PROCEEDINGS SERIES

THERMODYNAMICS
OF NUCLEAR MATERIALS
1979

PROCEEDINGS OF AN INTERNATIONAL SYMPOSIUM ON
THERMODYNAMICS OF NUCLEAR MATERIALS
HELD BY THE
INTERNATIONAL ATOMIC ENERGY AGENCY
IN JÜLICH, FEDERAL REPUBLIC OF GERMANY,
FROM 29 JANUARY TO 2 FEBRUARY 1979

In two volumes

VOL. I

INTERNATIONAL ATOMIC ENERGY AGENCY
VIENNA, 1980

EDITORIAL NOTE

The papers and discussions have been edited by the editorial staff of the International Atomic Energy Agency to the extent considered necessary for the reader's assistance. The views expressed and the general style adopted remain, however, the responsibility of the named authors or participants. In addition, the views are not necessarily those of the governments of the nominating Member States or of the nominating organizations.

Where papers have been incorporated into these Proceedings without resetting by the Agency, this has been done with the knowledge of the authors and their government authorities, and their cooperation is gratefully acknowledged. The Proceedings have been printed by composition typing and photo-offset lithography. Within the limitations imposed by this method, every effort has been made to maintain a high editorial standard, in particular to achieve, wherever practicable, consistency of units and symbols and conformity to the standards recommended by competent international bodies.

The use in these Proceedings of particular designations of countries or territories does not imply any judgement by the publisher, the IAEA, as to the legal status of such countries or territories, of their authorities and institutions or of the delimitation of their boundaries.

The mention of specific companies or of their products or brand names does not imply any endorsement or recommendation on the part of the IAEA.

Authors are themselves responsible for obtaining the necessary permission to reproduce copyright material from other sources.

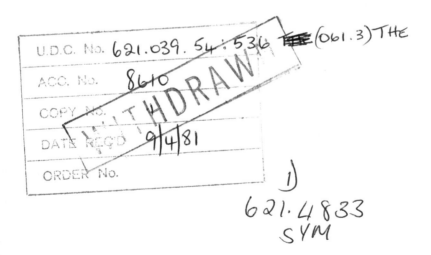
THERMODYNAMICS OF NUCLEAR MATERIALS 1979, VOL.I
IAEA, VIENNA, 1980
STI/PUB/520
ISBN 92—0—040080—9

FOREWORD

The design of nuclear fuels and their containment requires thermodynamic data on which to base estimates of fuel life, of cladding or coating performance and integrity, and of the behaviour of fuels under accident conditions involving high fuel temperatures and pressures. In the past 20 years, co-operation has been growing between the nuclear engineers and the scientists providing thermodynamic data. For example, nuclear engineers require such data on which to base their safety analyses and, because practical experience based on reactor accidents is almost totally lacking, they have to place considerable trust in the thermodynamicist's ability to provide them with accurate information. The cost of thermodynamics research in terms of materials, equipment, time and manpower is very high, and great experience is required in choosing the experiments or studies that will provide the necessary data. International meetings at which thermodynamicists and engineers meet to discuss data and to seek ways of obtaining information that is lacking serve to ensure that there is a minimum of unnecessary duplication of effort on a world-wide scale.

The International Atomic Energy Agency has played a major role in furthering the exchange of information on the thermodynamics of nuclear materials for over 17 years. In addition to the monographs being published on materials of importance in nuclear technology — for example the Atomic Energy Review Special Issues, and the series on The Chemical Thermodynamics of Actinide Elements and Compounds — and the holding of several small meetings on specific topics, the Agency has convened five symposia on the Thermodynamics of Nuclear Materials. The meetings of 1962, 1965 and 1967 covered the full range of basic thermodynamic studies and methodologies; the 1974 meeting continued the work of the earlier meetings and evidenced the growing interaction between thermodynamicists and engineers, with many studies initiated to obtain data that the engineers lacked. The 1979 Symposium, therefore, was aimed at 'applied thermodynamics' and at those basic studies of direct relevance to nuclear engineering. The very positive response engendered showed the value of supporting such co-operation between 'science' and 'technology'.

A significant portion of the symposium programme was devoted to oxide fuels, particular reference being made to fast breeder reactor applications. Oxygen diffusion and its effect on fuel/cladding interactions were discussed, while the behaviour of fission products, especially caesium, were considered in detail. Binary, ternary and even some quaternary phase diagrams were presented for fuel—fission product systems, these studies being based on both experimental and theoretical work; in one paper, the formation of two compounds was postulated that have yet to be experimentally identified. Sophisticated experiments had been undertaken to obtain experimental data at temperatures in excess of 7000 K, temperatures at which core components vaporize, and which could be expected to occur in core melt-down accidents. That such experiments were very necessary was supported by authors who showed that normal extrapolation of existing lower temperature data to temperatures above 2000 K could be seriously in error, if not completely valueless.

A new feature, one that will be of growing importance as moves are made towards setting up the first international fusion experiment (INTOR), is the presentation of work on the thermodynamics of materials required in fusion reactors — liquid lithium, solid lithium alloys and lithium-containing ceramics. The life of reactor blankets will depend on choosing suitable thermodynamic conditions for operation. Studies were also reported on diffusion of elements in glass matrices intended for fixing high-level radioactive waste. The stability of such vitreous storage media and the resistance of the glasses to leaching can be improved by considering the thermodynamic status of the major components.

The Proceedings of the 1979 Symposium is published in two volumes, and contains the texts of the 55 technical papers presented and of the resulting discussions. There is also a summary paper, and the publication includes a detailed subject index.

The Agency records with regret the death, shortly before the Symposium was held, of two eminent scientists who worked on the thermodynamics of nuclear materials, Dr. R.J. Ackermann (USA) and Professor O.S. Ivanov (USSR). They were both closely linked for many years with Agency activities on this subject, and their enthusiastic support will be sorely missed.

The Agency is indebted to the Government of the Federal Republic of Germany for the invitation to hold the meeting at the Jülich Nuclear Research Centre. The excellent facilities provided contributed in no small measure to the success of the meeting.

CONTENTS OF VOL. I

IN MEMORIAM

Section A: VAPORIZATION THERMODYNAMICS

Section B: EQUATION-OF-STATE STUDIES

Section C: SPECTROSCOPY

Section D: EMF STUDIES

Section E: DIFFUSION STUDIES

Section F: OXIDE FUELS

IN MEMORIAM

IN MEMORIAM

RAYMOND J. ACKERMANN

R.J. THORN, P.W. GILLES
Argonne National Laboratory,
Argonne, Illinois,
United States of America

Raymond J. Ackermann died on the 3rd of June, 1978, aged 48 years, following two heart attacks.

Among all the persons who have contributed to the success of the IAEA's activities involving the thermodynamics of nuclear materials, R.J. Ackermann is one of the few whose contributions have extended from the first symposium in 1962 to the present meeting. In addition to the Scientific Secretaries, a group of a few persons, through their sustaining efforts and results have made these activities unique in the transfer of basic scientific information into technology. Ray Ackermann was a recognized leader in these efforts. Beginning with papers presented at the symposium in 1962, he also contributed papers at the meetings in 1965 and 1974 and is a co-author of two papers published in these Proceedings. In addition, he has served on three panels dealing with thermodynamics of nuclear materials and was active on another at the time of his death.

He was a dedicated and enthusiastic participant in and supporter of these activities; he was frequently consulted on a variety of topics related to the subject. During 1970, he served as an IAEA fellow at the Bhabha Atomic Research Establishment, where he helped to initiate a programme in high temperature thermodynamics. With his death, the IAEA has lost what promised to be an even more productive contributor and leader in the future activities in thermodynamics of nuclear materials; such was his character that he had not reached his potential fulfillment.

As a graduate student at the University of Kansas, Ray Ackermann did the research for his thesis at Argonne National Laboratory. His thesis research, which was completed within one year, consisted of the high temperature thermodynamic properties of the U-O system derived from the measurement of the vapour pressure of uranium dioxide. In this research, he was the first in several respects: (1) He was the first student to be part of a long, continuing co-operative interaction between the high-temperature groups at Argonne National Laboratory and the University of Kansas; (2) His investigation of uranium dioxide was the first of several investigations of the gaseous oxides of uranium; these studies led to the interpretation by persons at the Atomic Energy Research Establishment, Harwell, UK, of the transport of oxygen, uranium and plutonium in the nuclear fuel; (3) His measurements were the first precise and accurate ones on any oxide phase above 2000°C; they extended over eight orders of magnitude in pressure and revealed a positive curvature which was the subject of several discussions. This research was a definitive study which has become a classic standard and reference. Subsequently, in 1955, after completion of his thesis, he continued the work as a staff member of Argonne National Laboratory.

During the next 23 years, Ray Ackermann studied several oxides, finding and quantitatively characterizing several phases having significant ranges of non-stoichiometry. A number of these involved phase equilibria ranging from pure metals to the lowest oxide. Among the phases of variable composition which he investigated are WO_{3-x}, La_2O_{3-y}, Y_2O_{3-x}, Ce_2O_{3-x}, U_3O_{8-z}, UO_{2-x}, ThO_{2-x}, ZrO_{2-x}. Because these phases cannot be retained through cooling to room temperature, i.e. they all disproportionate, their thermochemical properties must be determined directly at the high temperatures. His investigation of U_3O_{8-z} showed the presence of hysteresis in the pressure-composition isotherms, and therein he demonstrated the limitations of combustion to U_3O_8 in air as a gravimetric procedure. His study of UO_{2-x} established the lower phase boundary for this phase and the congruently subliming composition. In the most recent investigations, he determined the partial pressures of zirconium, ZrO, ZrO_2, and oxygen in equilibrium with ZrO_{2-x}. Undoubtedly, this is the first non-stoichiometric phase for which four partial vapour pressures have been determined at temperatures as high as 2500 K. In addition, these studies have produced a complete phase diagram for the $Zr-ZrO_2$ system and have established the extensive solubility (30 mol%) of oxygen in solid zirconium.

Through the direction which he gave to undergraduate students, an activity at which Ray was particularly outstanding, there were produced extensive measurements of the partial pressures of oxygen in equilibrium with wüstite, with magnetite and with both phases. The investigations of the wüstite phase included the determination of isotherms for the ternary systems of $Fe_{1+x}O$ containing magnesium, copper and nickel. The results are particularly significant in establishing the quantitative effects of redox potentials in solid oxides and the roles of valence-state energies and microdomains in solid-state chemistry.

In another systematic study, Ray determined phase equilibria in some binary metallic systems composed of tungsten with uranium and thorium, and of tantalum with these metals.

Ray's research was characterized particularly by his ability to determine with high reliability thermodynamic properties at extreme conditions of high temperature, by his ability to perform and conduct very effectively several investigations concurrently, by his versatility, and by his insights — which produced meaningful interpretations of several seemingly discordant results. He was particularly adept at identifying crucial experiments and at systematizing the results to produce interesting correlations. These served to establish the reliability of the results. His critical analysis of thermochemical data, an activity which interested him particularly, was just beginning. He had an innate feeling for the phenomenological concepts of thermodynamics. His knowledge of high-temperature thermochemical properties of nuclear materials, his dedication to the critical evaluation of data, his ability to co-operate with other scientists in several countries, and his enthusiastic personality made him the epitome for this international aspect of the IAEA's activity. In these endeavours, he exercised a mature judgement, an ability to select among discordant sources, and a willingness to express effectively his reasons.

In addition to the contributions to IAEA's programmes, Ray served as a member of the Libby-Cockcroft and the US-Euratom meetings. For the last several years, he had been a member of the committee on High Temperature Science and Technology of the National Research Council. Recently, he served as a co-chairman of the programme committee of a Conference on High Temperature Sciences Related to Open-Cycle Coal-Fired MHD Systems sponsored by Argonne National Laboratory and Argonne Universities Association. In the Chemistry Division of Argonne National Laboratory, he served on several committees and as a divisional representative to the Associate Laboratory Director for Physical Sciences. He worked closely with the visiting scientists from England and Japan.

Ray interacted with many people, and he did so with kindness, friendliness, sincerity, warmth, humility and generosity. He always had a genuine interest in his colleagues. He was frequently consulted and always responded with enthusiasm. Among those who knew him intimately he will be remembered both for the discussions of the philosophical implications of science and of life and for his witty, illustrative stories which brought a refreshing midwestern, rural perspective to the conversation. Some aspects of Ray's life, not known to many of us, reveal his dedication to the enrichment of the lives of others. We had known of his extremely worthwhile and unselfish service in India, but we were only vaguely aware of his early morning teaching at a local parochial high school. In his church, he was the choir director, and his wife, Irma, was the organist. At Ray's funeral, Ray's choir sang. In almost any kind of gathering, Ray was the leader for serious matters and for fun, with vigour, enthusiasm, stories, laughter, singing and ebullience. His approach to science and life was so exuberant that this memory will be long with us.

IN MEMORIAM

OLEG SERGEEVICH IVANOV

A.S. PANOV
A.A. Bajkov Institute of Metallurgy,
Academy of Sciences of the USSR,
Moscow,
Union of Soviet Socialist Republics

Oleg Sergeevich Ivanov died in Moscow on 20 September 1978 at the age of 65. He was well known both in the USSR and abroad as a leading expert who had made a significant contribution to the development of the science of reactor materials. His first work was published in the nineteen-thirties. He was one of the first scientists to study and measure the melting temperature and other properties of actinide metals. His works on thermodynamics and the phase diagrams of uranium, thorium and plutonium systems helped in the understanding of many aspects of the science of reactor materials and contributed much towards selecting and justifying reactor fuel options. During the last years of his life, he devoted considerable time to the thermodynamics of the uranium, thorium and plutonium systems. He headed the Soviet delegation to the 4th International Symposium on Thermodynamics of Nuclear Materials convened by the IAEA in Vienna in 1974, and was always heard with great interest at international meetings on this subject.

Oleg Sergeevich Ivanov was an admirable person, always cheerful and sociable, and had gathered around him a team of gifted scientists. The school of thermodynamics of nuclear materials which he set up at the Bajkov Institute of Metallurgy, USSR Academy of Sciences, is fruitfully carrying on the work which he started.

Section A

VAPORIZATION THERMODYNAMICS

A RE-DETERMINATION AND RE-ASSESSMENT OF THE THERMODYNAMICS OF SUBLIMATION OF URANIUM DIOXIDE

R.J. ACKERMANN[†], E.G. RAUH
Chemistry Division,
Argonne National Laboratory,
Argonne, Illinois,
United States of America

M.H. RAND*
Materials Development Division,
Atomic Energy Research Establishment,
Harwell, Didcot, Oxfordshire,
United Kingdom

Abstract

A RE-DETERMINATION AND RE-ASSESSMENT OF THE THERMODYNAMICS OF SUBLIMATION OF URANIUM DIOXIDE.

New mass-spectrometric measurements on the ion-intensity of UO_2^+ over urania from 1813 to 2463 K are reported. Although the mean value for the enthalpy of sublimation calculated from these measurements is close to previous values, a detailed examination of the results indicates that there is an appreciable curvature in the log p versus reciprocal-temperature curve for the process: $UO_2(s) \rightarrow UO_2(g)$. This is attributed to a large negative ΔC_p for the sublimation reaction, arising from the sharp increase in $C_p(UO_2(s))$ above ~ 1750 K. A thorough re-assessment of the previous studies on the sublimation of urania suggests an 'international' average value of $p_{UO_2} = (1.3 \pm 0.1) \times 10^{-6}$ atm at 2150 K; Knudsen effusion measurements above 2450 K ($p > 1 \times 10^{-4}$ atm) are thought to be in error due to departures from molecular flow. Thermal functions for $UO_2(g)$ have been calculated, assuming a linear molecule and electronic contributions to the partition function based on those of $ThO(g)$. Anharmonicity corrections have been included. When these functions are combined with the thermal functions for $UO_2(s)$, recently assessed, the third law heat of sublimation at 298.15 K becomes 147.8 kcal·mol^{-1} with a trend of only 0.2 kcal·mol^{-1} across the temperature range 1800 to 2400 K.

1. INTRODUCTION

A reliable equation of state for UO_2 and other nuclear fuel materials up to 6000 K is needed for the safety analysis of hypothetical core-disruptive accidents in fast-breeder reactors. The vapour pressure of the oxide fuel is an important parameter in the energy release mechanism and requires either a long

---

* Reprints, if required, should be requested from M.H. Rand.

TABLE I. THERMODYNAMICS OF SUBLIMATION OF URANIUM DIOXIDE

Investigators	Method	Temperature range (K)	ΔH (kcal·mol^{-1})	ΔS (cal·mol^{-1}·K^{-1})	Vapour pressure at 2150 K (10^{-6} atm)
Ackermann et al. (1956) [1]	mass effusion (total U)	1758–2378 (1600–2809)	142.5 ± 1.4	39.2 ± 0.7	1.2_1
Ivanov et al. (1962) [2]	mass effusion (total U)	1930–2160	150.0 ± 4.0	43.6 ± 1.9	1.9_1[a]
Voronov et al. (1962) [3]	Langmuir	1723–2573	140.5 ± 1.4	36.8 ± 0.2	0.57_7
Ohse (1966) [4]	mass effusion (total U)	2200–2800	151.8 ± 3.3	43.7 ± 1.4	1.3_2
Gross (1966) [5]	mass effusion (total U)	2021–2963	108.1	26.3	5.7_5[b]
Alexander et al.[c] (1967) [6]	transpiration	2090–2900	140.9 ± 1.9	37.9 ± 0.8	0.91_4
Gorban et al. (1967) [7]	mass effusion (total U)	1873–2573	147.5 ± 0.5	42.5 ± 0.5	1.9_7
Pattoret et al. (1968) [8]	mass spect. UO$_2$(g)	1890–2420	141.2 ± 1.1	39.4 ± 0.5	1.8_1
Tetenbaum and Hunt (1970) [9]	transpiration	2080–2705	143.1 ± 3.0	39.4 ± 1.2	1.1_6

Average = $1.36 ± 0.1_8$

[a] A recalculation of the tabulated data yields the present values.
[b] Not included in the average.
[c] Systematic temperature corrections obtained from Alexander (private communication) were applied to the originally reported values and the lowest temperature point has been disregarded.

extrapolation of the lower temperature data (T < 3000 K) or new methods of measurement at the temperatures of interest. Both options have certain limitations peculiar to the chemical thermodynamic behaviour of all materials at high temperatures.

The vapour pressure and enthalpy of sublimation of UO_2 have been reported by many investigators using the Knudsen (mass effusion), Langmuir (surface evaporation), mass spectrometric and transpiration methods. The results of the most extensive measurements to date are given in Table I. The results of Ackermann et al. [1] have been separated traditionally into low and high-temperature regions as a result of the observed positive curvature in the plot of log p against 1/T. It is necessary to re-examine these data in the light of new measurements and to re-interpret the high and low extremities of the original curve. Hence, in Table I are given the thermodynamic quantities derived from the most reliable central region (1758–2378 K) of the temperature range that is defined by no net statistical deviation from linearity of log p versus 1/T. From Table I, one can conclude that the variety of methods of measurement yield at 2150 K, a temperature common to nearly all of the studies, values of the vapour pressure which agree within a factor of approximately two, except that calculated from the anomalously high results of Gross [5]. Only the mass-spectrometric results of Pattoret et al. [8] correspond explicitly to the gaseous molecule, $UO_2(g)$; all of the other measurements involve the total uranium transport as $UO_2(g)$ and lesser amounts of $UO(g)$ and $UO_3(g)$. At temperatures near 2150 K, the combined partial pressures of the latter two are relatively small [8, 10] so that $p_{UO_2} \simeq 0.94\, p_{tot}$. Hence, the respectively derived enthalpies and entropies of sublimation should closely approximate to that of $UO_2(g)$ if the temperature range is centred around this temperature.

A further inspection of Table I indicates a spread of values for ΔH ranging from 140 to 152 kcal/mol. This variation adds considerably to the uncertainty resulting from an extrapolation of the vapour pressure to much higher tempera-tures. Furthermore, the very recent measurements of the vaporization rates of $UO_2(\ell)$ above 4000 K by laser irradiation and momentum detection or mass transfer methods [11–13] correspond to total vapour pressures (of many atmospheres) that appear to be significantly higher than the values for $UO_2(g)$ calculated from many of the results given in Table I. The molecularity of vaporization at these high temperatures has not been identified and, therefore, it will be elucidated from a more thorough assessment of the results given in Table I, supplemented by the present study.

Several of the previous measurements of the vapour pressure of UO_2 cited in Table I (Ackermann et al. [1], Ohse [4]), were carried out by extending the effusion method to pressures approaching 10^{-2} atm. The upper pressure limit of the effusion method must be more carefully examined in the light of some recent studies [14–16], which clearly show a breakdown of effusive flow for pressures exceeding 10^{-4} atm.

It is, therefore, the purpose of this study to improve the reliability of the thermodynamic data for the vaporization of uranium dioxide by: (i) presenting precise mass spectrometric measurements of the enthalpy of sublimation to $UO_2(g)$ over a wide (~ 700 K) temperature range; (ii) calculating the thermodynamic functions for UO_2(s and g) from the critical assessment of presently available data; and (iii) analysing the departure from molecular flow and the failure of the Knudsen effusion method at vapour pressures of uranium dioxide in excess of approximately 10^{-4} atm.

2.   EXPERIMENTAL METHODS AND RESULTS

Seven series of measurements of the temperature dependence of the partial pressure of $UO_2(g)$ over approximately stoichiometric UO_2(s) were made mass-spectrometrically by observing the UO_2^+ ion current as a function of the effusion cell temperature. The observations were obtained with a Bendix model 12–101 time-of-flight mass spectrometer equipped with a tungsten effusion cell heated by electron bombardment. Details of the cell assembly [17], power regulator [18] and the pyrometric measurement of temperature [19] have been reported elsewhere.[1] Three different samples of UO_2 (200–350 mg) were used and the overall instrumental sensitivity was changed from one series to another in an attempt to identify systematic instrumental errors that might involve a non-linear response. The possibility of error due to temperature gradients was investigated by altering the position of the cell with respect to the filament. When the cell was raised sufficiently, enough of a gradient could be introduced to cause a deposit to form within the orifice; at the other extreme the temperature measured through the orifice began to appear 'non-black-body'. Within these limits the position of the cell did not affect the reproducibility of the results within the experimental uncertainties. The possibility of temperature errors at the extremes of the pyrometer scales was minimized by making additional readings through a calibrated neutral-density filter, which shifted the readings down-scale. The linearity of the ion-detection system was established by measurements of the neon isotopes at varying gas pressures.

Each of the seven series of measurements consisted of 17–30 random observations of the UO_2^+ ion current-temperature relationship over the range 1813–2463 K. All the data normalized to a common sensitivity at 2150 K are plotted in Fig. 1. The maximum useful temperature was limited by the high rate of loss of sample and the rapid formation of deposits in the collimating slits, as well as by the possibility of exceeding molecular flow where ion currents are not

---

[1] Ref. [19] contains a description of the equipment and experimental procedure.

simply proportional to partial pressures. The combined results of all runs
(140 data points) were expressed as log $I(UO_2^+)T$ versus $1/T$ and treated by the
linear least-squares method. The normalization of all runs to an arbitrary ion
current at 2150 K yields the linear equation:

$$\log I(UO_2^+)T = (18.388 \pm 0.026) - (30\,927 \pm 60)T^{-1} \qquad (1)$$

from which the average enthalpy of sublimation is 141.5 ± 0.3 kcal·mol^{-1}, a
value in excellent agreement with the 141.2 ± 1.1 kcal·mol^{-1} reported by
Pattoret et al. [8].

A careful examination of the residuals shown in the upper plot of Fig. 2
indicates a curvature in the data beyond the experimental uncertainties and a
lack of a good statistical fit of a simple linear representation even though the
standard deviations in the slope and intercept are small. The results of each
series were then fitted to a three-parameter least-squares equation of the form:

$$\log I(UO_2^+)T = A + \frac{B}{T} + C \ln T \qquad (2)$$

and normalized to a common sensitivity at 2150 K. The residual pattern
resulting from this treatment is shown by the lower plot in Fig. 2. The plots
of the residuals clearly demonstrate that a linear least-squares representation
of the data is not quite adequate and shows a non-linear trend with temperature,
whereas the three parameter equation removes the residual curvature and shows
no such trend with temperature. All of the experimental precautions and
variations in the measurements of ion currents and temperatures did not remove
this curvature, nor did they reveal any readily identifiable systematic errors.
The solid line (Fig. 1) through the results will be identified and discussed in
much greater detail later. Attempts were made to extend the temperature range
beyond 2460 K into the region of non-effusive flow but the measurements were
severely limited by the rapid loss of sample. However, at the end of one of the
series, three observations were made in the range 2470–2580 K which showed
increased UO_2^+ ion intensities and a definite reversal of the downward curvature.
These are shown in Fig. 1 at the higher temperature end of the curve and will be
discussed in terms of the departure from molecular flow that follows.

3.   DISCUSSION

The constant C in Eq. (2) that has been used for the moment to fit and to
normalize the data shown in Fig. 1 is equal to the average value of the heat capacity
change divided by the gas constant, i.e.:

$$C = \frac{\overline{\Delta C_p}}{2.3\,R} \qquad (3)$$

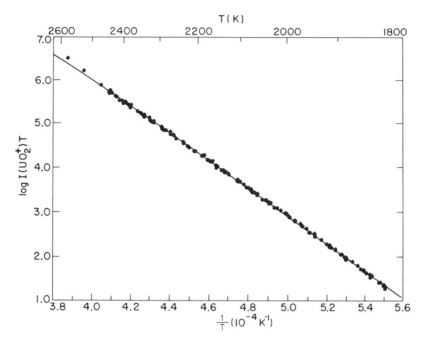

FIG.1.   Mass-spectrometric measurement of temperature dependence of $p(UO_2)$ over $UO_2(s)$, $log\ I(UO_2^+)\ T$ versus $1/T$.

Hence, values of $\overline{\Delta C_p}$ can be obtained from each individual series and from the combined normalized data; these values vary from extremes of -2.5 to -20 cal·mol^{-1}·K^{-1}. This variation is too large to obtain a reliable average value, and most certainly precludes the functional derivation of the nature of the dependence of $\overline{\Delta C_p}$ upon temperature. Therefore, the experimental data alone are inadequate as a result of the gentleness of the curvature compared with the existing random as well as unknown systematic errors. However, the results strongly indicate that $\overline{\Delta C_p}$ cannot be zero, but must be quite negative.

A small but insignificant systematic error is present at the higher temperatures for the data plotted in Fig. 1. The $UO_2(s)$ phase does not remain exactly stoichiometric over the large temperature range of this study. In fact, the congruently-vaporizing composition departs from the composition $UO_{2.00}$ above 2000 K and becomes progressively more substoichiometric, reaching a composition $UO_{1.98}$ near 2450 K [20]. The extent of this departure reduces the partial pressure of $UO_2(g)$ by less than 1% up to 2450 K and hence is considered insignificant.

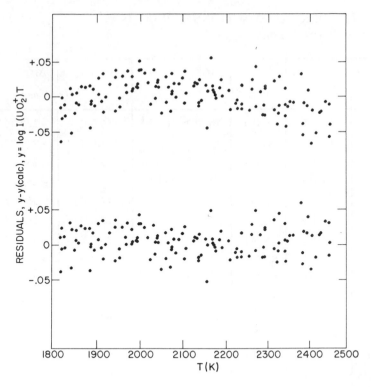

FIG.2. Residuals from least-squares analyses of mass-spectrometric results of Fig. 1. Upper: from linear least squares, log I(UO₂⁺) T = A + B/T. Lower: from three-parameter least squares, log I(UO₂⁺) T = A + (B/T) + C ln T.

In order to represent the vapour pressure of $UO_2(g)$ sufficiently accurately within the present temperature range and to permit extrapolations much beyond this range of measurements, a knowledge of the variation of ΔC_p with temperature is needed to complement the present results.

Uranium dioxide exists as a crystalline solid with the fluorite structure at all temperatures below the melting point. A number of compounds with this structure are known to possess excess contributions to the enthalpy and heat capacity at high temperatures, probably due to the formation of Frenkel defects in the anion sublattice [21, 22]. For such compounds, this is a co-operative process resembling fusion and is often described as pre-melting of the anion sub-lattice. In uranium dioxide, there is the additional possibility of an electronic contribution to the heat capacity, as discussed by MacInnes in these Proceedings [23]. Bredig [24] has analysed and discussed the excess enthalpy

TABLE II.   THERMAL FUNCTIONS FOR $UO_2(s, \ell)$

$H(298.15) - H(0) = 2696$ cal·mol^{-1}

T	C_p^0	S^0	$-(G^0 - H_{298}^0)/T$	$H^0 - H_{298}^0$	ΔH_f^0	ΔG_f^0
(K)		(cal·mol^{-1}·K^{-1})			(kcal·mol^{-1})	
298.15	15.200	18.410	18.410	0.000	− 259.30	− 246.60
400	17.008	23.159	19.033	1.650	− 259.07	− 242.29
600	18.661	30.414	21.681	5.240	− 258.50	− 234.03
800	19.543	35.912	24.579	9.067	− 258.07	− 225.94
1000	20.185	40.345	27.303	13.042	− 258.52	− 217.90
1500	21.528	48.784	33.137	23.471	− 260.55	− 197.24
2000	23.958	55.248	37.881	34.733	− 259.55	− 176.26
2500	31.423	61.258	41.951	48.268	− 256.41	− 155.73
2670 α	35.889	63.464	43.249	53.974	− 254.27	− 148.96
2670 β	39.923	63.627	43.249	54.410	− 253.83	− 148.96
3000	39.923	68.279	45.751	67.584	− 247.62	− 136.36
3120 β	39.923	69.845	46.648	72.375	− 245.38	− 131.95
3120(ℓ)	31.300	75.576	46.648	90.256	− 227.50	− 131.95
3500	31.300	79.173	49.987	102.150	− 223.70	− 120.53

in $UO_2(s)$, has related its similarity to that in other fluorite-structured materials and has concluded that this shows anomalous behaviour near 2670 K. Rand et al. [25] have separated the five sets of experimental measurements of enthalpy (675–3112 K) and the two sets of heat capacity (298–1006 K) into two regions, below and above 2670 K, the range below 2670 K including an exponential term to allow for the defect and electronic contributions. As a numerical convenience only, the discontinuity near 2670 K is represented by a first-order transition with an enthalpy change of 436 cal·mol^{-1}. The thermodynamic functions derived by Rand et al. [25] are given in Tables II and III.

     The heat capacity and other thermodynamic properties of $UO_2(g)$ have not been measured but can be calculated from known molecular parameters and some numerically consistent estimates of the electronic contribution to the partition function that can be closely fitted to the measured vapour pressure and enthalpy of sublimation over the large temperature range of this study. The need to know or to estimate reliably the low-lying electronic states and degeneracies in gaseous molecules is essential to any third-law treatment of the data that can be used for extrapolative purposes. At the present time, however,

TABLE III. THERMAL FUNCTIONS FOR $UO_2(g)$

$H(298.15) - H(0) = 2951$ cal·mol^{-1}

T	C_p^0	S^0	$-(G^0 - H_{298}^0)/T$	$H^0 - H_{298}^0$
(K)		(cal·mol^{-1}·K^{-1})		(cal·mol^{-1})
298.15	12.704	62.617	62.617	0.000
1000	17.942	82.369	70.529	11.840
1500	17.612	89.568	75.760	20.712
2000	17.635	94.631	79.875	29.512
2500	17.900	98.591	83.236	38.388
3000	18.303	101.889	86.078	47.435
3500	18.750	104.744	88.545	56.698

Electronic energy levels assumed:

Level (cm^{-1})	0	1500	4000	12 000
Degeneracy	1	6	6	18

there exists no reliable theoretical basis for predicting the term values and degeneracies of electronic states, particularly in a many-electron molecule such as $UO_2(g)$. Brewer and co-workers [26, 27] have attempted to approximate the electronic contribution in a gaseous molecule by using that for the related gaseous charged atom, e.g. the electronic contribution in a gaseous diatomic oxide, MO(g), is assumed to be the same as the electronic partition function for $M^{2+}(g)$ or its isoelectronic equivalent. These approximations in general are too large by several cal·mol^{-1}·K^{-1} in the entropy and free energy function in cases for which the electronic states in both the molecule and gaseous ion are known [28, 29].

At the mid-temperature (2150 K) of the present study the total vapour pressure of all uranium-bearing species from the experimental measurements given in Table I is 1.36×10^{-6} atm. The partial pressure of $UO_2(g)$ is about $(94 \pm 3)\%$ of this value, based on the mass spectrometric measurements of Ackermann et al. [10] and Pattoret et al. [8]. If this factor is applied to the total pressure measurements (both effusion and transpiration), an 'international average', $p_{UO_2} = 1.29 \times 10^{-6}$ atm, is obtained for the partial pressure of $UO_2(g)$

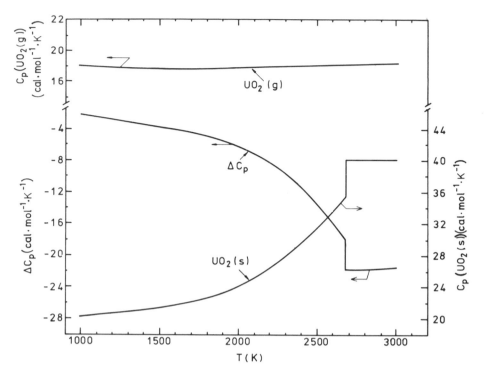

FIG.3. $C_p(UO_2(g))$, $C_p(UO_2(s))$ and ΔC_p versus T.

in equilibrium with virtually stoichiometric $UO_2(s)$ at 2150 K. The average enthalpy of sublimation of $UO_2(g)$ at 2150 K measured mass spectrometrically in the present study is 141.5 kcal·mol^{-1}, which agrees closely with that (141.2 kcal·mol^{-1}) reported by Pattoret et al. From these values of the pressure and its temperature dependence, an entropy of sublimation of 38.9 cal·mol^{-1}·K^{-1} at 2150 K is obtained. The absolute entropy of $UO_2(s)$ at 2150 K is 57.0 cal·mol^{-1}·K^{-1}; hence, the absolute entropy of $UO_2(g)$ is 95.9 cal·mol^{-1}·K^{-1} at 2150 K. It is possible to account for 91.5 cal·mol^{-1}·K^{-1} calculated from the known molecular parameters of $UO_2(g)$. The molecule is known to be linear in its ground state from matrix isolation spectroscopy, from which the measured stretching frequencies, $\nu_1 = 765.4$ cm^{-1} and $\nu_3 = 776.1$ cm^{-1}, were determined. A doubly degenerate bending frequency, $\nu_2 \simeq 190$ cm^{-1}, calculated from the valence force model applied to the uranyl ion, UO_2^{2+}, is consistent with the stretching frequencies. A previously observed absorption band at 81 cm^{-1} was tentatively assigned to the bending mode in $UO_2(g)$ but has not been confirmed and appears to be abnormally small. A U—O distance of 1.79 Å has been

used in these calculations. The breakdown of the Born-Oppenheimer approximation at high temperatures produces numerically significant anharmonicity
corrections which are included in these calculations.

Hence, the contribution to the entropy by unknown electronic states in
$UO_2(g)$ may be nominally as large as 4.4 cal·mol^{-1}·K^{-1}. The spectra for $U^{4+}(g)$
are not known, but those for isoelectronic $Th^{2+}(g)$ and $Pa^{3+}(g)$ [30] give rise to
electronic entropies of 7.1 and 6.2 cal·mol^{-1}·K^{-1}, respectively, which clearly
are too large.

Therefore we have constructed somewhat arbitrarily an electronic partition
function for $UO_2(g)$ using as a guide that for $ThO(g)$, which is the only gaseous
actinide oxide that has been so characterized. Both molecules have two unbonded
electrons but any further similarity is lacking. However, the thermodynamic
functions can be constructed in a consistent fashion from these electronic
estimates and the known parameters; they are given in Tables II and III. A plot
of the heat capacities of the gas and solid as well as ΔC_p is shown in Fig. 3. The
resultant ΔC_p is seen to be strongly temperature dependent and can be adequately
approximated by a parabolic dependence on temperature up to the transition
temperature of 2670 K.

An equation of the form:

$$\log p_{UO_2} = A + BT + CT^2 + \frac{D}{T} + E \ln T \tag{4}$$

which incorporates the parabolic dependence of ΔC_p on temperature:

$$\Delta C_p = 2.303 \, R \, [2BT + 6CT^2 + E] \tag{5}$$

has been fitted to the data of the present study. The constants B, C, and E
were fixed from the dependence of ΔC_p on temperature seen in Fig. 3, while
the remaining constants A and D were obtained by least-squares treatment of
all the ion-current data shown in Fig. 1 normalized to the value of
$p_{UO_2} = 1.29 \times 10^{-6}$ atm at 2150 K. With p measured in atmospheres and T in
kelvin, the numerical values of the constants are as follows:

A = 67.531
B = 4.382 × 10^{-3}
C = −4.411 × 10^{-7}
D = −3.709 × 10^4
E = −8.282

Equation (4) with the constants above gives rise to the curve drawn through the
data shown in Fig. 1. The enthalpy of sublimation of $UO_2(g)$ obtained by
differentiating Eq. (4) with respect to $1/T$ varies from 143.0 kcal·mol^{-1} at
1800 K to 138.5 kcal·mol^{-1} at 2400 K. The change of the enthalpy of

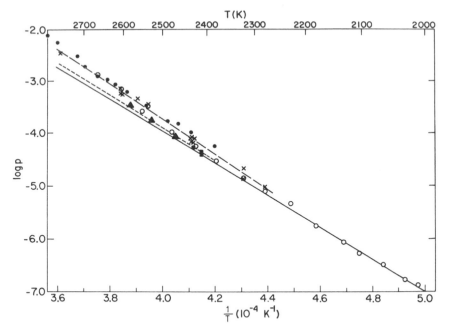

FIG.4. Comparison of the partial pressure of $UO_2(g)$ over $UO_2(s)$ from this study with total pressures of uranium-bearing species of other investigations (p in atmospheres). ○ Ackermann et al., Ref. [1], Series B-4 and B-8; ● ibid, Series H-1 and H-2; − − X − −, Ohse, Ref. [4]; ■ Cater, Ref. [32]; ▲ This study; − − − − Tetenbaum and Hunt, Ref. [9]; ——— $p(UO_2)$, Eq. (4), this study.

sublimation with temperature is dominated by the excess heat capacity of the solid phase as seen in Fig. 3. The uncertainties in the estimates of the bending vibrational constant and the electronic states in $UO_2(g)$ have only a very minor influence on the change in ΔC_p^0 and hence ΔH^0 with temperature. A third-law treatment of the data according to the equation:

$$\Delta H_{298}^0 = -RT \ln p - T\Delta \left(\frac{G_T^0 - H_{298}^0}{T}\right) \tag{6}$$

and the necessary quantities from Eq. (4) and Tables II and III yield a virtually constant value of $\Delta H_{298}^0 = 147.8$ kcal·mol^{-1} having an insignificant trend of only 0.2 kcal·mol^{-1} across the temperature range, 1800−2400 K. A plot of the residuals resulting from the least-squares treatment by means of Eq. (4) of all the data shown in Fig. 1 yields a distribution pattern that is virtually identical to that shown in the lower plot of Fig. 2 and, consequently, demonstrates the absence of any significant systematic errors.

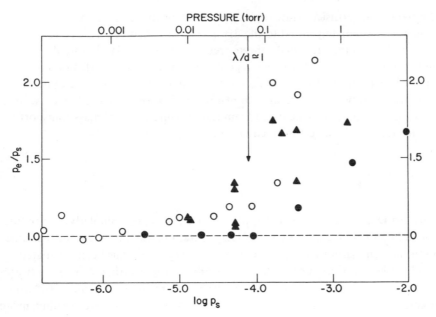

FIG.5. *Departure from molecular flow at high temperatures* (p_s *and* p_e *are in atmospheres).*
• *Hg, Carlson et al., Ref.* [14]; ○ UO_2, *Ackermann et al., Ref.* [1], *Series B-4;* ▲ UO_2,
Ohse, Ref. [4].

The partial pressure of $UO_2(g)$ above $UO_2(s)$ calculated from Eq. (4) is
plotted against reciprocal temperature in Fig. 4 and is compared with the total
pressure of uranium-bearing species reported in some of the earlier studies. The
results of Ackermann et al. [1] and Tetenbaum and Hunt [9] are virtually
indistinguishable up to 2450 K. Above this temperature the results of Ohse [4]
are in close agreement with the higher temperature measurements of
Ackermann et al. It is immediately clear that these latter two results are
substantially higher than the curve for $UO_2(g)$ that has been fixed at 2150 K
as described above. All previous mass spectrometric studies agree that $UO_2(g)$
is the major species vaporizing from congruently vaporizing $UO_2(s)$; the only
other observed molecules of any importance, $UO(g)$ and $UO_3(g)$, occur at
relatively smaller concentrations. Therefore, the ostensible higher pressures of
Ackermann et al. and Ohse above 2450 K cannot be explained by unobserved
gaseous species of sufficiently large partial pressures or by any reasonable
variation of the value of ΔC_p for the vaporization of $UO_2(g)$ that results from
an arbitrary adjustment of $C_p(UO_2(g))$.

The findings of Carlson et al. [14] describe quantitatively the increased
mass effusion rate of saturated mercury vapour that occurs from a thin-edged
orifice in the transition from molecular to hydrodynamic flow. The departure

from molecular (effusive) flow begins when the mean-free-path in the vapour becomes approximately equal to the diameter of the orifice. This condition will prevail when the saturated vapour pressure exceeds $\sim 10^{-4}$ atm. At higher temperatures, and hence for larger vapour pressures, the mass effusion rate increased to a limiting hydrodynamic value near 1.6 times the value expected for molecular flow. These results, shown in Fig.5, show a marked increase in the ratio of the pressure p_e calculated from the measured rate of mass transport of mercury vapour via the effusion equation:

$$p_e = \frac{m(2\pi RT)^{\frac{1}{2}}}{M} \tag{7}$$

to the saturated vapour pressure p_s plotted against log p_s. Similarly, the results of Ackermann et al. and of Ohse for UO_2 show a marked increase for the same condition. The saturated vapour pressure of $UO_2(g)$, p_s, has been calculated from Eq. (4). Inspection of the various results in Fig. 5 rather strongly indicates that the effusion method cannot be extended to pressures exceeding 10^{-4} atm for conventional size orifices, 0.05−0.1 cm diameter, and a corresponding mean-free-path of comparable magnitude. The deviation from molecular flow progressively increases with increasing pressures shown in Fig. 5 and hence, as seen in Fig. 4, causes the apparent enthalpy of sublimation to increase. Schulz and Searcy [15] have also shown in a convincing manner that molecular-flow conditions break down for $CaF_2(g)$ when the mean-free-path approaches the orifice diameter. However, Ewing and Stern [16], in a study of the vaporization of alkali halides, have proposed the unsuspected result that the transition from molecular to hydrodynamic flow depends on the mean-free-path and is independent of the diameter of the orifice. Although there still exists some lack of agreement among these investigations in the identification of the parameters explicitly associated with the break-down of molecular flow, it is clear that such occurs near 10^{-4} atm and has undoubtedly contributed an appreciable systematic error in the vapour pressure of UO_2 measured at the higher temperatures. Therefore, the values thereof derived from the high-temperature data of Ackermann et al. [1] and those of Ohse [4] exceeding 150 kcal·mol^{-1} are not valid quantities and should not be used to extrapolate the vapour pressure to higher temperatures [13, 31] even for comparative purposes.

The reasons for the breakdown of the effusion method at higher pressures do not apply to the transpiration technique employed by Tetenbaum and Hunt [9]. The small discrepancy between the transpiration results and those of p_{UO_2} in the present study is explainable in terms of the concentrations of $UO(g)$ and $UO_3(g)$ in the vapour, which increase relative to that of $UO_2(g)$ with increasing temperature. Also seen in Fig. 4 as ▲-points are the three highest temperature points of the present study shown previously in Fig. 1. These points very likely indicate the departure from molecular flow for $UO_2(g)$ observed mass spectrometrically. The ■-points are the effusion results reported earlier by Cater [32].

REFERENCES

[1] ACKERMANN, R.J., GILLES, P.W., THORN, R.J., J. Chem. Phys. 25 (1956) 1089; see also ACKERMANN, R.J., Argonne National Laboratory Rep. ANL-5482 (Sep. 1955).

[2] IVANOV, V.E., KRUGLYKH, A.A., PAVLOV, V.S., KOVTIN, G.P., AMONENKO, V.M., Thermodynamics of Nuclear Materials (Proc. Symp. Vienna, 1962), IAEA, Vienna (1962) 735.

[3] VORONOV, N.M., DANILIN, A.S., KOVALEV, I.T., Thermodynamics of Nuclear Materials (Proc. Symp. Vienna, 1962), IAEA, Vienna (1962) 789.

[4] OHSE, R.W., J. Chem. Phys. 44 (1966) 1375.

[5] GROSS, B., Institut für Kerntechnik der Technischen Universität Berlin Rep. TUBIK-4 (Oct. 1966).

[6] ALEXANDER, C.A., OGDEN, J.S., CUNNINGHAM, G.C., Battelle Memorial Institute Rep. BMI-1789 (Jan. 1967).

[7] GORBAN, Y.A., PAVLINOV, L.V., BYKOV, V.N., At. Ehnerg. 22 (1967) 465.

[8] PATTORET, A., DROWART, J., SMOES, S., Thermodynamics of Nuclear Materials, 1967 (Proc. Symp. Vienna, 1967), IAEA, Vienna (1968) 613.

[9] TETENBAUM, M., HUNT, P.D., J. Nucl. Mater. 34 (1970) 86.

[10] ACKERMANN, R.J., CHANDRASEKHARAIAH, M.S., RAUH, E.G., Argonne National Laboratory Rep. ANL-7048 (Jul. 1965).

[11] BOBER, M., BREITUNG, W., KAROW, H.U., SCHRETZMANN, J., J. Nucl. Mater. 60 (1967) 20.

[12] OHSE, R.W., BABELOT, J.F., BRUMME, G.D., KINSMAN, P.R., Ber. Bunsenges. Phys. Chem. 80 (1976) 780.

[13] ASAMI, N., NISHIKAWA, M., TAGUCHI, M., Thermodynamics of Nuclear Materials 1974 (Proc. Symp. Vienna, 1974) Vol. 1, IAEA, Vienna (1975) 287.

[14] CARLSON, K.D., GILLES, P.W., THORN, R.J., J. Chem. Phys. 38 (1963) 2725.

[15] SCHULZ, D.A., SEARCY, A.W., J. Phys. Chem. 67 (1963) 103.

[16] EWING, C.T., STERN, K.H., J. Phys. Chem. 78 (1974) 1998.

[17] RAUH, E.G., SADLER, R.C., THORN, R.J., Argonne National Laboratory Rep. ANL-6536 (Apr. 1962).

[18] CREMER, H.H., Z. Instrumentenkunde 7 (1964) 72.

[19] THORN, R.J., WINSLOW, G.H., Am. Soc. Mech. Eng. paper 63-WA-244 (1963).

[20] ACKERMANN, R.J., unpublished work.

[21] NAYLOR, B.F., J. Am. Chem. Soc. 67 (1945) 150.

[22] DWORKIN, A.S., BREDIG, M.A., J. Chem. Eng. Data 8 (1963).

[23] MacINNES, D.A., these Proceedings, paper IAEA-SM-236/37.

[24] BREDIG, M.A., in Colloq. Int. sur l'Etude des Trans. Cryst. à Haute Temp. au-dessus de 2000 K (Odeillo, 1971) C.N.R.S., Paris (1972) 183–191.

[25] RAND, M.H., ACKERMANN, R.J., GRØNVOLD, F., OETTING, F.L., PATTORET, A., Rev. Int. Hautes. Temp. Refract. (to appear in 1979).

[26] BREWER, L., CHANDRASEKHARAIAH, M.S., University of California, Radiation Laboratory Rep. UCRL-8713, Rev. (1960).

[27] BREWER, L., ROSENBLATT, G.M., *in* Advances in High Temperature Chemistry
      (EYRING, L., Ed.) Vol. 2, Academic Press, N.Y. (1969) 1–83.
[28] STEIGER, R.P., CATER, E.D., High Temp. Sci. **7** (1975) 288.
[29] ACKERMANN, R.J., RAUH, E.G., J. Pure Appl. Chem. (to appear).
[30] BREWER, L., J. Opt. Soc. Am. **61** (1971) 1666.
[31] FISCHER, E.A., KINSMAN, P.R., OHSE, R.W., J. Nucl. Mater. **59** (1975) 125.
[32] CATER, E.D., Argonne National Laboratory Rep. ANL-6140 (Mar. 1960) 86.

DISCUSSION

H.R. IHLE *(Chairman):* Did you try to apply corrections which take into
account viscous flow contributions at pressures exceeding about 10^{-4} atm?

M.H. RAND: No, we have not made such calculations since the precise
parameters governing the point at which such deviations begin do not seem to
be very firmly established. For example, there is no general agreement as to
whether the ratio of the mean-free-path to orifice diameter is critical or not.

A. SCOTTI: You refer to anharmonic contributions to C_p. Could you
please clarify how these are obtained?

M.H. RAND: The anharmonic contributions we have included are those
commonly used for diatomic molecules to describe the fact that the vibrations of
atomic molecules are not just simple harmonic motion[2], in particular constants
relating the energy level to $J^2(J + 1)^2$, where J = rotational quantum number,
and interactions between the vibrational and rotational modes.

E.A. FISCHER: If you extrapolate the curve which you assessed to the
liquid range, say up to 5000 K, will it agree with that recommended by the
International Working Group on Fusion Reactors (IWGFR) at Harwell? If not,
can you comment on the difference?

M.H. RAND: We have not made the extrapolation you suggest, but I think
we can predict that it will be distinctly lower than the curve for the *total* vapour
pressure of UO_2 recommended by the IWGFR meeting in June 1978. This
difference will be due to: (a) the neglect of any electronic levels above
12 000 cm^{-1} in $UO_2(g)$; (b) the fact that there is much more $UO_3(g)$ and
$UO(g)$ in the vapour over $UO_2(s)$ above 3000 K.

K.A. LONG: Could you please summarize your reasons for considering
that the increase in specific heat of UO_2 at low temperatures is due to defects
rather than to an electronic contribution?

M.H. RAND: I don't really want to enter into a discussion of the cause
of the C_p increase, which is treated more fully by McInnes in paper IAEA-SM-236/37
in these Proceedings. It is likely, in my opinion, that there are both electronic
and defect contributions — and probably others also, such as anharmonic effects.
However, the cause of the increase in C_p is not relevant to the arguments in our
paper; the sharp increase is not, I think, disputed.

---

[2] LEWIS, G.N., RANDALL, M., BREWER, L., PITZER, K.S., Thermodynamics,
McGraw Hill, New York (1961).

R.W. OHSE: I am pleased to note that your assessed 'international' average pressure value at 2150 K and heat of sublimation at 298.15 K agree best with our own data. As to the deviation at high temperatures, the disagreement with Ackermann's high-temperature data was pointed out as long ago as 1966 (your Ref. [4]). Extrapolations using Ackermann's and my low-temperature data showed reasonable agreement with theoretical assessment.

HIGH-TEMPERATURE VAPORIZATION
BEHAVIOUR OF OXYGEN-DEFICIENT THORIA*

R.J. ACKERMANN[†], M. TETENBAUM**
Argonne National Laboratory,
Argonne, Illinois,
United States of America

Abstract

HIGH-TEMPERATURE VAPORIZATION BEHAVIOUR OF OXYGEN-DEFICIENT
THORIA.
   The experimental results of the present study on the vaporization behaviour of oxygen-deficient thoria are directed toward a more precise and detailed study of the lower phase boundary (l.p.b.) and congruently vaporizing composition (c.v.c.), and intermediate compositions, and the corresponding oxygen potentials and total pressure at temperatures above 2000 K. The l.p.b. and c.v.c. values were found to fit an equation of the form $\log x = A + (B/T)$, where x is the stoichiometric defect in ThO_{2-x}. Oxygen potentials corresponding to the l.p.b. and c.v.c. have been estimated from vapour pressures and thermodynamic data. A very sharp decrease in oxygen potential occurs when thoria is reduced only slightly from the stoichiometric composition. In the temperature range from 2400 to 2655 K, the oxygen partial pressure dependency of x in ThO_{2-x} was found to be approximately proportional to $p_{O_2}^{-1/4}$. The small extent of reduction over a wide range of oxygen potentials at these temperatures is a clear illustration of the higher stability of the ThO_{2-x} phase compared with that of UO_{2-x}. Values of $\Delta \bar{H}_{O_2}$ and ΔS_{O_2} have been estimated for selected compositions from the dependence of the measured oxygen potential on temperature. Estimates of the standard free energy of formation of bivariant ThO_{2-x} compositions have been made. A substantial increase in the total pressure of thorium-bearing species occurs when stoichiometric thoria is reduced toward the lower phase boundary.

1.   INTRODUCTION

   The current interest in the United States in thorium-based fuels as alternative fuels in breeder reactor systems has led us to investigate the high-temperature vaporization behavior of oxygen-deficient thoria.

   The thermodynamic properties of the thorium-oxygen system at high temperatures have been determined principally from

---

 * Work performed under the auspices of the United States Department of Energy.
** Reprints, if required, should be requested from M. Tetenbaum.

effusion and mass-spectrometric measurements of vaporization of
the virtually stoichiometric dioxide [1] and the univariant system
$Th(\ell) + ThO_2(s) + vapor$ [2]. The most extensive studies of the
phase equilibria are reported by Benz [3] and clearly show that
the dioxide ThO_2 is the only stable oxide in the condensed state.
The vaporization studies, however, show that both $ThO_2(g)$ and
$ThO(g)$ are important molecules in the vapor phase. Rand [4]
has recently reviewed and critically assessed the physicochemical
and thermodynamic properties of thorium compounds and alloys
including a detailed analysis of the oxides.

The solid dioxide, thoria, is one of the most stable
refractory compounds known and for the most part is well characterized thermodynamically near the stoichiometric composition   [
The phase limits of thoria have not been defined precisely in
a chemical sense. The upper phase boundary is generally considered
to be ideally stoichiometric ThO_2 because a significantly oxidized
or hyperstoichiometric composition would require a valence of
thorium exceeding 4+ and, therefore, an associated departure
from the white color. Benz [3] has reviewed the earlier evidence
supporting the stoichiometric limit but cites several studies of
electrical conductivity and optical properties that suggest some
oxygen excess too small to be detected by ordinary chemical
methods of analysis. The lower phase boundary as a function of
temperature is reported by Benz [3] to vary nearly linearly from
$O/Th = 1.985 \pm 0.01$ at 1735°C to 1.87 ± 0.04 at 2740°C, the
monotectic temperature, and to 1.997 ± 0.01 at the melting point
of ThO_2, 3390°C. The rather large uncertainties in composition
probably result from the metallographic technique employed. Rauh
and Ackermann [5] accepted the composition data of Benz for the
lower phase boundary (l.p.b.) and calculated the standard free
energy of formation of $ThO_{2-x}(s)$ along the l.p.b. with increasing
temperature.

The vaporization of thoria in high vacuum produces a congruently vaporizing composition (c.v.c.) that is very close to the
stoichiometric composition. Ackermann *et al.*, [1] reported $O/Th =$
1.998 at approximately 2800 K but more recently [2] have reported
a c.v.c., $O/Th = 1.994 \pm 0.002$ at 2820 K, that was obtained from
both oxygen-rich and metal-rich samples. The combustion of
crystal-bar thorium metal at 1400°C in air yielded $O/Th = 2.000 \pm$
0.002. Aitken *et al.*, [6] have reported a hypostoichiometric
composition $O/Th = 1.96$ produced by heating thoria in "dry"
hydrogen at 2600°C.

Very little is known about the interdependence of oxygen
potential, total pressure of thorium-bearing species, composition,
and temperature in the bivariant region of $ThO_{2-x}(s)$ between the

phase limits. The few experimental data of Carniglia *et al.*, [7] suggest a dependence of $x \sim P_{O_2}^{-1/6}$, for $0 \leqslant x \leqslant 0.003$, $10^{-2} \geqslant P_{O_2} \geqslant 10^{-6}$ atm in the temperature range from 1400 to 1900°C. Hence, the experimental results of the present study of thoria are directed toward a more precise and detailed study of the l.p.b., the c.v.c., and intermediate compositions, as well as the corresponding oxygen potentials and total pressures at temperatures above 2000 K.

2.   EXPERIMENTAL

2.1 Measurements of the lower phase boundary of thoria

The lower phase boundary (l.p.b.) of ThO_{2-x} was measured in vacuum in the temperature range from 2005 to 2500 K by the vapor phase equilibration or isopiestic technique [5]. A mixture of mutually saturated phase, Th(ℓ) contained in a $ThO_{2-x}(s)$ cup, produces an equilibrium vapor at each desired temperature that reduces a physically separate disc or pellet of ThO_2 weighing about one gram. This equilibrated disc, out of contact with the liquid thorium, attains the composition of the lower phase boundary and, after cooling to room temperature, is analyzed by gravimetric combustion analysis in air at 1300 to 1400°C for the O/Th ratio even though disproportionation of the ThO_{2-x} may have occurred. All of the thoria discs were dark gray after the equilibration, but reverted to the original white color after combustion in air. Above 2500 K, a number of experimental difficulties became evident and precluded any reliable determinations of the l.p.b. The physical stability of the Th(ℓ)-$ThO_{2-x}(s)$ mixture began to deteriorate above this temperature, resulting in the rapid penetration of the thoria cup by the liquid metal and the subsequent creep of the liquid metal over the inside surface of the tungsten container.

2.2 Measurements of the congruently vaporizing composition

The congruently vaporizing composition (c.v.c) of thoria was determined over the range of temperature from 2400 to 2820 K by partially vaporizing samples of thoria from a tungsten effusion cell into a vacuum of $\sim 10^{-7}$ torr, following which the compositions of the residues were determined by combustion analysis. The thoria residues were a light to medium gray in appearance, but were not as dark as those obtained in the measurements of the l.p.b. The samples (1.5-2.0 g) reverted to the original white color in the combustion process. The weight gain was determined with a sensitivity of 0.01-0.02 mg. The overall precision of these measurements was somewhat better than that of the l.p.b. measurements because the total vapor pressure is relatively lower and creep of the metal component was not

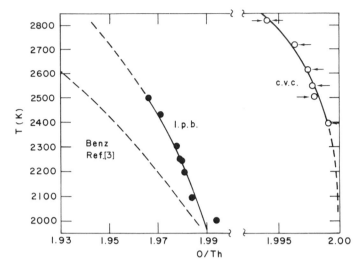

FIG.1.   Lower phase boundary (l.p.b.) and congruently vaporizing composition (c.v.c.) as a function of temperature for oxygen-deficient thoria.

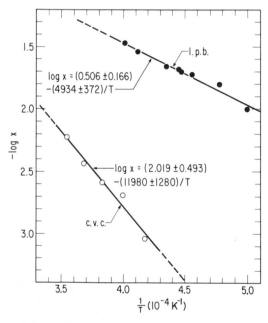

FIG.2.   Dependence of the stoichiometric defect x in ThO_{2-x} of the l.p.b. and c.v.c. on temperature.

involved.   The c.v.c. was approached from both the oxidized and
reduced condition.   In the former case, the sample was immediately
heated to the desired temperature for the duration (2-6 hours) of
the experiment, whereas in the latter case the sample was initially
heated to $\sim$2800 K to produce the substoichiometric composition at
that temperature, and then cooled to the desired temperature
(2504 K) in order to approach the c.v.c. from a "too reduced"
composition.

2.3   Measurements of oxygen potentials above oxygen-deficient thoria

The apparatus and general procedure used to measure oxygen
potentials and total pressure of thorium-bearing species above
bivariant ThO_{2-x} via gas equilibration techniques have been
previously described [8,9].   A small tapered tungsten crucible
having a perforated bottom was charged with approximately one
gram of sintered granules of stoichiometric thoria and press-fitted
into an open-ended tungsten condenser tube.   The carrier gas
consisted of hydrogen containing varying amounts of moisture;
the total pressure of carrier gas was approximately one atmosphere.
The oxygen-to-thorium ratio of the residue upon completion of a
run was derived from the continuous monitoring of the moisture
evolved during reduction of the initially stoichiometric thoria.
Oxygen potentials of the H_2-H_2O carrier gas were generally
chosen to be more negative than the values estimated for the
stoichiometric composition, but higher than oxygen potential
values estimated for lower phase boundary compositions.   However,
measurements were also carried out at 2500, 2600, and 2655 K
with carrier gas having oxygen potential values slightly higher
than the values estimated to establish the stoichiometric
composition.

Weight-loss measurements of the sample under investigation
served as a basis for estimating the total pressure of thorium-
bearing species.   Previous studies have shown that the agreement
between the values of the total pressure of metal-bearing species
derived from weight-loss measurements and those derived from
analysis of the sublimate in the condenser tube is very good [10,11].

3.   RESULTS AND DISCUSSION

The results of measurements of the lower phase boundary of
thoria are shown in Fig. 1 and compared with the metallographic
results of Benz [3] which are seen to indicate a more reduced
state.   The results of measurements of the congruently vaporizing
composition are also shown in Fig. 1.   The previous determination

of the c.v.c., O/Th = 1.994 at 2820 K reported by Ackermann and Rauh [2], is seen to fall on the curve in Fig. 1; therefore, the frequently cited earlier value [1] of 1.998 corresponding to ~2800 K was obtained on a sample that was apparently not fully reduced.

The extrapolation of the l.p.b. and c.v.c. to higher and lower temperatures can be accomplished by an equation of the form log x = A + (B/T), where x is the stoichiometric defect in ThO_{2-x}. An equation of this form adequately fits the extensive measurements of the l.p.b. of UO_{2-x} up to 2500 K, as pointed out by Winslow [12]. The results of the present study are shown in Fig. 2 and are seen to fit the linear dependence adequately.

Oxygen potentials corresponding to the univariant system (l.p.b.) Th(ℓ) + ThO_2(s) + vapor can be evaluated from the vapor pressure data given by Ackermann and Rauh [2] and the critical assessment of thermodynamic data reported by Rand [4] for the Th-O System. Ackermann and Rauh report data for the partial pressure of ThO(g) in the univariant system:

$$\frac{1}{2} Th(s/\ell, a \leq 1) + \frac{1}{2} ThO_2(s, a \leq 1) = ThO(g) \qquad (1)$$

The combined mass-effusion and mass-spectrometric data correspond to the temperature range from 2020 to 2420 K, but the latter data extend below the melting temperature of thorium (2023 K to as low as 1780 K. At 1800 K the activities of Th(s) and ThO_2(s) can be taken as unity with negligible error because the mutual solubilities of the two condensed phases are indeed small, as seen from the phase diagram of Benz [3] and from the present results for ThO_{2-x} (l.p.b.) seen in Fig. 1.

The equation for the partial pressure of ThO(g) that corresponds to Eq. (1), log p = 7.58 - (28630/T) [2], combined with the thermodynamic data for stoichiometric thoria and solid and liquid thorium [4] yields at 1800 K, ΔG_f°(ThO) = -38330 cal/mol and ΔH_f°(ThO, 298 K) = -6980 cal/mol. Hence, the value of the free energy of formation of ThO(g), ΔG_f°(ThO), can be generated at any desired temperature from the enthalpy of formation at 298 K and the free energy functions for ThO(g) [4]. The recalculation of the thermodynamic properties of ThO(g) differs by only ~500 cal/mol from those reported previously [2]; the latter were obtained from an equally weighted average of the results of the vaporization data corresponding to Eq. (1) and those of the isomolecular exchange reaction, YO(g) + Th(g) = Y(g) + ThO(g). A subsequent adjustment by Ackermann and Rauh [13] of their previous results [2] involved a revision of mass-spectrometric sensitivities and produced values for ΔG_f°(ThO) that are virtually identical with those obtained from Eq. (1). The numerical values

obtained in the present case are virtually identical to those given
by Rand [4], who based the third law assessment on the two
temperatures, 2080 and 2210 K, at which Ackermann and Rauh [2]
reported mass-effusion measurements. This fact supports the
notion that the activity of Th in the metal-phase and that of
ThO_2 in the oxide-phase are closely approximated by the atom
fraction, which departs only slightly from unity. Hence, the
oxygen potential at the lower phase boundary of the $ThO_{2-x}(s)$
can be evaluated from the equilibrium

$$Th(a_{Th} \simeq x_{Th}) + \tfrac{1}{2} O_2(g) = ThO(g) \tag{2}$$

for which

$$\overline{\Delta G}_{O_2}(l.p.b.) = 2\Delta G_f^\circ(ThO) + 9.152 \; T \log P_{ThO} \tag{3}$$

In the temperature range from 2200 to 2800 K, Eq. (3) is closely
approximated (within 50 cal/mol) by the linear equation

$$\overline{\Delta G}_{O_2}(l.p.b.) = -295870 + 44.95 \; T \tag{4}$$

By means of a similar calculation scheme and the vaporization data
for the congruently vaporization composition [1,2], one obtains
for the equilibrium at the c.v.c.

$$ThO_{2-x}(s, a \simeq 1) = ThO(g) + \left(\frac{1-x}{2}\right) O_2(g) \tag{5}$$

the equation for the oxygen potential

$$\overline{\Delta G}_{O_2}(c.v.c.) = -214600 + 37.14 \; T \tag{6}$$

Equations (4) and (6) generate the oxygen potential at each
desired temperature, and the corresponding compositions of the
l.p.b. and c.v.c. are given by the respective equations shown in
Fig. 2. The corresponding pairs of equations can be combined to
give the equations

$$\overline{\Delta G}_{O_2}(l.p.b.) = -318450 + 47.27 \; T - 4.576 \; T \log x \tag{7}$$

and

$$\overline{\Delta G}_{O_2}(c.v.c.) = -269420 + 46.38 \; T - 4.576 \; T \log x \tag{8}$$

Oxygen potentials above bivariant ThO_{2-x} compositions were
calculated from the moisture content of the carrier gas by means

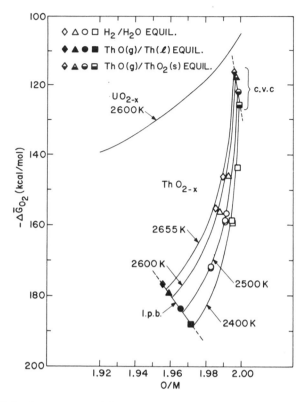

of the following expression

$$\overline{\Delta G}_{O_2} = 4.576T \log p_{O_2} \qquad (9)$$

where

$$\tfrac{1}{2} \log p_{O_2} = -\frac{12863}{T} + 2.88 - \log \frac{p_{H_2}}{p_{H_2O}} \qquad (10)$$

Equation (10) is derivable from well-known thermodynamic data between water vapor and its elements.

    The results of oxygen potential measurements above bivariant ThO_{2-x} compositions are shown in Fig. 3 for isotherms in the temperature range from 2400 to 2655 K.   The shapes of the isotherms shown in Fig. 3 were obtained from estimated values of the partial molar quantities $\Delta\overline{H}_{O_2}$ and $\Delta\overline{S}_{O_2}$ which were obtained as follows. Figure 4 shows a plot of $\overline{\Delta G}_{O_2}$ versus log x, where x is the

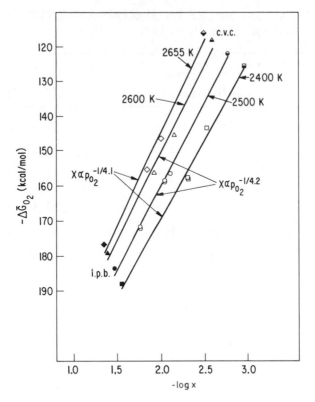

FIG.4. Plot of oxygen potential as a function of stoichiometric defect x in ThO$_{2-x}$.

stoichiometric defect in ThO$_{2-x}$ for the various isotherms investigated, and for O/Th ratios ranging from c.v.c. to the l.p.b. values. Assuming a linear fit, one obtains from a least-squares analysis of the data, the following expressions

$$2400 \text{ K} \qquad \Delta\overline{G}_{O_2} = (-258.9 \pm 5.8) - (44.9 \pm 2.2) \log x \qquad (11)$$

$$2500 \text{ K} \qquad \Delta\overline{G}_{O_2} = (-256.0 \pm 2.6) - (48.0 \pm 1.3) \log x \qquad (12)$$

$$2600 \text{ K} \qquad \Delta\overline{G}_{O_2} = (-250.9 \pm 8.8) - (50.3 \pm 4.3) \log x \qquad (13)$$

$$2655 \text{ K} \qquad \Delta\overline{G}_{O_2} = (-251.1 \pm 7.9) - (53.3 \pm 4.0) \log x \qquad (14)$$

From these equations, the oxygen partial pressure dependency of x in ThO$_{2-x}$ can be estimated and was found to be approximately proportional to PO$_2^{-1/4}$ over the temperature range from 2400 to

TABLE I. PARTIAL MOLAR ENTHALPIES AND ENTROPIES
OF SOLUTION OF OXYGEN IN ThO_{2-x}

O/Th	$-\Delta\overline{H}_{O_2}$ (kcal/mol)	$-\Delta\overline{S}_{O_2}$ (cal/mol · K)
1.955	380.6 ± 4.3	75.8 ± 1.7
1.960	381.9 ± 4.0	77.3 ± 1.6
1.965	383.6 ± 3.5	79.1 ± 1.4
1.970	386.1 ± 3.4	81.4 ± 1.3
1.975	388.0 ± 3.4	83.7 ± 1.4
1.980	391.5 ± 3.5	86.9 ± 1.4
1.985	395.0 ± 3.5	90.7 ± 1.4
1.990	400.5 ± 3.9	96.3 ± 1.5
1.995	410.5 ± 6.0	106.0 ± 2.4
1.998	421.5 ± 9.4	118.1 ± 3.7

2655 K. From classical defect theory [14], this dependency
suggests the possibility that singly charged oxygen vacancies
predominate under the conditions of this investigation.

By means of equations (11-14) and the thermodynamic relationships

$$\Delta\overline{S}_{O_2} = -\frac{\partial\Delta\overline{G}_{O_2}}{\partial T} \tag{15}$$

$$\Delta\overline{H}_{O_2} = \Delta\overline{G}_{O_2} + T\Delta\overline{S}_{O_2} \tag{16}$$

the partial molar enthalpy and partial molar entropy of solution
of diatomic oxygen in oxygen-deficient thoria can be estimated
for selected compositions. The results are given in Table 1.
The sharp changes in $\Delta\overline{H}_{O_2}$ and $\Delta\overline{S}_{O_2}$ suggest the possibility of the
occurrence of more than one type of defect for compositions very
close to stoichiometry [15]. It should be emphasized that the
accuracy of these values may be somewhat limited by the short
temperature range of these measurements (2400-2655 K).

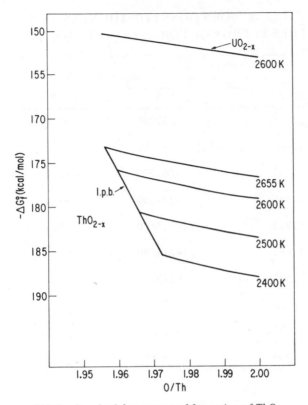

FIG.5.   *Standard free energy of formation of* ThO_{2-x}.

The $\Delta \overline{H}_{O_2}$ and $\Delta \overline{S}_{O_2}$ values given in Table 1 were used to generate the curves in Fig. 3. It is apparent from this study that a very sharp decrease in oxygen potential occurs when thoria is reduced only slightly from the stoichiometric composition. This small extent of reduction over a wide range of oxygen potential at these temperatures is a clear illustration of the higher stability of the ThO_{2-x} phase than that of UO_{2-x}. For comparison, oxygen potentials above UO_{2-x} at 2600 K are also included in Fig. 3.

Estimates of the standard free energy of formation of hypostoichiometric bivariant thoria compositions can be made from a knowledge of the relative partial molar free energy values of diatomic oxygen in thoria and the known standard free energy of formation for stoichiometric thoria. The free energy of formation of ThO_{2-x} compositions can be calculated from the Gibbs–Duhem equation in the form

$$\Delta G_f^\circ (ThO_{2-x},\ s) = \Delta G_f^\circ (ThO_2,\ s) - \tfrac{1}{2} \int_{2.00}^{2-x} \Delta \overline{G}_{O_2}\ d\left(\frac{O}{Th}\right) \qquad (17)$$

TABLE II. EQUATIONS FOR THE STANDARD
FREE ENERGY OF FORMATION OF ThO_{2-x}

| O/Th | $\Delta G_f^\circ = A + BT$ | |
	A	B
1.955	−282000	41.0
1.960	−283000	41.2
1.965	−283900	41.4
1.970	−284900	41.6
1.975	−287600	42.4
1.980	−288600	42.6
1.985	−289500	42.9
1.990	−290500	43.1
1.995	−291600	43.4
2.000	−292600	43.7

Calculated values of the standard free energy of formation as a
function of temperature for the various isotherms investigated
are shown in Fig. 5; the temperature coefficients for selected
compositions are given in Table 2. It is seen from Fig. 5 that
1) the standard free energy of formation of thoria becomes less
negative with increasing oxygen deficiency, and 2) for a given
composition, the stability of thoria decreases with increasing
temperature. It should be noted that, for a given oxygen-to-metal
ratio and temperature, the free energy of formation of thoria
is considerably more negative than that of urania [8]. This
again reflects the higher stability of the ThO_{2-x} phase compared
with that of UO_{2-x}.

The results of transpiration measurements of the total
pressure of thorium-bearing species above the bivariant Th-O
system in the temperature range from 2400 to 2655 K are shown
in Fig. 6. The total pressure of thorium-bearing species above
virtually stoichiometric thoria seen in Fig. 6 agree within 10%
with the effusion results of Ackermann et al. [1]. It is apparent
from Fig. 5 that a substantial increase in the total pressure of

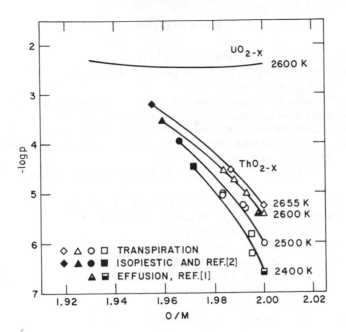

FIG.6. Total pressure of thorium-bearing species above the ThO₂₋ₓ phase (p in atmospheres: 1 atm = 101.325 kPa).

thorium-bearing species occurs when stoichiometric thoria is reduced toward the lower phase boundary. The increase in pressure can be attributed primarily to the increasing contribution of the ThO gaseous species. It should be noted that the isotherms can be smoothly extrapolated to the vapor pressure values of Ackermann and Rauh [2] above the lower phase boundary compositions. For comparison, the total pressure of uranium-bearing species above UO_{2-x} at 2600 K [9] is also included in Fig. 6. The striking difference in the curves for UO_{2-x} and ThO_{2-x} is related to the more complex vaporization of UO_{2-x} involving $UO(g)$, $UO_2(g)$, and $UO_3(g)$ [9]. Consequently, the congruently vaporizing composition for UO_{2-x} is significantly removed from the stoichiometric composition; however, such is not the case for ThO_{2-x}. From the results given above it follows that uranium-bearing species will predominate in solid solutions of UO_2 and ThO_2 of interest as alternative fuels.

Calculations are in progress to assess the interdependence of partial and total pressure of thorium-bearing species above bivariant ThO_{2-x} by means of our experimental oxygen potential values and the known or estimated free energy of formation of the gaseous species and condensed phase. The results of this assessment will be reported in a later paper.

REFERENCES

[1] ACKERMANN, R.J., et al., J. Phys. Chem. 67 (1963) 762.

[2] ACKERMANN, R.J., RAUH, E.G., High Temp. Sci. 4 (1972) 272.

[3] BENZ, R., J. Nucl. Mater. 34 (1970) 86.

[4] RAND, M., Thorium: Physico-Chemical Properties of Its Compounds and Alloys, Atomic Energy Review, Special Issue No.5, IAEA, Vienna (1975) 7.

[5] RAUH, E.G., ACKERMANN, R.J., Can. Metall. Q. 14 (1975) 205.

[6] AITKEN, E.A., BRASSFIELD, H.C., FRYXELL, R.E., Thermodynamics (Proc. Symp. Vienna, 1965), Vol.2, IAEA, Vienna (1966) 435.

[7] CARNIGLIA, S.C., BROWN, S.D., SCHROEDER, T.F., J. Amer. Ceram. Soc. 54 (1971) 13.

[8] TETENBAUM, M., HUNT, P.D., J. Chem. Phys. 49 (1968) 4739.

[9] TETENBAUM, M., HUNT, P.D., J. Nucl. Mater. 34 (1970) 86.

[10] TETENBAUM, M., High Temp. Sci. 7 (1975) 37.

[11] TETENBAUM, M., Trans. Am. Nucl. Soc. 23 (1976) 131.

[12] WINSLOW, G.H., High Temp. Sci. 5 (1973) 176.

[13] ACKERMANN, R.J., RAUH, E.G., J. Chem. Phys. 60 (1974) 2266.

[14] KOFSTAD, P., Nonstoichiometry, Diffusion, and Electrical Conductivity in Binary Metal Oxides, John Wiley and Sons, Inc., New York (1972).

[15] ROBERTS, L.E.J., MARKIN, T.L., Proc. Br. Ceram. Soc. 8 (1967) 201.

DISCUSSION

U. BENEDICT: What is the method which enables you to determine O/Th ratios with such high precision (i.e. to 4 decimal places)?

M. TETENBAUM: The weight gain of the 1.5–2.0 g samples of oxygen-deficient thoria from the c.v.c. measurements used for the determination of O/Th ratios (via oxidation to the stoichiometric composition $ThO_{2.000_0}$) was determined with a sensitivity of 0.01–0.02 mg. The fourth-place digit after the decimal point is written in these cases as a subscript to the oxygen-to-thorium ratios to indicate some uncertainty in the fourth-decimal-place digit.

A.S. PANOV: On what grounds do you assume that ThO_{2-x} has the same type of defects as UO_{2-x} (i.e. vacancies in the oxygen sub-lattice)? Is this based on neutron diffraction data or on the correlation between p_{O_2} and the entropy and enthalpy components of the oxygen chemical potential?

M. TETENBAUM: The oxygen partial pressure dependency of x in ThO_{2-x} was found to be proportional to $p_{O_2}^{-1/4}$. This dependency suggests that singly charged oxygen vacancies predominate in the fluorite lattice (based on classical defect theory) in the temperature range 2400–2655 K and over the composition range investigated. The rapid changes in oxygen potential over partial thermodynamic quantities $\Delta \overline{H}_{O_2}$, $\Delta \overline{S}_{O_2}$ very close to the stoichiometric composition suggest the occurrence of more than one type of defect with different energies of formation. Certainly, we would need sophisticated structural measurements at high temperatures in order to get a clue as to the nature of the defects in non-stoichiometric materials.

A.S. PANOV: How do you account for the considerably higher values of $|\mu_0|$ in ThO_{2-x} in comparison with those in UO_{2-x}, PuO_{2-x} and CeO_{2-x}? How would you explain this, using the energies of formation and interaction of defects?

M. TETENBAUM: The higher stability of the ThO_{2-x} phase as compared with the UO_{2-x}, PuO_{2-x} or CeO_{2-x} phases is reflected in the more negative values of $\Delta\bar{G}_{O_2}$ and particularly $\Delta\bar{H}_{O_2}$ at a given oxygen-to-metal ratio, and is therefore consistent with the increase in energy required to create an oxygen vacancy for the Th-O system. Interaction of defects should yield a decrease in $-\Delta\bar{H}_{O_2}$ with a corresponding levelling of the curve of $\Delta\bar{H}_{O_2}$ as the O/M ratio decreases (see Table I).

THERMODYNAMICS OF THE VAPORIZATION OF THORIUM AND URANIUM HALIDES

RAJENDRA PRASAD, K. NAGARAJAN, ZILEY SINGH,
M. BHUPATHY, V. VENUGOPAL, D.D. SOOD
Radiochemistry Division,
Bhabha Atomic Research Centre,
Trombay, Bombay,
India

Abstract

THERMODYNAMICS OF THE VAPORIZATION OF THORIUM AND URANIUM HALIDES.
The vaporization behaviour of thorium and uranium halides is being investigated in the solid and liquid regions by the boiling-temperature and transpiration techniques. The two techniques have been used for studies on $ThCl_4$, UF_4 and UCl_4 over the temperature ranges 880 to 1161 K, 1169 to 1427 K, and 763 to 1008 K, respectively. The data derived from the two techniques agree excellently with each other for these three halides, thereby establishing their vaporization behaviour. Boiling-temperature studies have been completed for ThF_4, $ThBr_4$, ThI_4 and UBr_4 over the temperature ranges 1395 to 1554 K, 842 to 1089 K, 790 to 1044 K, and 759 to 1004 K, respectively. In a number of cases, the present data are at variance with existing published data. The vapour pressure data were expressed as:

$$\log_{10} p = A - B\,T^{-1}$$

where p is in atmospheres and T in kelvin. The constants A and B for the vapour pressure equations are as listed below, and the data have been used to derive the relevant thermodynamic parameters.

	A	B		A	B
$ThF_4(s)$			$ThF_4(\ell)$	7.37	14154
$ThCl_4(s)$	10.10	11611	$ThCl_4(\ell)$	6.65	8023
$ThBr_4(s)$	10.65	11411	$ThBr_4(\ell)$	6.94	7813
$ThI_4(s)$	10.12	10316	$ThI_4(\ell)$	7.33	7915
$UF_4(s)$	10.03	15994	$UF_4(\ell)$	6.99	12014
$UCl_4(s)$	10.46	10443	$UCl_4(\ell)$	7.22	7627
$UBr_4(s)$	10.58	10122	$UBr_4(\ell)$	7.06	7287

1. INTRODUCTION

Knowledge of the vaporization behaviour of thorium and uranium halides is of importance in a number of areas, for example when considering halide volatility in the reprocessing of nuclear fuels and in molten salt reactor technology. Though a number of publications on this topic exist in the literature [1−19], it is clear from the compilations by Brown [20], Rand and Kubaschewski [21, 22], and Wagman [23] that the vaporization behaviour of these halides needs to be established. Various investigators have used different techniques for vapour pressure measurements, but it is rare that measurements by two independent techniques agree with each other and it is difficult to make a choice between the available sets of data. In order to resolve these discrepancies, studies on the vaporization behaviour of thorium and uranium halides are being carried out, both in the solid and in the liquid region by two independent techniques, namely boiling-temperature and transpiration. These sets of measurements should provide a consistent set of data from which the vaporization thermodynamics of these halides can be determined.

To date studies on the vaporization of $ThCl_4$, UF_4 and UCl_4 have been completed using both the techniques. For ThF_4, $ThBr_4$, ThI_4 and UBr_4, vapour pressure determinations by the boiling-temperature technique have been completed. The results for $ThCl_4$ by both the techniques and for UCl_4 by the transpiration technique have been communicated for publication [24, 25] but are presented in this paper together with the other data. Further investigations are in progress.

2. EXPERIMENTAL

2.1. Apparatus and procedure

The transpiration apparatus used in the studies on $ThCl_4$ and UCl_4 has been described elsewhere [25]. A molybdenum boat, a nickel reaction tube, a thermo-couple well and a condenser were used for studies on UF_4. The reaction tube's internal diameter had to be decreased from the 12 mm used earlier to 9 mm due to some unsaturation problems. In this assembly, the mass of UF_4 transported per unit volume of the carrier gas was independent of the flow rate in the region of 2×10^{-5} to 4.5×10^{-5} $m^3 \cdot min^{-1}$. The boiling-temperature apparatus used in the present study was the same as described previously [24] except that, in the case of UF_4 and ThF_4, the thermocouple well and reaction tube were made of nickel.

All halides except UF_4 and ThF_4 were handled inside a dry box having an argon-atmosphere with <5 ppm H_2O and <50 ppm O_2 as impurities. All thermo-couples used in the measurements were calibrated at the boiling temperature of water, and the melting temperatures of bismuth (544.5 K), antimony (904 K) and silver (1239 K) [26]. The furnaces used for the experiments had a constant temperature zone (±0.5 K) of 13 cm and the controlled temperature varied by no more than ±0.5 K.

2.2. Materials

All the halides except the fluorides were prepared by reacting the corresponding halogen with the thorium or uranium hydride at 570 to 700 K, followed by purification by sublimation. Fluorides were prepared by reacting active dioxides with anhydrous HF at 759—800 K.

2.3. Analysis

Thorium and uranium chlorides, collected in the condenser during each experiment, were removed by dissolution in distilled water. The quantities of uranium or thorium were determined gravimetrically by precipitating them as the hydroxide followed by conversion to U_3O_8 or ThO_2, respectively. The filtrate from the hydroxide precipitation step was analysed for chlorine gravimetrically as AgCl. In the case of UF_4, the nickel condenser was weighed before and after each experiment to obtain the quantity of material transported.

3.   RESULTS

The vapour pressure data from the present studies can be represented by the equation:

$$\log_{10} p = A - BT^{-1}$$

where p is measured in atmospheres and T in kelvin. The constants A and B for the thorium halides are listed in Table I and those for the uranium halides in Table II. In the case of $ThCl_4$, UF_4 and UCl_4, where studies were carried out by two techniques, it was observed that the agreement between the two sets of data was good, and hence the constants of the equation obtained by combining the two sets of data are also given. From these vapour pressure data, the melting temperature, boiling temperature, enthalpy of fusion and standard enthalpy and entropy of vaporization have been evaluated.

The values of ΔH^0 (g, 298 K) were calculated using both second-law (sigma method) and third-law methods for the thorium halides. The heat capacity data for solid and liquid thorium halides reported by Rand [22], and the heat capacity data for gaseous halides and free energy functions for condensed and gaseous

TABLE I. VAPOUR PRESSURE DATA ON THORIUM HALIDES

$\log_{10} p = A - BT^{-1}$; p in atmospheres, T in kelvin.

Compound	Temperature range (K)	A			B		
		Boiling temp.	Transpiration	Combined	Boiling temp.	Transpiration	Combined
$ThF_4(\ell)$	1395 to 1554	7.37 ± 0.20	–	–	14154 ± 306	–	–
$ThCl_4(s)$	880 to 1043	10.28 ± 0.12	9.93 ± 0.14	10.10 ± 0.08	11775 ± 122	11460 ± 130	11611 ± 80
$ThCl_4(\ell)$	1045 to 1161	6.52 ± 0.16	6.78 ± 0.24	6.65 ± 0.16	7866 ± 174	8177 ± 260	8023 ± 179
$ThBr_4(s)$	842 to 971	10.65 ± 0.13	–	–	11411 ± 119	–	–
$ThBr_4(\ell)$	971 to 1089	6.94 ± 0.16	–	–	7813 ± 162	–	–
$ThI_4(s)$	790 to 853	10.12 ± 0.33	–	–	10316 ± 267	–	–
$ThI_4(\ell)$	853 to 1044	7.33 ± 0.05	–	–	7915 ± 52	–	–

TABLE II. VAPOUR PRESSURE DATA ON URANIUM HALIDES

$\log_{10}p = A - BT^{-1}$; p in atmospheres, T in kelvin.

Compound	Temperature range (K)	A			B		
		Boiling temp.	Transpiration	Combined	Boiling temp.	Transpiration	Combined
$UF_4(s)$	1169 to 1307	10.13 ± 0.25	9.79 ± 0.19	10.03 ± 0.14	16120 ± 318	15714 ± 239	15994 ± 176
$UF_4(\ell)$	1318 to 1427	6.82 ± 0.25	7.21 ± 0.30	6.99 ± 0.24	11773 ± 344	12339 ± 413	12014 ± 335
$UCl_4(s)$	763 to 862	10.69 ± 0.24	10.43 ± 0.10	10.46 ± 0.12	10642 ± 203	10412 ± 82	10443 ± 100
$UCl_4(\ell)$	868 to 1008	7.21 ± 0.06	7.25 ± 0.13	7.22 ± 0.06	7625 ± 54	7649 ± 121	7627 ± 58
$UBr_4(s)$	759 to 790	10.58 ± 0.39	–	–	10122 ± 306	–	–
$UBr_4(\ell)$	800 to 1004	7.06 ± 0.05	–	–	7287 ± 44	–	–

TABLE III. THERMODYNAMIC PARAMETERS FOR THORIUM AND URANIUM HALIDES

Compound	Melting temperature (K)	Boiling temperature (K)	ΔH (fus, melt. temp.) ($kcal_{th} \cdot mol^{-1}$)	ΔH^0(g, 298 K) ($kcal_{th} \cdot mol^{-1}$) 2nd law	3rd law	ΔS^0(vap. 298 K) ($cal_{th} \cdot mol^{-1} \cdot K^{-1}$)
ThF_4	–	1921	–	80.90	78.71	45.4
$ThCl_4$	1043	1205	14.8	57.20	52.97	52.9
$ThBr_4$	971	1125	15.3	56.03	48.95	55.2
ThI_4	853	1079	10.7	50.46	49.70	54.1
UF_4	1309	1720	16.80	77.95	–	52.0
UCl_4	869	1057	11.87	50.86	–	53.9
UBr_4	806	1032	11.56	49.50	–	54.3

states reported by Wagman et al. [23] were used for the calculations. For ThF_4, measurements have so far been carried out in the liquid region only, and to calculate ΔH^0 (g, 298 K) the values of melting temperature (1383 K) and enthalpy of fusion at 1383 K (10.51 $kcal_{th} \cdot mol^{-1}$) were taken from an IAEA review [22].

In the case of the uranium halides, only second-law values for ΔH^0(g, 298 K) could be calculated due to the non-availability of free energy functions for these halides. For second-law calculations on uranium halides, heat capacity data reported by Barin and Knacke [27] were used. In the case of UF_4 and UBr_4, where the heat capacity data for the gaseous species were not available, the ideal value of 26.0 $cal_{th} \cdot mol^{-1} \cdot K^{-1}$ was used. The thermodynamic parameters calculated for various halides are given in Table III.

4.   DISCUSSION

4.1. Thorium halides

The vapour pressure of $ThF_4(\ell)$ has been determined by the boiling-tempera-ture method over the temperature range 1395 to 1554 K and the data from the present study are compared with those in the literature in Table IV. It is seen that the vapour pressure values from the present study are higher than those reported by Darnell and Keneshea [12] and the values estimated by Rand [22]. Transpiration studies on $ThF_4(s)$ and $ThF_4(\ell)$ are in progress to ascertain the vaporization behaviour of ThF_4.

The vapour pressure of $ThCl_4$ has been measured over the temperature range 880 to 1161 K both by boiling-temperature and transpiration methods and these data, which have been discussed in detail elsewhere [24], have established the vaporization behaviour of this compound.

The vapour pressure of $ThBr_4$ has been measured over the temperature range 842 to 1089 K by the boiling-temperature method and the data are compared with the only available data, those of Fischer et al. [13] in Table V. For $ThBr_4(s)$ the present data are much lower than those of Fischer et al., but for $ThBr_4(\ell)$ the two sets of data are in good agreement. There seems to be some systematic error in the data for $ThBr_4(s)$ reported by Fischer et al., since the value of ΔH^0 (fus, 952 K) reported by them is much lower than the value of ΔH^0 (fus, 971 K) observed in the present study. Transpiration studies, which are in progress, should resolve the discrepancy.

The vapour pressure of ThI_4 has been measured over the temperature range 790 to 1044 K by the boiling-temperature method and the data are compared with those available in the literature in Table VI. It is seen that for $ThI_4(s)$ the two sets of data reported in the literature are neither in agreement with each other nor with present study. For $ThI_4(\ell)$ the present data are in agreement with

TABLE IV. THERMODYNAMIC AND VAPOUR PRESSURE DATA ON $ThF_4(\ell)$, AND COMPARISON WITH LITERATURE

Investigators, temperature range and method	ΔH^0(g, 298 K) 2nd law ($kcal_{th} \cdot mol^{-1}$)	3rd law	ΔS^0(g, 298 K) ($cal_{th} \cdot mol^{-1} \cdot K^{-1}$)	Pressure (atm) 1400 K	1500 K	1600 K
Darnell and Keneshea [12] (1427–1595 K; boiling temp.)	81.49	80.79	46.1	1.08×10^{-3} extrapolated	5.75×10^{-3}	2.49×10^{-2} extrapolated
Rand [22] (suggested equation)	82.40	80.30	–	1.33×10^{-3}	4.75×10^{-3}	1.66×10^{-2}
Present study (1395–1554 K; boiling temp.)	80.90	78.71	45.4	1.81×10^{-3}	8.53×10^{-3}	3.32×10^{-2} extrapolated

TABLE V. THERMODYNAMIC AND VAPOUR PRESSURE DATA ON $ThBr_4$ AND COMPARISON WITH LITERATURE

Investigators, temperature range and method	ΔH^0(g, 298 K) 2nd law ($kcal_{th} \cdot mol^{-1}$)	3rd law	ΔS^0(g, 298 K) ($cal_{th} \cdot mol^{-1} \cdot K^{-1}$)	Pressure (atm) $ThBr_4$(s) 850 K	950 K	$ThBr_4$(ℓ) 1000 K	1100 K
Fischer et al. [13] (902–1124 K; boiling temp.)	48.98	48.18	47.9	3.45×10^{-3} extrapolated	5.34×10^{-2}	1.38×10^{-1}	6.69×10^{-1}
Present study (842–1089 K; boiling temp.)	56.83	48.95	55.2	1.68×10^{-3}	4.35×10^{-2}	1.34×10^{-1}	6.88×10^{-1} extrapolated

TABLE VI. THERMODYNAMIC AND VAPOUR PRESSURE DATA ON ThI_4 AND COMPARISON WITH LITERATURE

b: boiling temperature; d: damping of silica thread; e: effusion.

Investigators, temperature range and method	ΔH^0(g, 298 K) (kcal$_{th}$·mol^{-1}) 2nd law	3rd law	ΔS^0(g, 298 K) (cal$_{th}$·mol^{-1}·K^{-1})	Pressure (atm) ThI_4(s) 750 K	850 K	ThI_4(ℓ) 900 K	1000 K
Landis and Darnell [18] (unspecified; e)	50.88	49.3	51.4	1.13×10^{-4}	5.39×10^{-4}	–	–
Gerlach et al. [19] (550–690 K; d)	38.05	47.1	34.6	1.32×10^{-4} extrapolated	2.28×10^{-3} extrapolated	–	–
Fischer et al. [13] (856–1105 K; b)	–	47.1	–	–	–	3.64×10^{-2}	2.12×10^{-1}
Present study (790–1044 K; b)	50.46	49.7	54.1	2.33×10^{-4} extrapolated	9.68×10^{-3}	3.46×10^{-2}	2.62×10^{-1}

Fischer et al. [13] at 900 K but not at higher temperatures. Transpiration studies
on this sytem are nearing completion.

4.2. Uranium halides

The vapour pressure of UF_4 has been measured by the boiling-temperature
technique over the temperature range 1246 to 1424 K and by the transpiration
technique over the temperature range 1169 to 1427 K. The two sets of data agree
with each other to within 7% throughout the temperature range of investigation,
and hence the combined set of data is considered to provide a better representation
of the vaporization behaviour of UF_4. Results from the combined set are compared
with those in the literature in Table VII. It can be seen that the present data for
solid UF_4 agree excellently with those of Ryon and Twichell [1]. Three data
points for solid UF_4 reported by Langer and Blankenship [3] covering a range of
20 K are about 14% higher, while the data of Popov et al. [2] are up to 16% lower
than the present data. The agreement between the present data and those reported
by Akishin and Khodeev [4] and Chudinov and Chuprov [5] is poor. For liquid
UF_4, the data from the present study agree with those reported by Langer and
Blankenship [3], but the data of Popov et al. [2] are much lower. On the basis
of their limited data, Popov et al. have suggested the melting temperature of UF_4
is 1242 K, as against the experimentally determined value of (1309 ± 2) K by
Barton and Sheil [28], indicating some systematic error in Popov et al.'s measurements. The present data give 1309 K as the melting temperature, in good
agreement with the directly measured value. The values of $\Delta H^0 (g, 298 \text{ K})$ and
$\Delta S^0 (g, 298 \text{ K})$ are also compared with the literature data in Table VII, and it is
seen that the enthalpy and entropy data agree with the data of Ryon and
Twichell [1], Popov et al. [2] and Langer and Blankenship [3] to within
$0.75 \text{ kcal}_{th} \cdot \text{mol}^{-1}$ and $0.5 \text{ cal}_{th} \cdot \text{mol}^{-1} \cdot \text{K}^{-1}$, respectively.

The vapour pressure of UCl_4 has been measured by the transpiration technique
over the temperature range 763 to 971 K and by the boiling-temperature technique
over the temperature range 796 to 1008 K. The results from the transpiration
studies have been reported elsewhere [25], and these helped in solving the
discrepancy observed earlier while using this technique. The boiling-temperature
data are in good agreement with the transpiration data reported earlier, and with
the data reported by other workers on this system.

The vapour pressure of UBr_4 has been measured by the boiling-temperature
method over the temperature range 759 to 1004 K and the results are compared
with those in the literature in Table VIII. It is seen that the vapour pressure data
for solid UBr_4 from the present study lie between the data of Nottorf and
Powell [29] and Thomson [30]. For liquid UBr_4 the present data are in good
agreement with the data of Gregory [31] and Nottorf and Powell [29]. Further
confirmation will be obtained when the transpiration study is completed.

TABLE VII. THERMODYNAMIC AND VAPOUR PRESSURE DATA ON UF$_4$ AND COMPARISON WITH LITERATURE

b: boiling temperature; e: effusion; t: transpiration.

Investigators, temperature range and method	ΔH^0(g, 298 K) (kcal$_{th}$·mol^{-1})	ΔS^0(g, 298 K) (cal$_{th}$·mol^{-1}·K^{-1})	Pressure (atm)			
			UF$_4$(s)		UF$_4$(ℓ)	
			1200 K	1300 K	1350 K	1450 K
Ryon and Twichell [1] (1013–1133 K; e)	79.13	53.0	5.09X10^{-4} extrapolated	5.59X10^{-3} extrapolated		
Popov et al. [2] (1148–1273 K; t)	78.69	52.2	4.11X10^{-4}	4.45X10^{-3} extrapolated	5.18X10^{-3} extrapolated	1.68X10^{-2} extrapolated
Langer and Blankenship [3] (1314–1573 K; b)	79.40	53.1	—	—	1.26X10^{-2}	5.32X10^{-2}
Akishin and Khodeev [4] (917–1041 K; e)	75.43	52.1	1.50X10^{-3} extrapolated	1.46X10^{-2} extrapolated	—	—
Chudinov and Chuprov [5] (823–1280 K; e)	70.07	46.3	7.78X10^{-4}	6.38X10^{-3} extrapolated	—	—
Present study (1169–1427 K; t, b)	77.95	52.0	4.99X10^{-4}	5.28X10^{-3}	1.22X10^{-2}	5.02X10^{-2} extrapolated

TABLE VIII. THERMODYNAMIC AND VAPOUR PRESSURE DATA ON UBr₄ AND COMPARISON WITH LITERATURE

Investigators, temperature range and method	ΔH^0(g, 298 K) 2nd law (kcal$_{th}\cdot$mol^{-1})	ΔS^0(g, 298 K) (cal$_{th}\cdot$mol$^{-1}\cdot$K^{-1})	Pressure (atm) UBr₄(s)		Pressure (atm) UBr₄(ℓ)	
			700 K	800 K	850 K	950 K
Nottorf and Powell [29] (723–898 K; transpiration)	52.96	67.9	1.24×10^{-4} extrapolated	1.11×10^{-2}	3.39×10^{-2}	2.12×10^{-1} extrapolated
Gregory [36] (815–1033 K; diaphragm gauge, boiling temp.)	–	–	–	–	2.63×10^{-2}	1.33×10^{-1}
	–	–	–	–	3.09×10^{-2}	2.41×10^{-1}
Thomson [30] (573–723 K; effusion)	45.08	47.5	1.68×10^{-4}	8.01×10^{-3} extrapolated	–	–
Present study (boiling temp.)	49.50	54.3	1.31×10^{-4} extrapolated	8.40×10^{-3}	3.07×10^{-2}	2.45×10^{-1}

ACKNOWLEDGEMENT

The authors are grateful to Dr. M.V. Ramaniah, Head, Radiochemistry Division, BARC, for his keen interest in this work.

REFERENCES

[1]   RYON, A.D., TWICHELL, L.P., United States Report TL-7703 (1947).
[2]   POPOV, M.M., KOSTYLOV, F.A., ZUBOVA, N.V., Russ. J. Inorg. Chem. **4** (1959) 770.
[3]   LANGER, S., BLANKENSHIP, F.F., J. Inorg. Nucl. Chem. **14** (1960) 26.
[4]   AKISHIN, P.A., KHODEEV, Yu. Sr., Russ. J. Phys. Chem. **5** (1961) 574.
[5]   CHUDINOV, F.G., CHUPROV, D.Ya., Russ. J. Phys. Chem. **44** (1970) 1106.
[6]   MUELLER, M.E., USAEC Rep. AECD-2029 (1948).
[7]   JOHNSON, O., BUTLER, T., NEWTON, A.S., Chemistry of Uranium (KATZ, J.J., RABINOWITCH, E., Eds), USAEC Rep. TID-5290, Part 1 (1958) 1.
[8]   SHCHUKAREV, S.A., VASILKOVA, I.V., EFFINOV, A.I., USAEC Rep. AEC-tr-3870 (1956).
[9]   SAWLEWICZ, K., SIEKIERSKI, S., Radiochem. Radioanal. Lett. **23** (1975) 71.
[10] YOUNG, H.S., GRADY, H.F., Chemistry of Uranium (KATZ, J.J., RABINOWITCH, E., Eds), USAEC Rep. TID-5290, Part 2 (1958) 749.
[11] BINNEWIES, M., SCHAFFER, H., Z. Anorg. Allg. Chem. **407** (1974) 327.
[12] DARNELL, A.J., KENESHEA, F.J., J. Phys. Chem. **62** (1958) 1143.
[13] FISCHER, W., GEWEHR, R., WINGCHEN, H., Z. Anorg. Allg. Chem. **242** (1939) 161.
[14] YEN, Kung-Fen, LI, Sheo-Chung, NOVIKOV, G.I., Russ. J. Inorg. Chem. **8** (1963) 44.
[15] SU, Mein-Tseng, NOVIKOV, G.I., Russ. J. Inorg. Chem. **11** (1965) 270.
[16] KNACKE, O., MUELLER, F., VAN RENSEN, E., Z. Phys. Chem. (Wiesbaden), Neue. Folge. **80** (1972) 82.
[17] BINNEWIES, M., SCHAFFER, H., Z. Anorg. Allg. Chem. **410** (1974) 251.
[18] LANDIS, A.L., DARNELL, A.J., USAEC Rep. NAA-SR-5394 (1960) 15.
[19] GERLACH, J., KRUMME, J.P., PAWLET, F., PROBST, H., Z. Phys. Chem. (Frankfurt) **53** (1967) 135.
[20] BROWN, E.D., Halides of the Lanthanides and Actinides, John Wiley and Sons, London (1968).
[21] RAND, M., KUBASCHEWSKI, O., The Thermochemical Properties of Uranium Compounds, Oliver and Boyd, London (1963).
[22] RAND, M.H., "Thermochemical properties", Thorium: Physico-Chemical Properties of its Compounds and Alloys (KUBASCHEWSKI, O., Ed.), Atomic Energy Review, Special Issue No. 5, IAEA, Vienna (1975) 25.
[23] WAGMAN, D.D., SCHUMM, R.H., PARKER, V.S., US Dept. of Commerce Rep. NBSIR 77-1300 (1977).
[24] ZILEY SINGH, RAJENDRA PRASAD, VENUGOPAL, V., ROY, K.N., SOOD, D.D., J. Chem. Thermodynamics (in press).
[25] ZILEY SINGH, RAJENDRA PRASAD, VENUGOPAL, V., SOOD, D.D., J. Chem. Thermodynamics **10** (1978) 129.
[26] HULTEGREN, R., DESAI, P.D., HAWKINS, D.T., GLEISER, M., KELLEY, K.K., WAGMAN, D.D., Selected values of the Thermodynamic Properties of the Elements, American Society for Metals, Ohio (1973).

[27] BARIN, I., KNACKE, O., Thermochemical properties of Inorganic Substances, Springer-Verlag, Berlin (1973).

[28] BARTON, C.J., SHEIL, R.J., ORNL unpublished data quoted in Ref. [3].

[29] NOTTORF, R., POWELL, J., Ames Project Report for the Period May 10 to June 10, 1944, Rep. CC 1504 A-2087 (1955) *cited in Ref.* [6].

[30] THOMSON, R.W., SCHELBERG, A., OSRD Proj. SSRC-5, Princeton Rep. 24, A-809 (Oct. 1942) *cited in Ref.* [6].

[31] GREGORY, N.W., Rep. BC-3 (May 1946) *cited in Ref.* [6].

DISCUSSION

M.H. RAND: It is nice to see a systematic study of the vaporization of these halides by more than one technique.

There does seem to be a big problem with the vaporization data for all the thorium tetrahalides, since the agreement between the second- and third-law treatments is always very poor. It is a little difficult to see how the thermal functions for either the condensed or the gaseous phase can be as grossly in error as would be required to obtain agreement. This consideration and the fact that the entropies of vaporization are much larger than usual make one wonder whether the vaporization is as simple as is normally assumed. Is there a possibility of dissociation in the gas phase, for example?

Also, have you made any measurements on the volatility of UCl_3?

D.D. SOOD: If any dissociation or association effects were present during vaporization, the vapour pressure values as measured by the transpiration method would be quite different from those obtained by the boiling-temperature method. Since we do not observe any difference in the vapour pressure data measured by the two techniques for all the compounds, it is safe to assume that only undissociated monomolecular species are present in the vapour phase. Furthermore, in all our transpiration investigations the condensate was a beautiful crystalline material which on chemical analysis gave the expected halogen-to-metal ratio. Also, free chlorine, bromine or iodine was not observed in the gas phase either in the transpiration or in the boiling-temperature studies. So it is necessary to consider other mechanisms which might contribute to the high entropy of vaporization.

We are in the process of measuring the vapour pressures of other uranium halides. We are not sure whether it would be possible to overcome the problem of disproportionation of UCl_3 by using hydrogen carrier gas. However, studies in this direction are in progress.

E.F. WESTRUM Jr.: It is a pleasure to see that you have obtained such extensive data on an important series of actinide compounds. Since you have not stated the precision indices of your derived data on the entropies of melting, may I ask what is the reliability of these values and whether or not you have compared them with such literature values as are available?

D.D. SOOD: The primary data for these calculations are the enthalpies of vaporization of solid and liquid halides at the melting temperatures, which, in our studies, have a precision of approximately $\pm$ 0.5 kcal·mol^{-1}. Therefore, the derived enthalpies of fusion for the halides, which lie between 12 and 16 kcal·mol^{-1}, have a precision of $\pm$ 1 kcal·mol^{-1}. Calorimetric data on most of these compounds are not available and a comparison is not possible. However, the melting temperatures as derived from the data in the present study generally agree with the directly measured data, and this is an indirect proof of the good precision of our data.

M. TETENBAUM: How do you account for the significant difference in vapour pressure values obtained by means of the transpiration technique with UCl_4 systems as compared to the results obtained by other techniques and quoted in the paper?

D.D. SOOD: The vapour pressures determined using the transpiration technique by Johnson et al. and Sawlewicz and Siekierski do not agree with the data obtained by other techniques. However, our transpiration results are in excellent agreement with the results derived by methods such as effusion, boiling-temperature etc. The discrepancy in the case of the above authors probably arises because they did not establish the plateau conditions prior to their investigations.

THERMODYNAMICS OF COBALT-CONTAINING ALLOYS PROPOSED AS MATERIALS FOR HIGH-TEMPERATURE GAS-COOLED REACTOR SYSTEMS

K. HILPERT
Institute of Chemistry,
Kernforschungsanlage Jülich GmbH,
Jülich,
Federal Republic of Germany

Abstract

THERMODYNAMICS OF COBALT-CONTAINING ALLOYS PROPOSED AS MATERIALS FOR HIGH-TEMPERATURE GAS-COOLED REACTOR SYSTEMS.

Advanced high-temperature gas-cooled reactors demand appropriate high-temperature alloys as materials for the components situated outside the reactor core made of graphite. Some important candidate materials are the cobalt-containing alloys NIMONIC alloy PE 13, INCONEL alloy 617 and alloy IN-643. A contamination of the primary cooling cycle can occur if alloying constituents such as ^{59}Co, ^{50}Cr and ^{58}Fe get into the reactor core, where radioactive nuclides are generated by neutron capture. In order to predict the increase in radioactivity by deposition of the nuclides formed, the partial pressures of cobalt, chromium and iron over the alloys are of interest. Therefore, and also to get more data on the thermodynamics of the alloys, their vaporization has been investigated by high-temperature mass spectrometry. Chemical activities and activity coefficients of chromium, nickel, iron and cobalt in NIMONIC alloy PE 13 and IN-643 as well as of chromium, nickel and cobalt in INCONEL alloy 617 were determined. Polished ground sections were made of the measured samples; these indicated the presence of chromium carbides in the alloys. The chemical activities of chromium in Cr_3C_2, Cr_7C_3 and $Cr_{23}C_6$ were computed from the free energies of formation of these carbides given in the literature. In order to check these results the activity of chromium in Cr_7C_3 was determined by high-temperature mass spectrometry. The chemical activities and activity coefficients determined for the alloys are discussed with respect to the microstructure of the alloys and the chromium activities in Cr_3C_2, Cr_7C_3 and $Cr_{23}C_6$ that were obtained.

1. INTRODUCTION

The core gas-outlet temperature of a high-temperature gas-cooled reactor (HTR) using a steam cycle is about 750°C. One of the advantages of the HTR is its ability to provide outlet temperatures of >850°C, even reaching 1000°C. These temperatures are needed in the design of a direct-cycle (gas turbine) HTR, as well as in HTRs to be employed as sources of nuclear process heat, or as heat sources for long distance transfer of energy in the form of 'chemical enthalpy' [1].

TABLE I. NOMINAL COMPOSITION OF THE ALLOYS

In per cent by weight and by atomic fraction

	INCONEL alloy 617		NIMONIC alloy PE 13		alloy IN-643	
	wt%	N_i	wt%	N_i	wt%	N_i
Ni	54.40	0.5386	49.3	0.489	47.45	0.4801
Cr	22.31	0.2494	21.1	0.237	25	0.2856
Co	12.46	0.1229	1.5	0.015	12	0.1210
Fe	0.15	0.0016	18.1	0.189	3	0.0319
Mo	9.09	0.0551	9.13	0.056	0.6	0.0037
W	–	–	–	–	8.75	0.0283
C	0.07	0.0034	0.065	0.0032	0.5	0.0247
Nb	–	–	–	–	2	0.0128
Si	0.08	0.0017	0.35	0.0073	0.45	0.0095
Mn	0.02	0.0002	0.41	0.0043	–	–
Al	1.06	0.0228	–	–	–	–
Ti	0.35	0.0042	–	–	0.15	0.0019
P	–	–	0.01	0.0002	–	–
Zr	–	–	–	–	0.1	0.0007
S	0.007	0.0001	0.006	0.0001	–	–

The use of temperatures higher than 750°C gives rise to increased demands on the materials of the metallic components such as heat exchangers, gas turbine blades, hot-gas ducting and liners outside the graphite core [2]. Materials that might meet all requirements are the nickel-based alloys NIMONIC alloy PE 13 (alloy PE 13), INCONEL alloy 617 (alloy 617) and alloy IN-643 (IN-643). They were supplied by Henry Wiggin and Company Ltd., Hereford, England, International Nickel Company Inc., Huntington, USA, and International Nickel Ltd., England, respectively. The composition of the alloys is indicated in Table I.

Since the aforementioned alloys serve as materials for the components of an HTR primary cooling circuit, their constituents might get into the reactor core, where radioactive nuclides would then be generated by neutron capture. These radionuclides could later leave the core and increase the radioactivity of the components in the primary circuit above that due to the released fission products.

The transport of the alloying constituents can take place by vaporization and/or spalling of alloy particles due to corrosion [3]. The hazard involved in handling the components of the primary circuit resulting from neutron activation of the alloying constituents depends on the neutron activation cross-section of each stable nuclide in the alloy, the isotopic abundance of the nuclide, the activities of the radionuclides generated and their specific gamma-ray constants.

Taking these parameters into account, ^{59}Co — which is transformed into ^{60}Co — may be a dangerous alloying constituent since its isotopic abundance and neutron activation cross-section are comparatively large. The nuclides ^{59}Fe and ^{51}Cr, formed from ^{58}Fe and ^{50}Cr, respectively, are less important than ^{60}Co. This is due to the relatively small values for the corresponding isotopic abundances, neutron activation cross-sections and specific gamma-ray constants.

In order to predict the increase in radioactivity in the primary circuit due to the activation of the alloying constituents, their transport from the alloy through the reactor core has to be considered. In this context, the vaporization of the various alloys was studied using high-temperature mass spectrometry. The chemical activities obtained are needed to predict the formation of oxides or carbides at the surface of the alloys in the primary circuit from the partial pressures of the impurities in the cooling gas [4]. Additionally, the chemical activities of the alloys are of fundamental thermodynamic interest.

The purpose of this paper is to discuss the chemical activities obtained previously by our group [5, 6] for the alloys PE 13, alloy 617 and IN-643 with respect to their microstructure. Polished ground sections were, therefore, made from the alloys following rapid cooling from the temperature of the vapour pressure measurements. Since there is evidence for the formation of chromium carbides in the alloys and in order to explain the chemical activities of chromium determined for the alloys, the chemical activities of chromium in Cr_3C_2, Cr_7C_3 and $Cr_{23}C_6$ were computed. On account of the comparatively large error margins obtained for the computed activities of chromium in Cr_7C_3 and in order to check the computed data, the vaporization of the compound Cr_7C_3 was studied by high-temperature mass spectrometry.

2. EXPERIMENTAL

The experiments were carried out with the commercially available Knudsen cell/CH 5 mass spectrometer system supplied by Varian MAT, Bremen, Federal Republic of Germany. A special ion counting system was added by our group [7]. This was necessary to enable the vaporization of the alloys at temperatures present in a reactor system to be studied. More details about the instrument and other alterations that were carried out are given elsewhere [8, 9].

TABLE II. CHEMICAL ACTIVITIES AND ACTIVITY COEFFICIENTS IN ALLOY PE 13, ALLOY 617 AND IN-643

Alloy	Chemical activities[a]	Temp. range (K)	a_i at 1430 K	γ_i at 1430 K
alloy PE 13	$\log a_{Cr} = -1.378 + 0.137 \times 10^4\ T^{-1}$	1349–1493	0.380 ± 0.057	1.60 ± 0.24
alloy 617	$\log a_{Cr} = -0.981 + 0.062 \times 10^4\ T^{-1}$	1355–1527	0.284 ± 0.043	1.14 ± 0.17
IN-643	$\log a_{Cr} = -1.588 + 0.171 \times 10^4\ T^{-1}$	1350–1510	0.405 ± 0.061	1.42 ± 0.21
alloy PE 13	$\log a_{Ni} = -0.327 - 0.004 \times 10^4\ T^{-1}$	1349–1493	0.442 ± 0.066	0.904 ± 0.13
alloy 617	$\log a_{Ni} = -0.103 - 0.023 \times 10^4\ T^{-1}$	1355–1527	0.545 ± 0.082	1.01 ± 0.15
IN-643	$\log a_{Ni} = -0.405 + 0.010 \times 10^4\ T^{-1}$	1350–1510	0.462 ± 0.069	0.962 ± 0.14
alloy PE 13	$\log a_{Fe} = -1.330 + 0.086 \times 10^4\ T^{-1}$	1349–1493	0.187 ± 0.028	0.989 ± 0.15
IN-643	$\log a_{Fe} = -1.753 + 0.083 \times 10^4\ T^{-1}$	1350–1510	0.0672 ± 0.0101	2.11 ± 0.32
alloy PE 13	$\log a_{Co} = -1.736 + 0.040 \times 10^4\ T^{-1}$	1279–1500	0.035 ± 0.005	2.33 ± 0.33
alloy 617	$\log a_{Co} = -1.046 + 0.024 \times 10^4\ T^{-1}$	1214–1517	0.132 ± 0.020	1.09 ± 0.17
IN-643	$\log a_{Co} = -1.253 + 0.053 \times 10^4\ T^{-1}$	1215–1505	0.131 ± 0.020	1.07 ± 0.16

[a] T in kelvin.

Knudsen cells manufactured from tungsten were used for the vaporization of Cr_7C_3 to avoid the formation of carbides by interaction between the material of the Knudsen cell and the carbon in the samples. About 300 mg Cr_7C_3 were placed in the Knudsen cell for the vaporization measurements. The pulverized compound Cr_7C_3 (particle size 325 mesh, purity better than 99%) was supplied by CERAC, Milwaukee, USA. The main impurities, based on the manufacturer's analysis, are iron and oxygen.

The chromium partial pressures, p, observed over the compound Cr_7C_3 at the Knudsen cell temperature, T, can be computed from the ion intensity, I, of the most abundant chromium isotope from the equation:

$$p = f \cdot I \cdot T \tag{1}$$

The sensitivity:

$$f = \frac{p_c}{I_c \cdot T_c} \tag{2}$$

was determined in the course of a calibration measurement with pure chromium in the cell. The vapour pressure over pure chromium, p_c, was taken from the literature [10].

The measurement of the alloys has been described previously by Hilpert et al. [5, 6].

3.   RESULTS

3.1. Thermodynamic data

The following gaseous species could be observed in the gaseous phase over the alloys: Cr, Ni, Fe, Co for alloys PE 13 and IN-643; Cr, Ni, Co for alloy 617. The chemical activities obtained from the vapour pressure are given in Table II, together with the temperature range of the measurements [5, 6]. Additionally, chemical activities and activity coefficients, $\gamma_i = a_i/N_i$, at the mean temperature of 1430 K were computed using the atomic fractions given in Table I.

It could be shown that the compound Cr_7C_3 vaporizes incongruently according to the reaction:

$$2\,Cr_7C_3\,(s) \rightleftharpoons 5\,Cr(g) + 3\,Cr_3C_2\,(s) \tag{I}$$

TABLE III. STANDARD FREE ENERGIES OF REACTION BETWEEN 298 AND 1673 K, AND CHEMICAL ACTIVITIES AT 1430 K DERIVED THEREFROM AND FROM OUR EXPERIMENTS

Reaction (cf. §3.1) and reference	ΔG_T^0 ($J \cdot mol^{-1}$)	ΔG^0 at 1430 K ($kJ \cdot mol^{-1}$)
(II) [11]	$\Delta G_T^0 = 68534 + 6.44\,T$	77.74 ± 12.55
(III) [11]	$\Delta G_T^0 = 42049 + 11.9\,T$	59.07 ± 12.55
(IV) [11]	$\Delta G_T^0 = 13389 + 0.84\,T$	14.59 ± 12.55

Activity of interest	Activity equation	a_{Cr} at 1430 K
a_{Cr} in $Cr_{23}C_6(s)$ (calculated)	$\log a_{Cr} = 7.443 \times 10^{-2} - 361.02\,T^{-1}$	$6.637 \times 10^{-1} \left\{ \begin{matrix} +3.0 \times 10^{-1} \\ -1.6 \times 10^{-1} \end{matrix} \right.$
a_{Cr} in $Cr_7C_3(s)$ (calculated)	$\log a_{Cr} = -1.733 \times 10^{-1} - 1002.8\,T^{-1}$	$1.335 \times 10^{-1} \left\{ \begin{matrix} +9.1 \times 10^{-2} \\ -6.2 \times 10^{-2} \end{matrix} \right.$
a_{Cr} in $Cr_3C_2(s)$ (calculated)	$\log a_{Cr} = -2.025 \times 10^{-1} - 1469.2\,T^{-1}$	$5.890 \times 10^{-2} \left\{ \begin{matrix} +2.8 \times 10^{-2} \\ -2.2 \times 10^{-2} \end{matrix} \right.$
a_{Cr} in Cr_7C_3 (experimental)	$\log a_{Cr} = -0.523 - 617\,T^{-1}$	$1.11 \times 10^{-1} \left\{ \begin{matrix} +3 \times 10^{-2} \\ -2 \times 10^{-2} \end{matrix} \right.$

The following, typical chromium vapour-pressure curve was obtained from our measurements, carried out over the temperature range 1398–1758 K:

$$\log p_{Cr} = 6.655 - 20787\, T^{-1} \tag{3}$$

where p is in atmospheres and T is in kelvin. It was computed by making a least-squares fit for 14 vapour pressure points. Chemical activities (see Table III) were derived from Eq. (3) by employing the vapour pressures of pure chromium:

$$\log \dot{p}_{Cr} = 7.178 - 20170\, T^{-1} \tag{4}$$

given in Ref. [5]. The stated error of the chemical activity thus determined was estimated from the reproducibility of the measurements.

The chemical activity of chromium in Cr_7C_3 that was determined and the chromium activities in $Cr_{23}C_6$ as well as Cr_3C_2 can be computed from the standard free energies (see Table III) of the reactions:

$$\frac{1}{6} Cr_{23}C_6 \rightleftharpoons \frac{23}{6} Cr + C \tag{II}$$

$$\frac{23}{27} Cr_7C_3 \rightleftharpoons \frac{7}{27} Cr_{23}C_6 + C \tag{III}$$

$$\frac{7}{5} Cr_3C_2 \rightleftharpoons \frac{3}{5} Cr_7C_3 + C \tag{IV}$$

given in Ref. [11]. By suitably adding and subtracting these equations the reactions:

$$Cr + \frac{2}{9} Cr_7C_3 \rightleftharpoons \frac{1}{9} Cr_{23}C_6 \tag{V}$$

$$Cr + \frac{3}{5} Cr_3C_2 \rightleftharpoons \frac{2}{5} Cr_7C_3 \tag{VI}$$

$$Cr + \frac{2}{3} C \rightleftharpoons \frac{1}{3} Cr_3C_2 \tag{VII}$$

are obtained. The standard free energies of the reactions (V)–(VII) can, therefore, be computed from the data given in Table III for the reactions (II)–(IV). Chemical activities of chromium (see Table III) are then obtained using the equation:

$$a_{Cr} = RT \ln \Delta \bar{G}_{Cr} \tag{5}$$

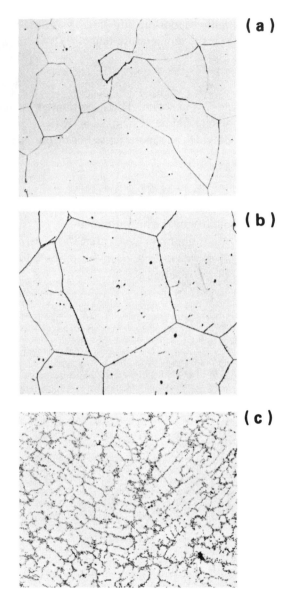

FIG.1.   Microstructures of the samples studied: (a) NIMONIC alloy PE 13; (b) INCONEL
alloy 617; (c) alloy IN-643. The magnification is X 63. (The polished ground sections were
made in the Institute of Reactor Materials, Kernforschungsanlage Jülich GmbH.)

In order to enable an easy comparison to be made of the chemical activities of chromium in the alloys and in the carbides, the chemical activity of chromium at 1430 K was also computed. The stated errors of the chemical activities are probable errors derived from the given errors in the free energies of the reactions (II)–(IV).

Obviously the chemical activity of Cr_7C_3 determined experimentally in this work is in good agreement with the value computed from the data of Ref. [11], indicating that the error margins for the computed activities are too large.

3.2. Microstructure

After the vaporization measurements of PE 13, alloy 617 and IN-643, the alloys in the Knudsen cell, at a temperature of about 1100°C, were cooled down as rapidly as possible. Polished ground sections were made from these alloys. The photomicrographs obtained are depicted in Fig. 1. They show a recrystallized matrix grain structure which represents a solid solution.

The photomicrographs of the alloys PE 13 and alloy 617 (Figs 1a, b) are similar. The intragranular areas contain a few widely scattered carbide precipitates. In addition, parts of the grain boundaries suggest the presence of some very fine carbide precipitates. The photomicrographs in Figs 1a, b resemble those given in the literature [12] for alloy 617. According to this reference, it could be shown by X-ray diffraction that $M_{23}C_6$ (where M is Cr and Mo) is a very abundant phase in the precipitates of alloy 617 at exposure temperatures between 649 and 1093°C. The phases $Cr_{23}C_6$, TiN and γ' (Ni_3Al) observed in addition are comparatively rare. The microstructure of the alloy IN-643 differs from those obtained for alloy PE 13 and alloy 617 (Fig. 1). It contains comparatively large amounts of precipitates, owing to the higher carbon content of the alloy IN-643 (see Table I). According to Ref. [13] the precipitates in alloy IN-643 consist of eutectic carbides and niobium-rich carbides. Tungsten, molybdenum and chromium are distributed between the two types of carbides [13].

These results are in agreement with the general rules concerning nickel-based alloys. Accordingly, these alloys contain an fcc nickel-based austenitic phase which is formed by a high proportion of the solid solution elements nickel, chromium, iron, cobalt, molybdenum and tungsten. Niobium, titanium, chromium, molybdenum and tungsten form carbides [14].

4.   DISCUSSION AND CONCLUSION

The chemical activities of nickel in the three alloys, PE 13, alloy 617 and IN-643, investigated by our group were found to agree to within the precision of measurement. The corresponding activity coefficients are, practically, equal to unity. This can be explained by the microstructure of the alloys, showing that nickel is only present in the solid solution of the fcc matrix, and by the nickel concentrations in the various alloys, which differ by only small amounts. The measurements also show that different compositions of the fcc matrix (see Table I and §3.2) exert no influence on the chemical activity of nickel in the various alloys (to within the accuracy of measurement).

Comments similar to those made for nickel are valid for the activities and activity coefficients of cobalt in alloy 617 and IN-643. The activity coefficient of iron in the alloy PE 13 is also practically unity. The activity coefficients of iron in IN-643 and of cobalt in alloy PE 13 are 2.11 and 2.33, respectively, indicating a tendency to immiscibility.

In contrast to nickel, cobalt and iron, which are essentially only present in the fcc matrix of the alloys investigated, chromium is present in the fcc matrix and in the various carbide precipitates. By comparing the chromium activities given in Tables II and III, it follows that the activity in the alloys is less than in $Cr_{23}C_6$ and greater than in Cr_7C_3 and Cr_3C_2.

This shows that practically no $Cr_{23}C_6$ is contained in the alloys at the temperature of the Knudsen measurements, since the chromium pressures over the alloys were determined with two effusion orifices, 1 and 0.3 mm diameter, yielding the same results. In addition, it follows that the carbides Cr_3C_2 and Cr_7C_3 can be present in the alloys but do not determine the chromium vapour pressures. These pressures may be determined by mixed carbides such as $(Cr, Mo)_{23}C_6$ and/or the fcc matrix.

ACKNOWLEDGEMENTS

The author thanks Professor Dr. H.W. Nürnberg, Director of the Institute of Applied Physical Chemistry, for the kind support of this work. He is also indebted to Dr. I. Ali-Khan and Dr. D.F. Lupton for discussions regarding the microstructure of the alloys. The making of the measurements by Mr. H. Gerads is gratefully acknowledged.

REFERENCES

[1] NÜRNBERG, H.W., WOLFF, G., Naturwissenschaften **4** (1976) 190.
[2] NICKEL, H., High Temp. – High Pressures **8** (1976) 123.
[3] INIOTAKIS, N., Kernforschungsanlage Jülich GmbH Rep. Jül-1394 (1977).
[4] WEDEN, G., NAOUMIDIS, A., NICKEL, H., Kernforschungsanlage Jülich GmbH Rep. Jül-1548 (1978).
[5] HILPERT, K., ALI-KHAN, I., J. Nucl. Mater. **78** (1978) 265.
[6] HILPERT, K., GERADS, H., LUPTON, D.F., J. Nucl. Mater. **80** (1979) 126.

[7] HILPERT, K., Advances in Mass Spectrometry, Vol. 7A, Heyden Son Ltd., London (1978) 584.

[8] HILPERT, K., GERADS, H., High Temp. Sci. **7** (1975) 11.

[9] HILPERT, K., Ber. Bunsenges. Phys. Chem. **83** (1979) 161.

[10] HULTGREN, R., DESAI, P.D., HAWKINS, D.T., KELLEY, K.K., WAGMAN, D.D., Selected Values of the Thermodynamic Properties of the Elements, American Society for Metals, Metals Park, Ohio (1973).

[11] KUBASCHEWSKI, O., EVANS, E.Ll., ALCOCK, C.B., Metallurgical Thermochemistry, Pergamon Press, Oxford (1974).

[12] MANKINS, W.L., HOSIER, J.C., BASSFORD, T.H., Metallurg. Trans. **5** (1974) 2576.

[13] FONTAINE, P.I., PENRICE, P.J., J. Inst. Met. **101** (1973) 15.

[14] DECKER, R.F., SIMS, C.T., *in* The Superalloys (SIMS, C.T., BASSFORD, T.H., Eds), J. Wiley, New York (1972) 33.

DISCUSSION

M. TETENBAUM: How do you relate the results of your measurements to actual reactor conditions?

K. HILPERT: In order to predict the contamination of the primary circuit of a gas-cooled high-temperature reactor due to deposition of ^{60}Co, Iniotakis from the Institute of Reactor Components of the Kernforschungsanlage (KFA) Jülich carried out some theoretical studies. He used the cobalt vapour pressures determined by us for his computations, from which it could be inferred that the increase of radioactivity in the primary circuit due to ^{60}Co could be neglected in comparison with the activity of the fission products.

Various experiments carried out at the Institute of Reactor Materials of the KFA revealed the formation of compounds such as Cr_2O_3 at the surface of alloys exposed to helium that contained impurities (e.g. H_2, H_2O, CO, CO_2 and N_2) of the type expected in the helium atmosphere of HTRs.

These compounds are formed by chemical reactions of the alloying components with the impurities. The chemical activities of the alloying components determined by us make it possible to predict the formation of compounds at the surface of the alloys from the partial pressures of the cooling gas impurities.

M.G. ADAMSON: In your presentation you discussed the possible influence of grain-boundary carbides on the chromium partial pressures measured over the three alloys. Could you please clarify whether or not the chromium partial pressure is determined by a Cr(carbide)/Cr(matrix) equilibrium in these measurements?

K. HILPERT: Our measurements show that the chromium pressures over the alloys are not determined by pure chromium carbides ($Cr_{23}C_6$, Cr_7C_3, Cr_3C_2). It follows from the foregoing and from the known microstructures of the alloys that the chromium pressures over the alloys may be dependent upon chromium dissolved in the fcc matrix of the alloys and/or in the mixed carbides containing, besides chromium, other metals such as molybdenum. The question whether it is chromium in the fcc matrix, or in the mixed carbides or in both that determines the pressure cannot be answered by our studies.

M.G. ADAMSON: What influence, if any, did surface oxidation have on the chromium partial pressure measurements?

K. HILPERT: At the beginning of the investigations the chromium partial pressures determined over the alloys increased by a factor of up to two with an increasing number of measurement runs. A similar effect was observed for the cobalt and nickel partial pressures over the alloys. This was due to oxidation of pure metals observed on their surfaces if the same metal lumps were used for several calibration measurements, and if the Knudsen cell chamber with the hot sample was aerated at comparatively high cell temperatures in order to save time. Oxidation causes a decrease of the Cr, Co and Ni partial pressures during the calibration measurement. This increases the sensitivity, E, used for the pressure computation, leading to the increased vapour pressures over the alloys. This effect could be avoided by using new material at each calibration run and by aerating the Knudsen cell chamber only if the Knudsen cell was cooled down to room temperature. No oxidation of the alloys was observed. During measurement of the alloys the cells containing the alloys were in each case cooled down to room temperature before aeration.

THERMODYNAMICS OF TiB$_2$
FROM Ti-B-N STUDIES*

T.J. YURICK, K.E. SPEAR
Materials Research Laboratory,
Pennsylvania State University,
University Park, Pennsylvania,
United States of America

Abstract

THERMODYNAMICS OF TiB$_2$ FROM Ti-B-N STUDIES.

Thermodynamic data for the formation of TiB$_2$ have been extracted from equilibrium pressure-temperature-composition investigations in the ternary Ti-B-N system. The value $\Delta H_{f,298}^0(TiB_2) = -72.7 \pm 1.6$ kcal/mol determined from a 3rd Law analysis of the present results does not agree either with standard reference book values of about -67 kcal/mol or with later calorimetric results of about -77 kcal/mol. The uncertainty listed with the value reported in this paper is a result of a detailed error analysis on possible random and systematic errors encountered in the investigation.

1. INTRODUCTION

Titanium diboride is a well-known high-temperature material which is extremely hard, resists oxidation and chemical attack, is a good neutron absorber, and has good thermal and electrical conductivity. Its current and projected uses are mainly as crucibles, electrode materials and protective coatings.

The present paper describes investigations on the thermodynamic stability of TiB$_2$ by means of equilibrium measurements in the ternary Ti-B-N system. This research was initiated as a result of our recent studies on the chemical vapour deposition of TiB$_2$ [1, 2]. Thermodynamic data for the diboride were critical in our equilibrium analysis of the Ti-B-Cl-H deposition system, and the $\Delta H_{f,298}^0$ (TiB$_2$) values reported in the literature differed by as much as 40 kcal/mol. A difference of 10 kcal/mol existed even after eliminating the obviously poor values. These large uncertainties prompted us to redetermine the enthalpy of formation by an equilibrium study of the Ti-B-N system which involved few experimental uncertainties.

---

* This research was supported by the United States National Science Foundation, Division of Materials Research.

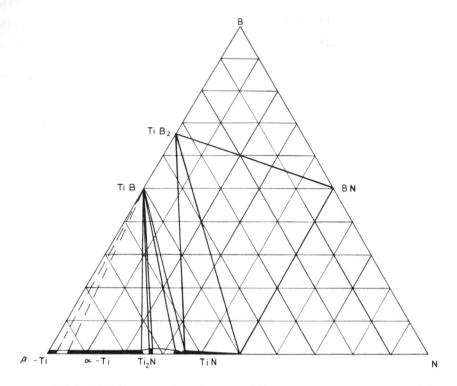

FIG.1.   *Ti-B-N ternary phase diagram at 1773 K as given by Nowotny et al.* [6].

      The method of using ternary equilibrium studies to obtain thermodynamic stabilities of binary phases was first illustrated by Brewer and Haraldsen [3] for boride, carbide and nitride phases. Spear et al. [4] later obtained precise thermo-dynamic values for the vanadium boride phases from investigations of the V-B-N system, and Baehren and Vollath [5] used similar methods in determining W_2B, Mo_2B and Re_3B stabilities.

      Studies by Nowotny et al. [6] on the condensed phase equilibria in the Ti-B-N system are illustrated by the diagram in Fig. 1. They showed that equilibrium could exist among TiN, BN, TiB_2, and $N_2(g)$, and determined the solubility of nitrogen in the titanium borides and of boron in the TiN phase to be very small. Fourteen boron-containing samples of TiN_{1-x} were found to have almost the same lattice constants as those of the pure TiN_{1-x} phase. We therefore decided to study the reaction:

$$TiN(s) + 2BN(s) = TiB_2(s) + \tfrac{3}{2}N_2(g) \qquad\qquad (I)$$

as a function of temperature and nitrogen pressure. By fixing both the temperature
and the nitrogen pressure in an experiment, only two solid phases plus the gas phase
can be present at equilibrium in a ternary system. Using this technique, we carried
out series of experiments at a given constant temperature to determine close limits
on the equilibrium nitrogen pressure for the reaction. Both 2nd and 3rd Law
treatments of the equilibrium data were carried out to obtain the 298 K enthalpy
change. Utilizing auxiliary formation data for TiN and BN, a value for the enthalpy
of formation of TiB_2 was calculated.

2.   BACKGROUND DATA

2.1. Ti-B system

The phase behaviour of the Ti-B system has been summarized by a number
of authors, including Rudy [7] and Hultgren et al. [8]. The congruently melting
(3498 K) TiB_2 phase with the hexagonal AlB_2-type of structure dominates the
diagram. An orthorhombic TiB phase melting by a peritectic reaction at 2463 K
is also firmly established. A Ti_3B_4 phase was reported by Fenish [9], but its
existence has not been confirmed. Little solid solution has been observed in the
two intermediate borides or in the end member Ti(s) and B(s) phases.

2.2. Ti-N system

The phase equilibria in the Ti-N system have been reviewed by Toth [10]
and by Hultgren et al. [8]. The cubic TiN phase of the NaCl-type is best known,
and melts congruently at 3223 K, according to the phase diagram given by
Hultgren et al. [8]. This cubic phase exists over a composition range of about
$TiN_{0.4}$ to $TiN_{1.0}$ at temperatures between about 1400 and 2600 K. A low
temperature tetragonal Ti_2N phase is also known, and α-Ti is stabilized up to
2623 K by the dissolution of up to 20 at.% nitrogen.

2.3. B-N system

Both cubic and hexagonal mononitride phases have been reported for the
B-N system. The cubic phase reported by Wentorf [11] is only stable under very
high pressures. The hexagonal phase with a graphite-type of structure reported
by Pease [12] is the stable form of BN under the conditions of the present
experiments. At 1 atm N_2, BN decomposes to boron and nitrogen at 2600±100 K
(JANAF [13]).

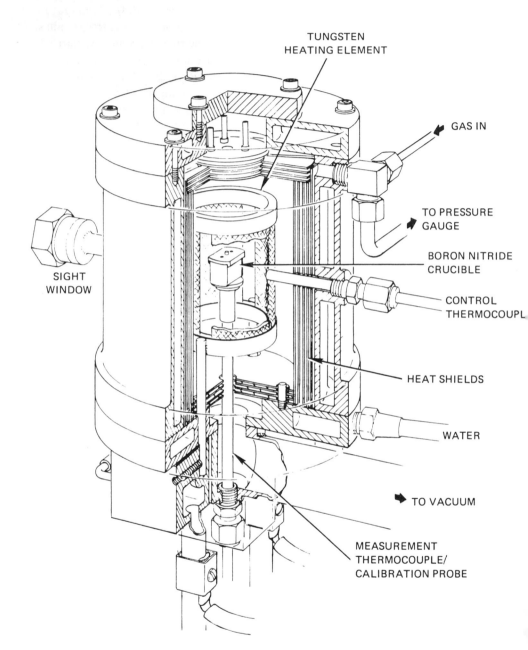

FIG.2.   Schematic of modified Centorr furnace and experimental configuration used in present investigation.   The heating element is about 5 cm dia.

2.4. $\Delta H^0_{f,298}$ (TiB$_2$)

The enthalpy of formation of TiB$_2$ has been reported in the literature by a number of investigators, and values have ranged from -32 to -77.4 kcal/mol. Critical assessments of the TiB$_2$ thermodynamic properties were completed by JANAF [13] in June 1966, and by Hultgren et al. [8] in April 1969. The respective $\Delta H^0_{f,298}$ values for TiB$_2$ listed by the references are -66.8 ± 4.0 kcal/mol and -66.9 kcal/mol. These values were derived after assessing measurements of calorimetric heats [14, 15], vapour pressures [16−18] and the equilibrium positions of chemical reactions involving the diboride [3, 19, 20]. Two calorimetric results have been reported since these assessments. Huber [21] performed oxygen bomb calorimetry on TiB$_2$ and calculated $\Delta H^0_{f,298} = -77.4 \pm 0.9$ kcal/mol. Akhachinskij and Chirin [22] used the method of direct synthesis from the elements in a Calvet calorimeter to obtain a value of -76.14 ± 0.85 kcal/mol.

3. EXPERIMENTAL

3.1. Materials

The materials used consisted of TiB$_2$ samples from Kawecki Berylco Industries, BN for crucibles from Union Carbide Corporation, and research grade nitrogen gas. The TiB$_2$ was in the form of -44 μm powder, and it contained less than 0.03 wt% metallic impurities according to analyses furnished by the supplier and our own spectrochemical laboratory. X-ray diffraction analysis produced sharp TiB$_2$ lines with d-spacings identical to those in patterns taken of diboride samples we produced by arc-melting the elements and by chemical vapour deposition. The only diffraction line not attributable to TiB$_2$ was a very weak line with the d-spacing of the strongest TiN diffraction line.

The (TiN-BN) mixtures were produced by nitriding several grams of TiB$_2$ at temperatures ranging from 1200 to 2000 K. All but one of the samples were nitrided in the same furnace that was used for the equilibrium experiments. The fact that a spectrochemical analysis of the nitrided sample gave identical results to the raw diboride indicates that impurities were not being introduced from the furnace or boron nitride crucible during the heating process.

The crucibles were machined from dense rods of boron nitride, and were then outgassed in vacuum to about 1400 K and in nitrogen to over 2100 K, temperatures greater than were used in the equilibrium experiments. The few per cent boron oxide used as a binder in the fabrication of the dense rods was removed during this process.

The research grade nitrogen gas was purified before being introduced into the furnace chamber by passing it through a heated copper purification unit.

Observations of the furnace chamber after experimental runs indicated very clean conditions, and were in marked contrast with observations after a run when the system contained a very small leak. Oxygen impurity in the system reacted with either the hot furnace assembly or the outside of the BN crucible to produce gaseous oxides, which condensed in the cooler regions of the furnace.

3.2. Equipment

A Centorr (Series 15) high-temperature resistance furnace, equipped with a tungsten mesh heating element was modified for use in the present experiments. A cutaway view of this furnace is shown in Fig.2, and requires little explanation. We connected this furnace to a vacuum system capable of routine operation at pressures of 10^{-8} to 10^{-9} atm.

A Bourdon-type of pressure gauge which could be read to about 1×10^{-3} atm was connected to the furnace in such a manner that the furnace and gauge could be isolated under static nitrogen pressures during an experiment, and pressures could be continuously monitored. This gauge was calibrated against an absolute mercury manometer between 0 and 1 atm, and found to be accurate to about 10^{-3} atm, the reading uncertainty of the gauge. The experiments were carried out under nitrogen pressures of 0.02 to ~0.6 atm.

Temperatures used in obtaining equilibrium data ranged from about 1850 to 2100 K, and could be easily obtained with the above furnace. The temperature of the furnace was controlled during an experiment by means of a proportional controller connected to a W-5%Re/W-26%Re thermocouple inserted through the side of the furnace. The thermocouple was shielded with an alumina sheath, and the tip of the couple was positioned just inside the tungsten mesh of the heating element. Temperature fluctuations as monitored by a thermocouple inserted into the bottom of the sample crucible were ±1 K or less during an experiment.

3.3. Temperature measurement and calibration

The actual temperature measurement was accomplished with the help of a vertical thermocouple probe/crucible assembly (see Fig.2). Details of this assembly are shown in Fig.3. The thermocouples were made from W-5%Re and W-26%Re wires, and each thermocouple was calibrated against the melting points of palladium (1827 K) and platinum (2045 K) before and after a series of experiments. The respective melting points used are those recommended by the International Practical Temperature Scale of 1968 [23]. During each experiment, two calibrated thermocouples were compared as a check against changes in EMF versus temperature characteristics. Measurements of the melting temperatures of Pd and Pt could be repeated to within 0.5 K (0.007 mV) during a given set of calibration experiments, and thermocouples did not change calibration by more than 1 to 2 K during their

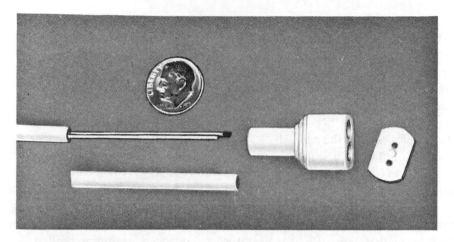

FIG.3. Measurement thermocouple/calibration probe assembly, with twin-cavity crucible
and lid. The top section of the outer sheath for the thermocouple and calibration probe is
lying below the assembly. This sheath fits into the bottom of the crucible and supports it in
the furnace (see Fig.2). The coin is about 18 mm dia.

lifetimes. The 99.8% pure Pd and the 99.99+% pure Pt were in the form of 0.25 mm
diameter wire.

Figure 3 shows the details of the calibration configuration. The thermocouple
and precious metal wire each had its own alumina protection sheath. These were
held in a larger 2-hole alumina sheath that also served as a support for the crucible.
The only difference between the experimental configuration used during calibration
and equilibrium runs was that the precious metal wire was replaced by another
calibrated thermocouple.

During a calibration experiment, the free ends of the precious metal were
connected in series with a small electric light and a variable DC power source.
The power was decreased until the light just barely glowed. The temperature
of the furnace was then slowly raised and the emf of the thermocouple noted
just as the glow disappeared, indicating the metal had melted and broken the
electrical circuit. The temperature of the furnace was always stabilized at about
25 K below the melting temperature of the precious metal to ensure thermal
equilibrium during the calibration. Our estimated absolute uncertainty in our
experimental temperatures is 3 K.

3.4. Procedure and characterization

A typical experimental run involved heating samples at constant temperature
and nitrogen pressure for a given length of time, and then examining the quenched

samples to determine if reaction (I) had proceeded toward the right or toward the left. In other words: Was the nitrogen pressure at the given experimental temperature high enough to nitride the TiB_2? We chose five temperatures at $\sim 50\,K$ intervals between 1875 K and 2082 K, and determined limits on the equilibrium nitrogen pressure for reaction (I) at each temperature by performing experiments at a number of different nitrogen pressures.

A double-cavity BN crucible with a lid , as shown in Fig.3, was used to hold the samples. About 0.5 g of TiB_2 was inserted into one cavity, and a similar quantity of $(TiN + 2BN)$ into the other. The sealed furnace containing a loaded crucible was first heated under vacuum to 1200–1300 K. Upon cooling following this bake-out, the pressure was $<10^{-8}$ atm. Slightly less than the desired amount of purified nitrogen was then introduced into the system, and the furnace temperature was increased automatically over a 1 to 2 hour period to a temperature about 50 K below that desired. The temperature and nitrogen pressure were then increased manually to the desired values, making certain that they did not increase above these values. The duration of the isothermal, isobaric heating ranged from 6 h at the lowest temperatures to 0.5 h at the highest temperatures.

Following the heating, the furnace was turned off and the samples cooled in nitrogen to <1500 K in a few minutes. In order to test if this quench rate was fast enough to avoid reaction with nitrogen while cooling, a few experiments were run in which the system was evacuated at the same time as the furnace was turned off. The altered procedure had no apparent effect on the samples.

After a sample had cooled, it was removed from the crucible and its colour and homogeneity noted. X-ray powder diffraction patterns were taken of each sample and compared with those from previous runs and with standard patterns of the nitrides and diboride. Colour changes from steel grey for TiB_2 to a dark golden-brown for a very nitrided sample, changes in TiN and TiB line intensities, and close monitoring of the strongest BN peak proved to be reliable techniques for determining the direction of reaction during an experiment.

4.   RESULTS AND ANALYSIS

4.1. Experimental results

The limits determined for the equilibrium nitrogen pressure for reaction (I) at each of five different temperatures are shown in Table I. Any pressure equal to or less than the listed P_{min} value at a given temperature resulted in reaction of the nitrides to form TiB_2; any pressure equal to or greater than P_{max} resulted in the nitriding of the diboride to form a mixture of the nitrides.

Ternary solutions in the TiN-BN-TiB_2 region of the system are concluded to be slight, based on the facts that (i) no two phases in this system are isomorphous,

TABLE I. LIMITS ON THE EQUILIBRIUM NITROGEN PRESSURE AND
ln K AS A FUNCTION OF TEMPERATURE FOR THE REACTION[a]
$TiN(s) + 2BN(s) = TiB_2(s) + \frac{3}{2}N_2(g)$

T (K)	N_2 pressure limits (atm)		ln K (footnote b)
	P_{min}	P_{max}	
1875	0.0513	0.0658	−4.2054
1926	0.1000	0.1039	−3.3733
1979	0.1737	0.1829	−2.5352
2029	0.2882	0.2976	−1.7913
2082	0.4842	0.4895	−1.0284

[a] A graph of ln K versus $1/T$ is given in Fig.4.

[b] $K = \{[a_{TiB_2}P_{N_2}^{3/2}]/[a_{TiN}a_{BN}^2]\};\ a_{TiB_2} = a_{BN} = 1;\ a_{TiN} = 0.95 \pm 0.05,\ P_{N_2} = \frac{1}{2}(P_{min} + P_{max}).$
See text for details.

(ii) our X-ray diffraction data gave no evidence for shifts in cell spacings from those
for the pure binary phases, and (iii) Nowotny et al. [6] reached similar conclusions
for this system.

Debye-Scherrer photographs taken with Cu Kα radiation always gave sharp
TiB_2 patterns which never showed even the slighest shifting of 2θ values, even at
2θ angles greater than 140°. The strongest BN line, which was the only one which
could always be easily observed for this phase, never shifted from one experiment
to the next. Also, the fact that the crucibles were made of boron nitride helped
to ensure unit activity for BN(s) in the system.

Titanium mononitride is the most uncertain of the phases in terms of possible
deviations from stoichiometry. The strong low-angle X-ray diffraction lines never
shifted from one experiment to the next, and always appeared at 2θ values corresponding to $TiN_{1.0}$. Thus, large deviations from stoichiometry are unlikely.
However, the high angle lines were not very sharp in that $(\alpha_1-\alpha_2)$ diffraction lines
for (hkl)s at 2θ values greater than 140° appeared together as a broad diffuse line
rather than as two sharp lines as they did in TiB_2 patterns. This lack of sharpness
could be caused by a lack of homogeneity in terms of the N/Ti ratio of the phase.
However, we feel certain that this ratio was always within the range from 0.90
to 1.00.

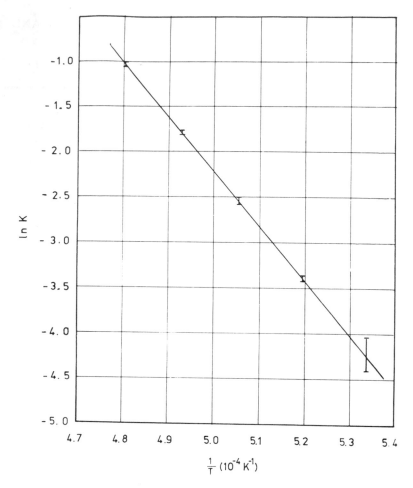

FIG.4.   Plot of ln K versus 1/T for reaction (I). The uncertainties in ln K at each temperature are indicated.

Based on the above results and conclusions, we assumed activities for the phases and calculated the equilibrium constants for reaction (I) at the five experimental temperatures. The ln K values are listed in Table I. The activities of TiB_2 and BN were assumed to be equal to unity; the activity of TiN was assumed to be equal to 0.95±0.05; the activity of nitrogen was assumed to be equal to the average of P_{min} and P_{max}, as listed in Table I. A plot of ln K versus 1/T is given in Fig.4.

4.2. Auxiliary thermodynamic data employed

The auxiliary thermodynamic data used in this study are listed in Table II together with the reported uncertainties in the $\Delta H^0_{f,T}$ and S^0_T values. In choosing which data to use, standard reference sources of JANAF [13], Hultgren et al. [8], Kubaschewski et al. [24] and NBS Technical Note 270–5 [25] were examined together with literature from the past ten years. Little information has been produced in the recent years except for the above-mentioned calorimetric results on the enthalpy of formation of TiB_2.

General agreement among the standard reference sources is quite good for the TiN(s), BN(s) and N_2(g) data listed, and for the S^0_{298} value for TiB_2. The listed uncertainties appear to be quite reasonable. The TiB_2 entropy has been measured by Westrum [26], but has not been confirmed by measurements from other laboratories. However the TiB_2 value is particularly consistent with values for other transition metal diborides (Spear et al.[4]), and is believed to be accurate within the uncertainty limits given.

The high-temperature heat content values for TiB_2 listed by JANAF [13] and by Hultgren et al. [8] differ appreciably. Therefore values from both sources are listed in Table II for comparison. The Hultgren et al. [8] values were used in analysing the experimental data since they give the best agreement between the 2nd and 3rd Law enthalpies and entropies of reaction.

4.3. Calculated thermodynamic properties

4.3.1. Reaction (I): TiN(s) + 2BN(s) = TiB_2(s) + $\frac{3}{2}N_2$(g)

A 2nd Law treatment of the data involves using the equation:

$$\ln K = -\frac{\Delta H^0_T}{RT} + \frac{\Delta S^0_T}{R} \tag{1}$$

and assuming that ΔH^0_T and ΔS^0_T are constant over the temperature range of the experiments. Thus a plot of ln K versus $1/T$ has a slope equal to $-\Delta H^0_T/R$ and an intercept equal to $\Delta S^0_T/R$. The temperature these quantities refer to is the reciprocal of the median of the $1/T$ values, which is 2000 K in the present experiments. The 1875 K experimental temperature was not considered since its equilibrium pressure limits have such a wide spread.

We calculated the slope of a line passing through the 1926 K and 2082 K values of ln K in a 2nd Law plot. These values, listed in Table I, gave:

$$\Delta H^0_T = 119.8 \text{ kcal}$$

for reaction (I). From the intercept we calculated:

$$\Delta S^0_T = 55.5 \text{ cal} \cdot K^{-1}$$

TABLE II. AUXILIARY THERMODYNAMIC DATA EMPLOYED[a]

T (K)	298	298	2000	2000	1875	1926	1979	2029	2082
Function	$\Delta H^0_{f,T}$ (kcal/mol)	$\Delta S^0_{f,T}$ (cal·mol⁻¹·K⁻¹)	$(H^0_T - H^0_{298})$ (kcal/mol)	$(S^0_T - S^0_{298})$ (cal·mol⁻¹·K⁻¹)	$-(G^0_T - H^0_{298})/T$ (cal·mol⁻¹·K⁻¹)				
TiN(s)	−80.7 ± 1.0	7.23 ± 0.05	21.523	22.775	18.552	18.836	19.126	19.394	19.675
BN(s)	−59.97 ± 0.37	3.536 ± 0.05	17.370	17.379	11.677	11.905	12.139	12.355	12.582
N₂(g)	–	45.77 ± 0.01	13.418	14.452	53.084	53.261	53.442	53.610	53.784
TiB₂(s)[b]	footnote c	6.808 ± 0.1	31.726	32.307	22.242	22.659	23.085	23.481	23.895
TiB₂(s)[d]	footnote c	6.81	30.633	31.479	21.995	22.399	22.811	23.189	23.577

a All data taken from JANAF [13] unless otherwise indicated.
b Not used in data analysis.
c Reported values for $\Delta H^0_{f,298}$(TiB₂) are discussed in §2.4.
d Data in this column are from Hultgren et al. [8], and are used in analysing the experimental data.

Using the $(H_T^0 - H_{298}^0)$ and $(S_T^0 - S_{298}^0)$ values listed in Table II, the following 2nd Law values have been calculated for reaction (I):

$\Delta H_{298}^0 = 125.3$ kcal

$\Delta S_{298}^0 = 59.9$ cal·K^{-1}

A 3rd Law treatment of the data uses the equation:

$$\Delta H_{298}^0 = -T \left(R \ln K + \frac{\Delta(G_T^0 - H_{298}^0)}{T} \right) \tag{2}$$

The data in Table I and auxiliary data in Table II yield:

T (K)	1875	1926	1979	2029	2085
ΔH_{298}^0 (kcal)	127.6	127.8	127.9	128.0	128.0

The average of the four highest temperature values is 127.9 kcal.

4.3.2. TiB₂ values

Using the 298 K ΔH^0 and ΔS^0 values for reaction (I) and the auxiliary data in Table II, the $\Delta H_{f,298}^0$ and S_{298}^0 values for TiB_2 can be calculated. These values are:

	2nd Law	3rd Law
$\Delta H_{f,298}^0(TiB_2)$ (kcal·mol^{-1})	−75.3	−72.7
$S_{298}^0(TiB_2)$ (cal·mol^{-1}·K^{-1})	5.5	6.81

The 3rd Law entropy is of course the listed value in Table II. The more negative 2nd Law enthalpy of formation and smaller 2nd Law entropy could both be

TABLE III. UNCERTAINTIES IN ΔH^0_{298} VALUES FOR REACTION (I) CAUSED BY EXPERIMENTAL UNCERTAINTIES

Source of uncertainty	$\delta \Delta H^0_{298}$ (kcal)	
	2nd Law	3rd Law
δP_{N_2} (± 0.005 atm)	± 1.89[a]	± 0.15
δT (± 3 K)	± 2.55[b]	± 0.20
δa_{TiN} (± 0.05)	± 1.33[c]	± 0.21
Total $= \sqrt{(\Sigma x^2)}$	± 3.44	± 0.33

[a] P_{max} (2082 K) and P_{min} (1926 K) were used to calculate the maximum change in slope which could be caused by pressure uncertainties.

[b] δT of $+2$ K at 2082 K and -2 K at 1926 K were used to calculate the effect of a systematic temperature error on the slope.

[c] δa_{TiN} of 0.0 at 2082 K and 0.05 at 1926 K were used to calculate the effect of a systematic error in a_{TiN} on the slope.

TABLE IV. UNCERTAINTIES IN $\Delta H^0_{f,298}(TiB_2)$ VALUES

Source of uncertainty	$\delta \Delta H^0_{f,298}(TiB_2)$ (kcal/mol)	
	2nd Law	3rd Law
Experimental[a]	± 3.44	± 0.33
$\Delta H^0_{f,298}$ nitrides[b]	± 1.24	± 1.24
C^0_p data[c]	± 2.0	± 1.0
Total $= \sqrt{(\Sigma x^2)}$	± 4.17	± 1.63

[a] See Table III.

[b] Uncertainties in TiN and BN values.

[c] Estimated uncertainties related to free energy functions, and enthalpy and entropy increments.

brought into agreement with the 3rd Law values if the slope of the ln K versus
1/T plot for reaction (I) were slightly steeper. The uncertainties in these values
are discussed in the next section.

4.4. Error analysis

The experimental uncertainties in T,P(N_2) and a_{Tin} were translated into
uncertainties in the ΔH^0_{298} values for reaction (I) and are listed in Table III. The
total uncertainty is calculated from the formula in the table.

The uncertainties resulting from the auxiliary data in Table II can be divided
into: (i) the formation data for the nitrides, (ii) the entropy data, and (iii)
functions derived from the high-temperature C^0_p data. The enthalpy of formation
of TiN(s) has an uncertainty of ±1.0 kcal and 2BN(s) have ±0.74 kcal. Their use
to calculate the TiB_2(s) enthalpy of formation from reaction (I) produces a
±1.24 kcal uncertainty.

The difference between the JANAF [13] and Hultgren et al.[8] thermal
function data for TiB_2 causes 2nd and 3rd Law enthalpy differences of about
1.1 and 0.6 kcal respectively, and the total uncertainty caused by possible high-
temperature C^0_p errors is estimated to be no more than ±2.0 kcal and ±1.0 kcal,
respectively.

All of the estimated uncertainties in $\Delta H^0_{f,298}$ (TiB_2) are summarized in
Table IV. It can be seen that experimental uncertainties dominate the total
uncertainty in the 2nd Law value, while the uncertain auxiliary values dominate
the total 3rd Law uncertainty. The uncertainty in the 2nd Law S^0_{298}(TiB_2) is
estimated at about ±2.0 cal·mol^{-1}·K^{-1} by a similar analysis.

5.   SUMMARY AND DISCUSSION

The values for $\Delta H^0_{f,298}$ and S^0_{298} for TiB_2 as measured in this investigation
are summarized below together with their uncertainties.

TiB_2	2nd Law	3rd Law
$\Delta H^0_{f,298}$ (kcal·mol^{-1})	−75.3 ± 4.2	−72.7 ± 1.6
S^0_{298} (cal·mol^{-1}·K^{-1})	5.5 ± 2.0	6.81 ± 0.1

The 3rd Law values are preferred since systematic errors in T,P(N)$_2$ or a_{TiN} could
easily cause the slope and intercept for reaction (I) to be too small. The a_{TiN} is
particularly suspected as a possible cause of the differences between 2nd and 3rd

Law values since a lower temperature $a_{TiN} = 1.00$ and a higher temperature $a_{TiN} = 0.90$ would change the 2nd Law enthalpy by about 2.7 kcal in the direction needed for agreement.

The recent calorimetric $\Delta H^0_{f,298}(TiB_2)$ values of -76.14 ± 0.85 kcal/mol [22] and -77.4 ± 0.9 kcal/mol [21] are both significantly more negative than the present 3rd Law value of -72.7 ± 1.6 kcal/mol. A reason for this disagreement is not known. The error analysis performed in the present study is believed to be quite reasonable, and it is very unlikely that the uncertainties are larger than given. Analytical chemistry uncertainties or incomplete reaction are the most likely sources of error in the calorimetric values. Small uncertainties in these areas can cause errors of several kcal/mol. In Huber's [21] oxygen bomb-calorimetry experiments, for example, combustion of only 99% as much sample as is believed to have burned would change the resulting $\Delta H^0_{f,298}(TiB_2)$ value from -77.4 to -72.9 kcal/mol.

Two other papers have examined reaction (I) in somewhat less detail than the present investigation. Brewer and Haraldsen [3] used a constant N_2 pressure of 0.5 atm, and from two experiments estimated the approximate equilibrium temperature for reaction (I) to be 1820 K. Performing a similar 3rd Law treatment of these data as was used in the present study yields for reaction (I) $\Delta H^0_{298} = 112.4$ kcal, and $\Delta H^0_{f,298}(TiB_2) = -88.2$ kcal/mol. This value is very negative, but then Brewer and Haraldsen [3] made no attempt to quantify their results. Williams [19] also used 0.5 atm nitrogen pressure and determined an equilibrium temperature of 2150 ± 25 K. This gives for reaction (I) $\Delta H^0_{298} = 131.8$ kcal, and $\Delta H^0_{f,298}(TiB_2) = -68.8$ kcal/mol. The listed ± 25 K uncertainty in temperature causes about a ± 1.5 kcal uncertainty in the calculated ΔH^0 values. Williams [19] states he could not show the reversibility of the reaction of the nitrides to form TiB_2, so said the 2150 K value is an upper limit. As an upper limit, his temperature is in agreement with the present results.

Besmann and Spear [1, 2] also concluded that $\Delta H^0_{f,298}(TiB_2) = -66$ kcal/mol was the most positive value that could still be consistent with their chemical vapour deposition results in the Ti-B-Cl-H system.

In order to further reduce the uncertainty in the presently determined enthalpy of formation for TiB_2, three types of improved data are needed. First, and most important, is the need for accurate high-temperature heat-content data for TiB_2. A marked reduction in the uncertainties in the free energy functions, enthalpy increments, and entropy increments would result.

Second, a re-determination of the TiN enthalpy of formation would be helpful. The BN value may be about as accurate as can be determined without extreme effort. Third, the stoichiometry of TiN in equilibrium with TiB_2, BN and $N_2(g)$ at temperatures above 1800 K should be examined in detail, and compared with the TiN stoichiometry in equilibrium with $N_2(g)$ at the same pressures and temperatures. This information would reduce the uncertainties regarding the TiN activity in the present studies. We believe the temperature and pressure uncertainties have been reduced to a minimum in the present studies.

ACKNOWLEDGEMENTS

This research is abstracted in part from the M.S. Thesis of T.J. Yurick submitted to The Pennsylvania State University (1979). Kawecki Berylco Industries kindly donated the TiB_2 powder used.

REFERENCES

[1] BESMANN, T.M., SPEAR, K.E., J. Cryst. Growth **31** (1975) 60.

[2] BESMANN, T.M., SPEAR, K.E., J. Electrochem. Soc. **124** 5 (1977) 786; J. Electrochem. Soc. **124** 5 (1977) 790.

[3] BREWER, L., HARALDSEN, H., J. Electrochem. Soc. **102** (1955) 399.

[4] SPEAR, K.E., SCHAFER, H., GILLES, P.W., "Thermodynamics of vanadium borides," High Temperature Technology, Butterworths, London (1969) 201.

[5] BAEHREN, F.D., VOLLATH, D., Planseeber. Pulvermetall. **17** (1969) 180.

[6] NOWOTNY, H., BENESOVSKY, F., BRUKL, C., SCHOB, O., Monatsh. Chem. **92** (1961) 403.

[7] RUDY, E., USAF Materials Laboratory Rep. AFML-TR-65-2, Part V (1969) 198.

[8] HULTREN, R., DESAI, P.D., HAWKINS, D.T., GLEISER, M., KELLEY, K.K., Selected Values of Thermodynamic Properties of Binary Alloys, American Society for Metals, Metals Park, Ohio (1973).

[9] FENISH, R.B., Trans. Metall. Soc. AIME **236** (1966) 804.

[10] TOTH, L.E., Transition Metal Carbides and Nitrides, Academic Press, New York (1971).

[11] WENTORF Jr., R.H., J. Chem. Phys. **26** (1957) 956.

[12] PEASE, R.S., Acta Crystallogr. **5** (1952) 356.

[13] STULL, D.R., PROPHET, H., JANAF Thermochemical Tables, 2nd ed., US Government Printing Office, Washington, DC (1971).

[14] LOWELL, C.E., WILLIAMS, W.S., Rev. Sci. Instrum. **32** (1961) 1120.

[15] EPEL'BAUM, V.A., STAROSTINA, M.L., Bor. Trudy Konf. Khim., Bora i Ego Soedinenii (1958) 97.

[16] KIBLER, G.M., LYON, T.F., LINEVSKY, M.J., DESANTIS, V.J., Technicals Rep. No. WADD-TR-60-646, Part III, Vol.2, General Electric Co., Evandale, Ohio (1964).

[17] SCHISSEL, P.O., TRULSON, O.C., J. Phys. Chem. **66** (1962) 1492.

[18] FESENKO, V.V., BOLGAR, A.S., Sov. Powder Metall. Met. Ceram. **1** (1963) 11.

[19] WILLIAMS, W.S., J. Phys. Chem. **65** (1961) 2213.

[20] SAMSONOV, G.V., J. Appl. Chem. USSR **28** (1955) 975.

[21] HUBER, E.J., Jr., J. Chem. Eng. Data **11** 3 (1966) 430.

[22] AKHACHINSKIJ, V.V., CHIRIN, N.A., "Enthalpy of Formation of Titanium Diboride," Thermodynamics of Nuclear Materials 1974 (Proc. Symp. Vienna, 1974) Vol.2, IAEA, Vienna (1975) 467.

[23] International Practical Temperature Scale of 1968 (IPTS-68), Metrologia **5** (1969) 35.

[24] KUBASCHEWSKI, O., EVANS, E.Ll., ALCOCK, C.B., Metallurgical Thermochemistry, 4th ed., Pergamon Press, Oxford (1967).

[25] WAGMAN, D.D., EVANS, W.H., PARKER, V.B., HALOW, I., BAILEY, S.M., SCHUMM, R.H., CHURNEY, K.L., NBS Technical Note 270-5, US Dept. of Commerce, Washington, DC (1971).

[26] WESTRUM Jr., E.F., Technical Documentary Report No. RTD-TDR-63-4096, Part I
     (KAUFMAN, L., CLOUGHERTY, E.Y., Eds), Manlabs, Inc., Cambridge, Massachusetts
     (1963) 239.

DISCUSSION

P.A.G. O'HARE *(General Chairman)*: It seems to me that fluorine bomb
calorimetry could provide a reliable value for $\Delta H_f^0(TiB_2)$ and thus remove the
discrepancies between the various values for this quantity. Would it be possible
to obtain a sufficiently pure specimen of TiB_2 for this purpose?

K.E. SPEAR: It would be fairly easy to obtain a sufficiently pure specimen
of single phase TiB_2; however, it would be difficult to determine the exact
composition of such a sample. A knowledge of the exact stoichiometry is
extremely important, as is illustrated by a previous calculation of mine, which
was related to the fluorine bomb calorimetry of ZrB_2. If a sample burned in
fluorine is thought to be $ZrB_{2.00}$, and is actually $ZrB_{1.98}$, the error in the resulting
$\Delta H_f^0(ZrB_2)$ is of the order of 4 kcal/mol.

Section B

EQUATION-OF-STATE STUDIES

MEASUREMENT AND ANALYSIS OF TRANSIENT VAPORIZATION IN OXIDE FUEL MATERIALS*

D.A. BENSON, E.G. BERGERON
Sandia Laboratories**,
Albuquerque, New Mexico,
United States of America

Abstract

MEASUREMENT AND ANALYSIS OF TRANSIENT VAPORIZATION IN OXIDE
FUEL MATERIALS.

This paper describes a series of experiments in which samples are heated to produce high-vapour-pressure states in times of 10^{-6} to 10^{-3} seconds. Experimental measurements of vapour pressures over fresh UO_2 from the pulsed electron beam and pulsed reactor heating tests are presented and compared with other high-temperature data. The interpretation of the vapour pressures measured in the tests is discussed in detail. Effects of original sample stoichiometry, chemical interactions with the container and non-equilibrium evaporation due to induced temperature gradients are discussed. Special attention is given to dynamic behaviour in rapid heating and vaporization of the oxide due to chemical non-equilibrium. Finally, similar projected reactor experiments on irradiated fuel are described and vapour-pressure predictions are made using available equilibrium models. A discussion of information accessible from such future tests and of its importance is included.

I. INTRODUCTION

In the production of temperatures above 3000 K the radiative loss of energy and the difficulty of finding suitably stable sample containers complicate the use of steady state high temperature experiments [1]. Even for situations where large radiative losses can be overcome with high power heating, the associated temperature gradients make analysis difficult. In experiments which require containment of the sample, the interdiffusion of sample and container materials at these extreme temperatures is capable of altering the sample properties, thus producing errors. Our approach in dealing with these problems has been to develop methods based upon the rapid transient heating of materials. These methods have enabled us to perform measurements of oxide reactor fuel properties at temperatures previously unattainable by steady state heating.

---

* This work was sponsored by the United States Nuclear Regulatory Commission.
** A United States Department of Energy Facility.

Recent analysis [2,3] of fast reactor power transients and hypothetical disassembly events has focused attention on the vaporization of reactor fuels undergoing rapid rates of heating. Vaporization properties in the temperature range from 4000 to 7000 K have been used in these analyses while the experimental data available to support these calculations had until recently been limited to temperatures well below this level. Transient vaporization data for fuel materials are required for two main purposes in the present safety analysis program. First, the vapor pressure data are needed to interpret the results of more complex transient heating experiments on complete fuel pin-coolant channel assemblies. Such data permit one to separate the pressure contribution due to the fuel from that due to vaporizing coolant or cladding material. This distinction is particularly important in developing models to describe fuel pin behavior during a transient event. Second, these data are needed to determine the adequacy of high temperature fuel constitutive relations being used in accident analysis. For example, the importance of rate dependent effects associated with diffusion processes and the departure from equilibrium due to a high vaporization flux must be considered.

The present paper will describe the methods which we have developed for oxide fuel vaporization, as well as means used to analyze these experiments. Pulsed electron beam and pulsed reactor measurements made at Sandia Laboratories will be described. Comparison of these results for fresh oxide material to other high temperature experiments will be presented. The potential importance of dynamic effects on the experiment and on reactor analysis will be discussed. Finally, an analysis extending these methods to the study of pre-irradiated mixed oxide fuels will be presented.

II. ELECTRON BEAM VAPORIZATION

The first high vapor pressure UO_2 data [4,5] from our work resulted from pulsed electron beam experiments. Unlike previous vaporization work, at a given pressure the energy state of the vaporizing liquid is determined rather than the temperature. This equation of state data format, i.e. $P(\rho,E)$, is the natural format for transient analyses in which thermal conduction is of secondary importance in determining the state of the heated material. This format may be used directly for the early time transient analysis of prototypic fuel pin tests in fast reactor safety analysis.

The Sandia Relativistic Electron Beam accelerator has been used [4] to heat samples of UO_2 in a time of ~ 1 µs using electrons in a beam with a 1.2 MeV average energy spectrum. Vapor pressure measurements are carried out with the apparatus [6]

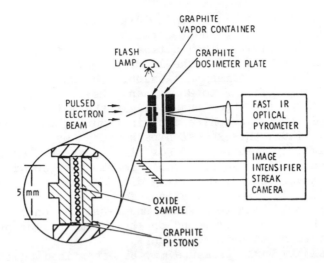

FIG.1.   Schematic view of the pulsed vaporization experiment. A thin UO₂ vapor sample is confined between pistons in the graphite vapor container. Pulsed-electron-beam vaporization of the sample induces container expansion which is monitored by a streak camera. An optical pyrometer measurement and other beam diagnostics determine the vapor sample energy state.

shown schematically in Fig. 1.   The cross sectional area of the electron beam may be varied to produce a specific energy of absorption in the target which varies from low levels to more than 3000 J/g.   Each target is composed of a thin vaporization sample sandwiched between two thin graphite pistons pressed into a graphite cylinder, thus forming a crucible which confines the vapor.   The target is positioned in the electron beam path where it is heated by an electron pulse.   After the sample is heated, the pistons which inertially confine the mixed liquid-vapor material are accelerated in opposite directions by the vapor pressure.   Two types of observations are made during each experiment.   First, the normal displacement of each target piston is measured by streak camera photography.   The piston expansion trajectories obtained from these records together with simple equations of motion are used to determine the pressure of the vapor sample.   Second, the specific energy absorbed at the center of a calorimeter plate positioned behind the target is measured with a time resolving optical pyrometer [7].   This specific energy is closely related to the specific energy of the vapor sample since the target and dosimeter are thin compared to the range of the electrons. These two measurements give the total vapor pressure as a function of energy along the liquid saturation curve.   The energy state, which is volume dependent, closely corresponds to a value on the liquid saturation curve.

The function of the graphite crucible in this experiment
is important. The large heat capacity per gram of carbon,
relative to UO_2, forces the initial graphite temperature to
remain below 2000°C, while the liquid-vapor sample reaches a
temperature several times that value. The short duration of
the experiments (from 2 to 20 μs) and the low thermal
conductivity of the vapor phase, limit thermal loss from the
vapor sample to the graphite container during the vapor driven
expansion. Thus, the expansion path traced by an experiment
is approximately isentropic. For mixed phase states away from
the critical region, an isentrope closely corresponds to
isobaric expansion thereby giving displacement profiles which
are parabolic in time. Deviations from isobaric and adiabatic
conditions could lead to a small under-estimation of the vapor
pressure; hence, a lower bound on the dynamic vapor pressure
is determined by these measurements.

The sample vapor pressure may be determined from the
graphite crucible expansion measurements by a simple differentiation analysis as follows. Since the pressurized vapor
sample is confined only by the inertial forces of the graphite
pistons and sample materials, the container expansion and
internal pressure may be related using the equations of motion
for the crucible walls. The compressibility of graphite is
orders of magnitude smaller than that of the expanding mixed-
phase sample so that the graphite can be approximated as a rigid
body and the equations of motion reduce to a simple form. The
pressure at the interior face of the piston as a function of
time, $P(t)$, is related to the piston displacement x by $P(t) = M\ddot{x}$,
where M is the mass per area of the piston and $\ddot{x}$ is the piston
acceleration.

In addition to the vapor driven acceleration of the piston,
there is a short duration impulse from the impact of condensed
phase liquid on the piston immediately following the pulsed
heating. This initial acceleration is larger than the purely
vapor driven acceleration. It is the later time acceleration
occurring after the condensed phase impulse which is used to
determine the UO_2 vapor pressure.

The energy density in the mixed phase sample is determined
by calorimetry and calculations based upon electron beam
diagnostic information. During each vapor experiment, the
temperature rise at the center of a graphite calorimeter plate
positioned behind the sample is measured. This temperature rise
is determined by infrared optical pyrometry after the electron
pulse, but before thermal diffusion alters the initial temperature distribution. The measurement must also be completed
before the calorimeter plate is destroyed by the rapidly
expanding vapor container. Enthalpy data [7] for the graphite
dosimeter material are used to relate the measured temperature
rise to the absorbed energy density at the plate surface.

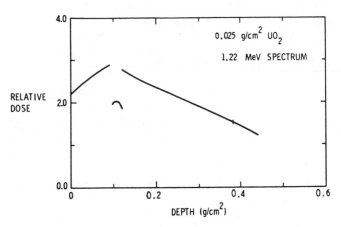

FIG.2. *Calculations of the electron energy absorbed for the target axis as a function of mass depth, X, for one vapor sample configuration used in these tests. The position of the UO₂ layer between layers of graphite is indicated by the curve discontinuities. The point at approximately 0.4 g/cm² depth is the point corresponding to the optical pyrometer measurement.*

The relation between the specific energy measured at the calorimeter plate position and the UO_2 sample specific energy is based upon electron transport calculations. An energy absorption profile in the target-calorimeter assembly is calculated by the electron-photon Monte Carlo method [8] using the experimentally determined electron beam spectrum as the input [9]. A typical absorption profile calculated for the UO_2 vapor experiments is shown in Fig. 2. The ratio of energy in the vapor sample to energy measured in the calorimeter plate is taken from this figure to deduce the energy in the vaporization sample for each experiment. The validity of the transport calculation for layered geometries of graphite and uranium bearing materials has been independently verified [10].

The energy state of the sample after heating at an average specific volume V_0 defined by the graphite container is taken to be the same as the energy of the liquid phase. Actually, a small energy difference occurs due to the vaporization which fills the crucible void space with vapor. This energy increment ΔE can be computed from $\Delta E = (\partial E / \partial V)_S (V_0 - V) = P(V - V_0)$, where E is the mixed-phase internal energy. Values of ΔE for the present experiments are less than 1% of the energy absorbed from the electron pulse and thus are negligible. Moreover, appropriate experiment design can be used to minimize $(V - V_0)$ to any desired level thereby reducing ΔE further if necessary. Thus, the energy absorbed from the electron beam can be directly identified as the internal energy on the liquid phase boundary, relative to the internal energy of the sample prior to heating.

This energy state and the vapor pressure corresponding to that
state are the quantities determined by the measurement method.

III. ANALYSIS OF VAPORIZATION PHENOMENA

Application of pulsed electron beam and pulsed reactor
heating to measurement of vapor equations of state at high
temperature is relatively new. Thus, much consideration has
been given to the interpretation of results from these experiments and toward examining possible sources of anomalous results.
In this section we will review some assumptions which are
implicit in the analysis of the electron beam heating tests.

CO Formation

One such question relates to the possibility of significant
carbon monoxide formation from interaction of UO_2 vapor species
with the graphite container walls. At the high temperature of
these transient vaporization experiments it is probable that a
fraction of the vapor will be in the form of atomic oxygen [11].
This is able to react according to:
$$C(s) + O(g) \rightarrow CO(g) \quad .$$
While some molecular oxygen may also be present in the vapor it
has a much lower concentration than atomic oxygen [12] and also
exhibits a substantially lower [13] probability of undergoing
combustion at the graphite surface.

For the CO reaction to proceed and create significant
quantities of CO, the corresponding amount of oxygen must
diffuse to the graphite surface. This occurs as a binary
process in the presence of high pressure oxide vapor species.
The binary diffusion coefficient for dissociated oxygen in UO_2
may be calculated by usual methods [14]. Similar published
calculations for binary oxygen diffusion have been published
which show diffusion coefficients in the range of 30 to
100 cm^2/s at 1 atmosphere and temperatures which range from
5000 to 7500 K [15]. If the partial pressure of oxygen is
assumed to be about 10% of the total pressure, the diffusion
of oxygen across a boundary layer of 0.1 mm at the face of the
target will result in an oxygen transfer of 0.1 to 0.5 $g/cm^2 \cdot s$
at pressures involved in these experiments. For a 10 μs experiment this corresponds to a loss of not more than 0.04% of the
sample mass or conversion of less than 0.3% of the total oxygen
present to CO. Since the predominant reaction shown above does
not change the net number of moles of gas in the vapor state,
this small alteration of the gas content cannot significantly
affect the total pressure transiently measured over the UO_2.
More detailed calculations including the rediffusion of CO and
the finite CO reaction rates will further reduce the effects
of oxygen diffusion and thus may be neglected for this problem.

Stoichiometry

The oxide material used for the electron beam experiments reported in Section IV has an O/M ratio of 2.08. It is well known that the vapor pressure over UO_2 at lower temperatures can vary with stoichiometry [16] by as much as two orders of magnitude because of variation in the oxygen partial pressure. Applying the calculations of Gabelnick and Chasanov [17] to our material we estimated that the vapor in equilibrium with our liquid has a higher pressure than that of perfectly stoichiometric UO_2 by a factor of approximately 1.3 at higher temperatures (6000 K), and about 2.0 at lower temperatures (3000 K). This dependence of the pressure-temperature relationship on stoichiometry is caused by altered vapor species as the temperature is varied. In their calculation the major contributor to the vapor pressure over stoichiometric UO_2 at lower temperatures is $UO_3(g)$, whose relative concentration is highly dependent on the oxygen potential and thus is affected by the oxide stoichiometry. On the other hand, at high temperatures the main contributor to the vapor pressure is $UO_2(g)$ whose concentration is relatively insensitive to oxygen potential and fuel stoichiometry.

Forced Congruent Evaporation

Related to the possibility of observing anomalous vapor pressures due to non-stoichiometry of the sample, is the possibility of producing forced congruent evaporation conditions if the evaporation from the sample surface is sufficiently rapid. Such evaporation could produce vapor pressures about an order of magnitude different from equilibrium evaporation.

The criteria suggested by Breitung [16] for determining whether or not evaporation is steady state "forced" congruent evaporation are that (a) the dimensions of the evaporating sample be larger than the depth of the (steady state) depleted oxygen zone, typically, D/v, where D is the diffusion coefficient of oxygen in liquid UO_2 and v the rate of evaporation of the surface, and (b) that the evaporation take place over time scales larger than those required to reach a forced congruent steady state, typically, D/v^2.

For our experiment v is at most 1 cm/s for $P = 10^3$ atm (a rate which approaches that of the laser pulse heating experiments). D is estimated by Breitung to be 10^{-2} cm^2/s for oxygen in liquid UO_2. Thus, the depleted zone depth, D/v, is approximately 2×10^{-2} cm, and requires a time, D/v^2, of at least 4×10^{-2} s for this zone to be depleted. The vaporizing liquid droplets in our experiments are approximately 2 μm (microns) in diameter, i.e. $\ll D/v$, and the measurements are taken in the first 10 or 20 microseconds of evaporation, i.e.

$<< D/\nu^2$. Consequently, we do not expect to be observing steady state forced congruent evaporation. The subject of non-equilibrium oxygen states will be discussed further, however, in Sec. IV as part of a non-equilibrium enthalpy model.

Adsorbed Gas

One further point which will be addressed as a source of error in the measurement of pressure is the possibility of an anomalous pressure contribution due to gas adsorbed on the surface of the UO_2 granules prior to heating. This is potentially serious due to the large sample surface area and is independent of the original chemical purity of the sample. During preparation of the powdered UO_2, the material was heated in a reducing atmosphere followed by a high vacuum, thereby producing urania with an O/M of 2.08. However, the samples were subsequently exposed to air during the experiment sample loading. The samples were then evacuated prior to the electron beam heating, but some additional gas may have remained adsorbed on the oxide granules. To examine the significance of this adsorbed gas we will assume that a complete monolayer of gas molecules covers the oxide surface and that this gas is released upon heating. For the electron beam sample geometry it is found that less than 1% of the vapor at pressures above 50 atm can come from this source. Since this concentration of uncondensable gas affects the pressure by a negligible amount, the adsorbed gas does not appear to be a significant source of error in the measurement of pressure.

Heat Losses

The total uncertainty in energy deposition of the present experiment is approximately ± 5% due to experimental uncertainties of pyrometry and carbon enthalpy. No correction of the energy state has been made for heat transfer losses by radiation and conduction between the high temperature (4000 K to 6000 K) UO_2 sample and the much cooler (< 2000 K) graphite crucible. If we assume that the graphite faces on both sides of the UO_2 sample present perfectly absorbing surfaces for radiation from the UO_2 samples, the heat loss from the UO_2 is 0.034 J in 10 μs for a 5000 K sample. If, as indicated by Tills et al. [18] conduction by the UO_2 vapor is present there may be added half again as much energy to this amount, or 0.051 J in 10 μs. This is 0.3% of the energy input of a typical sample and adds little to the previous 5% estimate for the energy uncertainty.

Due to the large inertial forces on the UO_2 liquid relative to the vapor and the rapid expansion of the graphite crucibles, it is unlikely that direct UO_2 liquid/carbon contact dominates the thermal transfer. However, as a worst case if one assumes perfect liquid graphite contact and a UO_2 conductivity [19] of 2×10^5 erg/s·K·cm an additional heat transfer of up to 0.2 J (approximately 1% of total) could be transferred from the sample. Thus, it is unlikely that any kind of thermal loss to the crucible can significantly affect the energy uncertainty.

Thermal Equilibrium

In any transient experiment, the existence of thermal equilibrium is of major importance. The present experiment may not have the same collection of chemical species present as in an ideal steady experiment because of the limited time for oxygen redistribution. Thus, while the sample may not exist in a state of chemical equilibrium, the existence of thermal equilibrium can be shown to exist for the rapid electron beam experiments examined. By thermal equilibrium we mean that the vapor and liquid phase have the same temperature and pressure. This will be the case if the net vaporization flux is small compared to absolute vaporization and recondensation fluxes. To estimate the recondensation flux we will use rates derived from ideal gas kinetics [20]. This analysis leads to vapor pressures, P, which depart from the equilibrium, P_e, by,

$$\Delta P = P_e - P = J \sqrt{2\pi mkT}$$

where m is the molecular mass, T the temperature and J the vaporization flux (molecules/cm^2·s).

For a vaporization container expanding with a rate $\dot{x}$ and for G the area ratio of UO_2 powder sample to the area of the expanding pistons, this becomes

$$\frac{\Delta P}{P} = \frac{\dot{x}}{GkT} \sqrt{2\pi mkT}$$

A maximum expansion rate for the UO_2 tests was found to be 10^5 cm/s using samples of 0.025 g/cm^2 of UO_2 powder with a diameter of ~ 2 μm. For the worst case in these tests the vapor flux produces a pressure error of 0.1%. Thus these vaporization rates produce no significant error in the measured pressure. This is a consequence of the large surface area in the powder sample. This argument is also consistent with the more extensive non-equilibrium modeling of Refling et al. [21].

IV. ELECTRON BEAM VAPOR DATA

The results of a sequence of electron beam vaporization experiments are presented in this section and analyzed with

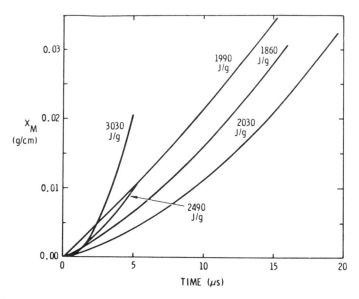

FIG.3. *Values measured for displacement as a function of time for five vaporization experiments. The indicated specific energy level for each experiment corresponds to the peak value in the mixed phase sample layer and is deposited in a time of 1 µs. The values of displacement have been scaled by multiplying each with the value of mass per area of that graphite plate.*

the methods described in the previous two sections of this paper. Expansion profiles from five electron beam tests are shown in Fig. 3. In each case these profiles have been expressed in reduced form by multiplying the displacement by the mass per area of the graphite plate which is being displaced by the vapor pressure of the UO_2 sample. Plate masses of 0.092 and 0.184 g/cm^2 have been used for these tests. Peak values for the energy deposited in the UO_2 sample layers are indicated for each expansion profile. The UO_2 sample was placed in the crucible as a powder with a nominal diameter of 2 µm. Prior to the tests, the region surrounding the powder granules is evacuated.

Values for the vapor pressure are obtained through the double numerical differentiation of the displacement profiles described in Sec. II. The results of the first differentiation for the five tests are shown in Fig. 4. There is some scatter associated with the velocity values determined by this means due to the finite resolution of the displacement photos. To indicate the resolution available in the current experiment, numerical velocity data points corresponding to the highest and lowest velocity tests are shown. The solid lines are linear fits to these points. In each case, the linear curves give an

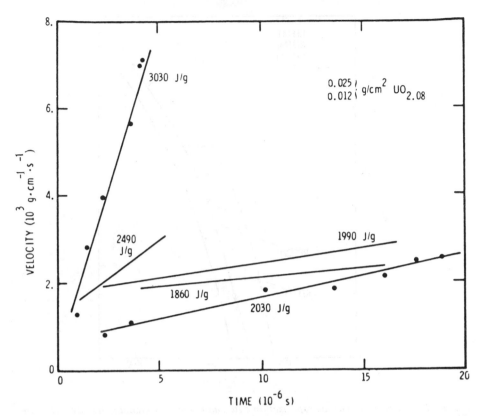

FIG.4. *Numerical differentiation of the information in Fig.3 results in this representation of velocity as a function of time. The specific energy levels corresponding to each experiment are shown for each curve. To illustrate the current resolution of the method, the numerically determined velocity values are shown for the upper and lower trace. Linear fits to these points are indicated by the solid lines. The slope of each of these curves is then proportional to the sample vapor pressure.*

accurate representation of the velocity points to within the data scatter and indicate no large negative curvature of the velocity data with time. Such a negative curvature would indicate a drop in vapor pressure during the observed expansion. The slope of these lines and, hence, the time averaged pressure during the expansion may be determined to within 10 to 20%.

The offset of these curves from the origin at t = 0 is similar to the behavior previously reported for gold vaporization [6]. There, the offset was found due to transfer of momentum from the condensed phase sample material. Possible contributions may also be present from small volume fractions of material having higher volatility than the UO_2 (adsorbed gas, free oxygen, as discussed in Sec. III).

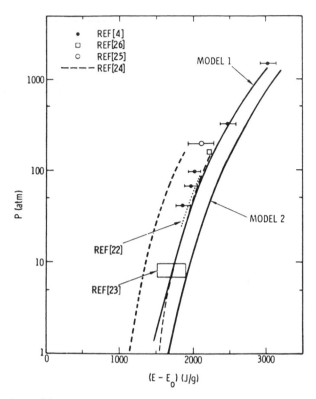

FIG.5. Comparison of dynamic vaporization data and models for uranium dioxide. The vapor pressure as a function of internal energy (relative to energy at room-temperature conditions) is shown. The data generated in the present investigation, and described in detail for the first time in this paper, are the black dots of Ref. [4].

    The vapor pressure which can be identified with each of
the linear fits in Fig. 4 is proportional to the slope of these
curves. In this way, values of vapor pressure corresponding
to the adsorbed energy are obtained. Pressure versus energy
(measured relative to energy at room temperature) data from
these experiments are shown in Fig. 5. The uncertainty
associated with the energy measurement is estimated to be ± 5%
and is shown by the bars on the data points in the figure.

V. COMPARISON OF RESULTS

    In addition to the electron beam results, data from pulse
reactor experiments have become available. These also represent
the vapor pressure as a function of internal energy. High

temperature experiments based on the laser heating of surfaces have also produced a collection of high vapor pressure data. In this section we compare the results of these tests.

Two groups have reported UO_2 vapor measurements for pressures from a few atmospheres to 65 atm using laser-induced heating of sample surfaces [22,23]. These data relate temperature and pressure. Heat capacity information is needed to place this information in a dynamic, or energy-dependent, format. The heat capacity described later as Model 1 has been used to place these data on Fig. 5. Electron beam data for UO_2 discussed in Sec. IV are shown by the points from Ref. [4]. Also shown are the results of Reil [24] generated by fission heating of UO_2 over millisecond times. Uncertainties in the extent of thermal diffusion during the longer experiment time in that work have led to indicated bounds for the vapor saturation curve (dashed lines in Fig. 5).

A third type of dynamic experiment that contains saturation curve information is a full fuel pin test. In these experiments the pressure of a coolant channel surrounding the pin is measured as the cladding ruptures during a reactor heated transient. The pressure and peak energy density of the fuel can be determined and are assumed for these tests to be dominated by the fuel vapor pressure. Results from the 5S test of the PBE series [25] and the TREAT (S-11) experiment [26] with unirradiated UO_2 pins are shown in Fig. 5. The error bars on the data indicate the authors quoted uncertainty in energy scale.

If limited thermal loss is assumed for the measurements of Ref. 24 then the data from all of the dynamic experiments agree to within their respective measurement uncertainties. Thus, a consistent result is found for the dynamic work. On the other hand, if one uses the measured heat capacity of Kerrisk and Clifton [27] and Leibowitz [28], together with the vapor pressure curve described by Booth [29], a theoretical curve may be constructed from low temperature data to compare with the dynamic results. This is labeled as Model 2 on the figure. Model 2 differs by more than a factor of 2 in pressure or nearly 15% in energy from the dynamic experiments.

Several reasons have been proposed for the difference between the dynamic experiments and extrapolated vapor data. One of these is that the vapor pressure of the hyperstoichiometric UO_2 used for several of the dynamic tests has a vapor pressure which is larger than for $U_{2.00}$. This does not appear consistent with the analysis in Section III. A second explanation which is discussed in more detail later in this section on possible rate effects in UO_2 suggests that the heat capacity should be altered for those experiments occurring on a millisecond time scale. Thus, we have selected a different heat capacity model for the high temperature material which agrees

with heat capacity measurements below 1500 K but excludes a
large fraction of the excess heat capacity occurring below the
melt temperature, a procedure which may be appropriate for
rapid events. This is described by

$$C_p(J/g \cdot K) = 0.276 + 3.78 \times 10^{-5} \, T(K) \text{ for } T < 3150$$

and a value of 0.51 J/g·K above that point [28], as well as a
heat of fusion of 278 J/g. This model, Model 1 in Fig. 5,
produces satisfactory agreement with all of the pieces of
data and gives a satisfactory comparison with the laser
heating data as well.

At this point it seems appropriate to explain why we have
chosen a value for the heat capacity which has been modified
from that measured by Kerrisk and Clifton and detail the
considerations we have taken in arriving at that value (Model I).

The possibility of observing various 'dynamic' or rate
dependent effects resulting from the lack of equilibrium,
particularly with respect to chemical species in experiments
on UO_2 and even in a reactor power excursion has been raised by
a number of people [30]. Questions of surface area limited
evaporation in the electron beam experiments have been addressed in previous sections and do not lead to alterations of the
measured results. The effects of chemical non-equilibrium in
the bulk UO_2 may, however, affect the measured pressure. We
have described the material in terms of a rate dependent
enthalpy state. The difference in enthalpy between an equilibrium and this non-equilibrium state is just the energy
resulting from oxygen redistribution.

Microscopic analysis of non-equilibrium in the solid phase
of UO_2 provides a model for non-equilibrium behavior in other
phases. In the high temperature solid phase of UO_2 the
decrease in heat capacity due to non-equilibrium effects can be
estimated by that normally attributed to oxygen Frenkel defects
[31]. The resulting heat capacity is expressed by Model I.
The time of relaxation to equilibrium in this model is just the
time of formation of oxygen Frenkel defects, estimated to be
between 50 μs and 20 ms. In contrast to our argument,
MacInnes has suggested the increase in solid heat capacity near
melt is an electronic contribution [32] for reasons presented
with a later paper in this conference. Such solid phase effects
provide a possible explanation for the difference between static
heat capacity measurements near the melt of UO_2 and the pulse
heating experiments of Affortit and Marcon [33].

While the energetics of oxygen redistribution have been
based upon solid phase structure the equivalent redistribution
in the liquid phase may be as important in reactor accident
analysis. Projecting the solid state analysis into the liquid

and vapor phases may be more difficult conceptually although no
less valid. Changes of stoichiometry with increases in temperature (associated with oxygen Frenkel defect formation in the
solid phase) will be suppressed in the liquid phase and along
the saturation curve if the material is heated sufficiently
rapidly. Such a suppression of oxygen redistribution in
rapidly heated UO_2 leads to the fixed congruent evaporation
model of Breitung [16]. In the event that the redistribution
occurs in the liquid, we can only estimate that a reduction
in the liquid phase enthalpy will be similar to that for the
solid phase because the basic oxygen redistribution process is
similar in both phases. Ultimately the magnitude and
relaxation time scale of rate effects in the heat capacity
which we postulate will have to be determined by experiment.

VI. IRRADIATED FUEL

Having gained reasonable confidence in our knowledge of
the high temperature vapor properties of fresh UO_2 from the
electron beam tests, it seems appropriate to extend our
investigations to irradiated fuel. Previous experiments
designed to measure vapor pressures due to fission products
as well as fission product release rates have not explored the
high temperature and pressure range or the high evaporation
rates typical of the fresh oxide work. Previous fission product
release experiments and theories deal almost entirely with slow
release from the solid fuel. Thus, a very high temperature-
pressure experiment offers the opportunity to test the theoretical predictions of Gabelnick and Chasanov [17], whose
calculations of high temperature fission product vapor pressures
are widely accepted. It also allows us to test the commonly
used hypothesis of instant release upon melt of all fission
products. This latter process has great significance to a
reactor accident sequence, since rapid fission product release
provides a convenient mechanism for early fuel dispersal and
accident termination.

New experiments to probe the behavior of fission product
release from fuel samples are being carried out in the pulsed
reactor at Sandia and at other facilities [34]. To demonstrate
the time and pressure resolution in such an experiment, samples
of fresh UO_2 have been pulse heated in the SPR III reactor
using the pressure vessel of Fig. 6. SPR is a fast-pulsed
reactor consisting of a stack of metal alloy fuel plates at the
center of which is an experiment cavity [35]. This cavity is
lined with a suitable moderator and the vessel of Fig. 6
secured inside the moderator. The reactor produces peak
specific energies of 2500 to 3500 J/g in fractional gram

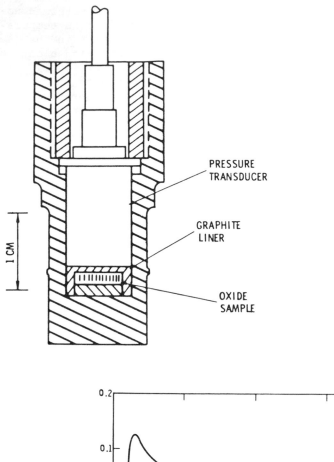

FIG.6.   Diagram of the
experimental vessel for
measuring vapor pressures
over oxide sample pulse
heated in the reactor.

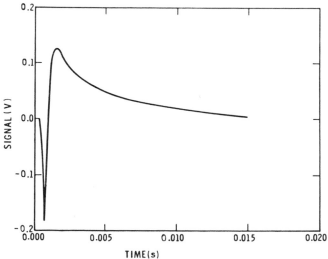

FIG.7.   Pressure transducer signal of a vapor pressure transient generated by the SPR III
reactor.   The negative signal at the beginning of the pulse is 'noise' induced by the neutron
pulse while the positive excursion describes the pressurization over a 0.35 g oxide sample.   The
pressure calibration for this transducer is 690 atm/V.

samples of fully enriched oxide within the vessel. This
extreme thermal state is achieved in a pulse having a 300 μs
FWHM and 90% of the pulse energy absorbed in less than 1 ms.
These samples are contained in a graphite liner within a zirc-
alloy pressure vessel. Vapor pressure generated over the sample
is measured with a pressure transducer designed to minimize its
sensitivity to the neutron flux. An example of a pressure
signal from this gage is shown in Fig. 7. The initial negative
excursion is a neutron induced "noise" signal which matches the
reactor flux history. The positive signal is a vapor pressure
pulse having a peak amplitude of 85 atm. The peak pressure
capability of this gage and pressure vessel is 700 atm.

There are a number of differences between the electron-
beam pulse heating experiments on fresh fuel and the proposed
SPR III neutron pulse heating experiments on irradiated fuel
which will significantly change the details of the experiment
analysis. One of these is that the energy is deposited in a
time on the order of a millisecond, rather than a microsecond.
This increased time may emphasize heat transfer and CO
formation. More sophisticated heat transfer and oxide diffusion analyses than those used for the e-beam experiments are
currently underway to explore these problems for the reactor
experiments.

Another difference may have to do with the type of non-
equilibrium effects present in the sample. In these
experiments the samples may fragment to produce a nearly
powdered material; however, the time scale of the experiment
is 10^3 times longer than for the e-beam experiment. In this
case, dynamic effects may or may not be evidenced for the base
material. For the irradiated fuel, however, there may be
additional dynamic effects associated with the release and
chemical recombination of fission products in the mixed oxide.
The plethora of previous studies on fission gas release from
solid fuel [36] is of little use to us here. If separation
occurs along grain boundaries, the products can be treated as
though those collected on the boundaries are released instanta-
neously as the material is heated through melt. However, the
proportion trapped at interstitial sites or as very small gas
bubbles may require longer to diffuse to the surface of the
liquid fuel. For example, the diffusion coefficient of xenon
or krypton in liquid fuel is closely approximated by $10^{-5} cm^2/s$
Based on this value it would take approximately 0.2 ms for a
xenon atom to diffuse from the center of a 2 micron diameter
droplet. This time delay in release of at least part of the
fission gas products may be observable in the experiments on
irradiated fuel.

It appears, also, that contamination of the irradiated
fuel during sample preparation and loading by N_2, CO, H_2 and

TABLE I. EQUILIBRIUM ESTIMATES OF PRESSURE DEVELOPED OVER
SAMPLES OF OXIDE FUEL DUE TO FISSION PRODUCTS (5% BURNUP)
AND SAMPLE CONTAMINATION

Specific Energy (Temp.)	Smear Density (g/cm^3)	Pressure (Including Fission Products) (atm)	Pressure (Pure UO_2) (atm)	Contaminant Pressure (atm)
806 J/g (2560 K)	3	93	3.7×10^{-4}	8
	4	135		16
	5	189		19
	6	267		23
1150 J/g (3150 K)	3	127	4.1×10^{-2}	14
	4	186		19
	5	287		24
	6	393		29
2000 J/g (4207 K)	3	248	5.4	19
	4	354		26
	5	519		32
	6	857		38
2500 J/g (5250 K)	3	691	94	24
	4	970		33
	5	1446		41
	6	2647		49

CH_4 may exist. A calculation of the partial vapor pressure of
18.5 μ moles (ideal gas) contaminant per g fuel may reasonably
represent the effect of these contaminants on the total vapor
pressure [34]. In Table I we have used the theory of Gabelnick
and Chasanov [17] to show the contribution of these contaminants
to the total vapor pressure for an equilibrium mixture of
irradiated fuel in which all fission products have been released.
The values of vapor pressure in the absence of fission products
are estimated by taking the UO_2 equilibrium values to approximate a mixed oxide sample $(U_{0.75}, Pu_{0.25}) O_{1.97}$. Specific energy
levels are tabulated which correspond to these vapor pressure
levels in Table I. If one assumes total release of the volatile
fission products as an ideal gas from samples of 5% burnup
material, the pressures as a function of smear density in
Table I are found. In addition, the effect of contaminant gas
at the indicated concentration is also shown. This concentration of contaminant gas adds only a small fraction to the

total pressure. This is in contrast to the results of experiments done at HEDL [37] and the Viper [34] experiments, where the pressure due to contaminants was found to be of the order of 50% or more of the total vapor pressure produced. The difference arises mainly because the lower temperature experiments at HEDL and in the Viper series do not yield complete release of all fission products, but probably do yield release of all contaminants.

The predictions displayed in Table I show that 8 to 20 times the fresh oxide pressure may be produced by the fission products in these samples. Thus, it appears that experiments probing such thermal states in fuels will be sensitive to the fraction and rates of fission product release. Heat transfer, hydrodynamics, fission gas release rates and non-equilibrium effects will be considered in the analysis of such experiments as they become available.

VII. SUMMARY

In this paper, an overview of the electron beam pulse heating of fresh oxide material has been presented. Comparison of data from the electron beam work with other pulsed heating studies shows a consistent representation of the vapor properties. However, a lack of agreement between extrapolations, based on low temperature static vapor pressure and heat capacity studies with the pulsed heating data indicate that fundamental differences in material properties may exist for rapidly heated samples. An explanation based upon chemical redistribution of the oxygen within the sample has been proposed which explains these observations. Direct measurements of such rate dependent behavior may be possible. Such measurements to define the magnitude and relaxation time scales would be useful in verifying the high temperature process involved. Finally, the possibility of extending such studies to irradiated fuel material has been discussed. Examples of pulsed neutron heating experiments have been described which should be useful in studying fission product release rates at high temperatures and pressures.

REFERENCES

[1]   NELSON, L.S., "Techniques for studying liquids and solids at extreme temperatures", Adv. in High Temp. Chem. 4 (1971) 171.

[2]   NICHOLSON, R.B., "Methods for determining the energy release in hypothetical fast reactor meltdown accidents", Nuc. Sci. and Eng. 18 (1964) 207.

[3]   DIETRICH, L.W. and JACKSON, J.F., "The role of fission products in whole core accidents", Proc. of Specialists Meeting at Harwell, IAEA 77-05900 (1977) 66.

[4]   BENSON, D.A. and BUCKALEW, W.H., "The uranium dioxide vapor equation of state", Trans. Am. Nucl. Soc., 23 (1976) 325.

[5]   BENSON, D.A., "Application of pulsed electron beam vaporization to studies of UO_2", Sandia Laboratories Rep. SAND77-0429.

[6]   BENSON, D.A., "Vapor properties of gold at extreme temperatures", J. Appl. Phys. 47 (1976) 4873.

[7]   BENSON, D.A. and OUELLETTE, A.L., Rev. Sci. Inst. 47 (1976) 291.

[8]   HALBLEIB, J.A. and VANDEVENDER, W.H., "CYLTRAN", Sandia Laboratories Rep. SAND74-0030 (1975).

[9]   BUCKALEW, W.H. and POSEY, L.D., "E-beam diagnostics", Sandia Laboratories Rep. SC-RR-69-512 (1969).

[10] LOCKWOOD, G.J., HALBLEIB, J.A. and MILLER, G.H., To be published, IEEE Trans. on Nucl. Sci., Dec. 1978.

[11] ACKERMANN, R.J. and THORN, R.J., Thermodynamics, Vol. 1, IAEA (1966) 243.

[12] KAROW, H.U., "Proc. of 7th Symp. on Thermophys. Prop., Nat. Bur. of Standards", Washington DC, 1977.

[13] ALLENDORF, H.D. and ROSNER, D.E., "Primary products in the attack of graphite by atomic and diatomic oxygen", Carbon 7 (1969) 515.

[14] BIRD, R.B., et al., "Transport Phenomena", John Wiley & Sons, New York (1960) 505.

[15] YUN, K.S., et al., "High temperature transport properties of dissociating nitrogen and oxygen", Phys. of Fluids 5 (1962) 672.

[16] BREITUNG, W., "Calculation of vapor pressures of oxide fuels up to 5000 K in equilibrium and non-equilibrium evaporation", Official Report USAEC-GFK, EURFNR-1283 (1975).

[17] GABELNICK, S.D. and CHASANOV, M.G., "A calculational approach to the estimation of fuel and fission product pressures and oxidation states to 6000 K", Argonne National Lab. Rep. ANL 7867 (1972).

[18] TILLS, J.L. and CRONENBERG, A.W., "A prediction of the thermal conductivity of UO_2 vapor and its application to LMFBR core disruptive analysis", Trans. Am. Nuc. Soc. 23 (1976) 326.

[19] FERRER, A.C. and BENACH, M.G., "Conductividad térmica de elementos combustibles", Energie Nucleare (Madrid) 17 (1973) 343.

[20] HIRTH, J.P., "Evaporation and sublimation mechanisms", Ch. of The Characterization of High Temp. Vapors, ed. by J.L. MARGRAVE, John Wiley, New York (1967).

[21]   REFLING, J.G., REYNOLDS, A.B., GARNER, P.L. and RAO, S.P.,
        "Non-equilibrium evaporation and condensation in liquid-
        metal fast breeder reactor fuel expansion", Nuc. Tech. 33
        (1977) 275.
[22]   OHSE, R.W., et al., "Application of laser pulse heating
        for the study of high temperature vapors", Proc. of 10th
        Materials Research Symposium on High Temp. Vapors and
        Gases, Gaithersburg, MD (1978).
[23]   BOBER, M., BREITUNG, W., KAROW, H.U. and SCHRETZMANN, K.,
        "Evaporation studies of liquid oxide fuel at very high
        temperatures using laser beam heating", Bericht KFK 2366
        (1976).
[24]   REIL, K.O. and CRONENBERG, A.W., "Effective equation of
        state experiments on uranium dioxide", Trans. Am. Nuc.
        Soc. 27 (1977) 576.
[25]   REIL, K.O., YOUNG, M.F., and SCHMIDT, T.R., "Prompt burst
        energetics experiments with fresh oxide", Sandia Laborato-
        ries Rep. SAND78-1561 (1978).
[26]   ZIVI, S.M., et al., Nucl. Sci. Eng. 56 (1975) 229.
[27]   KERRISK, J. and CLIFTON, D., Nuclear Technology 16 (1972)
        531.
[28]   LEIBOWITZ, L., et al., J. Nucl. Mat. 39 (1971) 115.
[29]   BOOTH, D.L., UKAEA Report TRG 1759 (1968).
[30]   OHSE, R.W., BERRIE, T.G., BOGENSBERGER, H.G., and
        FISCHER, E.A., "Extension of vapor pressure measurements
        of nuclear fuels (U, Pu) O_2 and UO_2 to 7000 K for fast
        reactor safety analysis", J. Nucl. Mat. 59 (1976) 112.
        Also see, BERGERON, E.G., "Theoretical considerations of
        rate effects in uranium dioxide", Trans. Am. Nucl. Soc.
        27 (1977) 575.
[31]   CATLOW, C. and LIDIARD, A., "Theoretical studies of point
        defect properties of uranium dioxide", Thermodynamics of
        Nuclear Materials, VOL. II (1975) 27, IAEA Vienna (1975).
[32]   MacINNES, D.A., "The electronic contribution to the
        specific heat of UO_2", These proceedings, paper IAEA-SM-236/37.
[33]   AFFORTIT, C. and MARCON, J.P., "Chaleur Spécific a haute
        température des oxydes d'uranium et de plutonium", Rev.
        Int. Hautes Tempér. et Réfract. 1 (1970) 236.
[34]   McTAGGART, M. and FINDLAY, J., "Progress report on the
        VIPER measurements of fission product pressure generation,
        Mar. 76-Sept. 77", AWRE/44/96/3 (1978).
[35]   BONZON, L.L. and SNYDER, J.A., "The Sandia Pulsed Reactor",
        Sandia Laboratories Rep. SLA-73-0551 (1973).
[36]   OLANDER, D.R., "Fundamental Aspects of Nuclear Fuel
        Elements", TID-26711-P1 (1976).
[37]   HINMAN, C.A. and SLAGLE, O.D., Private communication.

DISCUSSION

J. MAGILL: The central problem in dynamic heating experiments seems to be the interpretation of equilibrium data, i.e. saturated vapour pressure, from the measured quantities in the particular dynamic experiments. How valid is this and, in particular, with what accuracy do you expect to obtain the saturated vapour pressure?

D.A. BENSON: The saturated vapour pressure measured in the dynamic heating experiments is characteristic of a thermal equilibrium state, i.e. one in which the temperature and pressure of the liquid equal those of the vapour. On the other hand, the details of the chemical states which exist in these short-duration dynamic experiments may differ from states occurring in long-duration high-temperature events. Thus the dynamic saturation pressures are characteristic of microsecond to millisecond thermal transients. The accuracy of these results is bounded by a factor of ±5% on the energy co-ordinate and approximately ±50% in the pressure co-ordinate.

J.R. FINDLAY: I would like you to comment on the calculation of the impurity contribution to the estimates of pressures developed in the reactor-based experiments? Is a monolayer assumption a realistic method of evaluation?

D.A. BENSON: The impurity contributions to estimates of pressures over pre-irradiated oxide samples are based upon the results of chemical analysis on typical fuel-pin material and not on a monolayer model.

P.E. POTTER (Chairman): In the electron beam heating experiments, did you determine whether carbon monoxide was produced?

D.A. BENSON: No. Destruction of the rapidly expanding target in the electron beam accelerator diode makes such a measurement impractical. Carbon debris and residue from dielectric oil present in the diode after a test will mask the desired measurement.

P.E. POTTER (Chairman): Have you made or do you intend to make measurements on urania of lower oxygen contents?

D.A. BENSON: No, we have not done this, and we have no plans to do additional electron beam work with oxides. The emphasis of our present work is on the determination of fission-product release rates and fractions using reactor heating tests.

EQUATION OF STATE FOR SUBSTOICHIOMETRIC URANIA USING SIGNIFICANT STRUCTURES THEORY

E.A. FISCHER
Institut für Neutronenphysik und
  Reaktortechnik,
Kernforschungszentrum Karlsruhe,
Karlsruhe,
Federal Republic of Germany

Abstract

EQUATION OF STATE FOR SUBSTOICHIOMETRIC URANIA USING SIGNIFICANT
STRUCTURES THEORY.
  The Significant Structures Theory (SST) of Eyring was successfully used to predict
the equation of state in the liquid range for a variety of materials, including UO_2. However,
all these applications assumed that the liquid evaporates congruently, i.e. that the composition
of the vapour phase is identical to that of the condensed phase. In this paper an attempt is made
to apply SST to non-congruently evaporating materials, using hypostoichiometric urania as an
example. To this end, hypotheses additional to those of the original SST must be made. In
SST it is assumed that the partition function of the liquid can be expressed by suitably combining that of 'solid-like molecules' with that of 'gas-like molecules'. In the present work,
starting from the fact that the non-stoichiometry of solid urania is connected with lattice
defects, e.g. oxygen interstitials or oxygen vacancies, it is assumed that a simple oxygen-defect
model can be extrapolated into the liquid state. Thus, the solid-like partition function includes
a defect term, which determines the O/U ratio; the defect concentration depends on the
absolute activity of oxygen. The gas-like partition function allows for UO(g) and UO_2(g),
the ratio depending also upon the oxygen activity. The parameters of the theory are selected so
as to obtain agreement with experimental data at the melting point. The physical requirement
that the difference between liquid and gas disappears at the critical temperature necessitates
an adjustment of the solid-like partition function at high temperatures.

1. INTRODUCTION

For the analysis of hypothetical accidents in fast reactors, a knowledge of
the equation of state of oxide fuel at high temperatures is essential.

Equation-of-state data for UO_2 up to the critical temperature have been
evaluated by several authors, using essentially two different extrapolation
methods, namely the principle of corresponding states and the Significant
Structures Theory of Liquids (SST) of Eyring [1]. Probably the best known

data obtained by the former method, published by Menzies [2] in 1966, were used in several laboratories for many years. Later, it was noted that SST was rather successful for materials with ionic binding. In addition, it is more flexible than the principle of corresponding states, e.g. it allows one to account for dimerization, electronic excitation, etc. These observations encouraged the application of SST to UO_2 by Gillan [3] and by Fischer et al. [4].

Both of these extrapolation techniques can only be used directly for materials which evaporate congruently, where the composition of the gas phase can be assumed to be the same as that of the condensed phase. However, in the cases of urania and of mixed (U, Pu) oxide, this is only approximately true over a limited range of O/U ratios, and it is rather unrealistic for some other materials of interest, for example advanced nuclear fuels.

It is, therefore, desirable to develop one of these techniques so that a modified version can be used for non-congruently evaporating materials. In this paper, a possible procedure for UO_2 will be suggested.

Starting from the observation that non-stoichiometry in the solid material is connected with lattice defects, it will be hypothesized that a simple oxygen defect model can be extrapolated into the liquid range. Clearly, it is necessary to examine whether SST allows such an extension to be undertaken in a meaningful way. So far, a procedure has been formulated only for hypostoichiometric urania; however, it would easily be possible to use it for hyperstoichiometric material also, and probably for mixed oxide, too. It should be noted that the data used in this work should be considered as preliminary, and a re-examination and systematic adjustment to match experimental results is planned for the near future.

2.   SIGNIFICANT STRUCTURES THEORY OF LIQUIDS

The SST of liquids of Eyring [1] was quite successful in reproducing equation-of-state data for a number of liquids, including molten salts, which have ionic binding [5]. This fact encouraged its use for UO_2 [3], which is also believed to be ionic. As SST is used in this paper, the principles of the model will be outlined.

The liquid is pictured as being obtained from a crystal by removing molecules from a number of lattice sites, and placing them at the edge of the crystal. The 'fluidized vacancies' thus produced behave like a gas, whereas the molecules at the remaining lattice sites behave in a solid-like manner. Thus the total partition function, f_ℓ, of the liquid is given by the equation:

$$f_\ell(T,V) = f_s^{NV_s/V} f_g^{N(V-V_s)/V} \tag{1}$$

where f_s and f_g are the partition functions of the solid-like and gas-like molecules, V_s is the volume of the solid at the melting point, and V is the volume of the liquid.

The partition function of the gas is given by the product of translational, rotational, vibrational and electronic partition functions for the molecule under study:

$$f_g = f_g^{tr} f_g^{rot} f_g^{vib} f_g^{el} \tag{2}$$

For f_s, the standard form used in the SST is:

$$f_s = \exp\left(\frac{E_s}{RT}\left(\frac{V}{V_s}\right)^\gamma\right) f_{cr} f_{deg} \tag{3}$$

where f_{cr} is the partition function of a crystal in the Einstein approximation:

$$f_{cr} = \frac{1}{(1 - \exp(-\Theta_E/T))^3} \tag{4}$$

and f_{deg} accounts for positional degeneracy, which arises because fluidized vacancies provide additional sites for the solid-like molecules:

$$f_{deg} = 1 + n\,\frac{V-V_s}{V_s}\,\exp\left(\frac{-a\,E_s\,V_s}{RT(V-V_s)}\left(\frac{V}{V_s}\right)^\gamma\right) \tag{5}$$

where E_s is the binding energy, Θ_E the Einstein temperature, R the gas constant and where a, n, γ are parameters of the model.

The partition function f_ϱ is related to the Helmholtz free-energy function A by the equation:

$$A(T,V) = -kT \ln f_\varrho \tag{6}$$

The other thermodynamic functions are obtained from A in the usual way.

3. COMMENTS ON DEFECT MODELS

Defect theory was first applied by Thorn and Winslow [6] to describe the non-stoichiometric behaviour of urania. The authors assumed that oxygen interstitials and oxygen vacancies are the only defects. However, it became apparent later on by comparison with oxygen potential measurements that the defects responsible for hypostoichiometry are probably more complicated. Winslow developed a model which incorporates U^{3+} and U^{5+} ions, in addition to oxygen vacancies and interstitials. This was quite successful in reproducing experimental oxygen potential data [7].

As mentioned before, the main assumption involved in the extension of SST, as investigated in this paper, is that a defect model can be extrapolated into the liquid range to describe the behaviour of the solid-like lattice. It is not intended to assess the validity of defect models in this paper. Therefore it seems justified, for the purpose of studying the feasibility of such an approach, to ignore the theories of complex defects and to go back to a rather simplistic model. Hence it will be assumed that the only defects present in hypostoichiometric urania are oxygen vacancies. Although the equations of such a model can be found in the literature, it might be useful to summarize them here for later reference.

The composition variable x is connected with the density of vacancy defects, θ_v $(= N_v/2N)$, by the relation

$$x = 2 - (O/U) = 2\theta_v \tag{7}$$

The usual defect theory leads to the equation:

$$\ln \lambda_0 = \ln \frac{1 - \theta_v}{\theta_v} - \frac{E_v}{RT} - \ln q(T) \tag{8}$$

where λ_0 is the absolute activity of oxygen, E_v is the energy needed to create a vacancy, and $q(T)$ is the partition function of a three-dimensional harmonic oscillator.

From this model, the Helmholtz free energy associated with the vacancies is:

$$A_v = 2 RT \ln (1 - \theta_v) = -2 RT \ln \left(1 + \frac{\exp(-E_v/RT)}{\lambda_0 \, q(T)} \right) \tag{9}$$

It is easily seen that the partial pressure of UO_2 (g) over solid UO_{2-x} depends on the composition variable only through the free energy A_v; one has

$$RT \ln p_{UO_2} = \Delta G_f^0 \langle UO_2 \rangle - \Delta G_f^0 (UO_2) + A_v \tag{10}$$

Therefore, the partial pressure of UO_2 depends only weakly on O/U, in a way described by Eq.(10).

4.   OUTLINE OF THE EXTENDED SST

In the previous applications of the SST to urania, the assumption was always made that the material evaporates congruently. In this paper, an attempt will be made to extend it to non-congruently evaporating material. It is not obvious how such an extension can be accomplished, and it is necessary to make additional hypotheses to those of the original SST. Furthermore, it is not clear a priori whether the SST, being a rather simplistic model of a complicated physical system, is at all capable of describing the physical behaviour of a bivariate system correctly. One essential requirement for a correct description is that the O/M ratio is the same in the liquid and in the gas at or near the critical temperature. It will be seen that this requirement can only be fulfilled in an approximate way by the method proposed in this paper; it necessitates certain restrictions on the choice of the partition functions.

The equations will be developed, as an example, for the case of hypo-stoichiometric UO_{2-x}. It will be assumed that the gas phase contains only UO_2, UO and O. The condensed material is a homogeneous phase, and the oxygen deficiency in the solid material is caused by lattice defects, which may be assumed, as discussed above, to be oxygen vacancies. It will be postulated that these vacancies are also present in the solid-like lattice which, according to SST, characterizes the state of part of the molecules in the liquid. Thus, the solid-like partition function should allow, at least at temperatures near the melting point, for a defect concentration which depends on the absolute activity of oxygen.

Therefore, the partition function f_{cr} for the crystal must be modified in a suitable way. It is assumed that it can be expressed as:

$$f_{cr} = f_s^2 + f_s^1 \frac{\exp(-E_v/RT)}{\lambda_0} \tag{11}$$

where f_s^2 refers to the UO_2 formula, and f_s^1 refers to UO.

We now observe that the Einstein oscillator model for the crystal partition function is certainly not valid at high temperatures, because it is based on the assumption of a potential of infinite height. Following Ree, Ree and Eyring [8], it is assumed that the oscillator potential terminates at a certain level; beyond this level, the oscillator sublimes and behaves like a gas.

More specifically, it is assumed that f_s^2 can be constructed from contributions of different degrees of freedom in the following way.

(a)   *3 translational degrees of freedom*

$$f_s^{2,tr} = \left(\frac{1 - e^{-1\theta/T}}{1 - e^{-\theta/T}} + e^{-1\theta/T} \frac{(2\pi mkT)^{1/2} v_f^{1/3}}{h} \right)^3 \tag{12}$$

The Einstein oscillator terminates at the level 1, and then behaves like a gas confined to a volume v_f per molecule.

(b) 2 rotational degrees of freedom

$$f_s^{2,\text{rot}} = \left(\frac{1 - e^{-1\theta/T}}{1 - e^{-\theta/T}} + e^{-1\theta/T} \left(\frac{8\pi^2 I_2 kT}{h^2 \sigma} \right)^{1/2} \right)^2 \tag{13}$$

The Einstein oscillator associated with these two degrees of freedom behaves like a rotating linear molecule at high temperatures.

(c) 4 vibrational degrees of freedom

$$f_s^{2,\text{vib}} = \prod_{i=1}^{4} \frac{1}{1 - \exp(-\vartheta_i/T)} \tag{14}$$

This function is simply the vibrational partition function for UO_2, with three different vibrational modes, one of them with the degeneracy two. Thus, the partition function f_s^2 is given by:

$$f_s^2 = f_s^{2,\text{tr}} \, f_s^{2,\text{rot}} \, f_s^{2,\text{vib}} \tag{15}$$

Obviously at high temperatures this function goes over into the partition function of UO_2 gas, but without the electronic contribution.

The partition function f_s^1 is built up in the same way; the only difference is that there is just one vibrational degree of freedom.

The O/U ratio associated with (11), is then given by:

$$(O/U)_s = 2 - \frac{\dfrac{f_s^1}{\lambda_0 f_s^2} \exp(-E_v/RT)}{1 + \dfrac{f_s^1}{\lambda_0 f_s^2} \exp(-E_v/RT)} \tag{16}$$

We now examine the behaviour of this solid-like partition function at the melting point, assuming that Einstein oscillator terms dominate at this temperature. Then, the partition function (11) can be written, approximately, as:

$$f_{cr} = \frac{1}{(1 - e^{-\theta/T})^9} \left(1 + \frac{e^{-E_v/RT}}{\lambda_0 \, q(T)} \right) \tag{17}$$

$$q(T) = \frac{1 - \exp(-\vartheta/T)}{\displaystyle\prod_{i=1}^{4}(1 - \exp(-\vartheta_i/T))} \cong \frac{T^3}{\vartheta_{eff}^3} \tag{18}$$

which is indeed the partition function of a UO_2 crystal containing oxygen vacancies. This may be easily verified by comparing the term in brackets with the defect-free energy in Eq.(9).

It is then suggested by Eq.(10) that the UO_2 gas should be in equilibrium with the defect-containing crystal. Thus the gaseous partition function, to be used in the SST Eq.(1), is simply the one of $UO_2(g)$. This means, in terms of the physical model, that only molecular fluidized vacancies with the composition UO_2 are considered. The pressure of UO_2 is then determined as usual in the SST; clearly, it depends on the oxygen activity because of Eq.(11). Clearly, the O/U ratio of the liquid is then given by Eq.(16), assuming $(O/U)_\varrho = (O/U)_s$.

In addition, it is assumed that the gas phase also contains $UO(g)$, which is in equilibrium with O and UO_2. The density ratio is given by the equation:

$$\frac{n_1}{n_2} = \frac{f_g^1}{\lambda_0 f_g^2} \exp(\Delta H_0/RT) \tag{19}$$

where the subscripts 1 and 2 refer to UO and UO_2, and ΔH_0 is the reaction enthalpy at absolute zero. The exponential factor is necessary because the energy normalization in f_g^1 and f_g^2 must be consistent.

The O/U ratio in the gas phase is then:

$$(O/U)_g = 2 - \frac{n_1}{n_1 + n_2} = 2 - \frac{\dfrac{f_g^1}{\lambda_0 f_g^2}\exp(\Delta H_0/RT)}{1 + \dfrac{f_g^1}{\lambda_0 f_g^2}\exp(\Delta H_0/RT)} \tag{20}$$

The physical requirement that $(O/U)_\varrho \simeq (O/U)_g$ near the critical temperature can be fulfilled with the model suggested here if the following conditions are met:

(i)   The electronic partition function of the two gases are equal, $f_1^{el} = f_2^{el}$;
(ii) The energy, E_v, required to create a vacancy is of magnitude similar to the negative reaction enthalpy, $-\Delta H_0$;
(iii) The 'cut-off level' l of the oscillator can be chosen such that the gaseous behaviour dominates in the solid-like partition function at the critical temperature.

TABLE I. INPUT PARAMETERS FOR SST

E_s	720.5 kJ/mol		n	10.8
Θ_{UO_2}	1000 K		a	0.00352
Θ_{UO}	238 K		γ	-0.2602
V_s	27.9 cm^3/mol		l	7
E_v	837 kJ/mol		ΔH_0	725 kJ/mol

TABLE II. DATA OF THE UO AND UO$_2$ MOLECULES

	UO	UO$_2$ (linear molecule)
Bond length (Å)	1.764	1.79
Moment of inertia (g/cm^2)	0.774	1.70×10^{-38}
Vibrational frequencies (cm^{-1})	825	765 1
(degeneracy)		150 2
		776 1
Ground state multiplicity	5	5
Density of electronic levels (mol/J)	0.239×10^{-4}	0.239×10^{-4}
Energy of the lowest level (J/mol)	33490	33490

    Condition (i) will simply be postulated in the following calculations. This appears justified because the electronic partition functions of both gases are not well known. However, conditions (ii) and (iii) can only be fulfilled approximately. Thus, the model proposed is not inherently consistent. However, it will be shown that, by a suitable choice of the parameters, the relation $(O/U)_\ell \simeq (O/U)_g$ holds, at least approximately, over the most important range of O/U ratios.

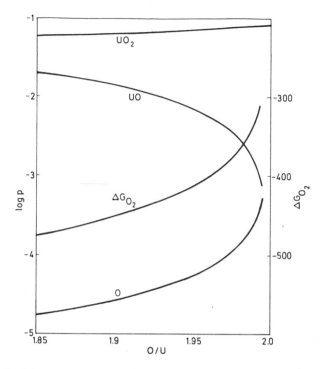

FIG.1. *Partial vapour pressures and oxygen potential at the melting temperature.*

TABLE III. THERMODYNAMIC DATA OF UO$_2$ AT THE MELTING POINT

Melting temperature	T_m	3120 K
Liquid specific volume	V_ϱ	30.87 cm^3/mol
Vapour pressure of UO$_2$		0.0715 bar
Entropy of fusion, $\Delta S/R$		2.44

5.   INPUT DATA AND RESULTS

The SST, as described above, was applied to liquid UO$_{2-x}$ in the O/U range 1.81 to 1.99. The input data used in the evaluation are listed in Tables I and II. There are several ways of determining the parameters a, n and γ of the model. In this paper, we follow essentially the method suggested by Gillan [3].

TABLE IV. VAPOUR PRESSURE OF URANIUM-BEARING SPECIES

O/U ratio Temp. (K)	1.810	1.911	1.964	1.983	1.992	1.810	1.911	1.964	1.983	1.992	1.810	1.911	1.964	1.983	1.992
	P_{UO_2} (bar)					P_{UO} (bar)					P_{tot} (bar)				
3120	0.0604	0.0661	0.0695	0.0709	0.0715	0.0243	0.0124	0.0053	0.0025	0.0012	0.0847	0.0785	0.0748	0.0734	0.0727
3500	0.486	0.531	0.558	0.568	0.574	0.170	0.087	0.037	0.018	0.008	0.656	0.618	0.595	0.586	0.582
4000	3.70	4.04	4.24	4.32	4.36	1.14	0.58	0.25	0.12	0.05	4.84	4.62	4.49	4.44	4.41
4500	16.53	18.00	18.88	19.21	19.37	4.61	2.35	0.99	0.48	0.22	21.14	20.35	19.87	19.68	19.59
5000	51.6	56.1	58.8	59.8	60.3	13.3	6.8	2.8	1.4	0.6	65.0	62.9	61.6	61.1	60.9
5500	126.0	136.6	142.9	145.3	146.5	30.4	15.4	6.5	3.1	1.4	156.4	152.0	149.4	148.4	147.9
6000	263	284	297	302	305	59	30	12	6	3	322	314	309	308	308
6500	462	499	522	530	534	101	51	21	10	5	563	550	543	540	539
7000	754	813	848	861	868	157	79	33	16	7	911	892	881	977	875

TABLE V. O/U RATIO IN THE GAS PHASE AS A FUNCTION OF TEMPERATURE AND O/U RATIO IN THE CONDENSED PHASE

$(O/U)_\ell$ Temp. (K)	1.810	1.911	1.964	1.983	1.992
			$(O/U)_g$		
3120	1.713	1.842	1.930	1.965	1.984
4000	1.764	1.874	1.945	1.973	1.988
5000	1.795	1.892	1.954	1.978	1.990
6000	1.812	1.904	1.959	1.980	1.991
6500	1.821	1.908	1.961	1.981	1.991
7000	1.827	1.911	1.962	1.982	1.992
7500	1.832	1.914	1.963	1.983	1.992

Thus, for nearly stoichiometric UO_2, the parameters are chosen such that the values for the volume change, and for the entropy change on melting, as predicted by the theory, are compatible with experimental values. Note that, in an earlier evaluation for UO_2 [4], the concept of representing the 'excess enthalpy' of solid UO_2 as a Frenkel defect term in the solid-like partition function was employed. This concept was abandoned in this work because it is now believed that the excess enthalpy is due to a λ-transition, rather than to Frenkel defects on the oxygen sublattice.

According to Eyring [1], the parameter γ accounts for the dependence of the binding energy on the average interionic distance, in the presence of long-range ionic forces. One expects, therefore, that for substoichiometric UO_{2-x} this parameter changes with O/U ratio. If γ is assumed to be a function of O/U, it is possible to reproduce a known dependence upon O/U of the liquid density at the melting point by the SST. However, to the author's knowledge, no information is available on the variation in density with O/U ratio. Therefore, γ was selected ad hoc such that the liquid density at the melting point is constant. The parameters a and n of the theory and also the melting temperature were assumed to be independent of O/U ratio.

The molecular data for UO and UO_2 were mainly taken following the suggestions of the Argonne group [10]; they are based on absorption spectroscopy data by Gabelnick et al. [11]. However, the electronic partition function was approximated by introducing a density, D, of electronic levels per unit energy, as suggested earlier [4]. The vapour pressure of UO_2 at the melting temperature was adjusted to the value obtained in the earlier evaluation [4], by a suitable choice of the binding energy.

It was the aim to reproduce the oxygen potential of UO_{2-x} as obtained from Blackburn's model [9] at the melting temperature by SST. For this purpose, two different 'effective' Einstein temperatures for UO and for UO_2 were introduced. They should be regarded as adjustable parameters rather than physical data. By a suitable selection of these parameters, and of E_v, Blackburn's oxygen potential curve could be well approximated.

It should be noted that the SST model does not explicitly account for the partial pressure of O_2. Thus, to adjust the results to Blackburn's data, it was necessary to relate the oxygen potential to the oxygen activity at the melting temperature. An analytical relation, suggested by Thorn and Winslow [6], was used for this purpose.

The important results are shown in Fig.1 and in Tables III–V. Figure 1 shows the partial pressures of UO_2, UO and O at the melting temperature, as functions of O/U ratio. The UO_2 pressure follows Eq.(10), UO increases as O/U decreases. The oxygen potential and the partial pressure of monotonic oxygen are in good agreement with Blackburn's model.

Table III shows the thermodynamic data of UO_2 at the melting point, Table IV the vapour pressures of the uranium bearing species at different temperatures. Up to about 6000 K, the total vapour pressure for O/U = 1.99 is in good agreement with the earlier evaluation [4]. At the melting temperature, the vapour pressure increases as O/U decreases. Clearly, the minimum, which is expected at the congruently evaporating composition, cannot be reproduced by the present version of the model, because UO_3 is not included in the gas phase.

Table V shows the O/U in the gas phase as a function of temperature and the O/U in the condensed phase. The requirement $(O/U)_g \simeq (O/U)_\varrho$ near the critical temperature is approximately fulfilled for $(O/U)_\varrho \geqslant 1.9$, but a difference of 0.02 exists for (O/U) = 1.81.

The present version of the SST provides estimated values for the critical constants, which depend slightly on O/U ratio. For 1.992 and 1.911, the critical temperatures are 7970 K and 7990 K, the critical pressures 1780 and 1820 bar, and the critical densities 2.08 and 2.10 g/cm^3, respectively. The differences from the earlier evaluation (7560 K, 1210 bar, 1.66 g/cm^3) are mainly due to a different selection of parameters, and reflect the uncertainties present in these predictions. In view of these uncertainties, the variation of T_c with O/U ratio is insignificant.

6.   CONCLUSIONS

It was shown that the Eyring's Significant Structures Theory of liquids can be extended by introducing a defect model into the theory to describe the thermodynamic behaviour of liquid UO_{2-x}. Thus a physical model has been presented, which was used to extrapolate partial vapour pressures into the liquid range, up to the critical temperature. However, the model is based on a hypothesis, and its utility still has to be confirmed by experiment. In addition, the model is not inherently consistent because the physical relation $(O/U)_g \simeq (O/U)_\varrho$ near the critical temperature is not automatically fulfilled. However, with the partition function suggested in this paper, this relation can be approximately fulfilled in the important range $O/U \geqslant 1.9$.

At the melting temperature, the present method reproduces the oxygen potential of the Blackburn model rather well, given a suitable choice of the parameters. The estimated critical constants differ from those of an earlier evaluation, due to the different selection of the parameters.

REFERENCES

[1] EYRING, H., JHON, M.S., "Significant Liquid Structures", John Wiley, New York (1969).
[2] MENZIES, C., UKAEA Rep. TRG-1119 (D) (1966).
[3] GILLAN, M.J., "Derivation of an equation of state for liquid UO_2 using the theory of significant structures", Thermodynamics of Nuclear Materials (Proc. Symp. Vienna, 1974), Vol.1, IAEA, Vienna (1975) 269.
[4] FISCHER, E.A., KINSMAN, P.R., OHSE, R.W., Critical assessment of equation of state data for UO_2, J. Nucl. Mater. 59 (1976) 125.
[5] LU, W.C., REE, T., GERRARD, V.G., EYRING, H., Significant structure theory applied to molten salts, J. Chem. Phys. 49 (1968) 797.
[6] THORN, R.J., WINSLOW, G.H., Nonstoichiometry in uranium dioxide, J. Chem. Phys. 44 (1966) 2632.
[7] WINSLOW, G.H., An examination of the total pressure over hypostoichiometric uranium dioxide, High Temp. Sci. 7 (1975) 81.
[8] REE, T.S., REE, T., EYRING, H., Proc. Natl. Acad. Sci. USA 48 (1962) 501.
[9] BLACKBURN, P.E., Oxygen pressures over fast breeder reactor fuel, J. Nucl. Mater. 46 (1973) 244.
[10] Chemical Engineering Division, Reactor Safety and Physical Property Studies Annual Report, Argonne National Laboratory Rep. ANL-8120 (1974).
[11] GABELNICK, S.D., REEDY, G.T., CHASANOV, M.G., J. Chem. Phys. 58 (1973) 4468.

DISCUSSION

J. MAGILL: Don't you think that in extending Significant-Structures Theory (SST) to non-congruent evaporation you are in a way fine-tuning the physical basis of the model unnecessarily? The sensitivity of the results of SST to the particular choice of the free parameters is such that any small variation of the parameters will swamp the effects of non-congruent evaporation.

E.A. FISCHER: I think the parameters E_v and the extra Einstein temperature for UO are necessary to reproduce the behaviour of the oxygen potential. However, your comment may be appropriate as regards the cut-off level 1, which influences the predicted critical constants. These constants are definitely liable to be in error.

M. HOCH: The critical point is reached when the number of atoms in gas per unit volume equals that in the liquid. As the density of the liquid changes much more slowly with temperature than that of the gas (vapour pressure), the vapour pressure is much more important than the liquid density in determining the critical temperature and pressure.

K.A. LONG: As input data for the calculation you used data which are independent of the O/U ratio. In calculations which we are currently performing on uranium carbide, using significant structure theory, it has become apparent that the variation of, for instance, ΔH_{ev} and ΔH_{sub} is substantial when the

C/U ratio varies. For example, ΔH_{ev} varies from 118 kcal/mol for C/U = 1.0 to 147 kcal/mol for C/U = 1.08. Do you not, therefore, consider that it is necessary to include variations of input data with U/O ratio in UO_2?

E.A. FISCHER: In the model, neither ΔH_{ev} nor ΔH_{sub} is an explicit input parameter. However, implicitly the enthalpy needed to evaporate a UO_2 molecule depends on the O/U ratio, because the pressure of UO_2 (g) depends on O/U in a manner given by the defect model.

The parameters n, a, γ introduced by Eyring in SST may all conceivably depend slightly on O/U. I chose to include this dependence only for γ, but there is no strong argument for this choice.

DO ELECTRONIC TRANSITIONS CONTRIBUTE
TO THE THERMODYNAMICS OF CONDENSED UO$_2$?

A review of the arguments

D.A. MacINNES
Safety and Reliability Directorate,
Culcheth, Cheshire,
United Kingdom

Abstract

DO ELECTRONIC TRANSITIONS CONTRIBUTE TO THE THERMODYNAMICS OF
CONDENSED UO$_2$? A REVIEW OF THE ARGUMENTS.

Recent analysis of the role of electronic transitions in the thermophysical properties
of UO$_2$ is surveyed. It is concluded to be highly likely that the $5f^2$ electrons on the U^{4+}
metal ion play a major role in both the specific heat and thermal conductivity, in that they
are primarily responsible for the large 'anomalous' increase displayed by each of these quantities
between T = 1600 K and T$_m$ = 3100 K. This has important implications for reactor analysis,
since to obtain the required data for molten fuel one must extrapolate existing data through
a wide range in temperature, and the behaviour of the electronic mechanisms may be expected
to extrapolate quite differently from that of the mechanisms in current use (Frenkel defect
generation and internal radiative heat transfer).

1. INTRODUCTION

It is well known [1, 2] that the electronic states of vapour molecules are
important in the thermodynamics of evaporated UO$_2$ and it has recently been
suggested [3–5] that thermally activated electronic transitions play a major role
also in the thermodynamics of condensed (both crystalline and molten) UO$_2$,
influencing in particular the specific heat $C_p(T)$ and the thermal conductivity K(T).
Both $C_p(T)$ and K(T) [6,7] of crystalline UO$_2$ show large 'anomalous' increases
as T increases from 1600 K to the melting temperature T$_m$ = 3100 K. The
mechanisms responsible for these increases have been taken to be, respectively,
Frenkel defect formation [6] and radiative heat transfer [8]. (It should be
noted [9] that electronic heat transfer was considered as an explanation of the
K(T) behaviour, but the majority view seemed to be that radiative transfer was
the correct mechanism.) It has recently been suggested [3, 4] that the increase in
$C_p(T)$ should be attributed to valence–conduction band electronic transitions
rather than Frenkel defect generation, and this re-interpretation will obviously
influence the choice of mechanism for the heat transfer process. This paper is
an attempt to summarize the main arguments concerning the interpretation of the

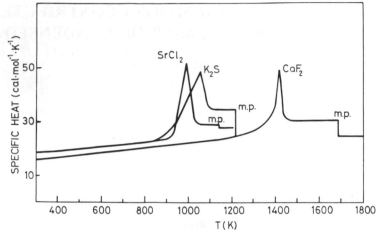

FIG.1.   The specific heat of CaF_2, $SrCl_2$ and K_2S in the crystalline and dense fluid regions
(m.p.: melting point).

behaviour of $C_p(T)$ and $K(T)$. From the point of view of reactor analysis, it is of
considerable importance to have a correct interpretation since most data on
$C_p(T)$ and $K(T)$ for molten fuel are obtained by extrapolation of data for
crystalline or just-molten fuel, and the different mechanisms can be expected to
extrapolate into the liquid region in widely different ways. The uncertainties thus
introduced may confidently be expected to affect markedly reactor analysis
parameters such as excursion yield, fuel/coolant interaction (FCI) initiation
criteria and fuel pin failure characteristics.

I have come to the broad conclusion that there is very strong evidence for
attributing a large part of the increase in $C_p(T)$ to electronic processes, although
one cannot rule out a significant contribution from Frenkel defects. This in turn
favours the interpretation of $K(T)$ behaviour in terms of electronic heat transfer,
and calls into question the extrapolation techniques used to obtain $C_p(T)$ and
$K(T)$ up to $T = 5000$ K.

2. SPECIFIC HEAT

Most of the attention has been focussed on $C_p(T)$ for two reasons. Firstly,
it is a simpler problem insofar as it does not involve electron dynamics, and
secondly, proper analysis of $C_p(T)$ is an essential forerunner to analysis of $K(T)$.
This section splits into two main parts, one concerned with the analysis of
available data for crystalline and just-molten UO_2, and the other containing a
few general comments about how one might expect the available data to extrapolate to $T = 5000$ K.

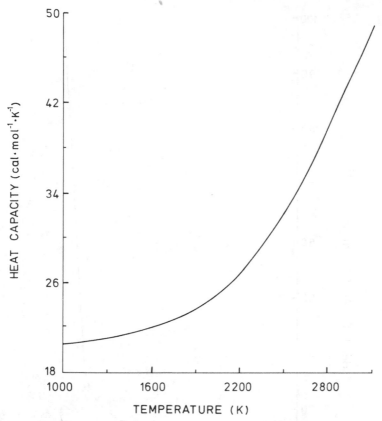

FIG.2.   *The specific heat of crystalline* UO_2 *(m.p. = 3100 K).*

2.1. Crystalline and just-molten UO_2

2.1.1. *Analysis of Frenkel defect model for $C_p(T)$ increase*

It is well known that $C_p(T)$ of UO_2 shows a large 'anomalous' increase as
T increases from 1500 K to 3100 K, and that this has been attributed to the
formation of anion Frenkel defects [6]. The most important point in favour of a
defect interpretation is that it is well known that cubic fluorite crystals such as
CaF_2 and BaF_2, which have a crystal structure identical to that of UO_2, show an
anomalous increase in $C_p(T)$ which can unambiguously be attributed to Frenkel
defect formation [10]. However, the $C_p(T)$ of these materials behaves in a well
established manner, i.e. as T increases, $C_p(T)$ rises sharply from the Dulong and
Petit value to a sharp peak at $0.85\ T_m$, and then decreases to a constant value
for $0.85\ T_m < T < T_m$ (see Fig.1). It has recently been pointed out [11] that this
behaviour is quite different from that observed in UO_2, and incidentally in other
actinide oxides such as ThO_2 and PuO_2 (Figs 2, 3 and 4). It is also noticeable that

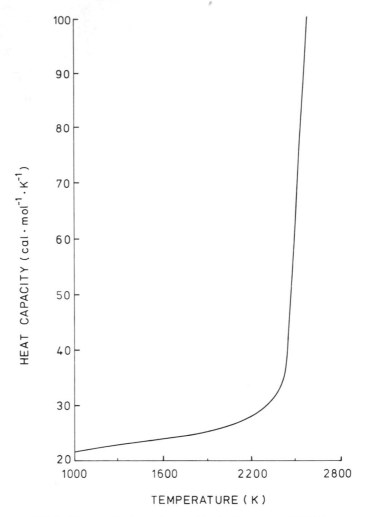

FIG.3. *The specific heat of crystalline* PuO_2 *(m.p. = 2650 K).*

the behaviour of different actinide oxides shows strong dissimilarities. From this it seemed apparent [11] that:

(a)   The mechanism causing the anomaly in UO_2 has different properties from that in CaF_2;

(b)   The mechanism causing the anomaly in UO_2 has different properties from that in PuO_2 and ThO_2.

This led to the conclusion that the anomaly in the actinides derived mostly from electronic properties.

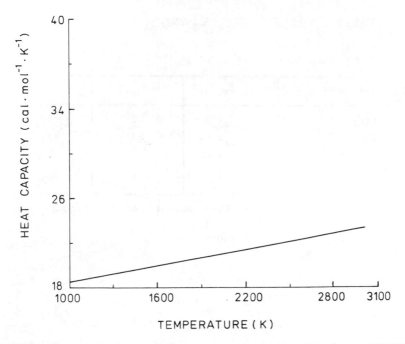

FIG.4. *The available data for the specific heat of crystalline* ThO$_2$ *(m.p. = 3600 K).*

Further comparison of the anomaly in UO$_2$ with those in CaF$_2$, SrCl$_2$ and K$_2$S is revealing. The thermal generation of Frenkel defects can be regarded as a premelting phase, in that increased disorder before melting reduces the disorder which occurs at melting, so that one can consider the total entropy required to melt a perfect crystal to be the premelting entropy plus the actual phase transition entropy. The premelting entropy, transition entropy and total entropy of melting for CaF$_2$, SrCl$_2$, K$_2$S and UO$_2$ are given in Table I. Clearly the UO$_2$ value of the total entropy is well in excess of that of the other materials. This is further indication that there is a process absorbing energy in UO$_2$ which does not occur in CaF$_2$, and the obvious choice is electronic transitions.

2.1.2. Small polaron model for electronic specific heat

It is well known from electrical conductivity studies [12] that there are a large number of electrons in the conduction band of UO$_2$ at T = T$_m$, and that the activation energy for a valence–conduction band transition is approximately 2 eV (see also Refs [3, 4] and the references therein). The most common model used to describe the behaviour of 5f^2 electrons on the U^{4+} ion is the small polaron

TABLE I. VALUES[a] OF S_p, S_m AND S_T

Entropy units: $cal \cdot mol^{-1} \cdot K^{-1}$

Compound	S_p	S_m	S_T
CaF_2	1.4	4.2	5.6
$SrCl_2$	4.4	3.4	7.8
K_2S	2.0	3.3	5.3
UO_2	4.4	5.8	10.2

[a] S_p : premelting entropy. This is calculated from the observed $C_p(T)$ behaviour.
      S_m : entropy change on melting.
      S_T : $S_p + S_m$.

model, in which an electron is assumed to be localized at any given time on a metal ion site. In this picture the valence–conduction band transition is represented by the reaction:

$$2U^{4+} \rightarrow U^{3+} + U^{5+} \tag{I}$$

In order to calculate the extent to which the reaction has proceeded at any given temperature, we need to know the entropy ΔS associated with the transition. This entropy has two components, each difficult to quantify:

(a) The entropy deriving from the degeneracy factor associated with the electronic configurations $2U^{4+}$ and $U^{3+} + U^{5+}$. This will depend sensitively on spin–orbit coupling and crystal-field splitting, and the appropriate data is not available for calculation.

(b) The electronic transition (I) alters the interionic forces and consequently the frequency spectrum and vibrational entropy of the crystal. (The entropy associated with this mechanism in the dense liquid has been calculated [13], via a model of charged hard spheres, to be between 2k and 3k per transition (k = Boltzmann's constant).)

It is possible to obtain indirectly an estimate of ΔS from work done on the oxidation properties of UO_2 [14] where it was argued that consideration of electronic reaction (I) was necessary to explain the behaviour of the partial molar

TABLE II. VALUES[a] OF K_e AND ΔS

K_e	10^{-7}	10^{-6}	10^{-5}
ΔS	4.94	7.24	9.55

[a]
  K_e : reaction equilibrium constant, Eq.(1).
  E_g : 2 eV.
  ΔS : increase of entropy $(cal \cdot mol^{-1} \cdot K^{-1})$ for
         electronic transition (I).

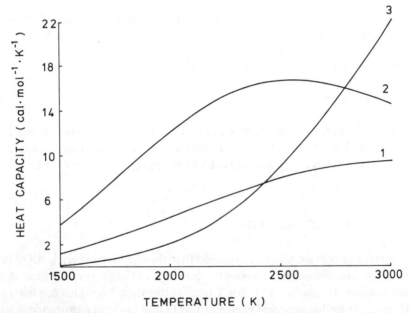

FIG.5. *The small polaron specific heat derived from reaction constants* $K_e = 10^{-7}$ *and* 10^{-6}.
The value of 10^{-5} *predicted* ΔS *to be too large and so was not plotted. The observed excess in*
$C_p(T)$ *is taken from Ref. [6]. Curve 1:* $K_e = 10^{-7}$; *curve 2:* $K_e = 10^{-6}$; *curve 3: experimental.*

entropy of oxidation of UO_2. This study predicted that the reaction would have
had to proceed appreciably at T = 1100 K in order to explain the observed
oxidation behaviour, and calculated the appropriate range of reaction constants

$$K_e = \exp \left(\frac{(T\Delta S - E_g)}{kT} \right) \tag{1}$$

The values of ΔS derived from the predicted values of K_e by assuming $E_g = 2$ eV
are given in Table II, and the specific heats calculated using $E_g = 2$ eV and

$K_e = 10^{-6}$ and 10^{-7} are displayed together with the observed 'anomalous' increase in Fig. 5. Two points are clear:

(i) The combination of $E_g = 2$ eV and the values of ΔS indicated by Catlow and Lidiard predict a major contribution to $C_p(T)$ from reaction (I).

(ii) The $C_p(T)$ thus predicted does not show exponential growth for all $T \leqslant 3000$ K. This is due to reaction saturation effects, and is important because recent work [15] has predicted the existence of a second-order phase transition on the basis of the observed deviation from exponential growth of $C_p(T)$ as T tends to 3000 K. This calculation shows electronic reaction saturation is a very plausible alternative explanation of the deviation.

It should be noted that other polarons are also possible. Catlow [16] has shown that the reactions:

$$2U^{5+} \rightarrow U^{4+} + U^{6+} \tag{II}$$

and

$$2U^{3+} \rightarrow U^{2+} + U^{4+} \tag{III}$$

have activation energies of less than 4 eV, and so may be expected to contribute significantly at $T = 3000$ K. However, I cannot at present calculate the extent of these reactions since we have no estimate of the appropriate values of ΔS.

2.1.3. Specific heat of molten UO_2

A further important piece of evidence that the $C_p(T)$ anomaly in UO_2 has a major electronic component is given by the value of $C_v(T)$ (specific heat at constant volume) for molten UO_2 for T just greater than T_m. This has been calculated [5] from the observed values of $C_p(T)$ and the other appropriate data via the standard thermodynamic relations. By comparison with $C_v(T)$ for other ionic liquids which are unlikely to have substantial electronic activity, $C_v(T)$ of UO_2 is shown to be appreciably in excess of the value one might expect on the basis of a simple ionic liquid containing U^{4+} and O^{2-} ions. This excess has been attributed to electronic transitions and is consistent with a large electronic component of $C_p(T)$ of the crystal.

2.1.4. Additional mechanisms possibly contributing to the anomaly

It appears that the major component of the anomaly is valence–conduction band electron transitions, with perhaps a contribution from anion Frenkel defects.

There are two other mechanisms which may be expected to contribute to a minor but perhaps non-negligible extent. These mechanisms are:

(a) anharmonic vibrations;

(b) intra-ionic electronic transitions in which an electron remains on a U^{4+} ion but is excited to an energy level higher up the spectrum created by spin–orbit coupling and crystal-field splitting.

It has been shown [17, 18] that (a) may be expected to be small and negative. It has been suggested that (b) makes a small contribution in PuO_2 [19], and one may expect this also in UO_2 [20].

2.2. Extrapolation of electronic $C_p(T)$ to $T = 5000$ K

The behaviour with increasing T of the electronic component of $C_p(T)$ is difficult to predict because the crucial parameters, E_g and ΔS, may be expected to vary with increasing temperature and specific volume. It has already been shown [13] that E_g may be expected to increase appreciably as T increases to 5000 K along the UO_2 saturation line. As in the crystal, the behaviour of ΔS is very difficult to obtain. The most promising source of data is via dynamic heating experiments in which the energy input and vapour pressure are measured. One can then use vapour pressure versus temperature data to extract $C_p(T)$. This approach has already been used [21] and one interpretation [5] of the results obtained is that the $C_p(T)$ of the liquid is much lower than is normally assumed. It is hoped to repeat this experiment, in the VIPER reactor, to gain greater confidence in this conclusion.

3. THERMAL CONDUCTIVITY

3.1. Current position

The thermal conductivity shows an increase [7] as T increases from 1600 K to T_m. This does not fit a phonon scattering picture which predicts a steady decrease with increasing T, so an additional mechanism of heat transport has to be assumed. Two mechanisms have been proposed, internal radiative transfer and electronic heat transfer, with the former model being preferred [8]. However, it can be argued strongly that the current predictions of extensive electronic activity in the crystal and dense fluid should for the following two reasons shift the balance in favour of the electronic heat transfer model:

(a) The radiative model relies on extrapolation of the near-infrared optical transmission properties of the crystal from temperatures below ~ 1500 K up to 3100 K. Since at $T = 3100$ K the maximum of the radiation spectrum

(a) cont.

corresponds to a wavelength of $\sim 10\,000$ Å (a photon energy of ~ 1 eV) the major changes which occur in the electronic properties between $T = 1500$ K and T_m render the simple extrapolation of the low temperature optical properties open to considerable doubt (the predicted increase in electronic activity should greatly decrease the optical transmission for 1 eV photons, perhaps making UO_2 as opaque as a metal).

(b)   The specific heat associated with the conduction electrons is much greater than one would normally expect (for a given electron and hole mobility one might expect the electronic thermal conductivity to be proportional to the electronic specific heat).

There is evidence [22] that $K(T)$ of UO_2 increases by a factor of more than three on melting. This cannot be explained by conventional (i.e. phonon) heat transfer characteristics but could be explained by a considerable increase in electron and hole mobility upon melting. Since the electrons and holes move by hopping between metal ions, the increased mobility could be a result of increased fluctuations in the relative positions of metal ions made possible by melting, allowing the metal ions to approach closer to each other and thus facilitate hopping.

3.2. Extrapolation of the electronic $K(T)$ to $T = 5000$ K

The possible range of behaviour of the electronic component of $K(T)$ makes simple extrapolation impossible, and so it seems unlikely that reliable theoretical modelling of $K(T)$ for molten UO_2 will be a viable proposition in the very near future. An experiment which would be very valuable and of considerable interest would be a measurement of the electrical conductivity of just-molten UO_2. If the jump in $K(T)$ on melting is a genuine effect and can be attributed to electronic heat transfer, one should see an increase, by approximately an order of magnitude, of the electrical conductivity as a result of melting.

4. CONCLUSION

There is very strong evidence for a major electronic component in $C_p(T)$ of UO_2, and this in turn is strong evidence for the interpretation of the 'anomalous' increase in $K(T)$ as electronic heat transfer. This leaves open to doubt the extrapolation of $C_p(T)$ and $K(T)$ to high temperatures (~ 5000 K). The best hope of obtaining data on $C_p(T)$ at these temperatures is from dynamic heating experiments in which input energy and resultant vapour pressure are the measured variables. The most promising experiment to obtain $K(T)$ up to $T = 5000$ K is based on electrical heating techniques and is currently under way [1]. A measurement of the electrical conductivity for just-molten UO_2 would be most revealing, since data and modelling combine to predict an increase of an order of magnitude over the crystal at its melting point.

ACKNOWLEDGEMENTS

The author wishes to record valuable conversations with Dr. C.R.A. Catlow. He also wishes to thank Dr. D. Martin and Mr. C.G. Theaker for assistance with the calculations.

REFERENCES

[1]   LEIBOWITZ, L., private communication.
[2]   BOGENSBERGER, H.G., FISCHER, E.A., BERRIE, P.G., KINSMAN, P.R., OHSE, R.W.,
      Gesellschaft für Kernforschung mbH, Karlsruhe, Rep. KFK 2272 (EUR 5501e)(1976).
[3]   MacINNES, D.A., UKAEA Rep. SRD R117 (1978).
[4]   MacINNES, D.A., J. Nucl. Mater. 78 (1978) 225.
[5]   MacINNES, D.A., UKAEA Rep. SRD R130 (1978).
[6]   KERRISK, G.F., CLIFTON, D.G., Nucl. Technol. 16 (1972) 531.
[7]   CHASANOV, M.G., LEIBOWITZ, L., GABELNICK, S.D., J. Nucl. Mater. 49 (1974) 129.
[8]   ANDERSON, E.E., Nucl. Technol. 30 (1976) 65.
[9]   HOBSON, I.C., TAYLOR, R., AINSCOUGH, J.B., J. Phys., D (London). Appl. Phys. 7
      (1974) 1003.
[10] DWORKIN, A.S., BREDIG, M.A., J. Phys. Chem. 72 (1968) 1277.
[11] MacINNES, D.A., CATLOW, C.R.A., UKAEA Rep. SRD R140 (1978).
[12] BATES, J.L., HINMAN, C.A., KAWADA, T., J. Am. Ceram. Soc. 50 (1967) 652.
[13] MacINNES, D.A., MARTIN, D., UKAEA Rep. SRD R145 (1978).
[14] CATLOW, C.R.A., LIDIARD, A.B., Thermodynamics of Nuclear Materials 1974 (Proc.
      Symp. Vienna, 1974), Vol.2, IAEA, Vienna (1975) 27.
[15] BREDIG, M.A., Coll. Int. L'étude des transformations crystallines à haute température,
      Odeille (1971) 183.
[16] CATLOW, C.R.A., Proc. R. Soc. (London), Ser. A 353 (1977) 533.
[17] STONEHAM, A.M., private communication.
[18] MacINNES, D.A., THEAKER, C.G., UKAEA Rep. SRD R142 (1978).
[19] MANES, L., Plutonium 1970 and Other Actinides (Proc. 4th. Int. Conf. Santa Fe, 1970),
      American Institute of Mining, Metallurgical and Petroleum Engineers, Inc., New York
      (1970) 254.
[20] BUTTERY, N.E., private communication.
[21] BENSON, D.A., Sandia Laboratories Rep. SAND-77-0429 (1977).
[22] KIM, C.S., BLOMQUIST, R.A., HALEY, J., FISHER, J., CHASANOV, M.G.,
      LEIBOWITZ, L., "Measurement of Thermal Diffusivity of Molten UO_2", 7th Symp.
      Thermophysical Properties (National Bureau of Standards, Gaithersburg, 1977), American
      Society of Mechanical Engineers, New York (1978).

HIGH-TEMPERATURE URANIUM DIOXIDE, SODIUM AND CAESIUM VAPOURS

The effect of thermal ionization and the plasma state on the heat- and radiation-transport properties of the vapour phase

H.U. KAROW
Institut für Neutronenphysik und Reaktortechnik,
Kernforschungszentrum Karlsruhe,
Karlsruhe,
Federal Republic of Germany

Abstract

HIGH-TEMPERATURE URANIUM DIOXIDE, SODIUM AND CAESIUM VAPOURS: THE EFFECT OF THERMAL IONIZATION AND THE PLASMA STATE ON THE HEAT- AND RADIATION-TRANSPORT PROPERTIES OF THE VAPOUR PHASE.

The paper considers how far the thermophysical state and the convective and radiative heat transport properties of vaporized reactor core materials are affected by the dense gaseous state and by the thermal ionization existing in the 'real' vapour state. The materials under consideration here are nuclear oxide fuel (UO_2), sodium (as LMFBR coolant material) and caesium (alkaline fission product, partly retained in the fuel of the core zone). For this purpose, the degree of thermal ionization and characteristic data describing the gaseous and plasma states of the saturated vapour of these materials have been calculated, in the case of UO_2 up to 5000 K, in the case of sodium and caesium up to their respective critical temperatures (T_c). The effect of the plasma state on convective thermal conductivity of the high-temperature vapour phases of UO_2 and sodium is discussed semiquantitatively. The contribution of the electron gas to the thermal conductivity of UO_2 vapour appears to be negligible up to ~4500 K. The same has been found for sodium vapour up to $T \simeq T_c$. Theoretical evaluations of the averaged radiative absorption coefficient and mean free path of saturated UO_2 vapour up to 5000 K show that uranium oxide vapour becomes optically thick, even for very thin vapour layers, its behaviour depending upon the temperature. In this problem the strongly plasma state of the vapour plays only a secondary role, i.e. in splitting and broadening the energy levels of the vapour molecules. The results for the radiative absorption of uranium oxide vapour are of relevance to problems in nuclear reactor safety analysis, when the radiative thermal heat exchange through a fuel-vapour layer has to be considered. The results are also relevant in experimental research involving pyrometric temperature measurements on liquid-oxide fuel.

1. INTRODUCTION

Safety analysis of sodium cooled fast reactors requires a knowledge of the thermal equations of state of the liquid/vapour equilibria of the fuel, coolant and structural materials over a temperature range from ~2000 to 5000 K.

TABLE I. CHARACTERISTIC DATA[a] DESCRIBING THE GAS KINETIC AND PLASMA STATES OF UO_2 VAPOUR BETWEEN THE BOILING POINT (3600 K) AND 5000 K

T (K) and kT (eV)	p (bar)	n_{tot} (cm^{-3})	χ_i (%)	n_0 (cm^{-3})	n_q (cm^{-3})	d_0 (Å)	d_q (Å)	λ_e (Å)	l_D (Å)	N_D	q_c	ln Λ	E_i^{p1} (eV)
3660 0.310	1	1.98 $\times 10^{18}$	0.5	1.97 $\times 10^{18}$	1.1 $\times 10^{16}$	49	280	22	280	1	0.2	2.8	5.2
4000 0.345	4	7.24 $\times 10^{18}$	0.7	7.19 $\times 10^{18}$	5.2 $\times 10^{16}$	32	165	21	135	0.5	0.3	2.2	5.1
4500 0.3	19	3.09 $\times 10^{19}$	1.2	3.05 $\times 10^{19}$	3.6 $\times 10^{17}$	20	87	20	54	0.2	0.7	1.4	4.8
5000 0.431	63	9.18 $\times 10^{19}$	1.9	9.0 $\times 10^{19}$	1.7 $\times 10^{18}$	14	52	19	26	0.1	1.3	0.7	4.5

[a] The following terms are used: T = vapour temperature; p = vapour pressure; n_{tot} = particle density; n_0 = neutral particle density; d_0 = average particle distance; χ_i = degree of ionization; E_i^{p1} = effective ionization energy; n_q = ion or electron density; d_q = mean ion distance; l_D = Debye shielding length; λ_e = de Broglie wavelength of the free electrons; N_D = charged particle number in a Debye sphere; q_c = parameter of plasma correlation [7]; lnΛ = Coulomb logarithm.

Information is also required on the relaxational and kinetic behaviour of the vapours of these materials during an adiabatic expansion. Furthermore, the momentum and heat-transport properties of the liquid and vapour phases should be known so that they can be applied in the modelling and numerical simulation of LMFBR hypothetical core accidents. To approach these questions experimentally, high-temperature experiments with nuclear fuel materials and fission-product materials using laser beam heating have been carried out at our laboratory [1−5]. In addition, for a correct evaluation of such experiments, the thermophysical properties of the materials under investigation must be sufficiently well known.

This paper deals theoretically with the thermal plasma state of the saturated vapour of liquid urania up to 5000 K, and those of sodium and caesium up to their respective critical points. The effect of the plasma state on the convective heat transfer and the thermal radiation transport in the vapour phases is discussed semiquantitatively.

2.   GASEOUS AND PLASMA STATE OF UO_2 VAPOUR

The gaseous and plasma states of the saturated vapour over liquid UO_2 are quantitatively characterized in Table I. It describes the gas kinetic state of oxide fuel vapour consisting of UO_2 or 'UO_2-like' molecules, ions and free electrons. The temperature and total vapour pressure data, p(T), given in the first two columns are based on our laboratory's own laser vapour-pressure measurements [5] and on the equation of state of UO_2 recently recommended by the International Working Group on Fast Reactors of the IAEA [6]. Calculation of the degree of ionization, χ_i, (third column) via the commonly used Saha equation by inserting the ionization potential of the undisturbed vapour molecule, E_i^0, does not give a useful result, because the effective ionization energy, E_i^{p1}, differs remarkably from the theoretical value for the isolated molecule under the actual conditions of UO_2 vapour above 4000 K.

As has been discussed in detail in a previous study [7, 8], there are several effects which lower the ionization energy of the molecules in the extremely dense UO_2 vapour system. In plasma physics, different approaches are available to calculate the lowering of the ionization potential in a 'common' kinetic plasma. Their validity, however, becomes questionable in the case of dense 'collective' plasmas [9−13]. The plausible approach of 'cut-off of highly excited valence electrons by nearest neighbour molecules' [9, 13] has been used to estimate the required effective ionization potential of the vapour molecules. The applicability of this method has been confirmed experimentally (e.g. Ref. [9], p.198 et seq.). In calculating the degree of ionization, χ_i, account has been taken of the fact that the fuel vapour over liquid UO_2 consists of various species, namely UO_2, UO_3, UO and O. The vapour composition calculated by Breitung [14] has been used in the evaluation of χ_i. Only the UO_2 and UO species have been regarded as ionizable ($E_i^0 = 5.5$ eV) because of the relatively high ionization potentials of UO_3 and O. For this reason, the values of χ_i, given in Table I, are somewhat lower than those obtained in the previous study mentioned above [7, 8].

As can be seen from Table I, oxide fuel vapour in the temperature range from the boiling point up to 5000 K represents both a dense gaseous system and a strong plasma. Up to 5000 K the neutral vapour component behaves as a perfect gas. This can be seen from an estimation of the van der Waals constants of UO_2 vapour. The ionized component of the fuel vapour, however, does not approach the state of being a perfect kinetic plasma. At temperatures around 5000 K, UO_2 vapour represents a strongly-coupled collective plasma ($q_c > 1$) with increasingly 'metallic' properties, shown by the fact that at temperatures between 4500 K and 5500 K the de Broglie wavelength, λ_e, of the free plasma electrons should reach the magnitude of the mean particle distance, d_0 and d_q, respectively. At this stage, the kinetic interaction is replaced by a wave-mechanical interaction between the neutral and ionized molecules and the conduction electrons in the UO_2 vapour.

3.   GASEOUS AND PLASMA STATE OF SODIUM AND CAESIUM VAPOURS

In the same way, the plasma state of sodium vapour has been evaluated up to the critical point. Caesium vapour has also been considered here, because

TABLE II. CHARACTERISTIC DATA DESCRIBING THE GAS KINETIC AND PLASMA STATES OF SATURATED SODIUM AND CAESIUM VAPOURS BETWEEN 1500 K AND THE CRITICAL TEMPERATURE[a]

T (K) and kT (eV)	p (bar)	Z	x_i	n_{tot} (cm^{-3})	n_q (cm^{-3})	d_0 (Å)	d_g (Å)	λ_e (Å)	l_D (Å)	N_D	q_c	$\ln \Lambda$	E_i^{pl} (eV)
SODIUM													
1500 0.129	10.9	0.83	2×10^{-7}	5.27×10^{19}	8.0×10^{12}	17	3050	34	6500	9.7	0.02	5.1	4.3
2000 0.172	78.3	0.62	3.5×10^{-5}	2.83×10^{20}	1.0×10^{16}	9.4	290	30	220	0.4	0.4	1.9	3.6
2506 0.216	257	0.12	8.2×10^{-4}	7.41×10^{20}	6.1×10^{17}	6.9	73	27	31	0.08	2.1	0.2	3
CAESIUM													
1500 0.129	27.7	0.74	4.3×10^{-5}	1.34×10^{20}	5.7×10^{15}	12	347	34	250	0.4	0.5	1.8	2.7
2050 0.177	118	0.23	2.4×10^{-2}	4.15×10^{20}	9.8×10^{18}	8.3	63	29	22	0.05	3.7	<0	2.2

[a] Terms as in Table I: in addition, Z = compressibility.

knowledge of its thermophysical properties could be of interest in studying fuel phenomena caused by the volatile fission products.[1] The equation-of-state data of sodium were taken from the KFK Sodium Property System Code of Thurnay (see Ref.[15]), while those for caesium were taken from Stone (see Ref.[16]).

Table II shows the numerical results. Sodium vapour at 1500 K appears to be a kinetic plasma. Between 2000 K and the critical temperature (2506 K) it becomes strongly coupled. At T_c, the quantitative validity of the Saha equation is questionable because the kinetic theory breaks down ($q_c \gtrsim 2$, $\ln \Lambda \simeq 0$). Similar results are obtained for the caesium plasma, which should already be strongly coupled at $T \gtrsim 1500$ K. At T_c (2050 K), the resulting figures should be viewed only as a semiquantitative proof of the existence of a metal-like plasma vapour.

Because of the limited quantitative validity of the analytical procedure, the effect of clustering of the alkali vapour atoms in the temperature range up to T_c has not been taken into account in the calculation of the degree of ionization, x_i. There is experimental evidence, however, that the ionization

---

[1] See SCHUMACHER, G., in Ref.[4], p.25 et seq.

TABLE III. CONVECTIVE THERMAL CONDUCTIVITY OF THE ELECTRON
GAS, κ_e, AND OF THE MOLECULAR GAS, κ_h, FOR SATURATED UO_2 VAPOUR

T (K)	3660	4000	4500	5000
Fully-ionized gas model				
$\ln \Lambda$ (calc.)	2.81	2.2	1.4	0.7
$\ln \Lambda$ (used)	2.81	2.2	1.4	1
κ_e $(mW \cdot cm^{-1} \cdot K^{-1})$	0.5	0.9	1.8	3.3
Lorentz gas model				
κ_e $(mW \cdot cm^{-1} \cdot K^{-1})$	1	1.3	2.2	3.8
Heavy particles [19]				
κ_h $(mW \cdot cm^{-1} \cdot K^{-1})$	3.4	5.4	8.1	9.3

potentials of clustered alkali atoms (X_2, X_3, X_4) are distinctly smaller than those
of the atoms [17, 18], which would mean an increase in the degree of ionization
of a vapour containing cluster molecules.

4.   CONVECTIVE THERMAL CONDUCTIVITY OF THE ELECTRON GAS
IN UO_2 VAPOUR AND IN SODIUM VAPOUR

What is the possible contribution of the free electron gas to the conductive
heat transfer in UO_2 vapour and in sodium vapour? While the heavy-particle
conductivity has already been evaluated [15, 19], the question posed has not yet
been considered. At present, a quantitative answer would be difficult to make
because of the existing uncertainties in the theoretical treatment of strongly-
coupled plasmas and of transition plasmas. An analytical kinetic theory of the
transport coefficients of strongly-coupled plasma has not yet been written [10−13].
In the case of weakly-coupled transition plasmas, two limiting approaches
are commonly used to evaluate semiquantitatively the thermal (and electrical)
conductivity of the electron gas − the Spitzer approach for the fully ionized gas
and the Lorentz gas approach (see, for example, Ref. [20]).

In the first case only the Coulomb interactions of the electrons with the ionized vapour molecules and with other electrons are considered, i.e. neglecting the neutral scatterers in the plasma. In the Lorentz gas approach for a weak plasma, the resistivity of the heavy-particle scatterers is assumed to be caused by neutral–electron scattering. Here a difficulty arises, namely in choosing correct values for an effective scattering cross-section, σ_{0e}, of the neutral 'lattice' molecules scattering the thermal electrons.

Table III gives the results for the electron thermal conductivity, κ_e, obtained from the two approaches, using the following two equations.[2]

Fully ionized gas

$$\kappa_{ei}\ (W \cdot cm^{-1} \cdot K^{-1}) = 1.85 \times 10^{-12}\ \frac{\sqrt{T^5}}{\ln \Lambda} \tag{1}$$

Lorentz gas

$$\kappa_{0e} = \frac{1}{3}\ \frac{1}{n_0 \sigma_{0e}}\ c_v\, n_e\, \overline{v}_e \tag{2}$$

where

$$c_v \simeq 3k \pm 1k$$

and

$$\overline{v}_e = \sqrt{\frac{8}{\pi}\ \frac{kT}{m_e}}$$

A value of $P_c \simeq 110\ cm^{-1}$ ($\triangle \sigma_0 \simeq 30\ Å^2$) of the electron–neutral collision probability has been used in Eq.(2). Table III gives, for comparison, the heavy-particle thermal conductivity, κ_h, calculated by Cronenberg et al. [19].

Comparing the different valves of κ_e obtained by the two approaches, it is felt that the lower values (fully ionized gas approach) will be nearer to reality. It has been found experimentally [20] that, owing to the long-range Coulomb interaction, the 'fully ionized' behaviour of a plasma appears progressively already at a degree of ionization of 1%. Thus, the comparison of the κ_h and κ_e values yields the result that the contribution of the electron gas to the thermal conductivity of UO_2 vapour is small in the temperature range below 4500 K, but that it can no longer be neglected at temperatures around 5000 K and above.

---

[2] Heat transport by ambipolar diffusion and recombination has been neglected.

TABLE IV. CONVECTIVE THERMAL CONDUCTIVITY OF THE
ELECTRON GAS, κ_e, AND OF THE ATOMIC GAS, κ_h, FOR
SATURATED SODIUM VAPOUR

T (K)	1500	2000	2506
Fully-ionized gas model			
$\ln \Lambda$ (calc.)	9.8	1.9	0.2
$\ln \Lambda$ (used)	9.8	1.9	1
κ_e (mW·cm^{-1}·K^{-1})	0.018	0.2	0.6
Heavy particles [19]			
κ_h (mW·cm^{-1}·K^{-1})	0.67	3.4	49

Table IV gives the results for saturated sodium vapour up to the critical
temperature. The contribution from the electron gas component to the total
thermal conductivity appears to be negligible over the whole temperature range
below the critical temperature.

5. RADIATIVE HEAT TRANSFER AND RADIATION MEAN FREE PATH
    IN UO$_2$ VAPOUR

The radiant heat exchange through a fuel-vapour layer is of importance in
certain problems occurring in nuclear safety research and in the safety analysis
of nuclear reactors. It is of interest to know under what conditions an oxide
fuel-vapour layer is optically thick or thin to thermal radiation. The same question
is relevant in pyrometric temperature determination in high-temperature experiments with liquid fuel materials [1—5]. In the cases mentioned, the spectral
range of interest is mainly the visible because, at temperatures above 4000 K,
the maximum spectral intensities of a black-body radiator lie in this range, as do
the working wavelengths of optical high-temperature pyrometers.

Various radiational interaction mechanisms occur in a high-temperature,
partially-ionized gas. In the visible spectral range, one has to consider electronic
bound—bound line absorption and bound—free absorption (photoionization).

Inverse bremsstrahlung absorption (free–free transitions in the electronvolt range) does not play an important role in UO_2-vapour plasma. Neither the absorption nor the emission spectrum of the multicomponent UO_2 vapour is known for the visible spectral range, nor are the line strengths known; these could be converted to oscillator strengths or spectral absorption cross-sections. The same lack of data holds for the uranium ions U^{2+}, U^{4+} and U^{6+}. The following theoretical approach has been undertaken to determine, at least semiquantitatively, the optical mean free path in saturated UO_2 vapour up to 5000 K.

Because of the relatively low value of the thermal energy quantum as compared with the ionization energy ($kT \ll E_i$), bound–free photoionization is neglected and only bound–bound absorption is considered.

The unknown energy-level spectra of the different UO_2 vapour species (UO_2, UO_3, UO) are approximated by that of a hydrogen-like model molecule. Its discrete energy spectrum is replaced by a smeared quasi-continuous spectrum. This is justified because, owing to the Stark effect, there is no energy degeneracy of the energy levels in a strongly bound polyatomic molecule which is, in addition, immersed in a dense and strong plasma.[3] Its partition function takes the form [7]:

$$Z = \int_{n_0}^{n_{max}} \left(\exp\left[-\frac{E_i - \{Ry/(n-\delta)^2\}}{kT} \right] \right) (2n^2 + 2n + \tfrac{1}{3})\, dn \qquad (3)$$

where n is the principal quantum number, δ the quantum defect constant, n_0 the ground state, n_{max} the cut-off orbit and Ry the Rydberg energy ($= 13.6$ eV). The absorption cross-section is determined for a molecule at a distinct excitation level n for spectral radiation of frequency ν and bandwidth $\Delta\nu$. The bound–bound absorption can excite the absorber molecule to an upper level within the range $m \pm \Delta m$, where Δm depends upon $\Delta\nu$. Integration over the continuous energy spectrum leads to an effective mean absorption cross-section, σ_ν, of the thermally excited absorber molecules of temperature T for radiation of frequency ν.

$$\sigma_\nu = \frac{I}{Z} \cdot \sqrt{\frac{Ry}{E_i}} \cdot \frac{\pi c r_0}{E_i/h} \cdot \bar{f}_\nu \qquad (4)$$

---

[3] The procedure is more suitable in the case of uranium oxide vapour, because the model absorber molecule is composed in reality of multielectron actinide atoms or ions. Even the undisturbed, non-excited molecules must be assumed to exhibit very line-rich dense spectra [21–24].

where

$$I = \int_{n_0}^{n_{max}} \left(\exp\left[-\frac{E_i - \{Ry/(n-\delta)^2\}}{kT} \right] \right) (2n^2+2n+\tfrac{1}{3})^2 \, dn$$

Here $\bar{f}_\nu$ is the mean (absorber) oscillator strength of a bound–bound transition of frequency ν, c is the velocity of light, and r_0 is the Bohr radius. The relation:

$$\int \sigma_\nu \, d\nu = \pi c r_0 f_{nn'}$$

for a bound–bound photoabsorption is used in Eq.(4).

The cut-off orbit n_{max} and the constant δ have been determined as shown in Ref.[7]. Combining the numerical results of (3) and (4) with the density data n_{tot} of Table I yields the absorption coefficient $K_\nu = n_{tot} \sigma_\nu$ and the radiation mean free path, $1/K_\nu$, in saturated UO_2 vapour. The results are given in Table V.

To avoid misinterpretations, the following comments should be borne in mind. The absorption coefficient, K_ν, and the radiative mean free path, L_ν, as calculated from Eq.(4), are not related to the 'black-line centre' of an absorbed spectral line of oscillator strength $\bar{f}_\nu$. They are averaged with respect to the absorption lines, which are assumed to overlap. In reality a certain line structure could still remain in the absorption spectrum of the UO_2 vapour molecules. This does not, however, alter the averaged absorption coefficient and radiative mean free path; there would only be a 'ripple' deriving from the non-averaged absorption coefficient varying with spectral frequency [9, 13].

Finally, there is the question of how a gap between the ground state and the lowest electronic states of the vapour molecules would alter the considerations discussed above. For this reason, the mathematical approach was modified, splitting the energy-level spectrum in the partition function into a term for the degenerate ground state of weight $2n_0^2$ and a quasi-continuous energy-level spectrum beginning at n_0, at ~1 to 2 eV excitation energy. The numerical results, however, turn out to be the same in this case (i.e. as given in Table V). Assuming such an energy gap would mean that there is a transmission window in the absorption spectrum of the vapour for infrared light of $h\nu < 1-2$ eV. (The remaining absorption by vibrational transitions is not discussed here.)

There is strong support from absorption experiments performed with pure uranium plasmas [21–23] for the quantitative correctness of the radiative absorption coefficients given in Table V. Conversion of the averaged absorption coefficient, 1.5×10^{-3} cm^{-1}, measured on a uranium plasma at T = 5100 K,

TABLE V. AVERAGED ABSORPTION COEFFICIENT, κ_ν, AND AVERAGED MEAN-FREE-PATH, L_ν, OF OPTICAL RADIATION IN SATURATED UO_2 VAPOUR UP TO 5000 K

Parameter: mean oscillator strength, $\bar{f}_\nu$, of a bound–bound transition between non-degenerate levels nn'.

$\bar{f}_\nu$	T (K) =	3660	4000	4500	5000
0.1	K_ν (cm) =	530	2×10^3	8.4×10^3	2.5×10^4
	L_ν (mm) =	0.019	5×10^{-3}	1.2×10^{-3}	4×10^{-4}
0.01	K_ν (cm) =	53	200	840	2.5×10^3
	L_ν (mm) =	0.19	0.05	0.012	0.004
0.001	K_ν (cm) =	5.3	20	84	250
	L_ν (mm) =	1.9	0.5	0.12	0.04

$p = 3.5 \times 10^{-4}$ bar, to the density conditions of saturated UO_2 vapour at 5000 K yields $L_\nu = 38\ \mu m$ at about 0.5 μm wavelength. This is in full agreement with the value $L_\nu = 40\ \mu m$, assuming weak spectral absorption lines in this spectral range with a mean oscillator strength of $\bar{f}_\nu = 0.001$. It is known from the spectroscopic literature that, as a rule, $f_\nu \simeq 10^{-2} - 10^{-3}$ is the typical order of magnitude of the oscillator strengths of the bound–bound transitions in the visible range, which is in accordance with the result obtained.

6.   CONCLUSIONS

The effect of the plasma state of saturated uranium oxide vapour on the convective thermal conductivity of the vapour phase appears to be negligible up to temperatures around 4500 K. A similar result has been found for saturated sodium vapour up to its critical temperature region (2500 K). It should be stressed however that evaluation of the transport coefficients of the dense vapour plasma has been more a qualitative than a quantitative one at the highest temperatures considered.

The evaluation of the averaged radiative absorption coefficient and mean free path of saturated UO_2 vapour up to 5000 K has shown that uranium oxide vapour becomes optically thick, even for very thin vapour layers, depending on the temperature. In this problem the dense gaseous and plasma states of the vapour play only a secondary role, i.e. in splitting, shifting and broadening the energy levels of the vapour molecules. The numerical results for K_ν and L_ν obtained from the theoretical approach have been found to be in quantitative agreement with experimental observations on high-temperature uranium plasmas.

The results for the radiative absorption of uranium oxide vapour are of relevance to problems in nuclear reactor safety analysis when radiative thermal heat exchange through a fuel vapour layer is involved. The mechanism of radiative heat transfer actually changes from the simple fourth-power law of Boltzmann in the case of the transparent vapour layer to the diffusion of trapped thermal radiation along the temperature gradient in the opaque vapour.

In the same way the results are relevant in experimental nuclear safety research involving pyrometric temperature measurements on liquid-oxide fuel. Similarly, the pyrometric temperature determination in experiments with liquid fuel material involving laser-beam-induced evaporation is affected because the open laser evaporation of fuel material, even into vacuum, proceeds in the form of a supersonic flow-off of a thin non-isothermal vapour layer at the boundary surface [1, 2]. If the vapour layer is opaque, it is the thermal radiation emitted by this non-isothermal vapour layer that the pyrometer sees instead of the thermal emission of the sample surface itself, and the vapour layer is at a lower temperature. The effect of radiation trapping would yield a measured value for the surface temperature which is too low. Consequently, a value for the vapour pressure would be assigned that is higher than that corresponding to the true sample surface temperature. Based on the numerical results for the radiation mean free path in UO_2 vapour, one would expect pyrometric temperature measurements to become unreliable at temperatures much exceeding ~4000 K during laser evaporation of oxide fuel material — unless there are ideal transmission windows for pyrometry in the greatly broadened line spectrum of the molecules in the fuel vapour. The latter seems improbable, as may be realized from observations on uranium plasmas [21–23].

Absorption experiments on laser-generated UO_2 vapour plumes, using a multiline laser as the reference light source, could provide an experimental proof of the conclusions. The intention of our group is to use a modified laser reflectometer [3, 4] with its modulated argon, krypton and HeNe laser beams to perform such absorption experiments. This device might also make it possible to measure the scattering of light passing through a UO_2 vapour plume.

ACKNOWLEDGEMENTS

I would like to thank Professor Hans Griem (presently with the Max Plank Institute, Munich), Professor H.P. Popp (Ruhr University of Bochum), Professor

F. Hensel (University of Marburg), and Drs H.J. Kusch and H. Ehrich (University of Kiel) for helpful discussions. The assistence of K.T. Müller in performing the computer calculations is greatly acknowledged.

REFERENCES

[1] BOBER, M., KAROW, H.-U., SCHRETZMANN, K., "Evaporation experiments to determine the vapour pressure of UO$_2$ fuel (3000–5000 K)", Thermodynamics of Nuclear Materials 1974 (Proc. Symp. Vienna, 1974) Vol.1, IAEA, Vienna (1975) 295–305.

[2] BOBER, M., BREITUNG, W., KAROW, H.U., SCHRETZMANN, K., "Evaporation studies of liquid oxide fuel at very high temperatures using laser beam heating", Gordon Research Conf. High-Temperature Chemistry (Tilton, USA, 1976); also Kernforschungszentrum Karlsruhe Rep. KFK-2366 (1976).

[3] BOBER, M., KAROW, H.U., "Measurements of spectral emissivity of UO$_2$ above the melting point", Proc. 7th Symp. Thermophysical Properties (National Bureau of Standards, Gaithersburg, 1977: CEZAIRLIYAN, A., Ed.), American Society of Mechanical Engineers, New York (1978) 344–350.

[4] KAROW, H.U., BOBER, M., "Experimental investigations into the spectral reflectivity and emissivity of liquid UO$_2$, UC, ThO$_2$ and Nd$_2$O$_3$", these Proceedings, paper IAEA-SM-236/22 Vol.1, p.155.

[5] BOBER, M., BREITUNG, W., KAROW, H.U., Thermodynamic Calculation and Experimenta Determination of the Equation of State of Oxide Fuels up to 5000 K, Kernforschungs-zentrum Karlsruhe Rep. KFK-2689 (1978).

[6] POTTER, P.E. (Ch.), International Atomic Energy Agency – International Working Group on Fast Reactors, Summary Report: Equations of State of Materials of Relevance to the Analysis of Hypothetical Fast Breeder Reactor Accidents (Specialists' Mtg Harwell, 1978), IAEA, Vienna, Rep. IAEA-IWGFR/26 (1978).

[7] KAROW, H.U., "Thermodynamic state, specific heat, and enthalpy function of saturated vapor over UO$_2$ between 3000 K and 5000 K", Proc. 7th Symp. Thermophysical Properties (National Bureau of Standards, Gaithersburg, 1977: CEZAIRLIYAN, A., Ed.), American Society of Mechanical Engineers, New York (1978) 373–378.

[8] KAROW, H.U., Thermodynamic state and gas kinetic relaxation behavior of saturated UO$_2$ vapor up to 5000 K", Rev. Int. Hautes Temp. Refract. 15 (1978) 347–354.

[9] ZELDOVICH, Ya.B., RAIZER, Yu.P., Physics of Shock Waves and High-Temperature Hydrodynamic Phenomena (HAYES, W.D., PROBSTEIN, R.F., Eds), Academic Press, New York (1966) 2 vols.

[10] KRASNIKOV, Yu.G., KULIK, P.P., NORMAN, G.E., "Non-ideal plasmas", Proc. 10th Int. Conf. Phenomena in Ionized Gases, Invited Papers (Oxford, 1971), Oxford University Press (1971) 405–435.

[11] DEUTSCH, C., GOMBERT, M.M., MINOO, H., Strongly coupled classical plasmas in laser fusion and astrophysics, Comments Plasma Phys. Controlled Fusion 4 1 (1978) 1–11.

[12] KALMAN, G., CARINI, P., (Eds), Strongly Coupled Plasmas (Proc. NATO Summer School Orléans, 1977), NATO Advanced Study Institutes Series, Ser. B (Physics) 36, Plenum Publ. Corp., New York (1978).

[13] GRIEM, H.R., private communication.

[14] BREITUNG, W., Berechnung der Dampfdrücke von oxidischen Brennstoffen bis 5000 K bei Gleichgewichts- und Nichtgleichgewichtsverdampfung, Kernforschungszentrum Karlsruhe Rep. KFK-2091 (1975).

[15] BOBER, M., BREITUNG, W., KAROW, H.U., KLEYKAMP, H., SCHUMACHER, G.,
      THURNAY, K., "Investigation of thermodynamic data of state of fast reactor core
      materials for hypothetical accident analysis: theoretical and experimental work at
      Karlsruhe (§7. The thermodynamic properties of sodium)", Summary Report: Equations
      of State of Materials of Relevance to the Analysis of Hypothetical Fast Breeder Reactor
      Accidents (Specialists' Mtg Harwell, 1978), IAEA, Vienna, Rep. IAEA-IWGFR/26 (1978)
      28–35.
[16] EWING, C.T., SPANN, J.R., STONE, J.P., MILLER, R.R., High temperature properties of
      Cs, J. Chem. Eng. Data 11 4 (1966) 473.
[17] ROBBINS, E.J., LECKENBY, R.E., WILLIS, P., The ionization potentials of clustered
      sodium atoms, Adv. Phys. 16 (1967) 739–744.
[18] LECKENBY, R.E., ROBBINS, E.J., The effect of molecular association on the electrical
      conductivity of sodium vapour, J. Phys., B (London), At. Mol. Phys. 1 (1968) 441–444.
[19] TILLS, J.L., CRONENBERG, A.W., SCHMIDT, T.R., A prediction of the thermal
      conductivity of UO_2 vapor, J. Nucl. Mater. 67 (1977) 67–76.
[20] HOYAUX, M.F., Arc Physics, Springer Verlag,Berlin (1968).
[21] SCHNEIDER, R.T., CAMPBELL, H.D., MACK, J.M., On the emission coefficient of
      uranium plasmas, Nucl. Technol. 20 (1973) 15–26.
[22] PARKS, D.E., LANE, G., STEWART, J.C., PEYTON, S., Optical Constants of Uranium
      Plasma, US National Aeronautics and Space Administration Rep. NASA CR-72348 (1968);
      also Gulf General Atomics Rep. 8244 (1968).
[23] POPP, H.P. private communication.
[24] RAND, M.H., "The thermodynamic functions of gaseous actinide elements", these
      Proceedings, paper IAEA-SM-236/39, Vol.1, p. 197.

DISCUSSION

J. MAGILL: I would like to point out that when we use the LTE or the
steady-state CORONA model for estimating the degree of ionization in the laser-
produced plasma and then calculate the amount of line radiation emitted, we find
this too low by as much as two orders of magnitude. When we account for transient
ionization, by solving the time-dependent rate equations and then using Griem's
formula for line radiation, agreement with experiment is obtained.

H.U. KAROW: I agree with your comment about the radiative line emission
of a common discharge plasma of moderate charged-particle density where the
Debye screening theory is applicable. As regards the strong quasi-continuous
absorption of the dense vapour of liquid-oxide fuel, however, I should like once
more to stress that in this case the collective plasma state of the vapour must be
considered to play only a secondary role, i.e. in splitting, shifting and Stark-
broadening the bound electronic states of the vapour molecules. This alters the
original line absorption spectrum of the dilute vapour giving, at particle densities
of $\gtrsim 10^{17}$ cm^{-3}, an overlapping quasicontinuous absorption spectrum of the saturated
fuel vapour. The exact values of the effective ionization potential and the degree of

ionization of the vapour are not of relevance in this context because the bound—free and free—free radiative absorptions in the vapour plasma can be neglected, as compared to the strong bound—bound absorption of the vapour.

K.A. LONG: In your calculations on gaseous UO_2, treating it as a plasma (due to thermal ionization), you appear to have completely ignored the effect of electron—electron and electron—ion interactions. In view of the fact that the inclusion of these interactions via the Debye theory is essential even for a plausible first-order theory, how can such calculations using the nearest-neighbour approach to the reduction of the ionization potential be credible?

Further more, recent experiments on alkali-metal vapours show a semi-conducting or insulator—metal transition around the critical density. Such transitions can be explained by the Mott-Hubbard theory, in which the electron—electron intra-atomic interaction plays an essential role. Thus since UO_2 at, say, 5000 K is well below the critical density, it would not seem possible for the conductivity of UO_2 vapour to be that high or that significant. Have you considered this in your calculations?

H.U. KAROW: I think that what you claim to be shortcomings have been clearly explained in the paper. You mention the findings, of Hensel and others, that in some cases metals lose their metallic conductivity near their critical point because of the decreasing particle density of the liquid phase with increasing temperature. An analogous phenomenon appears to occur in the vapour phase, but in the opposite direction. With increasing density of the vapour molecules the orbits of the higher-excited valence electrons progressively overlap the electron clouds of the neighbouring molecules. Such highly-excited electrons do not differ essentially from free conduction electrons. This effective cut-off of the upper bound excitation levels has been confirmed experimentally. It has been found by plasma diagnostic experiments that the Debye screening theory fails to calculate the effective lowering of the ionization potential of such highly dense vapour plasmas, while the nearest-neighbour approach yields reasonable results.

As for the problem of the heat transport properties of strongly-coupled vapour plasmas, I agree that there is as yet no complete theory of the transport coefficients that includes wave-mechanical interaction of the plasma particles.

EXPERIMENTAL INVESTIGATIONS INTO THE SPECTRAL REFLECTIVITIES AND EMISSIVITIES OF LIQUID UO_2, UC, ThO_2 AND Nd_2O_3

H.U. KAROW, M. BOBER
Institut für Neutronenphysik und Reaktortechnik,
Kernforschungszentrum Karlsruhe,
Karlsruhe,
Federal Republic of Germany

Abstract

EXPERIMENTAL INVESTIGATIONS INTO THE SPECTRAL REFLECTIVITIES AND
EMISSIVITIES OF LIQUID UO_2, UC, ThO_2 AND Nd_2O_3.

Safety research for fast reactors requires knowledge of emissivity data of nuclear fuel materials up to temperatures corresponding to the liquid state. A special integrating-sphere laser reflectometer has been used to measure the normal reflectivities and emissivities of UO_2, UC, ThO_2 and Nd_2O_3 in the solid state (premolten, refrozen material) and in the liquid state up to temperatures of 4000 to 4800 K. The measurement wavelengths were 0.63 μm and 10.6 μm. The emissivity curves, $\epsilon_\lambda(T)$, of the oxidic specimens measured at 0.63 μm show the same characteristic course: little temperature dependence below the melting point and a distinct increase of $\epsilon_{0.63\,\mu m}(T)$ in the liquid state. In the case of UO_2, the emissivity at the melting point (3120 K) is 0.84; at 4100 K it is 0.92. At 10.6 μm, a decrease of $\epsilon_{10.6\,\mu m}(T)$ has been measured for the liquid state of UO_2 and ThO_2. UC shows in both the solid and the liquid states only a small temperature dependence of $\epsilon_{0.63\,\mu m}(T)$, with a marked drop, however, at the melting point (2780 K) from 0.54 to 0.45. The results of the measurements are presented in diagrams and by 'fitted' equations related to the true and the black-body temperatures. The spectral emissivity values measured for UO_2 up to 4200 K have been applied in the evaluation of laser evaporation experiments. The vapour-pressure curve of liquid UO_2, as evaluated from these experiments, is presented.

1. INTRODUCTION

Emissivity data for the thermal radiation of nuclear fuel materials at temperatures above the melting point are of interest in safety research for nuclear reactors. The total hemispherical emissivity is relevant to problems in reactor safety analysis in which the thermal radiation of molten-core material is involved. Knowledge of the spectral emissivity makes it possible to derive reliable pyrometric temperature measurements of liquid fuel surfaces, e.g. up to temperatures of about 4000 K in the case of liquid oxide fuels. This is of interest when obtaining vapour pressure measurements using laser evaporation techniques [1, 2] and when carrying out high-temperature experiments with molten fuel material.

Knowledge of the wavelength and angular dependence of the emissivity and reflectivity with temperature could, in addition, lead to conclusions regarding the unknown physical structure and bonding character of the fuel materials in the liquid state. Furthermore, the radiative heat transfer in molten fuel could be evaluated [3].

Since no measured data existed on the thermal emissivity of nuclear fuel materials above 2500 K, we developed, three years ago (1976), a new technique for measuring the spectral reflectivity and emissivity of opaque ceramic materials to temperatures far above their melting points. With this experimental technique, the spectral emissivities of urania and of magnesia, alumina and tantalum carbide were measured up to 4000 K at wavelengths in the visible and infra-red spectral ranges [4, 5]. The emissivity data for urania were used when making measurements to determine the vapour-pressure curve of liquid UO_2 [2].

In the paper, the reflectivity measuring technique is described. The emissivity data of UO_2, remeasured with the improved apparatus, and the recently measured emissivity data of UC, ThO_2 and Nd_2O_3 are presented.

2.   MEASURING METHOD

Because of the relatively high opacity of these materials in the visible and infra-red spectral ranges at high temperatures, the spectral emissivity can be directly determined by measuring the spectral reflectance of the sample surface. The spectral directional-hemispherical reflectance $\rho_\lambda(\Theta; 2\pi)$ of the laser-heated, optically smooth (solid or liquid) sample surface is related to the required directional emissivity $\epsilon_\lambda(\Theta)$ by Kirchhoff's law:

$$\epsilon_\lambda(\Theta) = \alpha_\lambda(\Theta) = 1 - \rho_\lambda(\Theta; 2\pi) \tag{1}$$

For measuring the spectral reflectivity, ρ_λ, a special integrating sphere reflecto-meter has been developed using laser beam heating and monochromatic irradiation of the specimen with reference light (modulated laser light) of wavelength λ [4]. The intense local surface heating of the specimen allows of well-reproducible surface heating of UO_2 or of similar refractory materials to temperatures above 4000 K, with heating times of 1 to 100 ms per cycle within a surface area of ~ 0.5 mm in diameter. The reflected portion of the incident reference light is isotropically scattered by the diffuse-reflective, 4π integrating sphere and is measured with an appropriate spectral detector. A high-frequency modulation of the reference light beam allows of the detection of the reflected light in the presence of the intense thermal radiation emitted by the hot sample. The tempera-ture of the laser heated sample surface is determined using a fast micropyrometer looking into the integrating sphere.

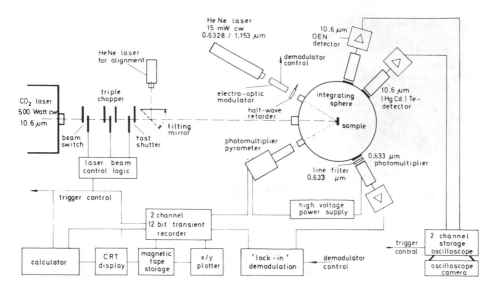

FIG.1. *Principle of measuring method and a schematic diagram of the experimental set-up.*

3.   APPARATUS

Figure 1 shows, schematically, the experimental assembly [4]. The small specimen is located in an integrating sphere of uniform, highly diffuse reflectance; it is 200 mm in diameter. Two spheres were prepared, with different surface coatings for measurements in the visible and near infra-red spectral ranges, and in the far infra-red range. During the measurements, pure argon gas is used as the atmosphere in the sphere, at a pressure of ~ 2 bar.

A 1 to 100 ms long, rectangular CO_2-laser pulse of 500 watt power level is focused on the specimen surface, which becomes locally heated to the liquid state. The temperature and the heating rate at the molten specimen surface are varied by varying the focused laser power density. The angle, Θ, of incidence or emission respectively, can be varied by altering the inclination of the sample. A special fast multiplier pyrometer is used to measure the spectral radiance temperature of the laser-heated samples up to temperatures of above 4000 K [2, 4]. It allows of measurements within a target spot of 40 μm. The working wavelength is 0.65 μm. From the digitally recorded pyrometer signal, the black-body or the true sample temperature, is evaluated by a computer.

The reflectivity measurements in the visible range are made in a white-coated integrating sphere. Simultaneously with the laser heating, the focus area on the specimen is irradiated by a modulated beam from either a 15 mW HeNe laser (at 0.6328 or 1.153 μm wavelengths) or a krypton-ion laser (at wavelengths in the

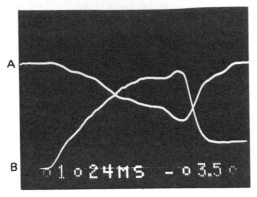

FIG.2. Digital oscilloscope recordings of reflectivity signal at 0.63 μm (A) and pyrometer signal (B) measured with a UO₂ sample Sweeptime 6 ms; $T_{max} \simeq 3650$ K.

T = const.

visible and near UV). The reflected reference laser light is measured with an appropriate photomultiplier detector. A spectral line blocking filter of matching wavelength is used in front of the photomultiplier to filter out the thermal radiation emitted by the hot sample. The directional-hemispherical reflectivity $\rho_\lambda(\Theta; 2\pi)$ of the sample is obtained by means of an additional calibration measurement in which the specimen is exchanged for a perfectly reflecting mirror of the same size. The directional emissivity is calculated with Eq.(1).

The reference light beam has been modulated to allow reflected light to be detected in the presence of residual thermal radiation passed by the blocking filter. An electro-optical ADP modulator[1] running at 200 kHz provides the modulation. The reflected light signal is demodulated by a fast lock-in demodulation system. The demodulated reflection signal is digitally recorded, and this is followed by an on-line computational evaluation of the spectral emissivity, $\epsilon_\lambda(T)$, as a function of the black-body or true sample temperature.

Figure 2 shows a typical measuring record, obtained with a UO₂ sample, as displayed on the digital oscilloscope. The upper trace (A) represents the reflectivity signal at 0.63 μm wavelength. The pyrometer signal is given by the lower trace (B), at the end of which the long constant-temperature portion represents the freezing point (f.p.) of the sample. The maximum temperature in this heating cycle is ~3650 K.

The reflection measurements at 10.6 μm wavelength are made in a gold-coated integrating sphere. The reflected portion of the incident CO_2-laser light is isotropically scattered and measured with special detectors for far IR radiation [4]. A similar reading is taken on a comparison standard using a fully reflecting mirror instead of the sample.

---

[1] ADP: ammonium dihydrogen phosphate.

TABLE I.   SPECIMEN MATERIAL SPECIFICATIONS

	Production	Per cent of theoretical density	Composition	Impurities
UO_2	sintering (NUKEM)	96%	O/U = 2.00	> 100 ppm C; 200 ppm Si + Cr + Fe
UC	sintering (NUKEM)	96.5%	4.66 wt.% C	180 ppm N_2; 3550 ppm O_2
ThO_2	sintering (Zircoa)	96%		200 ppm Ca; 150 ppm Si + Al + Co
Nd_2O_{3-x}	hot pressing (own lab.)	98.5%	x = 0.003	100 ppm Pr oxide + Sm oxide

Using this experimental device, we have made measurements of the normal
spectral emissivity, $\epsilon_\lambda(T)$, of urania and thoria at 0.63 μm and 10.6 μm wavelengths
to temperatures far into the liquid range. Measurements with uranium carbide
and neodymium oxide have, as yet, only been made at 0.63 μm wavelength.

4.   SPECIMENS

Specimens of UO_2, UC, ThO_2 and Nd_2O_3 were produced by sintering or
by hot pressing powder material. For the reflectivity measurements, discs of
6 mm in diameter and 1 mm in thickness were prepared. The sample surfaces
were polished to a surface roughness of $\lesssim$ 0.3 μm. The material specifications are
given in Table I.

5.   MEASUREMENTS AND RESULTS

Reflectivity measurements were carried out both on polished sintered sample
surfaces and, more extensively, on previously molten and refrozen sample surfaces.
The power density of the focused CO_2-laser beam was 0.1 to 0.4 MW/cm^2. In
the case of UO_2, for example, these power densities lead to heating rates between
1.5×10^6 and 2×10^6 K/s between 2000 K and the melting point (3120 K), and
of 0.1×10^6 and 0.5×10^6 K/s above the melting point.

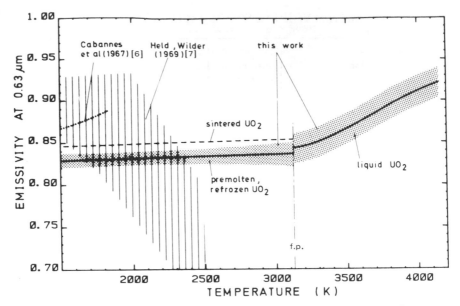

FIG.3. Normal spectral emissivity of UO_2 at 0.63 μm wavelength as a function of temperature.

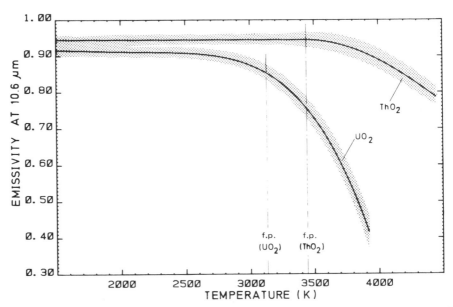

FIG.4. Normal spectral emissivity of UO_2 and ThO_2 at 10.6 μm wavelength as a function of temperature.

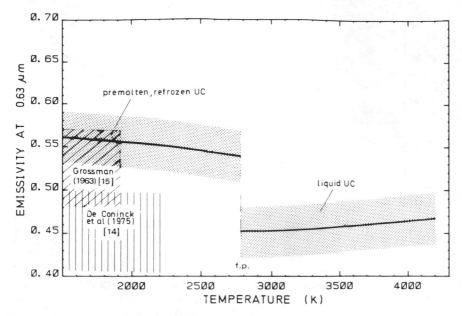

FIG.5.   Normal spectral emissivity of UC at 0.63 µm wavelength as a function of temperature.

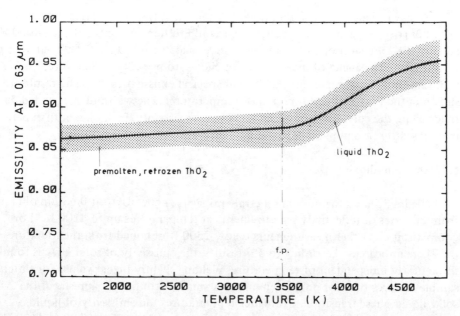

FIG.6.   Normal spectral emissivity of ThO₂ at 0.63 µm wavelength as a function of temperature.

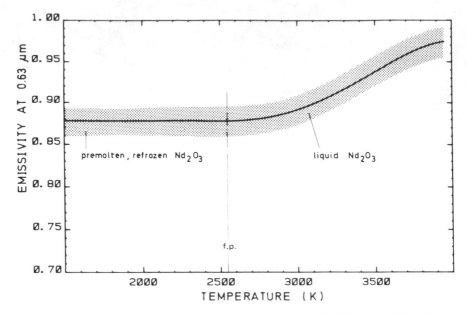

FIG. 7. *Normal spectral emissivity of Nd_2O_{3-x} at 0.63 μm wavelength as a function of temperature.*

At a wavelength of 0.63 μm, the directional-hemispherical reflectivity $\rho_\lambda(\Theta; 2\pi)$ of the specimens has been measured up to temperatures between 4000 K and 4800 K. In addition, measurements were made with UO_2 and ThO_2 at 10.6 μm wavelength. The angles of incidence were chosen to be < 30°. Therefore, the resulting emissivities represent the normal spectral emissivities, $\epsilon_\lambda^\perp$. The results are shown in the Figs 3–7 as a function of temperature. The width of the shaded data fields along the curves represent the ranges of uncertainty, including both experimental scattering and possible systematic errors.

5.1. Uranium dioxide, UO_2

Figure 3 shows the normal spectral emissivity, $\epsilon_\lambda^\perp$, of UO_2 at 0.63 μm obtained from 60 series of reflectivity measurements at temperatures up to 4200 K. For comparison two other measurements below 2500 K obtained from the literature [6, 7] are indicated. According to our results, the emissivity of solid UO_2 is found to be nearly independent of temperature, yielding slightly higher values for sintered sample surfaces than for premolten ones. A small discontinuity appears at the solid/liquid phase transition. At higher temperatures the emissivity of liquid UO_2 clearly increases from 0.84 at the melting point (3120 K) to 0.92 at 4100 K.

The course of the spectral emissivity of premolten and liquid UO_2 is represented by Eqs (2a, b):

300 K $<$ T $<$ 3120 K

$$\epsilon^{\perp}_{0.63\,\mu m}(T) = 0.8225 + 4.321 \times 10^{-6}\,T \tag{2a}$$

3120 K $<$ T $<$ 4200 K

$$\epsilon^{\perp}_{0.63\,\mu m}(T) = 0.843 + 1.4465 \times 10^{-5}\,(T{-}3120)$$

$$+ 1.64966 \times 10^{-7}\,(T{-}3120)^2 - 1.3136 \times 10^{-10}\,(T{-}3120)^3$$

$$+ 2.8990 \times 10^{-14}\,(T{-}3120)^4 \tag{2b}$$

For optical pyrometry purposes, $\epsilon^{\perp}_{0.63\,\mu m}$ is given in Eqs (3a, b) as a function of the radiance temperature (black-body temperature), T_r:

300 K $<$ T_r $<$ 3046 K

$$\epsilon^{\perp}_{0.63\,\mu m}(T_r) = 0.8225 + 4.9095 \times 10^{-6}\,T_r \tag{3a}$$

3046 K $<$ T_r $<$ 4200 K

$$\epsilon^{\perp}_{0.63\,\mu m}(T_r) = 0.843 + 1.8928 \times 10^{-5}\,(T_r{-}3046)$$

$$+ 1.5296 \times 10^{-7}\,(T_r{-}3046)^2 - 1.2347 \times 10^{-10}\,(T_r{-}3046)^3$$

$$+ 2.7849 \times 10^{-14}\,(T_r{-}3046)^4 \tag{3b}$$

The emissivity of UO_2 in the far IR at 10.6 μm was obtained from 50 series of measurements. The result is shown in Fig.4. No difference has been found in the reflectivity of sintered and premolten samples, and no discontinuity appears. However, in contradistinction to the course of the emissivity at 0.63 μm, at 10.6 μm the emissivity of liquid UO_2 has been found to decrease markedly with increasing temperature. In the numerical evaluation, the slope of the curve was corrected for isothermal reflectivity conditions in the measuring spot [5].

5.2. Uranium carbide, UC

The normal spectral emissivity of UC at 0.63 μm is given in Fig.5 up to 4200 K. In this case 30 series of reflectivity measurements made using both premolten and liquid sample surfaces were evaluated. Only a slight temperature dependence of the emissivity was detected, but a sharp drop appears at the melting point (2780 K), from 0.54 to 0.45. The course of $\epsilon^\perp_{0.63\,\mu m}(T)$ is represented by Eqs (4a, b):

300 K $<$ T $<$ 2780 K

$$\epsilon^\perp_{0.63\,\mu m}(T) = 0.566 - 2.7209 \times 10^{-6}\,T$$

$$+ 2.7697 \times 10^{-9}\,T^2 - 2.7102 \times 10^{-12}\,T^3$$

$$+ 2.8618 \times 10^{-16}\,T^4 \tag{4a}$$

2780 K $<$ T $<$ 4200 K

$$\epsilon^\perp_{0.63\,\mu m}(T) = 0.452 + 4.3247 \times 10^{-6}\,(T-2780)$$

$$+ 3.1967 \times 10^{-9}\,(T-2780)^2 + 1.6784 \times 10^{-12}\,(T-2780)^3$$

$$- 4.6641 \times 10^{-16}\,(T-2780)^4 \tag{4b}$$

For pyrometric temperature measurements, $\epsilon^\perp_{0.63\,\mu m}$ of UC is also given as a function of the radiance temperature, T_r:

300 K $<$ T_r $<$ 2580 K (solid UC)

$$\epsilon^\perp_{0.63}(T_r) = 0.566 - 3.244 \times 10^{-6}\,T_r$$

$$+ 1.982 \times 10^{-10}\,T_r^2 - 1.628 \times 10^{-12}\,T_r^3$$

$$- 7.976 \times 10^{-17}\,T_r^4 \tag{5a}$$

2527 K $<$ T_r $<$ 4000 K (liquid UC)

$$\epsilon^\perp_{0.63}(T_r) = 0.452 + 6.507 \times 10^{-6}\,(T_r-2527)$$

$$+ 4.695 \times 10^{-9}\,(T_r-2527)^2 + 9.876 \times 10^{-13}\,(T_r-2527)^3$$

$$+ 1.744 \times 10^{-16}\,(T_r-2527)^4 \tag{5b}$$

5.3. Thorium dioxide, ThO$_2$

Figure 6 shows the normal spectral emissivity of ThO$_2$ at 0.63 μm obtained from 40 series of reflectivity measurements at temperatures up to 4800 K. The course of $\epsilon^\perp_{0.63\ \mu m}(T)$ with temperature looks very similar to that of UO$_2$, showing the same increase in the liquid range. It is fitted by Eq.(6).

3440 K $<$ T $<$ 4800 K

$$\epsilon^\perp_{0.63\ \mu m}(T) = 0.876 - 3.515 \times 10^{-7}\,(T-3440)$$

$$+ 1.6 \times 10^{-7}\,(T-3440)^2 - 1.253 \times 10^{-10}\,(T-3440)^3$$

$$+ 2.853 \times 10^{-14}\,(T-3440)^4 \tag{6}$$

Again for optical pyrometry $\epsilon^\perp_{0.63}$ of liquid ThO$_2$ is given in Eq.(7) as a function of the radiance temperature T_r.

3370 K $<$ T$_r$ $<$ 4800 K

$$\epsilon^\perp_{0.63\ \mu m}(T_r) = 0.876 + 1.203 \times 10^{-6}\,(T_r-3370)$$

$$+ 1.573 \times 10^{-7}\,(T_r-3370)^2 - 1.251 \times 10^{-10}\,(T_r-3370)^3$$

$$+ 2.897 \times 10^{-14}\,(T_r-3370)^4 \tag{7}$$

The normal emissivity of ThO$_2$ at a wavelength of 10.6 μm was evaluated from 40 series of reflectivity measurements. The results are given in Fig.4. Above the melting point $\epsilon^\perp_{10.6\ \mu m}(T)$ decreases with increasing temperature, as found in the case of UO$_2$, but to a smaller extent.

5.4. Neodymium oxide, Nd$_2$O$_{3-x}$

The final material investigated was neodymium oxide. The normal spectral emissivity was obtained from 30 series of reflectivity measurements at 0.63 μm wavelength and temperatures up to 3950 K. The resulting curve is shown in Fig.7. The emissivity is found to be constant up to the melting point, whose value was obtained from the freezing constant-temperature plateau at 2540 K. The deviation from the melting temperature of 2598 $\pm$ 10 K measured by Coutures et al. [8] could be caused by a certain substoichiometry of the specimens used in our experiments. The deviation from stoichiometry, x, has not yet been determined. For liquid Nd$_2$O$_{3-x}$, the course of $\epsilon^\perp_{0.63\ \mu m}(T)$ with temperature is given by Eq.(8).

2540 K $<$ T $<$ 3950 K

$$\epsilon^{\perp}_{0.63\,\mu m}(T) = 0.878 + 1.1535 \times 10^{-6}(T-2540)$$

$$+ 3.7 \times 10^{-8}(T-2540)^2 + 7.005 \times 10^{-11}(T-2540)^3$$

$$- 4.4206 \times 10^{-14}(T-2540)^4 \tag{8}$$

For the purpose of optical pyrometry $\epsilon^{\perp}_{0.63\,\mu m}$ of liquid $Nd_2 O_{3-x}$ is given in Eq.(9) as a function of the radiance temperature:

2503 K $<$ T_r $<$ 3900 K

$$\epsilon^{\perp}_{0.63\,\mu m}(T_r) = 0.878 + 6.2866 \times 10^{-6}(T_r-2503)$$

$$+ 2.321 \times 10^{-8}(T_r-2503)^2 + 8.0226 \times 10^{-11}(T_r-2503)^3$$

$$- 4.653 \times 10^{-14}(T_r-2503)^4 \tag{9}$$

6.   DISCUSSION

It is the aim of these measurements to determine the reflectivity and emissivity of refractory ceramic materials in the liquid state. The validity of the measured liquid reflectivity data appears to be confirmed by the fact that the freezing temperatures (f.p.) evaluated with the measured emissivity from the radiance temperature of the freezing constant-temperature plateaux are in agreement with the melting point data derived from the phase diagrams of the substances. The applicability of the measuring technique has also been proved by the results of the comparison measurements performed with certain standard materials, e.g. MgO, $Al_2 O_3$, and UO_2 [5].

One limit of application of this measuring technique is reached when the material under investigation is not sufficiently opaque. This appears to be the case with ThO_2 in the solid state at low temperatures. The low temperature trans-lucency of this material (absorption coefficient ~ 30 cm^{-1}) could partly explain the disagreement between the lower temperature emissivity data published by Kneissl and Richmond [9] and our measurements (Fig.6).

The upper limit of application of the measuring technique occurs at the onset of appreciable evaporation of the specimen at temperatures when the sample vapour pressure becomes comparable with the inert gas pressure in the integrating sphere. In addition, the pyrometric temperature measuring method also ceases to be reliable when thermal radiation emitted by the sample is partly absorbed or scattered in the vapour layer above the sample surface [10].

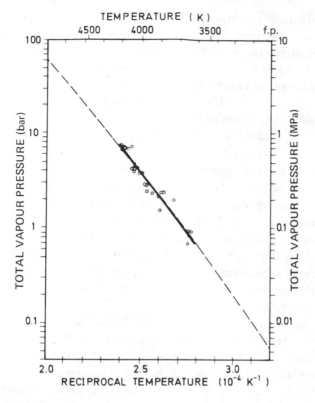

FIG.8.   *Vapour pressure of liquid UO₂.*
The circles represent experimental p-T values obtained in laser evaporation experiments with
pyrometric temperature measurements [2, 17] by using the emissivity data of Fig.3. The solid
line gives the fitted curve of the measured p-T values. The broken line is the vapour pressure
curve of liquid UO_2 recommended by the IAEA-IWGFR [17] (log p = 28.65−34 930 T^{-1}−5.64 log T:
p measured in MPa, T in K).

Considering the measured emissivity curves of the oxides investigated,
namely UO_2, ThO_2 and Nd_2O_3, it is seen that they show similar characteristic
courses at a wavelength of 0.63 μm (red region) with a distinct increase of
$\epsilon_{0.63\,\mu m}(T)$ in the liquid state of these materials. UO_2 and ThO_2 also behave
similarly in the infra-red at 10.6 μm wavelength. The decrease of $\epsilon_{10.6\,\mu m}(T)$
with temperature can be explained by a broadening of the 'reststrahlen' band
in the far infra-red region [11, 12], and by a possible contribution to the reflectivity
of free conduction electrons. Our measurements are in full agreement with the
lower temperature measurements performed on urania by Cabannes et al. [6],
and those of Clark and Moore on thoria [12].

The emissivity of uranium carbide in the visible red region (Fig.5) shows a different characteristic course to those of the oxidic materials investigated. The freezing-temperature value evaluated from the radiance temperature and the emissivity of the liquid UC surface at the freezing plateau coincides with the corresponding temperature value obtained from the radiance temperature and the emissivity of the frozen UC surface at the end of the freezing plateau. The resulting freezing temperature of UC is in full agreement with the melting temperature value (2780 K) published in Ref.[13]. No explanation can be given for the deviation of the emissivity values measured by De Coninck et al. [14] from our measurements. It might be caused by variations in stoichiometry of the specimen material or by oxygen impurity. There is good agreement, however, with the older emissivity measurements on UC of Grossman [15].

In the case of UO_2 (Fig.3), the emissivity measurements in the visible red region by Cabannes et al. [6], by Held and Wilder [7] and by Schoenes [16] are in agreement, to within their limits of error, with our measurements. Figure 8 represents a direct application of the liquid emissivity values of Fig.3 from the laser evaporation measurements to the evaluation of the vapour pressure curve of liquid UO_2 up to 4200 K [2, 17].

The measurements on ceramic materials will be continued and extended to wavelengths in the near UV, visible blue and green, and the near IR ranges. From these measurements, it will be possible to estimate the total thermal emittance of the ceramic melts. The dependence of the reflectivity on angle of incidence and polarization will be studied to get information on the optical constants $n_\lambda(T)$, $k_\lambda(T)$ and the radiative heat transfer in ceramic materials in the liquid state.

ACKNOWLEDGEMENTS

The measurements of this work have in part been performed by K.T. Müller. His excellent experimental assistance and theoretical support are gratefully acknowledged. The authors would also like to thank Professor F. Cabannes for helpful discussions.

REFERENCES

[1] BOBER, M., KAROW, H.-U., SCHRETZMANN, K., "Evaporation experiments to determine the vapour pressure of UO_2 fuel (3000–5000 K)", Thermodynamics of Nuclear Materials 1974 (Proc. Symp. Vienna, 1974) Vol.1, IAEA, Vienna (1975) 295–305.

[2] BOBER, M., BREITUNG, W., KAROW, H.U., Thermodynamic Calculation and Experimental Determination of the Equation of State of Oxide Fuels up to 5000 K, Gesellschaft für Kernforschung mbH, Karlsruhe, Rep. KFK 2689 (1978).

[3] KAROW, H.U., On the Evaluation of the Spectral Course of the Optical Constants $n_\lambda(T)$, $k_\lambda(T)$, and of the Radiative Heat Conductivity of Nuclear Fuel Materials in the Liquid Phase, Gesellschaft für Kernforschung mbH, Karlsruhe, Rep. KFK 2653 (1978).

[4] BOBER, M., KAROW, H.U., "Measurements of spectral emissivity of UO_2 above the melting point", Proc. 7th Symp. Thermophysical Properties (National Bureau of Standards, Gaithersburg, 1977: CEZAIRLIYAN, A., Ed.), American Society of Mechanical Engineers, New York (1978) 344–50.

[5] MÜLLER, K.T., Messung des spektralen Emissionsvermögens von keramischen Materialien im festen und flüssigen Zustand mit einem Laser-Reflektometer, Diploma Thesis, University of Karlsruhe, Kernforschungszentrum Karlsruhe, Rep. KFK-2803 (1979).

[6] CABANNES, F., STORA, J.P., TSAKIRIS, J., Facteurs de réflexion et d'émission de UO_2 à haute température, C.R. Hebd. Séances Acad. Sci., Ser. B 264 (1967) 45–8.

[7] HELD, P.C., WILDER, D.R., High-temperature hemispherical spectral emittance of uranium oxides at 0.65 and 0.70 μm, J. Am. Ceram. Soc. 52 (1969) 182–6.

[8] COUTURES, J.-P., VERGES, R., FOEX, M., Valeurs comparées des températures de solidification des différents sesquioxydes de terres rares; influence de l'atmosphère, Rev. Int. Hautes Temp. Refract. 12 (1975) 181–5.

[9] KNEISSL, G.J., RICHMOND, J.C., A Laser-Source Integrating Sphere Reflectometer, Technical Note 439, National Bureau of Standards, US Dept. of Commerce, Washington, DC (1968) 47–48.

[10] BOBER, M., BREITUNG, W., KAROW, H.U., SCHRETZMANN, K., On the interpretation of vapor pressure measurements on oxide fuel at very-high temperatures for fast reactor safety analysis, J. Nucl. Mater. 60 (1976) 20–30.

[11] MYERS, H.P., GYLLANDER, J.-Å., "The transmission of infra-red radiation through sintered uranium dioxide", New Nuclear Materials Including Non-Metallic Fuels (Proc. Symp. Prague, 1963) Vol. 1, IAEA, Vienna (1963) 323–9.

[12] CLARK, H.E., MOORE, D.G., A rotating cylinder method for measuring normal spectral emittance of ceramic oxide specimens from 1200 to 1600 K, J. Res. Natl. Bur. Stand., A 70 (1966) 393–415.

[13] FEE, D., JOHNSON, C.E., Phase Equilibria and Melting Point Data for Advanced Fuel Systems, Argonne National Laboratory Rep. ANL-AFP-10 (1975).

[14] DE CONINCK, R., VAN LIERDE, W., GIJS, A., Uranium carbide: thermal diffusivity, thermal conductivity and spectral emissivity at high temperatures, J. Nucl. Mater. 57 (1975) 69–76.

[15] GROSSMAN, L.N., Electrical Conductivity, Thermal Conductivity, and Thermal Emission for Fuel-Bearing Carbides: UC_2, UC, $U_{0.5}Zr_{0.5}C$, and ThC_2, General Electric Company (USA) Rep. GE-ST-2015 (1963).

[16] SCHOENES, J., Optical properties and electronic structure of UO_2, J. Appl. Phys. 49 3 (1978) 1463–65.

[17] POTTER, P.E. (Ch.), International Atomic Energy Agency − International Working Group on Fast Reactors, Summary Report, Equations of State of Materials of Relevance to the Analysis of Hypothetical Fast Breeder Reactor Accidents (Specialists' Mtg Harwell, 1978), IAEA, Vienna, Rep. IAEA-IWGFR/26 (1978).

VAPOUR PRESSURE MEASUREMENTS OF URANIUM CARBIDES UP TO 7000 K USING LASER PULSE HEATING

R.W. OHSE, J.F. BABELOT, K.A. LONG, J. MAGILL
Commission of the European Communities,
European Institute for Transuranium Elements,
Karlsruhe

Abstract

VAPOUR PRESSURE MEASUREMENTS OF URANIUM CARBIDES UP TO 7000 K USING LASER PULSE HEATING.

The laser pulse heating technique, originally developed for equation-of-state studies on UO_2 and the ternary fast breeder oxide fuel (U, Pu)O_2, has now been extended by additional high speed diagnostics to measurements of advanced carbide fuels, starting with UC. Vaporization studies over uranium monocarbide were performed over the temperature range from 6000 K to 8000 K. Various interaction phenomena with the laser-induced gas jet, such as optical absorption, backscattering and radial flow, were studied at laser power densities of $10^7 W/cm^2$. The average vapour pressure over the monocarbide at 6757 K was found to be 10.47 MPa. The final vapour pressure equation for the temperature range from the melting point at 2780 K, assuming a composition of $UC_{1.08}$, up to 7000 K, was determined to be:

$$\log p = 8.622 - (3.286 \times 10^4/T) - 0.715 \log T$$

where p is in megapascals and T is in kelvin. This yields a heat of evaporation within the measured range of temperature of $\Delta H_{evap} = 141$ kcal·mol^{-1} and entropy of evaporation of $\Delta S_{evap} = 67$ cal·mol^{-1}·K^{-1}. A spectral emissivity of 0.45 at the wavelength of 6540 Å was measured at the melting point of UC at 2780 K.

1. INTRODUCTION

The objective of this work was to make the first total pressure measurements on advanced fast-breeder carbide fuels far beyond the melting point in order to check the various total pressure estimations [1—4], and to provide the required input data for the equation-of-state and critical-point data prediction.

An equation of state for advanced fast breeder carbide fuels is required in order to predict the behaviour of the fuel in a hypothetical core disruptive accident (HCDA), and to estimate the energy release during the power transient of a prompt critical excursion. Carbides and nitrides are of interest as advanced fuels because they show better neutronic properties and have higher thermal conductivities than oxides, thus allowing of higher linear power ratings and a higher breeding ratio.

Direct measurements are essential since low-temperature data obtained by effusion, transpiration and mass spectrometric techniques are neither consistent, thus giving rise to large uncertainty limits on extrapolation, nor suited to extrapolation to extreme temperatures because of changes in the predominance of the different vapour species, and because of other effects caused by dissociation, excitation and ionization. Owing to the high rates of evaporation involved, direct measurements can only be accomplished by dynamic pulse heating techniques such as the exploding wire technique [5–7], neutron pulse heating [8], electron pulse heating [9], and laser pulse heating [10–12] as described earlier [13, 14]. In order to have access to the gas phase, the laser pulse heating technique previously applied to binary and ternary fast breeder oxide fuels [11, 14–16] was again chosen for the investigation of the uranium carbide system.

The problems of interpretation of vaporization experiments at extreme temperatures are (apart from the difficulties in making temperature measurements and in determining their 'meaning') to a large extent due to the temperature gradients, high rates of evaporation and hydrodynamic flow mechanism involved. These effects lead to radial displacement by recoil and expansion forces, to selective mass-dependent backscattering of evaporating molecules, and finally to composition changes at the surface which cannot be restored by the slower diffusion process.

In order to guarantee reliable vapour pressure data, the basic evaporation mechanism, including possible compositional changes at the surface, and the relationship between the measured rate of evaporation and the saturated vapour pressure has to be investigated as a function of increasing rate of evaporation.

2.   LITERATURE DATA ON THE VAPOUR PRESSURE OVER URANIUM CARBIDE

2.1.   Measured vapour pressure data below the melting point

Various experimental vapour pressure data measured over uranium monocarbide are presented in Table I and Fig. 1.

Good agreement exists between the two mass spectrometric studies of Storms [21] and Pattoret et al. [25], and the measurements of Vozzella et al. [19], Andrievskij et al. [27], and Tetenbaum et al. [29].

Some other measurements differ greatly from Storms' data, probably due to the rapid change of uranium and carbon activity with composition around $C/U = 1$, to interactions of uranium carbide with the crucible material, and to the presence of oxygen above the surface, leading to oxicarbide formation, with a corresponding CO partial pressure. Khromonojkin and Andrievskij [20, 27] obtained results in agreement with Storms' data [21] by improving their experimental technique. They concluded that the previously higher values found for the uranium pressure [17, 18, 20, 23, 24] were due to the use of tungsten Knudsen cells, leading to carbon depletion, and to non-stoichiometric specimens.

The various measurements have been discussed and compared by Storms [30], Pattoret et al. [25] and, more recently, during the IAEA Panel on the Assessment of Thermodynamic Properties of the U-C, Pu-C, U-Pu-C Systems held in Grenoble, France, in 1974 [1].

2.2. Assessment of thermodynamic data, and calculated vapour pressures below the melting point

Tetenbaum et al. [2] calculated the temperature dependencies of the total pressure and the partial pressures of UC_2, C_1, C_2, C_3, and U over UC_x for the composition range $x = 0.92-1.10$, from the thermodynamic data assessed by the IAEA Panel [1]. The given temperature range of validity of these equations is 2000–2500 K. Table II gives the vapour pressures over $UC_{1.00}$ calculated by Tetenbaum et al. [2].

For reactor safety analysis, data on the UC fuel are needed to much higher temperatures. An extrapolation of these low-temperature equations above the melting point requires, in the first place, a correction for the heat of fusion of UC; furthermore additional vapour species have to be taken into account, such as C_4 or possibly UC_4 at higher temperatures.

2.3. Thermodynamic extrapolation above the melting point

Sheth et al. first took an empirical approach [3, 32], by extrapolating the equations of Tetenbaum et al. [2] for the partial pressures of the individual species over $UC_{1.00}$, and by correcting each equation by a multiple of the heat of fusion of UC, depending on the number of UC molecules that are required to form the gaseous molecule. Additionally an equation was given for the partial pressure of C_4. The total vapour pressure was obtained by the addition of all partial pressures, leading to the following equation:

$$\log p = -25.0029 - (18\,132/T) + 7.5425 \log T \qquad (1)$$

for 2800 K $< T <$ 10 000 K, with p in megapascals.

By correcting these results using a self-consistent extrapolation method, new equations were published in 1977 [4], giving a total pressure over UC of:

$$\log p = 5.110 - (31\,704/T) + 0.197 \log T \qquad (2)$$

for 2800 K $< T <$ 10 000 K.

TABLE I. VAPOUR-PRESSURE DATA, MEASURED OVER URANIUM MONOCARBIDE

Measured quantity	C/U	Temperature range (K)	For [a] $\log p = A - B/T$ A	B	Method	Authors	Year	Ref.
Press. above UC	1	1948–2133	17.431	4.920×10^4	Effusion technique	Ivanov et al.	1962	[17]
U partial press.	1	1910–2660	5.11	2.780×10^4	Effusion technique	Alexander et al.	1963	[18]
U partial press.	1.1	2190–2525	7.197	3.624×10^4	Langmuir technique	Vozzella et al.	1965	[19]
U partial press.	1	1900–2400	6.179	2.809×10^4	Effusion technique	Khromonojkin et al.	1965	[20]
C partial press.	1	→	−1.360	1.324×10^4	→	→	→	→
U partial press.	0.995	2073–2373	5.775 ± 0.215	$(3.121 \pm 0.049) \times 10^4$	Mass spectrometry	Storms	1965	[21]
UC_2 partial press.	0.995	→	9.90 ± 2.77	$(4.628 \pm 0.644) \times 10^4$	→	→	→	→
U partial press.	1.078	→	7.046 ± 0.204	$(3.501 \pm 0.046) \times 10^4$	→	→	→	→
UC_2 partial press.	1.078	→	7.387 ± 0.812	$(3.993 \pm 0.189) \times 10^4$	→	→	→	→
U partial press.	1.03	–	6.225 ± 0.773	$(3.291 \pm 0.171) \times 10^4$	–	Krupka	1965	[22]
U partial press.	0.983	2250–2500	3.791	2.570×10^4	Effusion technique	Anselin et al.	1966	[23]
U partial press.	1.10	2250–2500	3.416	2.600×10^4	→	→	→	→
Press. above UC	1.03	2150–2500	7.8	3.500×10^4	Effusion technique	Gorban'yu et al.	1967	[24]
U partial press.	1.04	2250–2510	6.52	3.35×10^4	Mass spectrometry	Pattoret et al.	1967	[25]
C partial press.	1.1	2190–2525	7.347	3.608×10^4	Langmuir technique	Vozzella et al.	1968	[26]
U partial press.	1.01	2173–2573	3.69	2.85×10^4	Effusion technique	Andrievskij et al.	1969	[27]
U partial press.	0.97	–	4.86	2.700×10^4	Langmuir technique	Solov'ev	1971	[28]
Total U pressure		2255	(see Fig. 6 in Ref. [29])		Transpiration tech.	Tetenbaum,Hunt	1971	[29]

[a] Units of measurement: p in MPa; T in K.

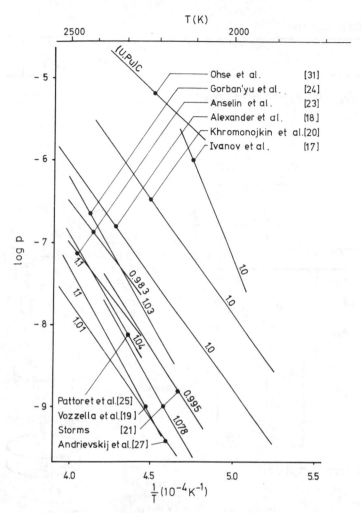

FIG.1.   Vapour pressure measurements below the melting point over $UC_{1\pm x}$ and, for comparison, the plutonium pressure over $(U_{0.80}Pu_{0.20})C_{1.01}$.

3.   EXPERIMENTAL APPROACH

3.1. Dynamic pulse heating

Direct pressure measurements up to the high critical temperatures of fast breeder oxide (approximately 7500 K) and carbide fuels (approximately 10 000 K) under the extreme conditions of high temperature gradients and high rates of evaporation require the development of dynamic pulse heating techniques.

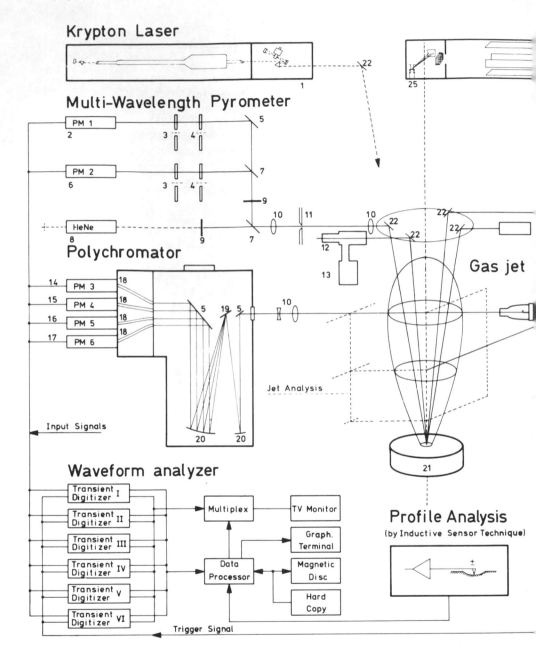

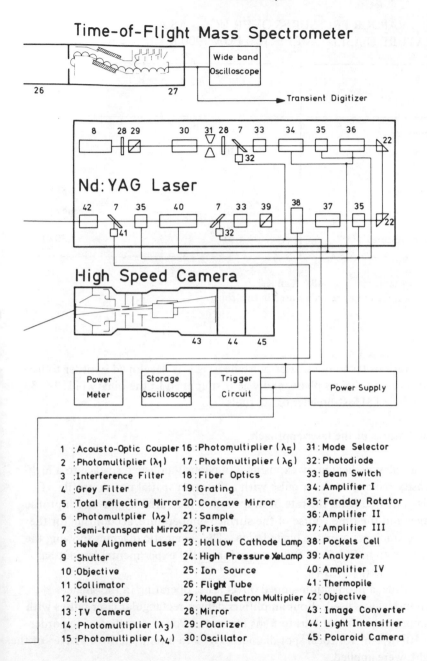

FIG.2. Schematic diagram of the lay-out of extended high speed diagnostics for thermophysical property measurements and the characterization of high-temperature vapour induced by laser pulse heating.

TABLE II.   VAPOUR PRESSURES OVER $UC_{1.00}$ AS A FUNCTION OF THE
TEMPERATURE (BASED ON REF. [2]) FOR $2000\ K \leqslant T \leqslant 2500\ K$

Calculated quantity	For[a,b] $\log p = A - B/T$	
	A	B
U partial pressure	4.51	-2.80×10^4
UC_2 partial pressure	7.30	-4.12×10^4
C_1 partial pressure	7.37	-4.01×10^4
C_2 partial pressure	9.14	-4.84×10^4
C_3 partial pressure	9.16	-4.87×10^4
Total vapour pressure	4.53	-2.80×10^4

[a] Units of measurement: p in MPa; T in K.
[b] Formula re-ordered to match form used in this paper.

Among the various techniques, laser pulse heating was developed in order to have
additional access to the gas phase, and was first applied to the binary [11, 14, 33]
and ternary [11, 15] fast-breeder oxide fuels.

3.2. Main features of the laser approach

The samples are heated by exposing a previously polished surface to a high
power density rectangular laser pulse with a Gaussian spatial power profile. The
pressure is evaluated from the rate of evaporation measured by the depth profile
of the target and the time course of the surface temperature, accounting for the
temperature gradients over the heated area. The schematic diagram showing the
technique is given in Fig. 2. A full description of the experimental lay-out has
been given elsewhere [34, 14].

For heating, a neodymium-doped YAG laser, operating in the single
transversal mode, was used. Four amplifiers deliver rectangular laser pulses with
a Gaussian power profile of 1 μs to 5 ms duration. Power densities of the order
of 10^6 to 10^7 W/cm^2 for focal spot diameters of 1 to 3 mm, defined as $1/e^2$ of the
peak height, were applied.

The temperature was measured by a high speed multi-wave-length pyrometer,
allowing a temporal resolution of 10 μs, a spatial resolution of 100 μm, and a
temperature resolution at 5000 K of the order of 5 K. A set of four interference
filters allowed temperature measurements to be made at 16 pairs of wavelengths.

The output signals of the photomultipliers were recorded on transient digitizers interfaced to a disc storage system and a digital computer. The pyrometer was calibrated against tungsten-ribbon standard lamps, calibrated by the Physikalisch-Technische Bundesanstalt (PTB) standards laboratory.

The laser target depth profile was measured by a magnetic sensor device of 0.1 μm depth resolution. It delivered a two-dimensional matrix of data points to be stored in the memory for profile and volume evaluation.

This technique has now been greatly extended [14, 34] using additional high speed diagnostics, such as multi-channel spectroscopy, high speed photography allowing of either a streak or framing mode, and time-of-flight mass spectrometry.

3.3. Problems arising at extreme temperatures

Classical Knudsen and Langmuir vaporization studies are performed well within the molecular flow region. Compositional changes at the surface of multicomponent systems are governed by diffusion processes. As the temperature is increased far above the melting point towards the critical temperature, the high rates of evaporation lead to hydrodynamic flow conditions, and to compositional changes at the surface which can no longer be restored by diffusion. The complex gas-dynamic expansion mechanism of the evaporating jet leads to selective back-scattering. Relaxation effects of excited internal degrees of freedom have to be taken into account.

The main problems associated with the increasing rates of evaporation and large temperature gradients are:

In the experiment
(A) Elimination of radial displacement and crack formation as a consequence of recoil and expansion forces;
(B) Measurement of temperature under optical absorption conditions (thermal ionization);

For evaluation and interpretation of results
(C) The concept of thermodynamic equilibrium, the basic evaporation mechanism and its consequence on surface composition, and the relationship between measured rate of evaporation and saturated vapour pressure;
(D) The effects of selective backscattering on surface compositional changes.

The considerations dealt with in §3.4 are mainly concerned with (A), the background to the radial-displacement problem. The theoretical backgrounds to (B), (C) and (D) will be the subjects of §4.

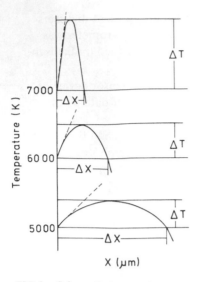

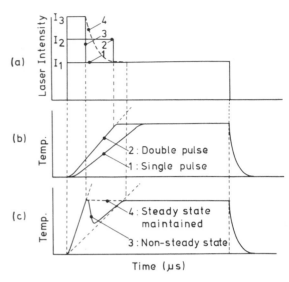

FIG.3. Schematic temperature
profile during steady-state evaporation using laser heating at
temperatures of the order of 5000
to 7000 K.

FIG.4. Reduction of temperature rise time by the
double-intensity laser pulse heating technique.
(a) Laser intensity versus time; (b) shortest rise
time for which a steady-state temperature profile
is established; (c) further reduction of temperature
rise time by programmed variation of laser intensity
with time.

3.4. Temperature profile and radial displacement

Figure 3 presents the temperature profiles in materials of low thermal
conductivity at temperatures of the order of 5000 to 7000 K, using the heat-
balance equations and assuming a constant absorption coefficient, heat conduction
inside the sample, and vaporization at the surface. The temperature gradient
caused by the high heat consumption of the evaporation process at the surface
increases with the surface temperature, whereas the width of the profile Δx decreases
It is the width of the profile down to the melting point which determines the
liquid layer thickness over the laser-heated target area. Due to the Gaussian power
and temperature profile across the laser-heated area, this liquid layer thickness
increases with radius toward the periphery, whereas the recoil forces caused by
the evaporating jet decrease.

In order to minimize or eventually to eliminate liquid or plastic radial
displacement, the penetration depth or width of the temperature profile has to
be limited to its minimum profile width at the final steady-state temperature.
To achieve this, the temperature rise time is shortened to the minimum time
required to ensure the build-up of the steady-state temperature profile. This was

TABLE III. SPECTROSCOPIC AND CHEMICAL ANALYSIS OF MOLTEN UC
(NUKEM)

Element	Concentration (ppm)	Element	Concentration (ppm)
Ag	< 1	Mn	< 1
Al	40	Mo	8
B	< 1	Na	8
Bi	< 1	Ni	< 1
Cd	< 1	Pb	< 1
Cr	3	Si	250
Cu	45	Sn	< 1
Fe	15	V	< 1
Mg	250	W	1500
		Zn	25

accomplished by a double-intensity laser pulse heating technique, described
previously [15], which is based on the reduction of the temperature rise time as
the intensity is increased (Fig. 4). The intensity of the pre-pulse is increased to the
maximum value I_2 (Fig. 4a) at which, after reaching the desired temperature T, the
steady-state profile is just established (Fig. 4b). The intensity of the pre-pulse is
then switched to the lower intensity I_1 of the main pulse (curve 2 in Fig. 4a).
A further increase in intensity to I_3 and rise time (3) leads to a temperature drop,
as indicated in Fig. 4c, since the heat consumption at the surface by evaporation
is fully imposed on the system although the steady-state profile has not yet been
established. (The additional heat loss by conduction could only be compensated
by a sophisticated programme involving a laser intensity change in the main pulse.)
In order to keep the width of the temperature profile as closely as possible down
to the unavoidable final steady-state profile width, the shortest temperature rise
time was chosen in all experiments. In addition to the axial temperature profile,
the radial temperature gradient, determining the gradients of the expansion and
recoil forces, was kept to a minimum by choosing the largest possible focal-spot
diameter.

3.5. Application to uranium carbides

On account of the expected complex evaporation behaviour, molten, almost-
oxygen-free, high-density UC samples from NUKEM were used. The solid UC rod

├──┤ 40 μm

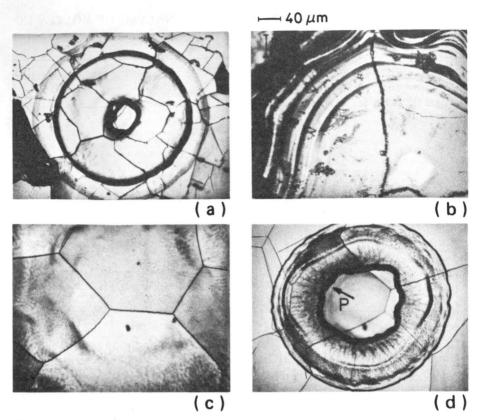

FIG.5. Micrographs of laser-heated uranium carbide targets, showing: (a) crack formation at
high quenching rates, and radial liquid displacement at the periphery of the heated area;
(b) ring structure; (c) elimination of cracks by keeping the sample temperature within the
plastic flow region (≥ 1900 K); (d) plastic flow (P) additionally influenced by the grain shape
acting as flow barrier. Crack formation in the liquid region occurs during quenching.

was cut into cylindrical sections and polished on both sides to better than 0.1 μm
surface roughness. A thorough sample degassing, using electron beam facilities,
was undertaken immediately before the laser heating experiments. The impurities
in the samples are listed in Table III.

The main problems arising in the first series of laser heating experiments were
crack formation and radial displacement (Fig. 5). Because of the plastic-like
properties of UC above 1800 K, the cylindrical UC discs were kept at 1800 K
throughout the laser heating experiments, using an additional electron bombardment
and were then slowly cooled down to room temperature. As shown in Fig. 5c,
this procedure allowed complete elimination of the previously observed crack
formation.

Various test series on liquid displacement, reported in §5, were necessary to investigate the reliability limits of measurements on uranium carbides. The superposed contribution of radial displacement, possibly due to broader initial temperature profiles during the rise time, to the target depth profile, obtained by evaporation, was tested by checking the time invariancy of the measurements, extrapolating to zero heating time.

The spectral emissivity measurements at the melting point of UC, 2780±25 K, applying a modified thermal analysis technique to the cooling curve after laser heating, confirmed the previously reported value of 0.45 at a wavelength of 0.65 μm [34]

4. THEORETICAL BACKGROUND

4.1. Temperature limit of pyrometry

The application of optical pyrometry to high-temperature laser evaporation experiments is limited by the fact that the radiation coming from the surface of the target, or from the laser itself, is absorbed in the gas jet. To estimate these effects a temperature criterion [34] for the limit of optical pyrometry, as a function of the material properties and the wavelength considered, is presented. The absorption mechanisms considered are photo-ionization and inverse bremsstrahlung, for which the absorption coefficient, b_ν (cm^{-1}), can be written as [35]:

$$b_\nu \cong \frac{10^{-7} n_g Z^2 \exp\{(h\nu - I_{eff})/kT_g\}}{T_g^2 (h\nu/kT_g)^3} \qquad (3)$$

where n_g is the gas number density, Z the ionic charge (= 1), I_{eff} (= $I - \Delta I$) the reduced ionization potential, T_g the gas temperature and ν the frequency of the radiation considered. Using the relations:

$$\left. \begin{array}{l} n_g = \dfrac{p_g}{kT_g} \\[2em] p_{sat} = 1.013 \times 10^6 \exp\{2.3 \, [A - (B/T_{sur})]\} \end{array} \right\} \qquad (4)$$

where p_g is the gas pressure and p_{sat} is the saturated vapour pressure at surface temperature T_{sur}, Eq. (3) can be rearranged to give the gas temperature limit T_g^{lim} provided that the absorption coefficient is known, i.e.:

$$T_g^{lim} = \frac{2.3 \dfrac{T_g^{son}}{T_{sur}} Bk - h\nu + I_{eff}}{k \left[\ln \left(\dfrac{0.1 k^2}{(h\nu)^3} \cdot \dfrac{p_g^{son}}{p_{sat}} \cdot \dfrac{1}{b_\nu} \right) + 2.3 \, A \right]} \qquad (5)$$

Here T_g^{son}/T_{sur} is the ratio of the gas temperature at the sonic point [14] divided by the surface temperature and p_g^{son}/p_{sat} is the gas pressure at the sonic point divided by the saturated vapour pressure at temperature T_{sur}. From gas dynamics [36, 37]:

$$\frac{T_g^{son}}{T_{sur}} = \frac{2}{3} \to 1 \tag{6}$$

and

$$\frac{p_g^{son}}{p_{sat}} = \frac{9}{2} \to 1 \tag{7}$$

depending on the number of de-excited degrees of freedom in going from the surface to the sonic point [14]. To obtain a value for b_ν, one must consider radiation from the laser passing through the gas to the surface and impose the condition that no more than 5% of the radiation should be absorbed, i.e. such that $b_\nu s = 0.05$, where s is the spatial coordinate. Taking s = 1 mm (the approximate focal-spot diameter), then $b_\nu = 0.5$ cm^{-1}.

From Table I average values of A and B were taken, 6.6 and 30 000 respectively, and extrapolated to above the melting point. Using the ratio T_g^{son}/T_{sur} = 0.667 and p_g^{son}/p_{sat} = 4.5, and considering 1 eV radiation, i.e. that of the neodymium laser, the limiting gas temperature was calculated as T_g^{lim} = 4415 K, corresponding to a surface temperature limit T_{sur}^{lim} = 6623 K. The calculation is not very sensitive to the laser wavelength; for a CO_2 laser the limiting temperature is lowered by approximately 400 K. Variation of the parameter b_ν by two orders of magnitude changes the limiting temperature by approximately 1000 K. Ionization potential reduction at the above gas temperature was of the order of 10^{-2} eV and can thus be neglected.

The absorbing region in the above calculation was taken to be the focal-spot diameter over which temperature and density are assumed constant. Gas dynamic calculations show, however, that the temperature drops by 65% and the density by 80% at one diameter away from the sonic point, thus allowing higher values of b_ν. This increases the limiting surface temperature to above 7000 K.

4.2. Basic evaporation process at high rates of evaporation

The fact that the equilibrium saturated vapour pressure, p_{sat}, and the rate of evaporation of a substance into vacuum are connected through the relation:

$$\dot{n}_+ = \frac{\alpha_\nu \, p_{sat}}{(2\pi m k T_{sur})^{\frac{1}{2}}} \tag{8}$$

where $\dot{n}_+$ is the rate of evaporation into a vacuum, α_ν the coefficient of evaporation,

T_{sur} the surface temperature, and m the molecular mass, was first shown by Hertz [38] and Knudsen [39] for the reaction $m_{solid} \rightarrow m_{gas}$ under the assumption of molecular flow. For evaporation from liquid surfaces, α_v is approximately equal to 1 [40].

At high rates of evaporation the assumption of molecular flow breaks down and hydrodynamic flow prevails in the gas phase, with associated density and temperature gradients. The rate of evaporation, however, is still given by relation (8), but the net flux transfer into the gas phase can be reduced due to the backscattering of particles from the gas phase onto the surface. On the basis of a one-dimensional steady-state kinetic description of the evaporation, this net flux transfer can be shown to be [36, 37] for $\alpha_v = 1$:

$$\dot{n}_+^{net} = \frac{p_{sat}}{(2\pi mkT_{sur})^{\frac{1}{2}}} (1-\beta) \tag{9}$$

where β, the total backscatter coefficient, can be interpreted in terms of the product of an equilibrium backscatter into a gas of density n_g, a non-equilibrium backscatter, and a factor which reduces the actual backscatter due to the forward motion of the gas phase [14]. For sonic evaporation into a vacuum it can be shown [14, 36, 37] that $\beta = 0.18$.

The above analysis is applicable to one-component systems only; for binary, single-phase systems such as UO_2 and UC showing a bivariant evaporation behaviour, one must consider, in addition, the composition of the evaporating surface.

4.3. Surface composition changes at high rates of evaporation

Surface composition changes follow a number of characteristic stages as the temperature and rate of evaporation increase. Within the molecular flow regime, the partial pressures and partial rates of evaporation are clearly defined by temperature and composition. Surface depletion tends to be restored by diffusion and the overall composition changes in the direction of the congruently evaporating composition. As the temperature and rate of evaporation are increased to such an extent that composition changes at the surface can no longer be restored by diffusion, the change of composition of the total sample is restricted to its surface layer and the evaporation enters into the forced congruency mode. Here the surface composition will change until the corresponding partial pressures lead to an overall gas composition which is equal to that of the bulk composition of the material. Assuming complete thermodynamic equilibrium, this composition change at the surface can be calculated applying the law of mass action as proposed by Breitung (see Ref. [41]), who reversed the approach taken in Rand's partial pressure calculation [42]. The higher partial pressure of uranium over the

monocarbide will lead to a carbon-rich phase, yielding a forced congruently
evaporating composition around the melting point of approximately $C/U = 1.1$.
As the temperature is increased towards 4000 K, law-of-mass-action calculations
indicate a change into substoichiometry. These calculations will eventually end
up showing extreme composition changes simply caused by the balanced requirements of rigorously extrapolated data. As the rates of evaporation are further
increased, gas dynamic effects of preferential backscattering of the lighter molecules
start to largely compensate these extrapolated surface composition changes.

5.   RESULTS AND DISCUSSION

5.1. Limit of measurement reliability

A critical analysis and appropriate theoretical models are necessary to check
the reliability of experimental results obtained under the extreme conditions of
high rates of evaporation (7000 K: > 1 m/s) and large temperature gradients
(70 000 K/cm), and to interpret these results in terms of equilibrium data. This
requires a systematic investigation of the observed and expected phenomena,
namely:

(a)   The basic evaporation mechanism at extreme rates of evaporation and its
       consequences on the relationship of measured rates of evaporation under
       hydrodynamic flow conditions to saturated equilibrium vapour pressures;
(b)   Computation of composition changes at the surface on the basis of appropriate
       thermodynamic models, accounting for the forced congruent evaporation
       mode and the gas dynamic effects of selective backscattering from the laser-
       induced gas jet;
(c)   Reliability and meaning of temperature measurement; possible effects of
       optical absorption leading to an upper limit of application for optical
       pyrometry;
(d)   Careful examination of the deviation of the measured depth profile from
       that due to pure evaporation only, due to additional radial plastic and liquid
       flow, caused by the large radial gradient of expansion forces within the
       heated area and recoil pressures of the evaporating vapour jet.

Of the above, (a) to (c) have been discussed in §4.2. The upper limit for optical
pyrometry (§4.1.) was found to be at least 7000 K. 10 series of test measurements
were made in the temperature range from 6000 K to 8000 K to investigate the
effects of optical absorption and a possible contribution of radial flow to the
depth profile across the laser-heated target area. Radial flow under steady-state
conditions can only be reduced or eliminated by reducing the radial temperature
gradient, i.e. enlarging the focal-spot diameter. The radial-flow contribution by an

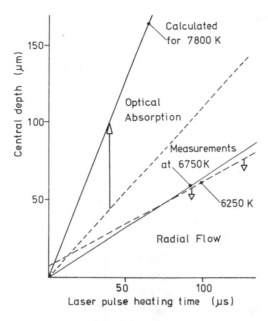

FIG.6. *Reliability limits for the vapour pressure measurements on UC; radial flow is used to define the lower temperature limit, and optical absorption the upper limit.*

initially large profile formation during the temperature rise was analysed by controlling the time invariancy of evaporation rate. Optimum laser heating pulse parameters, as described in §3.4., were selected for each temperature in order to guarantee the shortest temperature rise times. The time invariancy was checked up to 8000 K by 10 series of measurements, each at constant temperature, with pulse heating times up to 250 μs.

The central depths obtained at pulse durations from 50 μs to 250 μs were extrapolated to zero pulse time. Initial radial flow contributions, due to the larger temperature profiles formed at lower temperatures during the temperature rise time (Fig. 3) were found up to 6250 K, as demonstrated in Fig. 6 by the extrapolated 'zero-depth'. Measurements above 6300 K are thus considered as being undisturbed by initial radial flow phenomena.

The upper temperature limit of reliable measurements is mainly due to optical absorption of the light emitted from the laser-heated surface, leading to pyrometric measurements of too low a surface temperature, or to laser light absorption in the vapour jet, causing additional heating of the gas phase above the target, which can then affect the pyrometric measurements of the surface temperature. The start of optical absorption can be monitored (i) by its increasing effect during the build

OHSE et al.

TABLE IV. VAPOUR PRESSURE MEASUREMENTS OVER UC BY THE LASER PULSE HEATING TECHNIQUE OVER THE TEMPERATURE RANGE 6400 K TO 7000 K

T (K)	1/T (K^{-1})	t (μs)	d (μm)	p (MPa)	log p
6487	1.542×10^{-4}	236.4	130	8.73	0.94
6582	1.519×10^{-4}	128.0	81	10.10	1.00
6582	1.519×10^{-4}	128.2	85	10.60	1.03
6611	1.513×10^{-4}	224.9	152	10.83	1.03
6623	1.510×10^{-4}	171.2	120	11.25	1.05
6699	1.493×10^{-4}	176.0	122	11.18	1.05
6706	1.491×10^{-4}	180.4	116	10.38	1.02
6776	1.476×10^{-4}	169.1	126	12.09	1.08
6793	1.472×10^{-4}	96.0	60	10.15	1.01
6809	1.469×10^{-4}	98.9	61	10.03	1.00
6844	1.461×10^{-4}	137.9	93	11.00	1.04
6888	1.452×10^{-4}	94.1	51	8.87	0.95
6916	1.446×10^{-4}	140.8	87	10.12	1.01
6919	1.445×10^{-4}	95.0	69	9.63	0.98
6970	1.435×10^{-4}	94.1	69	12.06	1.08
6997	1.429×10^{-4}	139.9	98	11.55	1.06

up of the vapour jet, (ii) by the deviation from the power density/temperature relationship, and (iii) by a departure from the vapour pressure/temperature relationship, which is seen in the deviation of the measured depth from the expected depth. The time course of jet expansion investigated by high-speed photography in the streak and the framing mode still shows an appreciable increase in dimension during the first 50 μs. The expected depth values were calculated by plotting the log p versus 1/T relationship through the vapour pressure at the melting point of 2780 K, and the measured vapour pressure at 6757 K. The vapour pressure at the melting point was obtained by extrapolating the assessed data of Tetenbaum [2], which mainly relies on data from the IAEA Carbide Panel Meeting held at Grenoble in 1974 [1]. First deviations due to optical absorption (Fig. 6) were found to start at 7100 K.

The measurements were therefore considered to be reliable within the temperature range 6300 K to 7100 K.

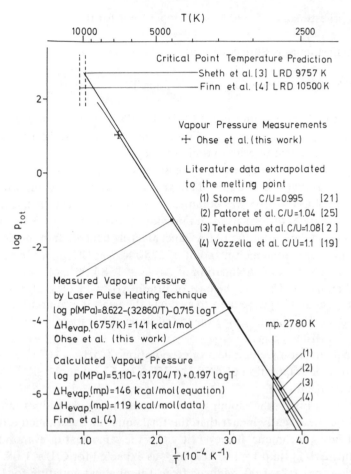

FIG.7.   *Total vapour pressure over uranium monocarbide, including extrapolations of low temperature data to the melting point, measurements at 6400—7000 K using the laser pulse heating technique, and prediction of critical-point data.*

5.2. Results and discussion

Table IV gives the results of five series of measurements within the temperature range from 6400 K to 7000 K. Figure 7 yields the overall pressure/temperature diagram, including vapour pressure data obtained by extrapolating vapour pressure measurements over solid $UC_{1 \pm x}$ up to the melting point, the average vapour pressure of 10.47 MPa at 6757 K, obtained from the five series of measurements given in Table IV, and finally the critical point temperature predicted by Sheth et al. [3, 43], and Finn et al. [4], applying the law of rectilinear diameter.

(Finn et al. [4] estimated 15 600 K as an upper limit for the critical temperature of UC.)

The final vapour pressure equation:

$$\log p = 8.622 - (3.286 \times 10^4 / T) - 0.715 \log T \tag{10}$$

was fitted to the average vapour pressure, 10.47 MPa at 6757 K, measured by the laser pulse heating technique over the liquid monocarbide (this work) and the vapour pressure and heat of evaporation at the melting point [2, 32, 44].

In order to be consistent with the measurements, the pressure chosen at the melting point was that over the forced congruently evaporating composition. A composition of C/U = 1.08 was obtained by a law-of-mass-action calculation, using the free energies of formation and uranium activity given by Tetenbaum et al. [2], which rely on data from the IAEA Panel Meeting held at Grenoble in 1974 [1]. The vapour pressure at the melting point of 2780 K over $UC_{1.08}$ was calculated to be 2.19 Pa. The heat of sublimation of ΔH_{sub} = 158 kcal/mol, calculated from Tetenbaum et al. [2], and the heat of fusion ΔH_f = 11.68 kcal/mol given by Sheth et al. [32] were used to evaluate the heat of evaporation ΔH_{evap} = 146 kcal/mol, thus giving the initial slope of the log p versus 1/T curve at the melting point.

The final equation yields a heat of evaporation of ΔH_{evap} (6757 K) = 141 kcal/m and an entropy of evaporation of ΔS_{evap} (6757 K) = 67 cal·mol⁻¹·K⁻¹ at the tempera ture of measurement.

As shown in Fig. 7, the vapour pressure equation calculated by Finn et al. [4] for C/U = 1.00, lies slightly higher than the final equation determined using the laser pulse heating technique; however, it yields the same heat of evaporation at the melting point, ΔH_{evap}(m.p.) = 146 kcal/mol, as expected for C/U = 1.08. The heat of evaporation for C/U = 1.00, deduced from Tetenbaum's equation for $UC_{1.00}$, yielding ΔH_{sub}(m.p.) = 128 kcal/mol, and from the heat of fusion, ΔH_f = 11.68 kcal/n [32], results in a lower value, ΔH_{evap} (m.p.) = 116 kcal/mol.

ACKNOWLEDGEMENTS

The authors would like to express their sincere thanks to Dr. Ronchi and Dr. Sari for helpful discussions and assistance in target evaluation, and to Messrs. R. Beukers, F. Capone, W. Heinz, G. Kramer and R. Selfslag for their assistance in performing the experiments.

REFERENCES

[1] ACKERMANN, R.J., POTTER, P., RAND, M., STORMS, E.K., Assessment of the Thermo-
    dynamic Properties of the U-C, Pu-C, and U-Pu-C Systems (IAEA Panel Mtg, Grenoble,
    France, 1974), (to be published).

[2] TETENBAUM, M., SHETH, A., OLSON, W., A Review of the Thermodynamic Properties
    of the U-C, Pu-C, and U-Pu-C Systems, Argonne National Laboratory, Illinois, Rep.
    ANL-AFP-8 (1975).

[3] SHETH, A., LEIBOWITZ, L., Equation of State and Transport Properties of Uranium
    and Plutonium Carbides in the Liquid Region, Argonne National Laboratory, Illinois,
    Rep. ANL-AFP-11 (1975).

[4] FINN, P.A., SHETH, A., WINSLOW, G., LEIBOWITZ, L., in Advanced LMFBR Fuels
    (Proc. ANS and USERDA Topical Mtg, Tucson, Arizona, 1977: LEARY, J., KITTLE, H.,
    Eds), American Nuclear Society, Illinois (1977) 189.

[5] GATHERS, G.R., SHANER, J.W., BRIER, R.L., Rev. Sci. Instrum. 47 (1976) 471.

[6] CEZAIRLIYAN, A., MORSE, M.S., BERMAN, H.A., BECKETT, C.W., J. Res. Natl.
    Bur. Stand., A 74 (1970) 65.

[7] SEYDEL, U., FUCKE, W., Z. Naturforsch., A 32 (1977) 994.

[8] REIL, K.O., CRONENBERG, A.W., Trans. Am. Nucl. Soc. 27 (1977) 576.

[9] BENSON, D.A., Application of Pulsed Electron Beam Vaporization to Studies of UO_2,
    Sandia Laboratories, Albuquerque, New Mexico, Rep. SAND-77-0429 (1977).

[10] ASAMI, N., NISCHIKAWA, M., TAGUCHI, M., Thermodynamics of Nuclear Materials 1974
    (Proc. Symp. Vienna, 1974) Vol. 1, IAEA, Vienna (1975) 287.

[11] OHSE, R.W., BERRIE, P.G., BOGENSBERGER, H.G., FISCHER, E.A., Thermodynamics
    of Nuclear Materials 1974 (Proc. Symp. Vienna, 1974) Vol. 1, IAEA, Vienna (1975) 307.

[12] BOBER, M., KAROW, H.-U., SCHRETZMANN, K., Thermodynamics of Nuclear Materials
    1974 (Proc. Symp. Vienna, 1974) Vol. 1, IAEA, Vienna (1975) 295.

[13] OHSE, R.W., v. TIPPELSKIRCH, H., The critical constants of the elements and of some
    refractory materials with high critical temperatures (29th IUPAC General Assembly
    Warsaw, 1977), High Temp.-High Pressures 9 (1977) 367.

[14] OHSE, R.W., BABELOT, J.-F., CERCIGNANI, C., KINSMAN, P.R., LONG, K.A.,
    MAGILL, J., SCOTTI, A.,"Application of laser pulse heating for the study of high
    temperature vapours, phase transitions and equation of state", Characterization of High
    Temperature Vapours and Gases (10th Materials Research Symposium Gaithersburg,
    Maryland, 1978), J. Nucl. Mater. 80 (1979) 232–248.

[15] OHSE, R.W., BERRIE, P.G., BRUMME, G.D., KINSMAN, P.R., "Advances in vapour
    pressure studies over liquid uranium plutonium oxides up to 5000 K", Plutonium 1975
    and Other Actinides (BLANK, H., LINDNER, R., Eds), North-Holland Publishing Company,
    Amsterdam (1976) 191–202.

[16] OHSE, R.W., KINSMAN, P.R., 5th Eur. Conf. Thermophysical Properties of Materials
    (Proc. Conf. Moscow, 1976), High Temp.-High Pressures 8 (1976) 209.

[17] IVANOV, V.E., KRUGLYKH, A.A., PAVLOV, V.S., KOVTUN, G.P., AMONENKO, V.M.,
    Thermodynamics of Nuclear Materials (Proc. Symp. Vienna, 1962), IAEA, Vienna (1962) 735.

[18] ALEXANDER, C.A., WARD, J.J., OGDEN, J.S., CUNNINGHAM, G.W., Carbides in
    Nuclear Energy (Proc. Symp. Harwell, 1963: RUSSEL, L.E., BRADBURY, B.T.,
    HARRISON, J.D.L., HEDGER, H.J., MARDON, P.G., Eds) Vol. 1, Macmillan and Co. Ltd.,
    London (1964) 192.

[19] VOZZELLA, P.A., DECRESCENTE, M.A., Thermodynamic Properties of Uranium
    Monocarbide, Pratt and Whitney Aircraft, Middletown, Connecticut, Rep. PWAC-478 (1965).

[20] KHROMONOZHKIN, V.V., ANDRIEVSKIJ, R.A., Thermodynamics (Proc. Symp. Vienna, 1965) Vol. 1, IAEA, Vienna (1966) 359.

[21] STORMS, E.K., Thermodynamics (Proc. Symp. Vienna, 1965) Vol. 1, IAEA, Vienna (1966) 309.

[22] KRUPKA, M.C., quoted in Ref. [21], p. 323.

[23] ANSELIN, F., POITREAU, J., Mesure de la pression de dissociation du monocarbure d'uranium entre 2250 et 2500 K par la méthode d'effusion de Knudsen, Commissariat à l'Energie Atomique, Centre d'Etudes Nucléaires de Fontenay-aux-Roses, France, Rep. CEA − R 2961 (1966).

[24] GORBAN'YU, A., PAVLINOV, L.V., BYKOV, V.N., Energ. Atom. (Paris) 23 (1967) 72.

[25] PATTORET, A., DROWART, J., SMOES, S., Bull. Soc. Fr. Ceram. 77 (1967) 75.

[26] VOZZELLA, P.A., MILLER, A.D., DECRESCENTE, M.A., J. Chem. Phys. 49 (1968) 876.

[27] ANDRIEVSKIJ, R.A., KHROMONOZHKIN, V.V., GALKIN, E.A., MITRIFANOV, V.I., At. Ehnerg. 26 (1969) 494.

[28] SOLOV'EV, G.I., given by SHETH, A., GABELNICK, S.D., FOSTER, M.S., CHASANOV, M., JOHNSON, C.E., in Argonne National Laboratory, Argonne, Illinois, Rep. ANL/CEN/AF-100 (1974).

[29] TETENBAUM, M., HUNT, P.D., J. Nucl. Mater. 40 (1971) 104.

[30] STORMS, E.K., The Refractory Carbides, Academic Press, New York, London (1967).

[31] OHSE, R.W., CAPONE, F., "Evaporation behaviour of the ternary uranium plutonium carbides", Plutonium 1975 and other Actinides (BLANK, H., LINDNER, R., Eds), North-Holland Publishing Company, Amsterdam (1976) 245.

[32] SHETH, A., TETENBAUM, M., LEIBOWITZ, L., Trans. Am.Nucl. Soc. 22 (1975) 233.

[33] BABELOT, J.F., BRUMME, G.D., KINSMAN, P.R., OHSE, R.W., Vapour pressure measurements over liquid UO_2 and (U, Pu)O_2 laser surface heating up to 5000 K, Atomwirtschaft 22 (1977) 387.

[34] OHSE, R.W., BABELOT, J.F., KINSMAN, P.R., LONG, K.A., MAGILL, J., High speed pyrometry and spectroscopy in laser pulse heating experiments at extreme rates of evaporation, 6th Eur. Conf. Thermophysical Properties of Materials − Properties and Application (Proc. Conf. Dubrovnik, Yugoslavia, 1978) High Temp. − High Pressures 11 (1999) 225−239.

[35] ZELDOVICH, Ya.B., RAIZER, Yu.P., Physics of Shock Waves and High Temperature Phenomena (HAYES, W.D., PROBESTEIN, R.F., Eds) Vol. 1, Academic Press, New York (1967) 131.

[36] ANISIMOV, S.I., Sov. Phys. − JETP 27 2 (1968) 182.

[37] YTREHUS, T., Von Karman Institute for Fluid Dynamics, Rhode Saint Genese, Belgium, Technical Note 112 (1975).

[38] HERTZ, H., Ann. Phys. 17 (1882) 117.

[39] KNUDSEN, M., Ann. Phys. 28 (1909) 75; 28 (1909) 999; 29 (1909) 179.

[40] POUND, G.M., J. Phys. Chem. Ref. Data 1 1 (1972) 135.

[41] BOBER, M., BREITUNG, W., KAROW, H.U., SCHRETZMANN, K., Kernforschungszentrum Karlsruhe, Rep. KFK 2366 (1976).

[42] RAND, M.H., MARKIN, T.L., Thermodynamics of Nuclear Materials 1967 (Proc. Symp. Vienna, 1967), IAEA, Vienna (1968) 637.

[43] SHETH, A., LEIBOWITZ, L., WINSLOW, G., Trans. Am. Nucl. Soc. 23 (1976) 130.

[44] FEE, D.C., JOHNSON, C.E., Phase Equilibria and Melting Point Data for Advanced Fuel Systems, Argonne National Laboratory, Argonne, Illinois, Rep. ANL-AFP-10 (1975).

DISCUSSION

H.U. KAROW: Referring to the measuring limits in the pyrometric temperature determination of laser evaporation experiments, you mentioned calculations yielding sample temperatures of $\gtrsim 5000$ K in the case of UO_2 samples and 7800 K in the case of UC as the upper measuring limits for pyrometry on these samples evaporating into vacuum. We have evaluated temperature values near 4000 K as the measuring limit in the case of UO_2 samples. Did you in your calculations take into account only the bound—free photoionization mechanism and free—free absorption (which we totally neglected), or did you also consider the more stringent bound—bound absorption of the fuel vapour?

R.W. OHSE: Let me first clarify the situation. Experiments on the reliability limit of optical pyrometry used in vapour pressure measurements over UC were performed over the temperature range from 6000 to 8000 K. The upper limit of reliability due to optical absorption was investigated as a function of jet expansion observed by high-speed photography, by the power density/temperature relationship and, lastly, by the turn-off from the expected pressure/temperature relationship, up to a surface temperature of 7100 K. According to our gas-dynamic studies on the axial pressure, density and temperature drop, this corresponds to a gas temperature of 4700 K at the sonic point, being only a very limited number of mean free paths away from the surface. The same applies to the uranium oxide measurements. Your assumption of an upper limit as high as 7800 K is apparently caused by misunderstanding. My co-authors, Mr. Long and Mr. Magill, would perhaps like to comment further.

K.A. LONG: Our calculations have considered bound—free and free—free transitions in the evaluation of the optical absorption, where bound states are considered to be the discrete states which lie below the effective ionization potential, thereby taking into account the reduction of the latter. These calculations on UO_2 yield a temperature limit for pyrometric temperature measurement in agreement with your own previous estimates. For UC the limit for the surface temperature was found to be about 1000 K higher, thus justifying the experimental measurements. However, it is difficult to give an exact limit since, as Mr. Karow has pointed out himself in his paper (IAEA-SM-236/21), such calculations are semiquantitative.

The use of a smeared-out continuum of levels for the bound states in UO_2 in the calculation of the bound—bound contribution to the optical absorption is questionable. This effectively reduces the ionization potential to zero, since all the states are then extended. This is not in accord with the observation of discrete states in the cores of atoms in solids, e.g. even the 4f levels in rare-earth metals are localized, as is seen from their magnetic properties. As was first pointed out by Mott, it is the electron—electron interaction that keeps the states localized, since the transfer of electrons to neighbouring atoms involves an increase in the electrostatic potential energy. The application of this theory has been confirmed by its explanation of insulator—metal transitions in alkali metal vapours, transitions which may become important in reactor safety analysis (above 2500 K for sodium).

Calculations need to be performed to estimate the oscillator strengths in UO_2, since these could be much lower than in simple atoms. It is necessary to carry out further calculations in which, firstly, a calculated discrete spectrum is used and, secondly, an 'approximate' continuum spectrum is used to calculate the absorption in order to evaluate the accuracy of such an approximation.

J. MAGILL: If any appreciable amount of radiation, coming either from the surface or from the laser itself, is absorbed in the gas jet, the temperature and pressure of the jet increase very rapidly, sending a shock wave back along the axis of irradiation. The electron density increases to above 10^{21} cm^{-3} (the critical density for neodymium laser radiation), preventing light from reaching the target surface, and the depth of the crater thus produced will be very small. In practice our depth measurements are very large, being consistent with a vapour phase rather than a plasma in the expanding jet.

Section C
SPECTROSCOPY

THE THERMODYNAMIC FUNCTIONS OF GASEOUS ACTINIDE ELEMENTS

Effects of unobserved energy levels

M.H. RAND
Materials Development Division,
Atomic Energy Research Establishment,
Harwell, Didcot, Oxfordshire,
United Kingdom

Abstract

THE THERMODYNAMIC FUNCTIONS OF GASEOUS ACTINIDE ELEMENTS: EFFECTS
OF UNOBSERVED ENERGY LEVELS.
    The actinide gases have large number of unobserved energy states — up to 3×10^6 for
Pu(g) — which could contribute to the partition function and its derivatives, from which the
thermal functions of these gases are calculated. Existing compilations have simply ignored
these levels. By making reasonable assumptions as to the distribution of these energy states,
their effect on the functions can be calculated. It is concluded that the existing compilations
will be inadequate above $\sim$2000 K. The effect is particularly marked on the heat capacity.
For example, when unobserved levels for Pu(g) are included, the heat capacity of Pu(g) reaches
a maximum value of more than 12R at 3200 K. Similar considerations will apply to the gaseous
actinide ions.

I.   Introduction

    In order to estimate the uncertainty in the thermal functions
of the actinide gases calculated from their electronic levels, it
is desirable to estimate the total number of energy states that
should be included in the partition function. The magnitude of the
error in omitting these from the calculations, as is done at present,
can then be estimated with various assumptions as to their distribution.

    Gurvich and Yungman[1] have shown that the inclusion of unobserved
levels for five electronic configurations of U(g) could have an
appreciable effect on the tempered Gibbs energy (free energy function,
$-(G_T^0 - H_{298}^0)/T$) at 5000K and above. However, the effects on S and
$(H_T - H_{298}^0)/T$ will be larger, and increase rapidly with temperature at
low temperatures. The effect on Cp may therefore be important at
more modest temperatures.

The problem of the divergence of the partition function due to the (in principle) infinite number of energy levels below the ionization limit, is inevitably involved in these considerations.

2.   Nomenclature

This will be made clear by a preliminary discussion of the uranium atom. This has a ground state configuration $5f^3 6d7s^2$. Each configuration corresponds to a number of terms (e.g. ^5L, ^5K) whose number and designation can be written down for each configuration. Each term (with given quantum numbers S and L) gives rise to a number of energy levels, of varying J values; each energy level has a degeneracy $(2J + 1)$. The total number of states of a level, term or configuration is the number of its contributions, weighted by the $(2J + 1)$ factor, to the partition function, $\sum_i (2J_i+1)\exp(-\varepsilon_i/kT)$. Thus the ^5L term corresponds to 5 levels with J = 6 to 10, and a total number of states of $13 + 15 + 17 + 19 + 21 = 85 = (2S + 1)(2L + 1)$

3.   Total number of states

The total number of states of any configuration is obtained simply, since it is the number of ways the electrons can be arranged in that configuration without violating the Pauli principle. If any electron orbit capable of containing n electrons actually contains m, this is $w_i = {}^n C_m = n!/m!\,(n-m)!$ The total number of states for a configuration is thus $W = \Pi_i\, w_i$, where the product is taken over all unfilled shells. This is illustrated in Table I which lists the seven terms of the f^2 configuration, with their S and L values. The total number of states for this configuration is thus 91, and equal to $^{14}C_2$.

The terms and thus the number of levels and total number of states for a given configuration are the same for L-S, J-j or mixed coupling; in the latter instance, the various levels correspond to mixtures of terms.

The total number of states for configurations containing no f electrons, is typically less than 1000 (e.g. 504 for d^5s for Mo) and the contributions of unobserved levels at 20000 cm^{-1} and above will be small, even at quite high temperatures.

TABLE I. THE SEVEN TERMS AND TOTAL NUMBER OF STATES OF THE f^2
CONFIGURATION

Term	^3P	^3F	^3H	^1S	^1D	^1G	^1I	Total
(2S+1)	3	3	3	1	1	1	1	
L	1	3	5	0	2	4	6	
Number of states (2S+1)(2L+1)	9	21	33	1	5	9	13	91

However, when f electrons are present, the total number of states
can be of the order of 10^6, and the effects of even high-lying levels
can become appreciable, as we shall see.

4. Application to actinide elements

The lowest-lying configurations of the gaseous actinide atoms and
ions have been given by Brewer[2], together with observed or estimated
positions of the lowest energy levels for these configurations. For
the present purposes, we are only interested in configurations which
have an appreciable extension below the ionization level for the ground-
state configuration (I). For the actinide elements up to Cm we have
therefore calculated the total number of states whose lowest energy
level lies somewhat below I. Of course, many of these energy states
will lie above the ionization level and are assumed, for the moment,
not to contribute to the partition function for the unionized gas
(see section 8).

These calculations are summarized in Table II. By contrast the
existing compilations of thermodynamic data[3] are based on a very small
fraction of the total levels. For uranium, for which the energy level
list is the most complete, 214 odd levels and 919 even levels were
available, with a total number of states of 12929, only 2.4% of the
total attributable to the lowest 19 configurations.

TABLE II. TOTAL NUMBER OF STATES FOR CONFIGURATIONS WHICH
HAVE SIGNIFICANT CONTRIBUTIONS TO THE PARTITION FUNCTIONS OF
ACTINIDE GASES

Element	No. of config. considered	Ground state ionization level (I) (cm^{-1})	Lowest level (cm^{-1}) of highest config. incl.	Total number of states (W)
Th	16	49 000	42 000	13 896
Pa	19	47 500	44 500	136 078
U	20	48 800	41 000	547 876
Np	19	49 900	44 000	1 676 402
Pu	18	48 900	47 000	3 499 496
Am	11	48 340	40 000	1 564 134
Cm	14	48 500	41 000	2 542 683

5.   Correction for unobserved levels

In order to calculate the thermal properties of gases the sum-
mations

$$Q'_0 = \sum_i (2J_i+1) \exp(-\varepsilon_i/kT)$$

$$Q'_1 = \sum_i (2J_i+1) \frac{\varepsilon_i}{kT} \exp(-\varepsilon_i/kT)$$

and
$$Q'_2 = \sum_i (2J_i+1)(\frac{\varepsilon_i}{kT})^2 \exp(-\varepsilon_i/kT)$$

need to be calculated. With the large numbers of energy levels which
occur for configurations containing f electrons, the mean spacing between
the levels becomes so small that the exact summation can be closely
approximated by integrals of the form

$$Q_m = N \int_{x=B}^{x=H} (ax)^m P(x) e^{-ax} dx \quad (m = 0, 1, 2) \quad \ldots\ldots (1)$$

where x = ε_i
     a = 1/kT
     P(x) is the fraction of energy states lying between x and x + dx
     N = $\Sigma(2J+1)$ is the total number of states,
and   B and H are the lower and upper energies to be included in the sum.

The problem is thus reduced to defining B, H and P(x) for each configuration.

For most configurations, B is the energy of the lowest level, which has been tabulated by Brewer[2]. However, for the ground-state and some other low lying configurations, the positions of a fair number of levels up to $\sim$ 15000 cm^{-1} have been identified, and the exact summation has been retained for these levels, the integration being started from some value of B above the lowest level, the value of N being adjusted accordingly.

If we assume that energy levels above the ground-state ionization level (I) should be excluded from the summation of the partition function,

$$H = min\ (I,Z)$$

where Z is the highest energy level of the configuration. We shall see shortly that for configurations containing more than 2f electrons, I < Z even for the ground-state configuration.

The main problem is thus the definition of P(x), the distribution of energy levels between B and Z. Gurvich and Yungman[1], in their treatment of five configurations of uranium, made the simplest assumption of a uniform distribution of levels, which as they stated probably leads to an overestimate of the corrections. Since then however substantial progress has been made in calculating, at least approximately, energy levels of some configurations containing 4f and 5f electrons. Judd's pioneering calculations[4] for the $5f^3ds^2$ ground state of U(g) were extended by Spector[5], and by Guyon et al[6]; however, none of these authors give all the calculated levels for this configuration. Cowan[7] has given schematic distributions of the energy levels for the lanthanide gases, and has given[8] brief details of a similar calculation for the $5f^3ds^2$ configuration of U(g). These latter show that the uppermost level for this configuration lies at $\sim$ 70000 cm^{-1}, way above the ionization level (48800 cm^{-1}). Cowan[9] has kindly supplied the author with all the calculated levels for this configuration and has pointed out that for lanthanide elements,

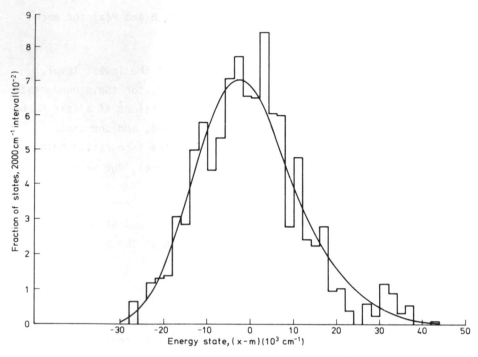

FIG.1.   Distribution of energy states for U(g) 5f³ds².

the energy levels follow fairly closely a skewed Gaussian distribution of the type

$$P(t) = \frac{1}{\sqrt{2\pi}} \left[1 - \frac{\alpha_3}{2}\left(t - \frac{t^3}{3}\right)\right] e^{-t^2/2} \, dt \qquad \cdots\cdots (2)$$

where $t = (x-m)/\sigma$, x being an energy state

$\quad\quad\quad m = \bar{x}$, the mean energy state

$\quad\quad\quad \sigma = \mu_2^{\frac{1}{2}}$, the standard deviation

$\quad\quad\alpha_3 = \mu_3/\sigma^3$, the momental skewness

where μ_k = the k th moment about m

$\quad\quad\quad = \sum_i (x_i - \bar{m})^k / \sum_i (2J_i + 1)$

Note that x_i is a value for an energy <u>state</u>, so that each energy <u>level</u>
has $(2J_i + 1)$ states of equal energy.

It will be seen shortly that a distribution function of this form
fits Cowan's calculated values very well. This form has therefore
been assumed for all the configurations.

With this form of distribution function, the integrals Q_m (m = 0, 1, 2) giving the contribution of each configuration to the partition function and its temperature derivatives can be evaluated without undue difficulty. The algebraic details are given in the Appendix.

The precise definition of B, H and P(x) for the configurations of U(g) and Pu(g) is now considered.

6. Calculations for U(g) and Pu(g)

6.1 Uranium

The histogram of Cowan's calculated energy states for the $5f^3ds^2$ configuration of U(g), is shown in Fig. 1, together with a skewed Gaussian distribution (equation 2) with

$$m = 27370 \text{ cm}^{-1}$$
$$\sigma = 11870 \text{ cm}^{-1}$$
$$\alpha_3 = 0.531$$

Since f-f interactions are stronger than other interactions, the width of a configuration will probably increase with the number of 5f electrons. There will be variations within such groupings depending on the distribution of the other electrons between the 6d, 7s, 7p and 8s shells, but the use of such fine detail is not warranted for the present exercise. Thus the following widths have been assumed:

No. of 5f electrons	1	2	3	4	5	6
Width of config. $(Z-L)(\text{cm}^{-1})$	60000	70000	80000	90000	100000	110000

where the width for most $5f^3$ configurations is assumed to be rather higher than that for the ground state, which contains paired $7s^2$ electrons. The distribution of states within all configurations is assumed to be the same as in the ground state configuration, so that the values of $m/(Z-L)$ (0.385), $\sigma/(Z-L)$ (0.167) and α_3 (0.53) have been assumed to apply to the other configurations. Thus the data used are summarised in Table III.

TABLE III. DATA FOR U(g)

Configuration 5f	6d	7s	7p	8s	Number of States	Lowest observed level (cm⁻¹)	B (cm⁻¹)	m (cm⁻¹)	σ (cm⁻¹)
Odd Levels									
3	1	2			3 640	0	14 000	27 370	11 870
3	2	1			32 760	6 249	14 000	37 100	13 400
3	3				43 680	–	21 000	51 800	13 400
1	4	1			5 880		22 000	45 100	10 000
4		1	1		12 012	22 792	22 792	57 400	15 000
2	1	2	1		5 460		24 000	51 000	11 700
2	2	1	1		49 140		25 000	52 000	11 700
2	3		1		65 520		31 500	58 500	11 700
3	1	1		1	14 560	32 857	32 857	63 700	13 400
4	1		1		60 060	34 161	34 161	68 800	15 000
Even levels									
4		2			1 001	7 021	14 000	37 800	13 400
2	2	2			4 095	11 503	14 000	38 500	11 700
3		2	1		2 184	13 463	14 000	44 300	13 400
3	1	1	1		43 680	14 644	14 644	45 400	13 400
2	3	1			21 840		15 500	42 500	11 700
4	1	1			20 020	14 840	14 840	49 500	15 000
3	2		1		98 280	27 887	27 887	58 700	13 400
5		1			4 004		28 000	66 500	16 700
4		2			45 045	34 000	34 000	68 700	15 000
4			2		15 015	41 000	41 000	75 700	15 000
Ionization limit						48 800			
Total number of states					547 876				

As will be seen, the integration for the lowest five configurations was started as 14000 cm⁻¹, the 78 levels (764 states) below this point having been summed exactly. Values for the lowest observed levels are taken from Steinhaus et al [8], updated by information from Blaise [10]; values for the configurations where no levels have been observed are from Brewer [2].

TABLE IV. DATA FOR Pu(g)

Configuration					Number of States	Lowest Observed level (cm^{-1})	B (cm^{-1})	m (cm^{-1})	σ (cm^{-1})
5f	6d	7s	7p	8s					
Odd levels									
5	1	2			20 020	6 314	12 000	44 800	16 700
5	2	1			180 180	14 912	14 912	53 400	16 700
6		1	1		36 036	15 449	15 449	57 800	18 400
5	3				240 240	-	27 000	65 500	16 700
5		1	2		60 060	-	30 000	68 500	16 700
6	1		1		180 180	33 070	33 070	75 400	13 400
5	1	1		1	80 080	39 618	39 618	78 100	16 700
4	1	2	1		60 060	-	47 000	81 700	15 000
Even levels									
6		2			3 003	-	14 000	38 500	16 700
6	1	1			60 060	13 528	13 528	55 900	18 400
5		2	1		12 012	17 898	17 898	56 400	16 700
5	1	1	1		240 240	20 402	20 402	58 900	16 700
6	1			1	60 060	31 573	31 573	73 900	18 400
6	2				135 135	31 710	31 710	74 100	18 400
6				2	45 045	-	36 000	78 400	18 400
4	2	2			45 045	36 051	36 051	70 700	15 000
5	2		1		540 540	-	38 000	76 500	16 700
4	3	1			240 240	-	42 500	77 200	15 000
Ionization limit						48 900			
Total number of states					3 499 496				

6.2 Plutonium

No calculation for f^6s^2 ground state of Pu(g) have been reported, but the calculations of Cowan[7] on the $4f^n$ states suggest a width of > 100000 cm^{-1}. The assumptions used for the widths and shape of the distribution of the U(g) configurations have therefore been applied for all the Pu(g) configurations, giving the data in Table IV. The observed levels are those characterized by Fred[11] with additions by

Blaise[10]; unobserved lowest levels are estimates by Brewer[2]. The
10 levels (56 states) of the f^6s^2 configuration below 14000 cm^{-1}, and the
known levels (43 states) of the f^5ds^2 configuration have been summed
exactly.

7.   Results

Table V gives the thermal values given by Oetting et al[3],
extended to 10 kK, calculated from the observed levels (1133 extending
to 38713 cm^{-1} for U(g) and 1075 extending to 43823 cm^{-1} for Pu(g))
and the values calculated from the data of Tables III and IV. We may
emphasize again here that to maintain accuracy at lower temperatures,
the lowest observed levels (< 14000 cm^{-1} in both cases) have been
summed exactly. Fig. 2 shows the differences in the heat capacity
curve.

Although they confirm the remarks in the introduction, the results
are still somewhat surprising, in the size, and for Pu(g), the location
of the massive increase in C_p due to the unobserved levels. Thus for
U(g), the peak which for the observed levels lies at 6.42 R at 4200K,
is displaced to 9.09R at 4400K, while for Pu(g), with many more
unobserved levels, the effect is even more dramatic. The peak from the
'observed' levels 7.45 R at 4400K is increased to no less than 12.34 R
at 3200K.

The other thermal functions, of course, reflect this large increase
in heat capacity, and even the tempered Gibbs energy $-(G^0_T - H^0_{298})/T$, which is
relatively insensitive to C_p, shows appreciable discrepancies for Pu(g).

The contributions of the various configurations to the partition
functions of U(g) and Pu(g) at temperatures close to that of the peaks
in the heat capacity are shown in Figs 3 and 4. The right hand portion
of these figures shows the relative position (vertically) of the lowest
level of the configuration, and horizontally the number of states in
each configuration. It is clear that at temperatures below 4000K, the
principal contributions are from configurations with a large number of
levels starting below $\sim$20 000 cm^{-1}.

TABLE V. THERMAL FUNCTIONS FOR U(g) AND Pu(g) CALCULATED
FROM ~1100 OBSERVED LEVELS ('obs') AND FROM ESTIMATED
DISTRIBUTIONS OF UNOBSERVED LEVELS

Uranium gas

T (K)	C_p 'obs' (cal·mol^{-1}·K^{-1})	calc (cal·mol^{-1}·K^{-1})	S_T 'obs' (cal·mol^{-1}·K^{-1})	calc (cal·mol^{-1}·K^{-1})	$(G_T-H_{298})/T$ 'obs' (cal·mol^{-1}·K^{-1})	calc	H_T-H_{298} 'obs' (cal·mol^{-1})	calc
1000	5.80	5.80	54.57	54.57	50.60	50.60	3 962	3 962
2000	8.66	8.82	59.37	59.39	53.84	53.84	11 050	11 090
3000	11.54	13.78	63.47	63.85	56.38	56.43	21 260	22 260
4000	12.72	17.69	66.99	68.44	58.60	58.86	33 540	38 320
5000	12.47	17.58	69.82	72.44	60.58	61.19	46 220	56 230
6000	11.62	15.63	72.02	75.48	62.31	63.33	58 290	72 890
8000	9.65	11.92	75.09	79.44	65.15	66.91	79 520	100 210
10000	8.15	9.54	77.07	81.82	67.35	69.67	97 220	121 460

Plutonium gas

T (K)	C_p 'obs' (cal·mol^{-1}·K^{-1})	calc (cal·mol^{-1}·K^{-1})	S_T 'obs' (cal·mol^{-1}·K^{-1})	calc (cal·mol^{-1}·K^{-1})	$(G_T-H_{298})/T$ 'obs' (cal·mol^{-1}·K^{-1})	calc	H_T-H_{298} 'obs' (cal·mol^{-1})	calc
1000	7.84	7.84	49.43	49.43	45.10	45.10	4 330	4 330
2000	11.77	13.57	56.21	56.46	49.04	49.06	14 340	14 800
3000	13.42	24.14	61.32	64.24	52.32	52.79	27 000	34 350
4000	14.68	20.76	65.37	70.98	55.09	56.55	41 120	57 690
5000	14.49	15.77	68.66	75.02	57.49	59.88	55 850	75 700
6000	13.06	13.20	71.19	77.64	59.57	62.63	69 680	90 060
8000	9.95	10.40	74.49	81.02	62.93	66.84	92 530	113 400
10000	8.042	8.63	76.49	83.13	65.45	69.90	110 330	132 300

These figures suggest that the contributions of configurations not
included in Tables III and IV are likely to be insignificant at temperatures
up to 10kK. Quantitative confirmation of this conclusion is presented
in the next section.

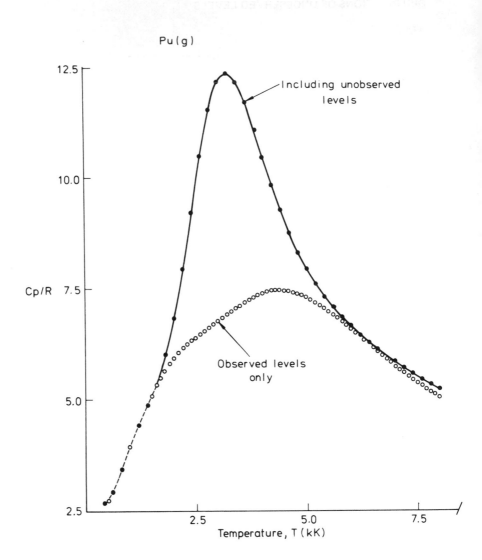

FIG.2.   *Heat capacities of U(g) and Pu(g).*

U(g)

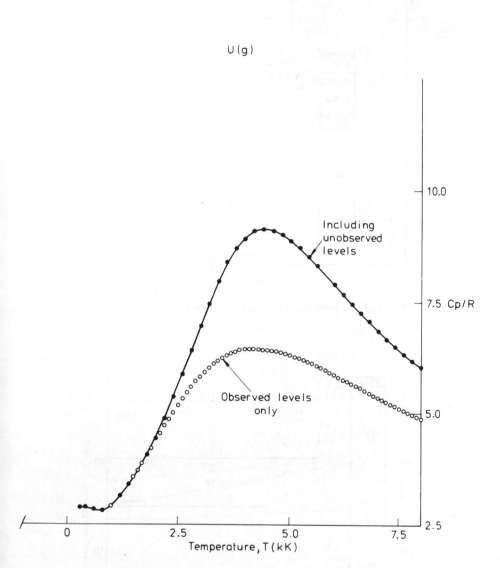

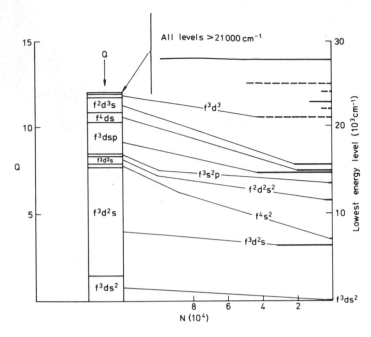

FIG.3.   U(g) at 4000 K; contributions of each configuration to the partition function.

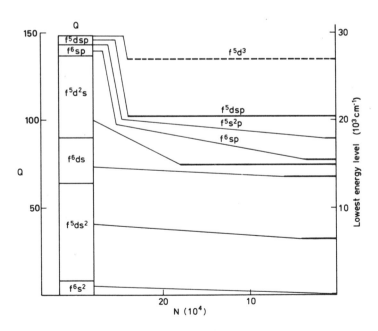

FIG.4.   Pu(g) at 3000 K; contributions of each configuration to the partition function.

8. Additional configurations

Consideration of corrections for additional configurations requires a discussion   of the inherent divergence of the electronic partition function, due to the fact that the number of energy levels even below the ionization limit, is in principle, infinite. Some method of limiting the number of levels contributing to the partition function and its temperature derivatives is therefore required. McChesney[12] summarizes the fairly arbitrary methods proposed, based on a truncation related to the size or principal quantum number of the orbit, or to a lowering of the ionization potential due to the presence of surrounding charged particles, usually nearest neighbours only.

For U(g) and Pu(g), all the configurations considered above involve only 5f, 6d, 7s, 7p or 8s electrons. Since these orbitals are occupied either in the ground state, or in low-lying levels, it seems unlikely that any of these configurations would have any levels lying beyond any spatial truncation limit, so that the only limitation of the states contributing to the partition function which has been applied in the foregoing treatment is to exclude all levels above the ground-state ionization level.

We can make an estimate of the contribution of additional configurations for U(g), as follows. There are a further 27 configurations involving 5f, 6d, 7s and up to two 7p electrons with a total number of states of 289 594. These presumably lie above the configurations listed by Brewer[2]. There will be many more configurations, lying wholly above the first ionization level, and possibly beyond a 'spatial' truncation limit. The effect of these additional configurations is therefore not likely to exceed that found by taking a further 300 000 states starting at 41 000 cm^{-1}. The width of such a conglomeration of states has been assumed to be 70 000 cm^{-1}, so that a further entry may be added to Table III with:

Number of states = 300 000

B = 41 000 cm^{-1}; m = 68 000 cm^{-1}; σ =11 700 cm^{-1}.

When such an addition is made, the thermal functions at 5000 and 10 000K become:

T (K)	C_p (cal·mol^{-1}·K^{-1})	S_T (cal·mol^{-1}·K^{-1})	$(G_T-H_{298})/T$ (cal·mol^{-1}·K^{-1})	H_T-H_{298} (cal·mol^{-1})
5 000	17.61	72.44	61.19	56 250
10 000	9.64	81.88	69.69	121 960

Comparison with the values given in Table V shows that the effects of configurations additional to those given in Table III for U(g) are negligible up to 10 000K within the uncertainties of the present assumptions.

9. Conclusions

The existing tabulations of the thermal functions of the actinide gases, apart perhaps from Th(g), are likely to be in error above about 2000K, owing to the neglect of unobserved electronic levels. Inclusion of these levels for configurations with a lowest level somewhat below the ground state ionization energy, together with reasonable assumptions concerning the width and distribution of levels, leads, for U(g) and Pu(g) to a massive peak in the heat capacity curve between 3500 and 4500K. The effect of any further configurations is likely to be negligible below 10 000K. Similar effects will undoubtedly occur for Pa, Np, Am and Cm gases and ions.

Although the assumptions underlying these calculations are thought to be reasonable, detailed calculations of the distribution of the unobserved levels, even to a low accuracy, are highly desirable to provide more reliable thermal functions for all the actinide gases and their ions.

The pronounced influence of the unobserved levels for the metallic gases re-emphasizes the necessity to include some estimates of the electronic contribution to the thermal functions for other actinide gaseous molecules particularly the oxides, where interest in the thermodynamic properties extends to at least 5000K, in reactor safety assessment studies.

The elegant and precise studies of Edvinsson and his colleagues on ThO(g) - for references and discussion see Rand [13] - are the only extensive data available on an actinide oxide. Clearly the participation

some of the electrons in covalent bonds will markedly decrease the number of
energy states contributing to the partition function. Theoretical studies
on the electronic spectra of such gaseous molecules would be most valuable.

REFERENCES

[1]   GURVICH, L., YUNGMAN, V.S., Thermodynamics (Proc. Symp. Vienna, 1965) Vol.2, IAEA,
      Vienna (1966) 613.
[2]   BREWER, L., J. Opt. Soc. Am. **61** (1971) 1101.
[3]   OETTING, F.L., RAND, M.H., ACKERMANN, R.J., The Chemical Thermodynamics of Actinide
      Elements and Compounds, Part 1, The Actinide Elements, IAEA, Vienna (1976).
[4]   JUDD, B.R., Phys. Rev. **125** (1962) 613.
[5]   SPECTOR, N., Phys. Rev., A **8** (1973) 3270.
[6]   GUYON, F., BLAISE, J., WYART, J.F., J. Phys. (Paris) **35** (1974) 929.
[7]   COWAN, R.D., Nucl. Instrum. Methods **110** (1973) 173.
[8]   STEINHAUS, D.W., RADZIEMSKI, L.J., COWAN, R.D., BLAISE, J., GUELACHVILI, G.,
      OSMAN, Z.B., VERGES, J., USAEC Rep. LA-4501 (1971).
[9]   COWAN, R.D., private communication, Dec. 1975.
[10] BLAISE, J., private communication, Dec. 1975.
[11] FRED, M., private communication to R.J. Ackermann, 1975.
[12] McCHESNEY, M., Can. J. Phys. **42** (1964) 2473.
[13] RAND, M.H., GOLDBECK, O. von, FERRO, R., GIRGIS, K., DRAGOO, A.L., Thorium: Physico-
      chemical Properties of its Compounds and Alloys (KUBASCHEWSKI, O., Ed.), At. Energy Rev.
      Spec. Issue No.5 (1975).

Appendix

Calculation of Contributions to the Partition Function for a Skewed
Gaussian Distribution

For any configuration, the fraction of states with energy x lying
between t and t + dt is assumed to be

$$\frac{1}{\sqrt{2\pi}} \left[1 - \alpha(t - t^3/3) \right] \, dt$$

where $t = (x-m)/\sigma$, and m and σ are the mean and standard deviation
of the distribution of energy states. The contributions of this
configuration to the partition function and its derivatives are then

$$\text{To } \frac{Q_o}{N} : q_o = \frac{1}{\sqrt{2\pi}} \int_{x=B}^{x=H} e^{-ax} \left[1-\alpha(t-t^3/3)\right] e^{-t^2/2} \, dt$$

$$\text{To } \frac{Q_1}{N} : q_1 = \frac{1}{\sqrt{2\pi}} \int_{x=B}^{x=H} e^{-ax} \cdot ax \cdot \left[1-\alpha(t-t^3/3)\right] e^{-t^2/2} \cdot dt$$

$$\text{To } \frac{Q_2}{N} : q_2 = \frac{1}{\sqrt{2\pi}} \int_{x=B}^{x=H} e^{-ax} \cdot (ax)^2 \left[1-\alpha(t-t^3/3)\right] e^{-t^2/2} \cdot dt$$

where a = 1/kT, and B and H are the lower and upper energy levels included in the integration.

If we define

$$S_n = \frac{1}{\sqrt{2\pi}} \int e^{-ax} (ax)^n e^{-t^2/2} \, dt$$

$$T_n = \frac{1}{\sqrt{2\pi}} \int e^{-ax} (ax)^n \cdot t \cdot e^{-t^2/2} \, dt$$

$$U_n = \frac{1}{\sqrt{2\pi}} \int e^{-ax} (ax)^n \frac{t^3}{3} e^{-t^2/2} \, dt$$

these contributions may be written

$$q_i = \left[S_i - \alpha(T_i - U_i)\right]_{x=B}^{x=H} \quad i = 0, 1, 2$$

If the multipliers t and $t^3/3$ are expressed in terms of x, we
find

$$T_n = \frac{1}{a\sigma} \cdot S_{n+1} - \left(\frac{m}{\sigma}\right) \cdot S_n$$

and $$U_n = \frac{1}{3(a\sigma)^3} \cdot S_{n+3} - \frac{1}{(a\sigma)^2} \cdot \frac{m}{\sigma} \cdot S_{n+2} + \frac{1}{a\sigma}\left(\frac{m}{\sigma}\right)^2 \cdot S_{n+1} - \frac{1}{3}\left(\frac{m}{\sigma}\right)^3 \cdot S_n$$

Thus the required contributions q_i can be expressed in terms of the
integrals S_n, n = 0 to 5.

A recurrence relation for S_n can be established by a further change of variable and consideration of the integrals

$$R_n = \int (u - \gamma)^n \, e^{-u^2/2} \, du$$

for which integration by parts leads directly to the recurrence relation:

$$R_{n+2} = -\gamma R_{n+1} + (n+1) R_n - (u-\gamma)^{n+1} \, e^{-u^2/2}$$

with

$$R_0 = \sqrt{\frac{\pi}{2}} \, \text{erf} \left(\frac{u}{\sqrt{2}} \right)$$

and

$$R_1 = -e^{-u^2/2} - \gamma R_0$$

However, with

$$u = {}^{x}/\sigma + \gamma$$

$$\gamma = a\sigma - m/\sigma$$

and

$$\beta = a^2\sigma^2/2 - am$$

$$S_n = \frac{e^\beta \, (a\sigma)^n}{\sqrt{2\pi}} \, R_n$$

so that the recurrence relation for S is

$$S_{n+2} = (am - a^2\sigma^2) S_{n+1} + (n+1) a^2\sigma^2 S_n - \frac{a\sigma}{\sqrt{2\pi}} \, (ax)^{n+1} \, e^{\{\beta - (u^2/2)\}}$$

with

$$S_0 = \frac{e^\beta}{2} \, \text{erf} \left(\frac{u}{\sqrt{2}} \right)$$

$$S_1 = (am - a^2\sigma^2) S_0 - \frac{a\sigma}{\sqrt{2\pi}} \, e^{\{\beta - (u^2/2)\}}$$

The required contributions q_i can thus be computed.

DISCUSSION

K.E. SPEAR: In cases like the calculation of the vapour pressure of UO_2 gas from condensed UO_2 at very high temperatures, won't the contributions of unobserved energy levels to the thermal functions of the gas be cancelled by similar contributions to the thermal functions of the condensed phase?

M.H. RAND: Yes, if the thermal functions are used for this purpose, this will be true as a zero'th approximation at least, as has been suggested by Mr. McInnes (paper IAEA-SM-236/37). However, this will of course not be true for decomposition reactions, such as the vaporization of uranium carbide.

J.D. DROWART: Can the treatment you described for U and Pu lead to general methods for predicting the electronic states (or substates) and their excitation energies in the gaseous molecules of these and related molecules?

M.H. RAND: No, we have not attempted to produce a general method for predicting energy levels. Such a method would indeed be most valuable. Our intention was to emphasize that, with a large number of energy levels, calculations of even modest accuracy are sufficient to provide a reasonable estimate of the electronic contribution to the thermal functions, and to encourage such computations.

K.A. LONG: The problem of extending your calculations on actinide metal gaseous species to molecular species such as UO_2 is complicated substantially by the reduction of symmetry. In the metal atoms one can use an average spherical potential, composed of the potential of the nuclear charge and of the core electrons. Since the potential is spherical, angular momentum quantum numbers can be used, and this makes the classification of states relatively simple, as is the calculation of the degeneracy factors. In a UO_2 molecule there are three overlapping potentials, which renders the solution very difficult, since no realistic spherical potential can be constructed.

J.D. DROWART: The group theory further predicts only the great variety of states formed upon modification of the symmetry, but not the energy of one state in relation to another. For most of the molecules of interest here, not even the spatial structure nor therefore the point group is known experimentally. Additional problems arise because it may reasonably be assumed that spin—orbit interaction causes splitting of the states expected under LS coupling, whereafter interaction of substates with similar symmetry species might cause one or more substates to become energetically isolated from the others. Estimated rotational, vibrational and electronic contributions to the thermodynamic functions can therefore be quite uncertain. A further complication in this respect is that certain thermodynamic quantities, owing to their functional dependence, are more sensitive than others to the choice of the term values and degeneracies of low-lying excited states.

H.U. KAROW: In your studies you have considered the electronic energy
level spectra of the isolated undisturbed actinide atoms. As regards the application
of these data to the thermodynamic description of dense high-temperature vapours
of these species, it should be noted that the electronic excitation levels can be split,
shifted and broadened by the Stark effect. This happens when the molecule is
immersed in a dense or even strongly-coupled vapour plasma as, for example, in
the case of liquid oxide fuel vapour above 4000 K.

PHOTOELECTRON SPECTROSCOPY OF
4f, 5f AND 2p ORBITAL ELECTRONS AND
THE THERMOCHEMISTRY OF URANIUM OXIDES*

G.E. MURCH, R.J. THORN
Chemistry Division,
Argonne National Laboratory,
Argonne, Illinois,
United States of America

Abstract

PHOTOELECTRON SPECTROSCOPY OF 4f, 5f AND 2p ORBITAL ELECTRONS AND THE
THERMOCHEMISTRY OF URANIUM OXIDES.

The basic reaction describing the partial pressure of oxygen in equilibrium with UO_{2+x} is resolved into a set of elementary reactions for which the energies can be obtained through the point-charge model with corrections for ionicity and through 5f and 2p orbital-electron energies measured with X-ray photoelectron spectroscopy (XPS). It is shown that the energies of the valence orbitals are a direct measure of the oxygen potential and that the orbital binding energies adjusted for the coulombic self-potential are linearly related to the ionicity derived from the optical dispersion theory of Phillips and van Vechten. Measurements of the O^{2-} ($1s^2$---) orbital energies in alkaline earth oxides and UO_2 and ThO_2 indicates an ionicity of 0.93 for the two actinide dioxides. From measurements of the intensities of the photoelectron emitted from the 5f orbital in uranium metal, uranium dioxide and uranium tetrafluoride, it is shown that the densities of electrons in the 5f orbitals are the same in U, UO_2 and UF_4.

I. INTRODUCTION

Through the use of the photoelectron-spectroscopic investigation of the 4f and 5f orbital electrons on uranium and the 2p orbital electrons on oxygen in uranium metal and its oxides, one can obtain measures of valence-state energies and densities of states in valence bands. Quantitative values for these fundamental properties can be used to understand and describe the thermochemistry of uranium and its compounds. Among the questions which can be answered through such investigations are those related to the ionic characteristics, the electron affinity of oxygen, the fifth ionization potential of uranium, the extent of localization of the 5f electrons,

* Work performed under the auspices of the Division of Basic Energy Sciences of the United States Department of Energy.

the valency of uranium in the metal, the nature of the orbitals on O^{2-} formed from the 2p orbitals on oxygen and the electrons originally on uranium, and the two valence states in a non-stoichiometric phase such as UO_{2+x} or U_3O_{8-z}. To demonstrate the extent to which questions concerning these topics are answered through spectroscopic studies, we discuss (1) the chemical reactions which interrelate partial molar thermodynamic quantities and the measured binding energies; (2) results obtained with alkaline earth fluorides to establish the quantitative basis for the relations, and (3) some of the results obtained with uranium metal and oxides. The discussion serves to establish a more complete description of the fundamental thermochemistry of nonstoichiometric phases and to identify the known and unknown thermochemical and auxiliary data.

II. THERMOCHEMICAL REACTIONS OF VALENCE STATES IN UO_{2+x}

The thermochemical reactions which relate the partial molar thermodynamic quantities to the valence states in a non-stoichiometric phase such as UO_{2+x} are:

$$2U^{4+}cs + O(g) + is \rightarrow 2U^{5+}cs + O^{2-}is; \quad -E_i \qquad (1)$$

and

$$2U^{4+}cs + O(g) + as \rightarrow 2U^{5+}cs + O^{2-}as; \quad -E_v \qquad (2)$$

In these reactions cs, is and as represent cation, interstitial, and anion sites in the cubic fluorite structure. When UO_2 is oxidized, the first reaction predominates; hence even though both reactions are present, the following discussion involves only the first one. In these reactions and the following ones, we assume that O^{2-} exists in the solid even though it is un-stable in the gas. Some difficulties have been encountered resolving this difference, but chemical evidence and calculated lattice energies are not inconsistent with the assumption. An examination of the results obtained with photoelectron spectroscopy and discussed subsequently are consistent with the existence of O^{2-} in the lattice.

The two reactions given above can be written as the sum of several elementary reactions which can be selected in several ways. To relate them to topics which will be apparent in discussions cited below we select the following.

For the cation one writes:

$$U^{4+}cs \rightarrow U^{5+}cs* + e(g); \quad E_{bg}^+; \tag{3}$$

$$U^{5+}cs* \rightarrow U^{5+}cs; \quad E_R^+ . \tag{4}$$

For the anion one writes:

$$O^{2-}(g) + is \rightarrow O^{2-}is; \quad -E_d^- \tag{5}$$

$$O^{1-}is* + e(g) \rightarrow O^{2-}is; \quad -E_{bg}^- \tag{6}$$

$$O^{2-}is + O^{1-}(g) \rightarrow O^{1-}is* + O^{2-}(g); \quad -(E_e^- + \delta^-(f_i)) \tag{7}$$

$$O(g) + e(g) \rightarrow O^{-1}(g); \quad A_1 . \tag{8}$$

In these reactions the asterisk indicates an unrelaxed state in the lattice. Thus reaction (4) corresponds to a slight relaxation after an electron is removed from $U^{4+}cs$. If reaction (3) and (4) are multiplied by two and if all the reactions are added, then reaction (1) is obtained. Hence

$$-E_i = 2E_{bg}^+ + 2E_R^+ + E_d^- - E_{bg}^- - E_e^- - \delta^-(f_i) - A_1 . \tag{9}$$

The first reaction (i.e., reaction (3)) is the one which occurs when one of the 5f electrons on $U^{4+}cs$ is emitted upon radiation with X-rays. Hence E_{bg}^+ is measured directly with photoelectron spectroscopy. If this reaction is combined with the reaction:

$$U^{5+}cs* + U^{4+}(g) \rightarrow U^{4+}cs + U^{5+}(g); \quad E_e^+ + \delta^+(f_i) \tag{10}$$

then the result is:

$$U^{4+}(g) \rightarrow U^{5+}(g) + e(g), \tag{11}$$

i.e., the fifth ionization of uranium. Because reaction (10) corresponds to the reverse of the process wherein an electron on $U^{4+}cs$ is removed to the gas without producing any ionization in the gas but with the result that a U^{5+} is left at the cation site in an unrelaxed state, the energy is calculable through the electrostatic potential in the point charge model with a small correction for the ionic character $\delta^+(f_i)$. Thus, one writes that:

$$I_5(U) = E_{bg}^+ U(5f) + E_e^- + \delta^+(f_i). \tag{12}$$

Similarly, one can combine the reaction for the photoionization of the anion:

$$O^{2-}is \xrightarrow{h\nu} O^{1-}is* + e(g), \quad E^{-}_{bg} \tag{13}$$

with the equation for the electrostatic potential at the anion
site, i.e., reaction (7), to obtain:

$$O^{2-}(g) \rightarrow O^{1-}(g) + e(g), \quad A_2 \tag{14}$$

for which one writes:

$$A_2 = E^{-}_{bg} (O^{2p}) - E^{-}_{e} + \delta^{-}(f_i). \tag{15}$$

If A_2 is substituted for the RHS of equation (5), the result
is

$$-E_i = 2E^{+}_{bg} + 2E^{+}_{R} - E_d - A_2 - A_1 . \tag{16}$$

Because both A_2 and A_1 are constant, the photoelectron-spectroscopic value, $^2E^{+}_{bg}$, is a direct measure of the oxidation potential
of the cation.

III.   DERIVATION OF VALENCE STATES ENERGIES AND IONICITIES
FROM X-RAY (XPS) PHOTOELECTRON SPECTROSCOPY

The potential, quantitative use of the reactions and the
accompanying energies presented above in solid state thermochemistry and the demonstration of the role of photoelectron
spectroscopy have been investigated through a study of
alkaline earth fluorides and oxides, and uranium and its
oxides. The alkaline earth fluorides were selected because
more of the energies associated with the reactions equivalent
to those cited above are known. For instance, the electron
affinity of fluorine is known; the cations existing in the
solids have inert-gas, closed shells with high energies for
the first excited electronic states; cohesive energies have
been calculated; and finally the ionic characters of them are
known.

The orbital binding energies have been measured in a
GCA/McPherson ESCA36. For all the values reported herein

MgK$_\alpha$ X-radiation was used and the energies are referred to the Fermi level of the instrument for which the work function is 4.3 eV. Hence values referred to the gaseous electron are obtained by adding 4.3 to the values for E_b, i.e., $E_{bg}^+ = E_b^+ + 4.3$. In our instrument the analyzer and sample chambers are evacuated with either an Airco Tempescal turbomolecular pump or an ion pump or both. A helium cryopump is used to differentially pump the sample chamber. During measurements the pressure at the turbo pump or ion pump is 2×10^{-7} torr, so the pressure in the sample chamber is at least an order of magnitude lower. Samples were prepared for mounting in the sample chamber by pressing powders (analytical reagents) onto stainless steel fritted plates in a hydraulic press (approximately 1000 lbf/in^2).

The principal variable which limits the accuracy and probably also the precision with which the orbital binding energies in nonconducting solids can be determined with photoelectron spectroscopy is the electrical charge which accumulates on the solid during X-radiation and consequent photoelectron emission. All the ionic materials acquire a charge varying from 2 to 6 eV. To obtain a measure, and hence a correction, of this charge two procedures are frequently used. In one, gold is sublimed onto the sample and the Au(4f$_{7/2}$) orbital energy at 83.8 eV from the Fermi level is used to obtain the correction. In another, the C(1s) orbital energy of the carbon contaminant which is always present in all samples is used. In the second case the chemical state of carbon is unknown, but its presence in all cases makes it a convenient reference. For the present, we assume that the state is elemental carbon, so the reference energy is 284.8 eV.

The values for the electrostatic potentials, E_e^+ and E_e^-, at the cation and anion sites were calculated through the use of a computer program furnished by van Gool and Piken [1]. For the ionicities or ionic characters of the solids we have employed the scale established by Phillips [2] and van Vechten [3] on the basis of the short and long wavelength limits of the index of refraction. The numerical values used were calculated by Levine [4].

For the alkaline earth and sodium fluorides, the values of $E_b^-[F(1s)] - E_e^- + 4.3$ are plotted in Figure 1 as a function of ionicities for the two cases: (1) gold sublimed onto the samples and (2) carbon C(1s) of the contaminant used as a reference, both used to correct for the electrical charge on the sample. The two plots are linear and parallel. Hence the functional dependence $\delta^-(f_i)$ on ionicity is linear.

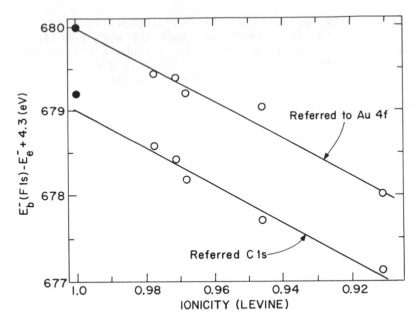

FIG.1. *Plot of energy for the reaction:*

$$F^-(1s^2\text{---})(g) \rightarrow F(1s\text{---})(g) + e(g)$$

derived from XPS with alkaline earth and sodium fluorides as a function of ionicities of the solids. E_b is the XPS measured energy; E_e^- is the coulombic, electrostatic potential; 4.3 is the spectrometer work function. To correct for sample charging gold deposits Au(4f) and carbon contaminants C(1s) were used. Value at 680.0 eV is from fluorine K_α and electron affinity. Value at 679.2 eV is from HF-SCF calculation by Bagus [5].

At unit ionicity the correctly adjusted binding energy should equal the energy to ionize the (1s) orbital electron from a gaseous fluorine ion. The linearly extrapolated value for the gold corrected values is 680.0 eV; that for C(1s) is 679.0 eV. The value derived from the K_α X-ray line and the electron affinity of fluorine is 680.0 eV. The value obtained through HF-SCF energies for F and F$^-$ is 679.2 eV [5]. The agreement in the case of the correction through Au(4f) with the experimental 680.0 eV appears to indicate that the gold deposits sublimed on the samples yield a reliable correction for electrical charge. If this is the case, then the chemical state of the carbon contaminant corresponds to a binding energy of 285.8 eV for C(1s). The agreement in either case with the value derived from HF-SCF calculations is within the error in

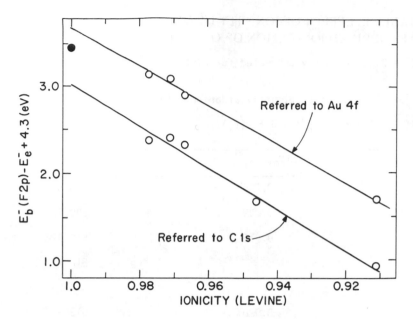

FIG.2.   *Plot of energy for the reaction*

$$F^-(1s^2\text{---}2p^6)(g) \rightarrow F(1s^2\text{---}2p^5)(g) + e(g)$$

derived from XPS with alkaline earth and sodium fluorides as a function of ionicities of solids.
Symbols have the same meaning as in Fig.1.   Value at 3.45 eV is the electron affinity of
fluorine (Berry and Reimann [6]).

the calculation. In Figure 2, are plotted the values for
$E_b[F(2p)] - E_e^- + 4.3$ for the F(2p) orbital electron as a
function of ionicities. In this case also the dependencies
are linear and the lines are parallel. The extrapolated value
for gold differs by 0.25 eV from the electron affinity for
fluorine at 3.45 eV [6]. For the carbon case, the difference
is 0.45 eV for C(1s) at 284.8 eV. Adjustment of the values
for C(1s) to the values of Au(4f$_{7/2}$) implies that the value
for carbon is at 285.5 eV.

The results presented in Figures 1 and 2 demonstrate that
although the precision in measuring binding energies and in
establishing their relation to ionicities is approximately
0.1 eV, the accuracy is only approximately 0.3 eV. If the
results in the two cases are averaged, then one obtains agreement within 0.1 eV.

TABLE I. EFFECT OF IONICITY OF OXIDE ON
XPS-DERIVED IONIZATION OF $O^{2-}(1s)(g)$

E_b^- = Binding energy corrected for electrical charge through $C(1s)$.

E_e^- = Electrostatic potential at anion site.

$I(O^{2-}1s) = E_b^- - E_e^- + 4.3 + \delta(f_i)$.

Oxide	Ionicity (f_i)	$E_b^- - E_e^- + 4.3$ (eV)
Al_2O_3	0.796[a]	508.8
MgO	0.839	509.4
CaO	0.916	512.0
UO_2		512.7
ThO_2		512.8
SrO	0.928	512.9
BaO	0.931	513.4

[a]Values from Levine [4].

An investigation of the alkaline earth oxides has produced the results displayed in Table I wherein values for
$E_b[O(1s)] - E_e^- + 4.3$ are listed as a function of ionicities of
the oxides. In this case, the measured binding energies were
corrected for electrical charge through the carbon contaminant
only. In this case, a comparison cannot be made between the
extrapolated value at $f_i = 1$ and a calculated value for the
ionization of $O^{2-}(g)$ because the latter has not been accomplished. Measurements of the binding energies of the $O(1s)$ orbital
electron in UO_2 and ThO_2 yield the values of 512.7 eV and
512.8 eV respectively for $E_b - E_c^- + 4.3$. In these two cases the
reference used for the electrical charge acquired by the sample
was also $C(1s)$ of the carbon contaminant. Comparisons in these
two cases serve to check the use of carbon to correct for the
electrical charge, because UO_2 has sufficient conductivity that
the correction is only 0.1 eV whereas ThO_2 is sufficiently
nonconducting that the correction is 3.5 eV. A comparison of
the values for $E_b(1s) - E_e^- + 4.3$ obtained with UO_2 and ThO_2
with those obtained with the alkaline earth oxides yields a
value of 0.93 (with an error of approximately 0.01) for the
ionicity of these actinide oxides.

IV. VALENCY OF URANIUM IN METAL FROM DENSITY OF 5f ELECTRONS

To complete the use of the formalism sketched above it is necessary to know the valence of uranium in the metal, the self-potential of the cation in a metal, and the oxidation-reduction potential of the free electrons. When such are known, then the gaseous atoms can be accepted for the zero of valency and one can write the condensation,

$$U(5f^3 6d7s^2)(g) \rightarrow U^{n+}(5f^{6n})e_n(s),$$

as a process of internal oxidation of U and reduction of bound electrons. Herein we address only the first item, but cite what might be possible for the other two.

The extent of the localization of the 5f electrons and the valence of uranium in the metal have been subjected to several studies since Zachariasen [7] first derived a value of 6+ for the valence on the basis of the systematic variation of the metallic radii. The subject has been summarized recently by Oetting, Rand and Ackermann [8]. The values reported vary from 0.9 to 6 as shown in Table II. Because all the various attempts to derive the value from the systematic variation of metallic radii depend upon some accepted reference, this procedure cannot be accepted in a more absolute sense. Thus it is not surprising that the derived values vary from 4.2 to 6. Hopefully, one might expect that band structure calculations would yield a more definitive value. However, such attempts have produced varying results also. The energetics of the internal oxidation-reduction which occurs upon condensation probably cannot be readily incorporated in the wave-mechanical calculations without including the ionization potentials and the self-potentials. Thus the exact location and density of charge in the 5f orbital is left uncertain because the 5f band is strongly overlapped and hybridized with the 7sp and 6d broad bands and because the uncertainty in the exchange parameter obscures the situation [14].

All the procedures cited above are either indirect or they do not contain sufficient information to define the situation existing in the metal. None of them consists of a direct measurement of the density of charge in the 5f orbital. To obtain a more direct measurement we have employed the intensities of electrons emitted during photoionization. Although an absolute measurement even with this technique is difficult, a relative measurement can be accomplished. Thus, through measurements of the relative intensities of the 4f and 5f orbital emissions in uranium metal, UO_2, UF_4, U_4O_9,

TABLE II. REPORTED VALUES FOR VALENCE OF
URANIUM METAL

Valence	Basis	Authors
6	Radii with Th at 6	Zachariasen [7]
4.2	Radii with Cm at 3	Cunningham & Wahlman [9]
6	Radii with Eu and Yb at 2	Sarkisov [10]
4.5	Radii with Ra at 2	Weigl and Trinkle [11]
0.9	APW Calculations	Kmetko and Hill [12]
5	Band structure with muffin-tin potential	Julien, et al. [13]

and U_3O_8, one can establish a more plausible value for the
valence in the metal [15]. Although a precision of only
approximately 30 percent is commonly attained in compositions
derived from intensities, a recent investigation [15] has shown
that a precision of 5 percent in the compositional ratios
derived from intensity ratios is possible provided the effects
of surface contamination are removed and correction is made
for the kinetic energy of the electrons. However, the effect
of surface contamination is somewhat but not entirely amelio-
rated by the fact that two orbitals on the uranium ion are
involved and the effect of kinetic energy is nil. The primary
problem is that uranium metal is so readily oxidized on its
surface that it must be established that the measurements are
in fact made on the metal free of oxide. To accomplish such
the metal, free of visible oxide, was cleaned in the electron
spectrometer through the use of argon ion bombardment. The
ratio of the intensities of the 5f and 4f orbital electrons
were then measured first within a few minutes and subsequently
as a function of time as the surface oxidized in the vacuum.
By extrapolating to zero time a value was obtained for the
pure metal.

The measured ratios of the intensities (5f/4f) are plotted
in Figure 3 as a function of the measured binding energies.
Therein, one observes that the ratio is the same for U, UO_2,
and UF_4. Hence, to the extent that the valence in the last
two cases is 4+ it is also 4+ in the metal. The value for
UO_2 and UF_4 is established through the calculation of the
fifth ionization potential. Since both of these compunds are

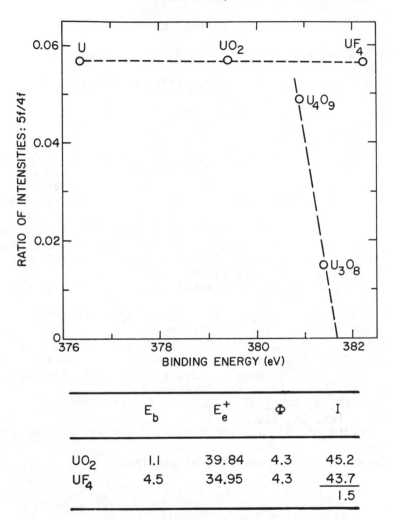

	E_b	E_e^+	Φ	I
UO_2	1.1	39.84	4.3	45.2
UF_4	4.5	34.95	4.3	43.7
				1.5

FIG.3. *Ratios of intensities of 5f to 4f orbital photoelectron emissions versus 4f orbital binding energies in U, UO_2, UF_4, U_4O_9, and U_3O_8.*

highly ionic, the values for the self-potentials can be calculated adequately from the point charge model. Within the uncertainty of the known or estimated value, both the derived values establish the presence of the U^{4+} cation. The small difference between the two values is insignificant with respect to this conclusion, but it is significant with respect to the measurements in that it exceeds the experimental errors.

REFERENCES

[1] VAN GOOL, W., and PIKEN, A. G., J. Mat. Sci. 4 (1969) 95, 105.

[2] PHILLIPS, J. C., Phys. Rev. Letters 20 (1968) 550; Rev. Mod. Phys. 42 (1970) 317.

[3] VAN VECHTEN, J. A., Phys. Rev. 182 (1969) 891.

[4] LEVINE, B. F., J. Chem. Phys. 59, (1973) 1463.

[5] BAGUS, P. S., Phys. Rev. 139 (1965) A619.

[6] BERRY, R. S., and REIMANN, J. Chem. Phys. 38 (1963) 1540.

[7] ZACHARIASEN, W. H., The Metal Plutonium (COFFINBERRY, A. S., MINER, W. N., Eds.) University of Chicago Press, Chicago, Illinois (1961) p. 99; J. Inorg. Nucl. Chem. 35 (1973) 3487.

[8] OETTING, F. L., RAND, M. H., and ACKERMANN, R. J., The Chemical Thermodynamics of Actinide Elements and Compounds, Part 1. The Actinide Elements, International Atomic Energy Agency, Vienna (1976) p. 41.

[9] CUNNINGHAM, B. B., and WAHLMAN, J. C., J. Inorg. Nucl. Chem. 26 (1964) 271.

[10] SARKISOV, E. S., Dokl. Akad. Nauk. SSSR 166 (1966) 627.

[11] WEIGL, F. and TINKL, A., Radiochem. Acta 10 (1968) 78.

[12] KMETKO, E. A., and HILL, H. H., Plutonium 1970 (Proc. 4th Int. Conf. Plutonium and Other Actinides, Santa Fe, New Mexico, 1970) 1, Metall. Soc. and Am. Inst. Min. Met. Petrol. Eng., Inc., New York, p. 233.

[13] JULLIEN, R, GALLEANI d'AGLIANA, E., AND COQBLIN, B., Phys. Rev. B6, (1972) 817.

[14] KOELLING, D. D., and FREEMAN, A. J., Phys. Rev. B7 (1973) 4454.

[15] THORN, R. J., unpublished manuscript.

DISCUSSION

L. MANES: Band calculations (LMTO-ASA method) have been performed by M.S.S. Brooks and R. Kelly for AnO_2 (actinide dioxides). As a parameter, the charge (ionicity) of the actinide was varied. The position of the 5f level with respect to the 2p-like valence band coincides with the photoemission spectra (Th to Cm) measured by B.W. Veal et al.[1] when An is given a charge of less than 4. These results, which were presented at the Third International Conference on Electronic Structure of the Actinides (Grenoble, 1978) are being published in the Journal de Physique.

---

[1] VEAL, B.W., et al., Phys. Rev. 15 (1977) 2929.

H. BLANK: May I ask why you compared UO_2 and ThO_2 with monoxides? Would it not be more reasonable to compare them with other dioxides of the same structure?

G.E. MURCH: Since we are comparing the sum of the measured binding energies of the oxygen 1s and the oxygen self-potential with the individual ionicities, both the ratio of M to O and the lattice structure should be automatically compensated for.

Section D

EMF STUDIES

ТЕРМОДИНАМИЧЕСКИЕ СВОЙСТВА ГАММА-УРАНА ПРИ ОБРАЗОВАНИИ ТВЕРДЫХ РАСТВОРОВ С АЛЬФА-ПЕРЕХОДНЫМИ МЕТАЛЛАМИ

Ю.В. ВАМБЕРСКИЙ, А.Е. ИВАНОВ, О.С. ИВАНОВ

Институт металлургии им. А.А. Байкова АН СССР,

Москва,

Союз Советских Социалистических Республик

Представлен В.В. Ахачинским

Abstract—Аннотация

THERMODYNAMIC PROPERTIES OF GAMMA-URANIUM IN THE FORMATION OF SOLID SOLUTIONS WITH ALPHA-TRANSITION METALS.

An electromotive force (EMF) method involving the use of a liquid chloride electrolyte was used for measuring the variations in the chemical potential of uranium in the formation of solid solutions: in the system U−Nb−Mo with a niobium and molybdenum atomic ratio of 1:1 and an (Nb + Mo) concentration of 5, 10 and 15 at.%; in the system U−Nb−Zr with a niobium and zirconium atomic ratio of 1:1 and an (Nb + Zr) concentration of 15 at.%; in the system U−Pd, with a palladium concentration of 2 and 20 at.%; in the system U−V, with a vanadium concentration of 4, 7, 30, 45, 55, 80 and 98 at.%. The measurements were performed in the 1048-1173 K range. The activity and the variations in entropy and enthalpy of uranium were calculated. Comparisons were made of the concentration dependences of variations in the chemical potential of γ-uranium when some α-transition metals were dissolved in it. In the case of the U−V system calculations were made of partial thermodynamic functions of vanadium, variations in the Gibbs free energy, the entropy and enthalpy of mixing, and also of excess values of variation in the Gibbs free energy and entropy. The thermodynamic properties of alloys in the U−Nb and U−V systems were compared. It was found that a solution of γ-uranium in vanadium is formed with the release of a large amount of heat and it has a negative entropy of mixing of considerable magnitude ($\Delta H = -1.419$ kcal·(g−at)$^{-1}$, $\Delta S = -1.056$ cal·(g−at)$^{-1}$·K^{-1} for a saturated solution concentration of $\sim$ 3 at.% U).

ТЕРМОДИНАМИЧЕСКИЕ СВОЙСТВА ГАММА-УРАНА ПРИ ОБРАЗОВАНИИ ТВЕРДЫХ РАСТВОРОВ С АЛЬФА-ПЕРЕХОДНЫМИ МЕТАЛЛАМИ.

Методом электродвижущих сил (ЭДС) с жидким хлоридным электролитом измерены величины изменения химического потенциала урана при образовании твердых растворов в системе U − Nb − Mo с соотношением атомных долей Nb и Mo 1:1 и концентрацией Nb + Mo 5, 10 и 15 ат%; в системе U − Nb − Zr с соотношением атомных долей Nb и Zr 1:1 и концентрацией Nb + Zr 15 ат%; в системе U − Pd при концентрации Pd 2 и 20 ат%; в системе U − V при концентрации V 4, 7, 30, 45, 55, 80 и 98 ат%. Измерения проведены в температурном интервале 1048-1173 К. Вычислены активность, изменения энтропии и энтальпии урана. Сопоставлены концентрационные зависимости изменения химического потенциала γ-урана при растворении в нем некоторых a-переходных металлов. Для системы U − V вычислены парциальные термодинамические функции ванадия, изменения термодинамического потенциала Гиббса, энтропии и энтальпии смешения, а также избыточные величины изменения термодинамического потенциала Гиббса и энтропии. Сопоставлены термодинамические свойства сплавов в системах U − Nb и U − V. Найдено, что раствор γ-урана в ванадии образуется с выделением большого количества тепла и имеет значительную по величине отрицательную энтропию смешения ($\triangle H = -1,419$ ккал · (г-ат)$^{-1}$, $\triangle S = -1,056$ кал · (г-ат)$^{-1}$· град$^{-1}$ при концентрации насыщенного раствора $\simeq$ 3 ат% U).

ТАБЛИЦА I. СОСТАВ ИССЛЕДОВАННЫХ СПЛАВОВ УРАНА, РЕЖИМЫ ПРЕДВАРИТЕЛЬНОЙ ТЕРМИЧЕСКОЙ ОБРАБОТКИ И ЧИСТОТА МЕТАЛЛОВ

№№ п.п.	Легирующий элемент	x, ат. доля	Температура отжига, К	Длительность отжига, ч	Чистота, вес %
1	Nb + Mo с соотношением атомных долей 1:1	0,05; 0,10	1273±30	72	Nb – 99,7
2	Nb + Zr с соотношением атомных долей 1:1	0,15	1273±30	38,5	Mo – 99,988
		0,15	1273±30	73	Zr – 99,98
3	Pd	0,02; 0,20	1173±10	96	99,9
4	V	0,04; 0,07; 0,30; 0,45; 0,55	1173±20	1248	99,45
		0,07; 0,30; 0,98	1273±30	72	
		0,80	–	–	
5	–	–	–	–	U – 99,8

I. Введение

Подавляющее большинство литературных данных о термодинамических свойствах сплавов урана относится к интерметаллическим соединениям. Это объясняется с одной стороны тем, что фазовые диаграммы большинства двойных систем с ураном имеют интерметаллические соединения; с другой – тем что энергетический эффект при образовании интерметаллических соединений значительно выше, чем при образовании растворов, поэтому системы с интерметаллическими соединениями предпочтительнее в качестве объектов исследования. Так как существует перспектива использования твердых растворов урана в качестве ядерного горючего, изучение их физико-химических свойств, в том числе термодинамических, является актуальной задачей.

Настоящая работа посвящена исследованию изменения химического потенциала урана при растворении в нем некоторых α - переходных металлов в температурном интервале 1048-1173 К. Использовался метод электродвижущих сил (ЭДС) с жидким электролитом, составленным из эквимолярной смеси хлоридов натрия и калия с добавкой 5-6 вес % трихлорида урана. Особенности применения метода для изучения урановых сплавов изложены в $[1]$. Измерения ЭДС и температуры производились цифровым вольтметром с входным сопротивлением 10^9 Ом или высокоомным потенциометром. Во всех опытах по окончании испытания гальванических элементов электролит исследовался на присутствие легирующих элементов сплавов. Таким образом было установлено, что равновесие реакции обмена

$$3Me(\text{сплав}) + nU^{3+}(\text{электролит}) \rightleftharpoons nU(\text{сплав}) + 3Me^{n+}(\text{электролит}) \qquad (1)$$

(Me - легирующий элемент сплава, n - валентность его ионов, U^{3+} - трехвалентные ионы урана)

в значительной мере смещено вправо для сплавов уран-цирконий и уран-титан, для которых не удалось получить равновесных значений ЭДС, и в меньшей степени для сплава уран-ниобий-цирконий, для которого удалось произвести измерения ЭДС. При работе с остальными исследованными сплавами переход легирующих элементов в электролит не обнаружен.

Гальванические элементы составлялись следующими способами:

$$\ominus \; U_{TB} \,/\, U^{3+}_{\text{ж}} \,/\, [(1-x)U + xMe]_{TB} \; \oplus \qquad (2)$$

$$\oplus \; [(1-x_1)U + x_1 Me]_{TB} \,/\, U^{3+}_{\text{ж}} \,/\, U_{TB} \,/\, U^{3+}_{\text{ж}} \,/\, [(1-x_2)U + x_2 Me]_{TB} \; \oplus \qquad (3)$$

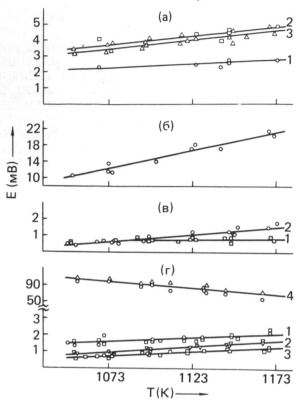

Рис. 1. *Температурные зависимости ЭДС для сплавов с концентрацией легирующих элементов в атомных долях. Для сплава Nb – Mo: 1 (а) – 0,05, 2 (а) – 0,10 (элемент 3); 3 (а) (△) – 0,15 (элемент 2). Для сплава Nb – Zr: (б) – 0,15 (элемент 2). Для Pd: 1 (в) (□) – 0,02 (элемент 2); 2 (в) (○) – 0,20 (элемент 2). Для V: 1 (г) – 0,30 (○) и 0,55 (□) (элемент 3); 2 (г) – 0,04 и 0,30 (элемент 4); 3 (г) – 0,04 и 0,07 (элемент 4); 4 (г) – 0,45 (△), 0,80 (○) и 0,98 (элемент 5).*

$$\ominus \left[(1-x_1)U + x_1 Me \right]_{TB} \big/ U_{ЖС}^{3+} \big/ \left[(1-x_2)U + x_2 Me \right]_{TB} \oplus \qquad (4)$$

$$\ominus \left[(1-x_1)U + x_1 Me \right]_{TB} \big/ U_{ЖС}^{3+} \big/ \overset{\oplus}{\left[(1-x_3)U + x_3 Me \right]_{TB}} \big/ U_{ЖС}^{3+} \big/ \left[(1-x_2)U + x_2 Me \right]_{TB}^{\ominus} (5)$$

(x – атомная доля легирующих элементов в сплавах; $x_3 > x_2 > x_1$).

Состав и режимы предварительной термической обработки исследованных сплавов приведены в таблице I. Все сплавы после отжига закаливались в холодной воде.

2. Парциальные термодинамические функции урана

На рис. I приведены температурные зависимости ЭДС гальванических элементов, рассчитанные по способу

ТАБЛИЦА II. КОЭФФИЦИЕНТЫ УРАВНЕНИЯ (6) И ОШИБКИ ОПРЕДЕЛЕНИЯ ЭДС

Система	x, ат. доля	№ прямой на рис.I	$a \cdot 10^6$, В·град$^{-1}$	$b \cdot 10^3$, В	$2G_\varepsilon \cdot 10^3$, В
U-Nb-Mo	0,05	Iа	4,396	-2,513	±0,22
	0,10	2а	9,127	-6,020	±0,39
	0,15	3а	9,564	-6,758	±0,43
U-Nb-Zr	0,15	б	87,298	-81,417	±1,58
U-Pd	0,02	Iв	1,658	-1,048	±0,17
	0,02; 0,20	2в	8,974	-8,982	±0,23
U-V	-	2г	7,088	-6,630	±0,13
	0,04	-	-1,812	2,531	±0,34
	-	3г	5,548	-5,220	±0,12
	0,07	-	3,736	-2,690	±0,46
	0,30 0,45 0,55 0,80	Iг	5,276	-4,099	±0,21
	0,98	4г	-314,000	436,000	±9,00

наименьших квадратов и описывающиеся уравнением

$$\mathscr{E} = aT + b \qquad (6)$$

Коэффициенты уравнения (6) и ошибки измерения ЭДС, определенные как удвоенные среднеквадратичные отклонения ($2G_\varepsilon$) от прямых (6) (что соответствует доверительной вероятности 0,95), представлены в таблице П. Для сплавов уран-палладий значения ЭДС, измеренные при $x = 0,02$ и $T = 1050{-}1077$ К, обрабатывались совместно с значениями ЭДС, измеренными при $x = 0,20$, что связано с пересечением фазовой границы $\gamma{-}U / \gamma{-}U + UPd_3$ диаграммы состояния

ТАБЛИЦА III. ПАРЦИАЛЬНЫЕ ТЕРМОДИНАМИЧЕСКИЕ ФУНКЦИИ УРАНА

$(\Delta\bar{G}_U$ и $\Delta\bar{H}_U$ в ккал·(г-ат)$^{-1}$; $\Delta\bar{S}_U$ в кал·(г-ат)$^{-1}$·град$^{-1}$)

Система	x, ат. доля	$\Delta\bar{G}_v$		a_v		$\Delta\bar{S}_v$	$\Delta\bar{H}_v$
		1048 K	1173 K	1048 K	1173 K		
U–Nb–Mo	0,05	-0,145±0,015	-0,183±0,015	0,933±0,007	0,924±0,006	0,304	0,174
	0,10	-0,245±0,027	-0,324±0,027	0,889±0,012	0,870±0,010	0,632	0,417
	0,15	-0,226±0,030	-0,309±0,030	0,897±0,015	0,876±0,012	0,664	0,470
U–Nb–Zr	0,15	-0,697±0,109	-1,452±0,109	0,715±0,039	0,536±0,026	6,040	5,633
U–Pd	0,02	-0,053±0,012[a]	-0,062±0,012	0,976±0,006[a]	0,974±0,005	0,108	0,065
	0,20	-0,029±0,016	-0,107±0,016	0,987±0,008	0,955±0,007	0,624	0,625
	0,04	-0,044±0,024	-0,028±0,024	0,979±0,011	0,988±0,010	-0,128	-0,178
	0,07	-0,085±0,032	-0,117±0,032	0,960±0,015	0,951±0,013	0,255	0,182
U–V	0,30 0,45 0,55 0,80	-0,099±0,015	-0,145±0,015	0,954±0,007	0,940±0,006	0,368	0,287
	0,98	-7,403±0,623	-4,705±0,623	0,028±0,010	0,133±0,040	-21,584	-30,023

а) Для 1090 K

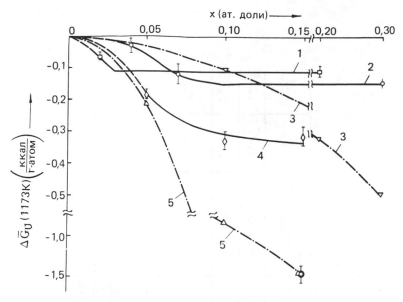

Рис. 2. Изменение химического потенциала γ-урана при растворении в нем: палладия (1), ванадия (2), ниобия (3), ниобия и молибдена с соотношением атомных долей 1:1 (4), молибдена (5), ниобия и циркония с соотношением атомных долей 1:1 (◉) для T = 1173 К.

[2] при изменении температуры сплава с $x = 0,02$ в пределах 1048-1173 К. Значения коэффициентов уравнения (6) и ошибок определения ЭДС для сплавов уран-ванадий при $x = 0,04$ и $x = 0,07$ вычислены из соответствующих величин для прямых 1(г), 2(г) и 3(г) рис.1; при $x = 0,98$ коэффициенты уравнения (6) приняты равными коэффициентам для прямой 4(г), так как поправки, равные величинам ЭДС для прямой 1(г), значительно меньше ошибки измерения для прямой 4(г).

Парциальные термодинамические функции урана определялись по следующим уравнениям:

$$\Delta \bar{G}_U = -3F\varepsilon = -69,186\,(aT+b) \quad \text{ккал·(г-ат)}^{-1} \qquad (7)$$

$$\Delta \bar{S}_U = -\left[\frac{\partial(\Delta \bar{G}_U)}{\partial T}\right]_P = 69186\,a \quad \text{кал·(г-ат)}^{-1}\text{град}^{-1} \qquad (8)$$

$$\Delta \bar{H}_U = \Delta \bar{G}_U + T\cdot\Delta \bar{S}_U = -69,186\,b \quad \text{ккал·(г-ат)}^{-1} \qquad (9)$$

$$a_U = exp\left(\frac{\Delta \bar{G}_U}{RT}\right) \qquad (10)$$

Результаты вычислений сведены в таблицу Ш.

ТАБЛИЦА IV. ПАРЦИАЛЬНЫЕ ТЕРМОДИНАМИЧЕСКИЕ ФУНКЦИИ ВАНАДИЯ

($\Delta\bar{G}_V$ и $\Delta\bar{H}_V$ в ккал·(г-ат)$^{-1}$; $\Delta\bar{S}_V$ в кал·(г-ат)$^{-1}$·град$^{-1}$)

x, ат.доля	a_V		$\Delta\bar{G}_V$		$\Delta\bar{S}_V$	$\Delta\bar{H}_V$
	1048 K	1173К	1048 К	1173К		
0,04	0,550	0,423	-1,244	-2,003	6,072	5,119
0,07	0,792	0,806	-0,485	-0,503	0,144	-0,334
0,30 0,45 0,55 0,80	0,858±0,015	0,925±0,013	-0,319±0,036	-0,181±0,033	-1,104	-1,476
0,965	—	0,925±0,013	—	-0,181±0,033	-1,104	-1,476
0,97	0,858±0,015	—	-0,319±0,036	—		
0,98	0,938±0,006	0,977±0,004	-0,133±0,014	-0,055±0,010	-0,624	-0,787

На рис.2 сопоставлены изменения химического потенциала μ - урана при растворении в нем ряда d - переходных металлов при II73 К.Кривые 3 и 5 построены по полученным ранее [I,3] данным для сплавов уран-ниобий и уран-молибден и приведены для сравнения с соответствующими данными для сплавов уран-ниобий-молибден и уран-ниобий-цирконий.

Изменение химического потенциала урана в сплавах уран-ниобий-молибден (кривая 4) при $x < 0,05$ близко к изменению этой величины в сплавах уран-молибден (кривая 5). При увеличении содержания ниобия и молибдена в тройных сплавах падение химического потенциала урана интенсивно замедляется, и при $x = 0,15$ кривая (4) сближается с кривой (3) для сплавов уран-ниобий. Замена в тройном сплаве уран-ниобий-молибден при $x = 0,15$ молибдена на цирконий приводит к резкому снижению величины химического потенциала урана - кривая (4), точка ◎. Химический потенциал урана мало отличается по величине в сплавах уран-ванадий и уран-ниобий при $x \leqslant 0,15$. Концентрации насыщенных μ - растворов уран-палладий и уран-ванадий взяты из работ [2] и [4] соответственно.

3.Термодинамические свойства системы уран-ванадий

Интегральные величины изменения термодинамического потенциала Гиббса при образовании сплавов находились с помощью интегрирования уравнения Гиббса-Дюгема:

$$\Delta G(x;T) = x \cdot \int_0^{1-x} \frac{\Delta \overline{G}_U(x;T)}{x^2} \, d(1-x). \tag{II}$$

Интегрирование производилось с использованием фазовых границ диаграммы состояния [4], поскольку характер изменения ЭДС в зависимости от концентрации не противоречит указанной диаграмме состояния.

$$\Delta G(0,04;T) = 0,04 \left[\int_0^{0,93} \frac{\Delta \overline{G}_U(x;T)}{x^2} d(1-x) + \Delta \overline{G}_U^* \cdot \int_a^{0,9} \frac{d(1-x)}{x^2} + \int_{0,9}^{0,93} \frac{\Delta \overline{G}_U(x;T)}{x^2} d(1-x) + \int_{0,93}^{0,96} \frac{\Delta \overline{G}_U(x;T)}{x^2} d(1-x) \right]. \tag{I2}$$

Здесь $(I - x) = a = 0,03$ при I048 К и $a = 0,035$

ТАБЛИЦА V. ИНТЕГРАЛЬНЫЕ ТЕРМОДИНАМИЧЕСКИЕ ФУНКЦИИ СИСТЕМЫ УРАН – ВАНАДИЙ
(ΔG и ΔH в ккал·(г-ат)⁻¹; ΔS в кал·(г-ат)⁻¹·град⁻¹)

x, ат. доля	ΔG 1048K	ΔG 1173K	ΔS	ΔH	ΔG^E 1048K	ΔG^E 1173K	ΔS^E
0,04	-0,092±0,025	-0,107±0,024	0,120	0,034	0,257±0,025	0,284±0,024	-0,213
0,07	-0,113±0,021	-0,144±0,020	0,248	0,147	0,414±0,021	0,446±0,020	-0,255
0,10	-0,121±0,017	-0,149±0,017	0,224	0,114	0,555±0,017	0,608±0,017	-0,421
0,30	-0,165±0,021	-0,156±0,021	-0,072	-0,240	0,425±0,026	0,514±0,026	-0,632
0,45	-0,198±0,024	-0,162±0,023	-0,288	-0,500	0,322±0,029	0,438±0,028	-0,788
0,55	-0,220±0,026	-0,165±0,025	-0,440	-0,681	0,260±0,031	0,385±0,030	-0,900
0,80	-0,275±0,031	-0,174±0,030	-0,808	-1,122	0,085±0,036	0,256±0,035	-1,168
0,965	–	-0,180±0,033	-1,056	-1,419	0,032±0,034	0,173±0,033	-1,339
0,97	-0,312±0,034	–					
0,98	-0,278±0,028	-0,148±0,022	-1,040	-1,368	0,074±0,028	0,081±0,022	-1,235

при II73 К - предельная растворимость урана в ванадии; (I - x) = 0,9 - предельная растворимость ванадия в уране, принятая независящей от температуры;
(I - x) = 0,93 и (I - x) = 0,96 - составы исследованных сплавов в области γ - раствора диаграммы состояния; $\Delta \bar{G}_U^* = const$ - изменение химического потенциала урана в двухфазной области
$\gamma + \delta$ при $T = const$.

В связи с тем, что при графическом интегрировании
уравнения (II) экстраполяция подинтегральной функ-
ции на значения $x \longrightarrow I$ неудобна, ΔG (0,98; T)
и ΔG (α ; T) определялись следующим образом:
графическим интегрированием уравнения Гиббса-Дюгема

$$ \lg a_V = -\int_0^{1-x} \frac{1-x}{x} \, d \lg a_U \qquad (13) $$

находились значения активности ванадия,изменения
химического потенциала ванадия

$$ \Delta \bar{G}_V = RT \ln a_V \qquad (14) $$

и изменения термодинамического потенциала Гиббса

$$ \Delta G(x;T) = (1-x)\Delta \bar{G}_U (x;T) + x \cdot \Delta \bar{G}_V (x;T). \qquad (15) $$

Экстраполяция подинтегральнои функции в уравнении
(13) на значения $x \longrightarrow I$ осуществляется легко.

Первый интеграл уравнения (12) определялся из
найденного с помощью равенства (15) значения
$\Delta G (\alpha;T)$:

$$ \int_0^\alpha \frac{\Delta \bar{G}_U (x;T)}{x^2} \, d(1-x) = \frac{\Delta G(\alpha;T)}{1-\alpha}, $$

второи интеграл вычислялся аналитически, третий и
четвертый - графическим интегрированием.

Интегрирование производилось для температур 1048
и II73 К.Значения изменения химического потенциала
и активности ванадия при $x < 0,I$ вычислялись по
уравнениям (15) и (14) из найденных значений ΔG
и определенных экспериментально значений $\Delta \bar{G}_U$.
Методика определения ошибок $\Delta G(x;T)$ описана
в [5].

Результаты вычислений парциальных термодинамических функций ванадия приведены в таблице IУ, а инте-
гральных термодинамических функции системы уран-ва-
надии - в таблице У.

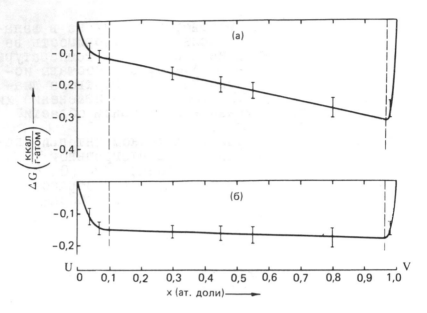

Рис. 3.   Изменение термодинамического потенциала Гиббса в системе уран – ванадий: (а) – при 1048 К; (б) – при 1173 К.   Пунктиром показаны фазовые границы диаграммы состояния [4].

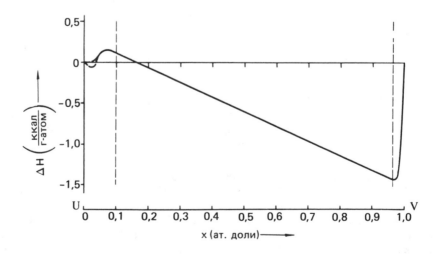

Рис. 4.   Энтальпия смешения в системе уран – ванадий.

На рис.3 изображены концентрационные зависимости изменения термодинамического потенциала Гиббса. По форме эти зависимости сходны с соответствующими зависимостями для системы уран-ниобий $[5]$ - минимальные значения ΔG смещены в сторону легирующего элемента на концентрационной оси. Численные значения ΔG в системе уран-ванадий примерно на порядок меньше, чем в системе уран-ниобий.

На рис.4 приведена концентрационная зависимость энтальпии смещения, которая имеет сходство с соответствующей зависимостью для системы уран-ниобий $[5]$. В обеих системах происходит изменение знака энтальпии смещения с положительного на отрицательный с увеличением концентрации легирующего элемента. Следует отметить большую экзотермичность реакции образования δ - раствора на основе ванадия. Этот эффект проявлялся при приготовлении сплавов с содержанием 98 ат.% ванадия. При сплавлении шихты в дуговой печи происходило чрезвычайно сильное разогревание, которое во многих случаях приводило к взрывообразному разбрызгиванию слитка.

Энтропия смещения в зависимости от состава изменяется симбатно с энтальпией смещения, как можно видеть по данным таблицы У.

ЛИТЕРАТУРА

[1] ВАМБЕРСКИЙ, Ю.В. и др., Thermodynamics of Nuclear Materials, 1974 (Proc. Symp. Vienna, 1974) Vol.2, IAEA, Vienna (1975) 321.

[2] CATTERALL, J.A., et al., J. Inst. Met. 85 (1956) 63.

[3] ВАМБЕРСКИЙ, Ю.В. и др., Физико-химический Анализ Сплавов Урана, Тория и Циркония, М., Наука (1974) 20; 37.

[4] SALLER, H.A., ROUGH, F.A., J. Met. 5 (1953) 545.

[5] VAMBERSKY, Yu.V., et al., J. Nucl. Mater. 55 (1975) 96.

THERMODYNAMIC STUDIES ON MOLYBDENUM/NOBLE-METAL ALLOYS

M. YAMAWAKI, Y. NAGAI, T. KOGAI, M. KANNO
University of Tokyo,
Tokyo,
Japan

Abstract

THERMODYNAMIC STUDIES ON MOLYBDENUM/NOBLE-METAL ALLOYS.

Electromotive force cells have been used to determine the activities of molybdenum for Mo-Pd, Mo-Rh and Mo-Ru-Pd alloys over the temperature range 1200–1300 K, and thermo-dynamic functions were derived from the results. Solid $ZrO_2 - 11$ mol.% CaO was used as the electrolyte. In ternary Mo-Ru-Pd alloy the atomic ratio Ru:Pd was fixed to 72.5:27.5 in order to simulate the white metallic inclusions in irradiated $(U,Pu)O_2$ fuel. Activities of molybdenum showed negative deviations from Raoult's law (ideal solution behaviour) in the composition range where the atomic fraction of molybdenum N_{Mo} is $\lesssim 0.3$, and then positive deviations in the intermediate N_{Mo} range. It was shown that the estimation of activities of molybdenum in the white metallic inclusions based on the regular-solution approximation was generally fairly good, but might lead to error in the intermediate and higher N_{Mo} ranges. Standard Gibbs free energy of formation of the Mo-Rh intermetallic ϵ-phase was shown to be more negative than those of some other molybdenum intermetallic phases, indicating a higher thermodynamic stability of the ϵ-phase.

1. INTRODUCTION

Molybdenum is one of the most plentiful and important of the fission products. A considerable part of this element in irradiated $(U,Pu)O_2$ fuel exists as a component of white metallic inclusions which are otherwise composed of such noble metals as technetium, ruthenium, rhodium and palladium [1 – 3]. The activity of molybdenum, a_{Mo}, in this noble metal alloy is believed to strongly affect the oxygen potential of the fuel as it determines the proportion of molybdenum in the irradiated fuel which is to be oxidized [4, 5]. Johnson et al.[3] developed a method for determining the oxygen potential gradients in $(U,Pu)O_2$ fuel, based on the Mo/MoO_2 redox couple. To apply this method, the knowledge of a_{Mo} as a function of alloy composition is required.

However, there seem to be no thermodynamic data available for molybdenum/noble-metal alloys. Johnson et al. [3] only estimated a_{Mo} of the metallic inclusions, based on the regular solution approximation after the method by Kaufman et al. [6]. In the present study, a_{Mo} and thermodynamic functions of Mo-Pd, Mo-Rh and Mo-Ru-Pd alloys have been determined by measuring the EMFs of galvanic cells using the solid electrolyte ZrO_2 (+ CaO).

2. PRINCIPLE

Under the conditions of, essentially, exclusively ionic conduction in the solid-oxide electrolyte, the following cell is constructed:

$$(-)Mo,MoO_2 \mid ZrO_2 (+ CaO) \mid Mo/\text{noble-metal alloy, } MoO_2 (+),$$

where 'noble metal' stands for either palladium, rhodium or Ru-Pd alloy. Electrode reactions at the two electrodes are as follows:

$$(+)MoO_2 + 4e^- \rightleftharpoons Mo \text{ (in alloy)} + 2O^{2-} \tag{I}$$

$$(-)Mo + 2O^{2-} \rightleftharpoons MoO_2 + 4e^- \tag{II}$$

The net reaction is:

$$Mo(s) = Mo\,(\text{in alloy}) \tag{III}$$

Under isothermal, isobaric and reversible conditions, the Gibbs free energy change for the reaction is given by:

$$\Delta \overline{G}_{Mo} = \overline{G}_{Mo} - \overline{G}^0_{Mo} = -4EF = RT \ln a_{Mo} \tag{1}$$

where E is the open-circuit cell potential, i.e. the EMF, F is Faraday's constant, R is the universal gas constant, a_{Mo} is the chemical activity of molybdenum in the alloy, referred to a standard state of oxygen-saturated pure molybdenum of unit activity, and T is the absolute temperature of the cell.

Using the measured value of E and Eq.(1), a_{Mo} and $\Delta \overline{G}_{Mo}$ for a specific temperature can be derived. From the temperature dependence of E, the relative partial molar entropy, $\Delta \overline{S}_{Mo}$, and enthalpy, $\Delta \overline{H}_{Mo}$, can be derived using the following equations:

$$\Delta \overline{S}_{Mo} = 4F \left(\frac{\partial E}{\partial T} \right)_{x,p} \tag{2}$$

$$\Delta \overline{H}_{Mo} = 4F \left(T \left(\frac{\partial E}{\partial T} \right)_{x,p} - E \right) = \Delta \overline{G}_{Mo} + T \Delta \overline{S}_{Mo} \tag{3}$$

Using the Gibbs-Duhem equation, relative partial molar quantities of either palladium or rhodium, and the relative integral molar quantities can be derived.

3. EXPERIMENTAL

The materials used in the preparation of alloys were molybdenum, palladium, rhodium and ruthenium powders of 99.95 wt% purity each. Weighed amounts of two elements for binary alloys or three elements for ternary alloys were pressed into tablets. These were then melted in a plasma-jet furnace on a water-cooled copper hearth, under an atmosphere of purified argon to produce an alloy button. To ensure homogenization of the alloy, the melting of the specimen was repeated several times, each melting being followed by inversion of the specimen after cooling. After the melting procedure had been completed, a further homogenizing heat treatment was also given to each button. Then the alloy button was milled off to produce alloy, powder, from which magnetizable impurities were removed with a magnet. The MoO_2 powder which was to be mixed with the alloy powder was prepared by heating ammonium molybdate at $550°C$ in a flow of an $H_2 + H_2O$ gas mixture of fixed composition. The MoO_2 powder was then adequately mixed with a specimen of alloy powder at a weight ratio of $1:7$ and pressed into a tablet of 10 mm diameter. The tablets were heated for 20 h at about $1000°C$ and served as the (alloy)-MoO_2 electrodes. For the Mo-MoO_2 electrodes, 99.95 wt% of molybdenum and reagent grade MoO_3 powders were mixed at a molecular ratio of $2:1$, and were pressed into tablets and heat-treated in silica tubes for 24 h at $1100°C$.

The solid electrolyte was in the form of stabilized zirconia crucibles consisting of ZrO_2 and 11 mol.% CaO. They were 12 mm i.d., 1 mm thick and 30 mm high, and were manufactured by Nihon-Kagaku-Togyo Co. Ltd. A schematic diagram of the cell arrangement is shown in Fig. 1. The electrodes were isolated from the environment by a gas-tight alumina-base cement. The temperature of the cell was measured with a Pt/Pt–13% Rh thermocouple spot welded to the platinum plate which was inserted between the alloy electrode and a spring-loaded alumina tube. Grounded nichrome strip was wound round the outer silica tube containing the cell assembly to provide electrical shielding of the cell system. The EMF measurements were carried out in a flow of purified argon.

Cell potentials were measured with a universal digital voltmeter (Yokogawa Electric Co. Ltd., Type 2501). The output from the digital voltmeter was printed out by a digital printer at one minute intervals during measurements. The temperature of the specimen was initially raised up to about $950°C$, and it was then maintained at a constant temperature until a constant EMF reading was obtained. Then the temperature was cyclically raised and lowered. This temperature cycling was repeated several times, during which about 10 EMF values were measured for each specimen. Reversibility of cell operation was checked by verifying that there was no significant variation with time of voltage at constant temperature, by noting the approach to the same open-circuit voltage from temperatures both above and below the temperature of measurement, and by

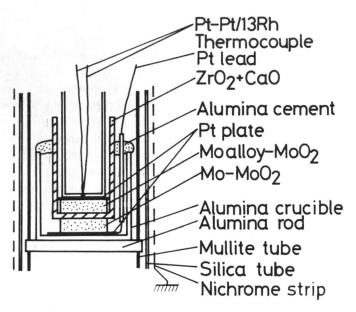

Pt–Pt/13Rh
Thermocouple
Pt lead
ZrO$_2$+CaO

Alumina cement
Pt plate
Mo alloy–MoO$_2$
Mo–MoO$_2$

Alumina crucible
Alumina rod

Mullite tube
Silica tube
Nichrome strip

FIG.1.   *Schematic diagram of the cell arrangement.*

noting recovery to the same voltage after a passage of current in either the forward
or the reverse direction.

EMF measurements were taken in the following cells:

$$Mo, MoO_2 \quad | \quad ZrO_2 + (CaO) \quad | \quad Fe, Fe_xO \tag{4}$$

$$Mo, MoO_2 \quad | \quad ZrO_2 + (CaO) \quad | \quad Mo{-}Pd \; alloy, MoO_2 \tag{5}$$

$$Mo, MoO_2 \quad | \quad ZrO_2 + (CaO) \quad | \quad Mo{-}Rh \; alloy, MoO_2 \tag{6}$$

$$Mo, MoO_2 \quad | \quad ZrO_2 + (CaO) \quad | \quad Mo{-}Ru{-}Pd \; alloy, MoO_2 \tag{7}$$

For all the Mo-Ru-Pd ternary alloys tested the atomic ratio of Ru : Pd was fixed
at 72.5 : 27.5, with the aim of simulating the white metallic inclusions in a fast
breeder reactor fuel. The composition of this hypothetical inclusion was determined on the basis of typical elemental yields of technetium, ruthenium,
palladium and rhodium evaluated for the Japanese experimental fast breeder
reactor MONJU (Tc : Ru : Pd : Rh ≃ 13 : 48 : 29 : 10). In simulating this quaternary
alloy with a Ru-Pd binary alloy, the following assumptions were used: with
regard to the transformation contribution to phase stability, technetium and
rhodium would be equivalent to Ru(hcp) and Pd(fcc), respectively, and, further,

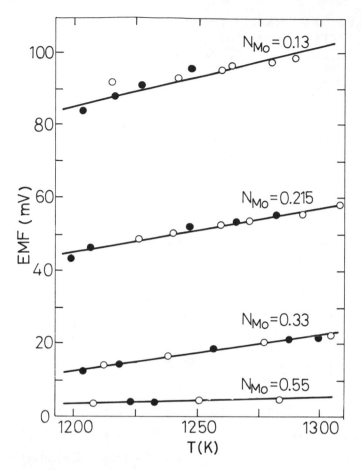

FIG.2. Electromotive force data as functions of temperature for Mo-Pd alloys.

the electronic contribution would be the dominant factor so that the phases that occur would be principally determined by the electron/atom ratios.

4.   RESULTS AND DISCUSSION

Initially, the EMF of cell (4) was measured as a function of temperature. The EMF values obtained were close to the mean values of those obtained by Rapp et al., Kozuka et al. and Drobyschev et al. (see Ref.[8]), and the slope of the EMF curve accorded well with those of these research workers. This result suggests that the measuring system of the present study is very reliable.

TABLE I. TEMPERATURE DEPENDENCES OF THE
EMF OF CELLS (5), (6) AND (7)

Comp. (at.%)	EMF (mV)
Pd-13Mo	$(-93.4 \pm 18.7) + (0.150 \pm 0.015)T$
Pd-21.5Mo	$(-103.6 \pm 3.9) + (0.124 \pm 0.003)T$
Pd-33Mo	$(-105.4 \pm 3.0) + (0.098 \pm 0.002)T$
Pd-55Mo	$(-10.0 \pm 0.4) + (0.011 \pm 0.000)T$
Rh-15Mo	$(-296.3 \pm 20.0) + (0.384 \pm 0.016)T$
Rh-20Mo	$(-157.7 \pm 63.1) + (0.264 \pm 0.050)T$
Rh-30Mo	$(-203.6 \pm 11.4) + (0.191 \pm 0.009)T$
Rh-50Mo	$(-32.8 \pm 9.0) + (0.031 \pm 0.007)T$
65.3Ru-24.7Rh-10Mo	$(62 \pm 2.2) + (0.0573 \pm 0.0013)T$
58Ru-22Rh-20Mo	$(-21.5 \pm 4.5) + (0.0754 \pm 0.0037)T$
50.8Ru-19.2Rh-30Mo	$(-29.1 \pm 4.1) + (0.0517 \pm 0.0034)T$
36.3Ru-13.7Rh-50Mo	$(-16.5 \pm 3.1) + (0.0266 \pm 0.0025)T$
21.8Ru-8.2Rh-70Mo	$(-6.8 \pm 3.9) + (0.0139 \pm 0.0032)T$
14.5Ru-5.5Rh-80Mo	$(-18.9 \pm 2.6) + (0.0242 \pm 0.0022)T$
7.3Ru-2.7Rh-90Mo	$(-35.1 \pm 4.2) + (0.0355 \pm 0.0035)T$

The temperature dependence of the EMF of cell (5) is shown in Fig.2. The relationships between the EMF and temperature for Mo-Pd, Mo-Rh and Mo-Ru-Pd alloys are shown in Table I. From these results a_{Mo} and the partial molar quantities of the molybdenum component, $\Delta \bar{G}_{Mo}$, $\Delta \bar{H}_{Mo}$ and $\Delta \bar{S}_{Mo}$, for the binary alloys were calculated, and these are shown in Table II. For these alloys

a_{Pd}, a_{Rh} and related partial molar quantities were obtained using the Gibbs-Duhem equation. In the Gibbs-Duhem integrations, the phase diagram of Mo-Pd alloy by Shunk [9] and that of Mo-Rh alloy by Elliott [10] were made use of, respectively. The partial molar quantities obtained for the palladium and rhodium components, as well as the integral quantities ΔG, ΔH and ΔS, are also shown in Table II. The error ranges for these values were not evaluated on account of the difficulty of estimating errors deriving from graphical integrations. For the ternary alloy, only a_{Mo} values were obtained from the EMF/T relations, and these are shown in Table III.

The values of a_{Mo} for the binary alloys are plotted in Fig.3. In the two systems, Mo-Pd and Mo-Rh, a_{Mo} showed negative deviations from ideal solution behaviour for $N_{Mo} \lesssim 0.3$. With increasing N_{Mo}, a_{Mo} increased steeply up to nearly unity and then remained constant over a two-phase region. In Fig.3, semi-theoretically calculated values of a_{Mo} are also plotted, using triangles. The calculation of these a_{Mo} values is based on the regular solution approximation for each phase, after the method by Kaufman et al. [6]. A similar trend of the a_{Mo}/N_{Mo} relation was observed between the experimentally obtained a_{Mo} values and the theoretically calculated ones for each binary alloy system studied, even though the match between the two kinds of a_{Mo} values for the Mo-Pd system was better than that for the Mo-Rh system.

The values of a_{Mo} for the Mo-Ru-Pd ternary alloys are plotted in Fig.4. A relation between a_{Mo} and N_{Mo} similar to those observed in the binary systems was obtained; however, in the ternary system, a_{Mo} increased with N_{Mo} even over two-phase regions, as the compositions of the respective co-existing phases changed with N_{Mo}. In the Mo-Ru-Pd system, the composition range of $N_{Mo} \lesssim 0.25$ and that between ~ 0.45 and ~ 0.95 are two-phase regions consisting of Ru(hcp) and Pd(fcc) and of Ru(hcp) and Mo(bcc), respectively. In the former region the Ru(hcp) phase is the dominant phase, so that the composition range of $N_{Mo} \lesssim 0.6$ of the ternary alloy could be simulated as a single hcp phase without large errors being introduced into the estimated a_{Mo} values. Johnson et al. [3] observed that metallic alloy inclusions in irradiated $(U,Pu)O_2$ fuels comprised a single hcp phase, and they estimated the value of a_{Mo} of these inclusions based on the regular solution approximation, after Kaufman et al. [6]. Their a_{Mo} values are also plotted in Fig.4, using squares. Both groups of a_{Mo} values in Fig.4 appear to be in fairly good accord with one another over the composition range $N_{Mo} \lesssim 0.6$, which is of significance in fast breeder reactor fuels, even though in the intermediate and higher N_{Mo} ranges the discrepancies tend to become larger than in the small N_{Mo} range. This result appears to support the appropriateness of the above estimation of a_{Mo} by Johnson et al.

The values of ΔG and ΔH at 1273 K and ΔS over the temperature range of 1200 – 1300 K obtained in this study are shown in Fig.5. For both the Mo-Pd and Mo-Rh alloys, ΔH and ΔS showed large positive values over the whole

TABLE II. THERMODYNAMIC QUANTITIES FOR Mo-Pd AND Mo-Rh ALLOYS AT 1273 K, CALCULATED FROM EMF DATA

N_{Mo}	a_{Mo}	a_{Pd} or a_{Rh}	$\Delta\bar{G}_{Mo}$ (cal·g-at⁻¹)	$\Delta\bar{G}_{Pd}$ or $\Delta\bar{G}_{Rh}$	$\Delta\bar{H}_{Mo}$ (cal·g-at⁻¹)	$\Delta\bar{H}_{Pd}$ or $\Delta\bar{H}_{Rh}$	$\Delta\bar{S}_{Mo}$ (cal·g-at⁻¹·K⁻¹)	$\Delta\bar{S}_{Pd}$ or $\Delta\bar{S}_{Rh}$	ΔG (cal·g-at⁻¹·K⁻¹)	ΔH	ΔS (cal·g-at⁻¹·K⁻¹)
(Mo-Pd Alloys)											
0.13	$0.029^{+0.001}_{-0.002}$	0.74	-9000±1700	-770	8600±1700	0	13.8±1.4	0.60	-1800	1100	2.3
0.215	$0.138^{+0.003}_{-0.002}$	0.52	-5000±360	-1600	9600±360	-110	11.4±0.3	1.16	-2300	2000	3.3
0.33	$0.494^{+0.002}_{-0.009}$	0.36	-1800±280	-2800	9700±280	-200	9.0±0.2	2.02	-2400	3100	4.3
0.55	$0.864^{+0.017}_{-0.017}$	0.25	-370±40	-3800	920±40	6350	1.0±0.0	8.0	-1900	3400	4.1
(Mo-Rh Alloys)											
0.15	$0.00089^{+0.00004}_{-0.00005}$	0.82	-18000±1800	-500	27000±1800	5600	35.4±1.5	4.82	-3100	8800	9.4
0.20	$0.0018^{+0.0001}_{-0.0001}$	0.72	-16000±5800	-820	15000±5800	10000	24.4±4.6	8.82	-3900	11000	11.9
0.30	$0.236^{+0.012}_{-0.002}$	0.136	-3600±1100	-5000	19000±1100	8400	17.6±0.8	10.5	-4600	12000	12.6
0.50	$0.784^{+0.020}_{-0.016}$	0.064	-610±830	-6900	3000±830	19000	2.9±0.6	20.1	-3800	11000	11.5

TABLE III. ACTIVITY OF MOLYBDENUM,
a_{Mo}, FOR Mo-Ru-Pd ALLOYS AT 1273 K,
CALCULATED FROM EMF DATA

Comp. (at.%)	a_{Mo}
65.3Ru-24.7Rh-10Mo	$0.0073^{+0.0002}_{-0.0002}$
58Ru-22Rh-20Mo	$0.066^{+0.009}_{-0.008}$
50.8Ru-19.2Rh-30Mo	$0.262^{+0.009}_{-0.008}$
36.3Ru-13.7Rh-50Mo	$0.531^{+0.015}_{-0.026}$
21.8Ru-8.2Rh-70Mo	$0.672^{+0.023}_{-0.022}$
14.5Ru-5.5Rh-80Mo	$0.648^{+0.019}_{-0.019}$
7.3Ru-2.7Rh-90Mo	$0.692^{+0.026}_{-0.025}$

composition range. The enthalpy of formation, ΔH, can be divided into three terms, i.e. the binding term, the strain term and the transformation term. The strain term for these binary alloy systems was estimated to be 27 and 137 cal/g-at. at $N_{Mo} = 0.50$ for Mo-Pd and Mo-Rh alloys, respectively, by using the values of the strain energy contribution to the interaction parameter evaluated by Kaufman et al. [6]. These values of the strain term are very small because the atomic size difference between Mo and Pd and that between Mo and Rh are as small as 0.9% and 3.2%, respectively. The transformation term for the primary solid-solution region was estimated to be about 250 cal/g-at. at $N_{Mo} = 0.10$ for both Mo-Pd and Mo-Rh alloys, based on the values of the transformation enthalpy for pure metals estimated by Kaufman et al. [6]. This term for the ϵ-phase of Mo-Rh alloy at $N_{Mo} = 0.50$ was estimated to be as large as 1000 cal/g-at.

These estimates show that the binding term is most probably the largest term in the enthalpy change on alloying, not only for Mo-Rh alloy but also for Mo-Pd alloy. However, the probable error ranges for ΔH and ΔS of this study were comparatively large owing to the narrow temperature ranges of the measurements, so that more detailed discussions based on the ΔH and ΔS values obtained might prove fruitless.

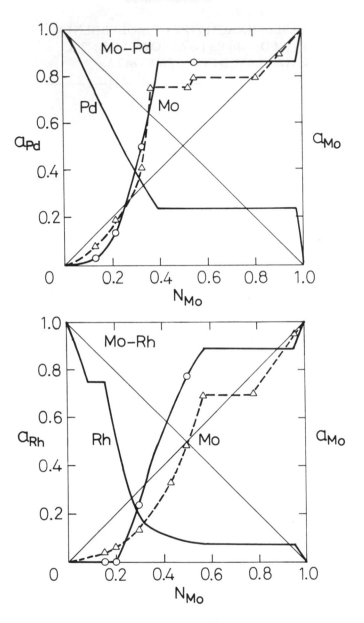

FIG.3. Activities of molybdenum and palladium or rhodium versus N_{Mo} at 1273 K.
$-\circ-$ or ———— : obtained from EMF values of this study; $-\triangle-$: calculated based on regular solution approximation, after the method by Kaufman [6].

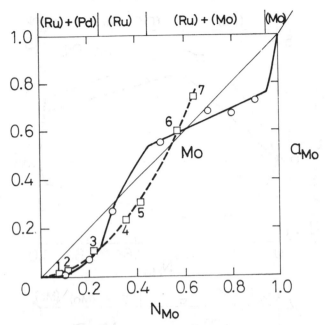

FIG.4. *Activities of molybdenum for Mo-Ru-Pd alloys at 1273 K obtained in this study*
(−○−) and those for white metallic inclusions in (U, Pu)O₂ fuels for various temperatures
estimated by Johnson et al. [3] (−□−).
Temperature and fuel pin code for each point is as follows (from Ref.[3]):

No.	Temperature (K)	Fuel pin
1	1943	C-15
2	1908	C-11
3	2358	C-11
4	2053	F2R
5	1723	F2Z
6	2953	F2R
7	2943	F2R

ΔG of Mo-Rh alloy showed a minimum in the ε-phase. The ΔG value for
the Mo-Rh ε-phase (30 at.% Mo) at 1273 K, i.e. − 4.6 kcal/mol, is compared with
ΔG values for the Mo-Co μ-phase (53 at.% Mo) and for the Mo-Ni μ-phase
(47 at.% Mo), i.e. − 1.4 and − 0.5 kcal/mol, respectively; the order coincides with
a decreasing order of melting points of these intermetallic phases, i.e. Mo-Rh,
2075°C; Mo-Co, 1510°C; Mo-Ni, 1362°C.
    A similar study to the present one is now under way in the authors' laboratory
for Mo-Ru alloys.

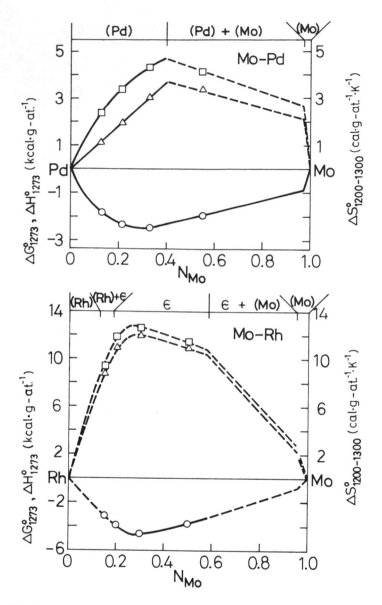

FIG. 5. *Relative integral molar Gibbs free energies, enthalpies and entropies of formation of Mo-Pd and Mo-Rh alloys.* —○- -: ΔG^0_{1273}; —△- -: ΔH^0_{1273}; —□- -: $\Delta S^0_{1200-1300}$.

ACKNOWLEDGEMENT

The authors wish to express their appreciation to Dr. T. Mukaibo for his warm encouragement during the course of this study. They also wish to acknowledge the aid of Messrs. T. Yoneoka, S. Kokubo and K. Shida in the experimental work, and Professor Z. Kozuka and Dr. I. Katayama of Osaka University for the valuable suggestions made.

REFERENCES

[1] JOHNSON, C.E., JOHNSON, I., BLACKBURN, P.E., CROUTHAMEL, C.E., React. Technol. **15** (1972–73) 303.

[2] JOHNSON, C.E., Physical Aspects of Electron Microscopy and Microbeam Analysis (SIEGAL, B., BEAMAN, D.R., Eds), John Wiley & Sons, New York (1975) 373.

[3] JOHNSON, I., JOHNSON, C.E., CROUTHAMEL, C.E., SEILS, C.A., J. Nucl. Mater. **48** (1973) 21.

[4] OLANDAR, D.R., Fundamental Aspects of Nuclear Reactor Fuel Elements, USERDA Rep. TID-26711-P1 (1976) 172.

[5] ADAMSON, M.G., private communication.

[[6] KAUFMAN, L., BERNSTEIN, H., Computer Calculation of Phase Diagrams, Academic Press, New York (1970).

[7] ZADOR, S., ALCOCK, C.B., J. Chem. Thermodyn. **2** (1970) 9.

[8] KATAYAMA, I., SHIMATANI, H., KOZUKA, Z., Nippon Kinzoku Gakkai-Shi (J. Japan Inst. Metals) **37** (1973) 509.

[9] SHUNK, F.A., Constitution of Binary Alloys, 2nd Suppl., McGraw-Hill, New York (1969) 520.

[10] ELLIOTT, P., Constitution of Binary Alloys, 1st Suppl., McGraw-Hill, New York (1965) 631, 630.

[11] KATAYAMA, I., AOKI, M., KOZUKA, Z., Nippon Kinzoku Gakkai-Shi (J. Japan Inst. Metals) **39** (1975) 1210.

DISCUSSION

P.E. POTTER: These measurements are of great significance in the calculation of equilibria in the alloys of molybdenum, technetium, ruthenium, rhodium and palladium. If only regular solution models are used, it is difficult to reproduce phase boundaries of many òf the binary systems, so that further measurements of activities as well as some more phase-diagram studies are required.

THERMODYNAMIC STUDIES ON
LIQUID Mg-In-Sn TERNARY SOLUTIONS

Z. MOSER
Institute for Metal Research,
Polish Academy of Sciences,
Cracow,
Poland

R. CASTANET
Centre de recherche de microcalorimétrie
et de thermochimie,
Centre national de la recherche scientifique,
Marseille,
France

Abstract

THERMODYNAMIC STUDIES ON LIQUID Mg-In-Sn TERNARY SOLUTIONS.

The thermodynamic properties of Mg-In-Sn liquid solutions were determined for t = 0.03, 0.07, 0.1, 0.2, 0.4 and 0.6 (where t = $X_{Sn}/(X_{Sn}+X_{In})$) and at various magnesium concentrations by the EMF method, using concentration cells with a solid magnesium reference electrode. In addition, in dilute magnesium solutions for t = 0.0881, 0.1613, 0.2250 and 0.2795, integral enthalpies obtained with a sensitive calorimeter were used in calculations of the partial enthalpy of magnesium. Data from EMF and calorimetric studies were analysed and used with previous results obtained for the Mg-In system for presentation of Wagner's linear equations:

$$\ln \gamma_{Mg(Mg\text{-}In\text{-}Sn)} = \ln \gamma^0_{Mg(In)} + X_{Mg(In)}\, \epsilon^{Mg}_{Mg} + X_{Sn}\, \epsilon^{Sn}_{Mg}$$

$$\Delta \bar{H}_{Mg(Mg\text{-}In\text{-}Sn)} = \Delta \bar{H}^0_{Mg(In)} + X_{Mg(In)}\, \rho^{Mg}_{Mg} + X_{Sn}\, \rho^{Sn}_{Mg}$$

1.   INTRODUCTION

In previous investigations [1] EMF and calorimetric techniques were analysed and compared in application to liquid Mg-In solutions; the magnitude of the respective errors suggested the acceptance of partial excess free energies from the EMF technique and partial enthalpies from the calorimetric method in subsequent thermodynamic calculations. It was decided that it would be of interest to extend such analyses to ternary magnesium-indium-tin liquid alloys.

TABLE I.   THERMODYNAMIC DATA OF LIQUID Mg-In-Sn SOLUTIONS FROM EMF STUDIES

X_{Mg}	X_{In}	X_{Sn}	t	$E = a + bT$ (see footnote a)		$\Delta \bar{G}_{Mg}^{*}$ (923 K)	$\Delta \bar{H}_{Mg}$
				a	b	(cal/mol)	(cal/mol)
0.03	0.94	0.03		103.727	0.164	− 5323.6	− 4785
0.05	0.92	0.03	0.03	105.074	0.129	− 4834.8	− 4847
0.07	0.90	0.03		98.668	0.115	− 4593.4	− 4552
0.10	0.87	0.03		101.028	0.090	− 4255.3	− 4661
0.03	0.90	0.07		100.643	0.172	− 5532.0	− 4643
0.05	0.88	0.07	0.07	91.049	0.154	− 5268.9	− 4200
0.10	0.84	0.06		77.086	0.126	− 4696.6	− 3556
0.03	0.87	0.10		77.230	0.203	− 5793.1	− 3563
0.05	0.85	0.10	0.10	73.743	0.176	− 5401.3	− 3402
0.10	0.81	0.09		97.510	0.106	− 4795.4	− 4498
0.03	0.78	0.19		96.194	0.190	− 6095.7	− 4438
0.05	0.76	0.19		93.227	0.161	− 5672.9	− 4301
0.10	0.72	0.18		83.474	0.134	− 5283.0	− 3851
0.20	0.64	0.16	0.2	73.492	0.098	− 4615.9	− 3390
0.30	0.56	0.14		58.604	0.084	− 4087.0	− 2703
0.40	0.48	0.12		53.846	0.060	− 3351.2	− 2484
0.50	0.40	0.10		45.759	0.038	− 2459.2	− 2111
0.03	0.58	0.39		122.014	0.175	− 6659.3	− 5629
0.05	0.57	0.38		106.395	0.166	− 6481.0	− 4908
0.07	0.56	0.37		94.887	0.157	− 6211.4	− 4377
0.10	0.54	0.36	0.4	104.091	0.126	− 5989.5	− 4800
0.20	0.48	0.32		86.620	0.098	− 5222.8	− 3996
0.30	0.42	0.28		82.081	0.072	− 4667.5	− 3787
0.40	0.36	0.24		71.511	0.050	− 3733.3	− 3299
0.03	0.39	0.58		135.127	0.171	− 7091.3	− 6234
0.05	0.38	0.57		125.980	0.155	− 6941.8	− 5812
0.07	0.37	0.56	0.6	116.654	0.144	− 6657.1	− 5381
0.10	0.36	0.54		120.239	0.119	− 6380.2	− 5547
0.20	0.32	0.48		100.944	0.103	− 6088.8	− 4657
0.30	0.28	0.42		90.329	0.082	− 5464.4	− 4167

[a]   Units of measurement: E(mV); T(K).

The aim of the present paper is to report thermodynamic data deriving from the two experimental methods, in order to calculate interaction parameters of both free energy and enthalpy of magnesium, and to apply Wagner's linear equations [2] for dilute magnesium solutions in indium when studying the effect of tin additions.

2.   EXPERIMENTAL RESULTS

EMF data were obtained with cells of the type:

$$Mg(s) \mid MgCl_2 \text{ in } LiCl-KCl_{(eut)} \, (\ell) \mid Mg\text{-}In\text{-}Sn \, (\ell) \tag{1}$$

The EMFs were measured against a solid reference magnesium electrode, but liquid magnesium was chosen as a standard state. Hence appropriate terms were introduced into the mathematical evaluation of the thermodynamic functions to compensate. Measurements were carried out at different t-values (where $t = X_{Sn}/(X_{Sn} + X_{In})$) at temperatures between 770 and 890 K. The experimental arrangements were the same as those described in Ref. [3].

For each composition a linear representation of the temperature dependence of the EMF was obtained by a least-squares fit to the data. From linear equations, EMF data were calculated to give the partial excess free energy of magnesium at 923 K $(\Delta \bar{G}^*_{Mg})$ and the partial enthalpy $(\Delta \bar{H}_{Mg})$ as presented in Table I.

In addition, for dilute magnesium and moderately dilute tin solutions, integral enthalpies were measured by a high-temperature Calvet calorimeter. These measurements were made at t = 0.0881, 0.1613, 0.2250 and 0.2795.

3.   DISCUSSION

Figure 1 shows the plotted values (points) of $\Delta \bar{G}_{Mg}$ versus t at 923 K. Values between $0 \leqslant t \leqslant 0.2$ refer to dilute tin and magnesium solutions and were worked out to estimate precisely the ternary interaction parameter ϵ^{Sn}_{Mg}. This was not possible with ϵ^{In}_{Mg} (measurements when $t > 0.6$), since at temperatures below the melting point of magnesium (reference electrode) some intermetallic compounds precipitate, as was seen from the EMF versus temperature curves. Experimental points for the binary system, t = 0 (Mg-In) and t = 1 (Mg-Sn) were taken from previous investigations. The data for Mg-In [1] at $X_{Mg} = 0.03$ to 0.6 were fitted by a polynomial, and the values for Mg-Sn [4] at $X_{Mg} = 0.03$ to 0.35 using Sharma's [5] results for higher magnesium concentrations were treated similarly.

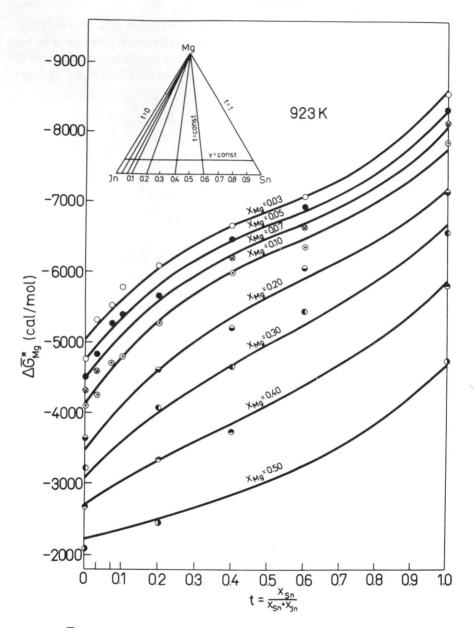

FIG.1. $\Delta \overline{G}_{Mg}^*$ values as function of t at 923 K in liquid Mg-In-Sn solutions from EMF studies (points) and from binary data from Refs [1, 4, 5]. Continuous lines are from Eq. (2), while the compositional triangle shows alloys investigated.

Binary data were worked out with 30 experimental $\Delta\bar{G}^*_{Mg}$ values obtained at 923 K with the Mg-In-Sn system (Table I). An analytical method of Pelton and Flengas [6] was employed. This method allows of the application of the Gibbs-Duhem equation analytically to establish equations for all partial and integral functions. As an example we present the polynomial equation for $\Delta\bar{G}^*_{Mg(Mg-In-Sn)}$, with two parameters t and y (where y = $1-X_{Mg}$) in the form:

$$\Delta\bar{G}^*_{Mg(Mg-In-Sn)}\ (923\ K) = (-10299.8 + 53379.3\ t - 45472.7\ t^2 - 6418.7\ t^3)y^2$$

$$+\ (-16461.6 - 205244.1\ t + 139097.5\ t^2)y^3$$

$$+\ (56239.4 + 196922.6\ t - 84663.1\ t^2)y^4$$

$$+\ (-35053.5 - 51267.0\ t)y^5 \tag{2}$$

A good fit was obtained with 12 coefficients (standard deviation, 95 cal). The $\Delta\bar{G}^*_{Mg}$ data calculated from Eq. (2) for binaries (t = 0 and t = 1) and for the Mg-In-Sn ternary systems are plotted in Fig. 1 as continuous curves. The experimental data of Mg-In [1] and Mg-Sn [4] from separate EMF studies are also plotted.

Experimental data for the Mg-In-Sn ternary system and those data from Eq. (2) are in agreement except at t = 0, where some differences appear between the $\Delta\bar{G}^*_{Mg}$ extrapolated from the ternary system and the experimental $\Delta\bar{G}^*_{Mg}$ of binary Mg-In. There is a similar tendency when determining the ternary interaction parameter ϵ^{Sn}_{Mg} from experimental data. To obtain this parameter, experimental values of $\Delta\bar{G}^*_{Mg}$ from Table I for dilute magnesium solutions (X_{Mg} = 0.03 − 0.1) were recalculated into $\ln \gamma_{Mg}$ for t = 0.03, 0.07, 0.1 and 0.2 and linearly extrapolated to X_{Mg} = 0. The values of $(\ln \gamma_{Mg})_{t\ =\ const.}$ thus obtained were plotted against X_{Sn} in Fig. 2 and

$$\epsilon^{Sn}_{Mg} = \left(\frac{\partial \ln \gamma_{Mg}}{\partial X_{Sn}}\right)_{\substack{X_{Sn}=0 \\ X_{Mg}=0}} = -3.66 \pm 0.5$$

was obtained as a slope. In Fig. 2, points refer to extrapolated experimental data, while the broken line results from an extrapolation from Eq. (2) at t ≤ 0.2, X_{Mg} = 0, the slope of which gives ϵ^{Sn}_{Mg} = 3.40, close to the experimental value. However, the value of $\ln \gamma^0_{Mg(In)}$ at 923 K deriving from binary studies, −2.72, is different from the value of −2.99 extrapolated from ternary experimental data and from −3.04 extrapolated from Eq. (2).

Wagner's linear equation for the ternary Mg-In-Sn system may then be presented in the form:

$$\ln \gamma_{Mg(Mg-In-Sn)}\ (923\ K) = -\ 2.72 + 4.92\ X_{Mg} - 3.66\ X_{Sn} \tag{3}$$

TABLE II. CALORIMETRIC INTEGRAL ENTHALPY IN LIQUID Mg-In-Sn SYSTEM AND CALCULATED PARTIAL ENTHALPY OF MAGNESIUM

X_{Mg}	X_{In}	X_{Sn}	t and ref. temp.	$\Delta\overline{H}'$ (cal/mol)	$\Delta\overline{H}_{Mg}$ (cal/mol)
0.009	0.904	0.087		− 62.3	− 5900
0.014	0.899	0.087		− 88.5	− 5302
0.018	0.896	0.086		− 109.1	− 5337
0.024	0.890	0.086	t = 0.0881	− 141.8	− 5259
0.029	0.886	0.085	947 K	− 165.9	− 5327
0.034	0.881	0.085		− 194.7	− 5420
0.040	0.875	0.084		− 227.1	− 5316
0.045	0.871	0.084		− 250.2	− 5183
0.048	0.868	0.084		− 271.0	− 5651
0.052	0.865	0.083		− 289.7	− 5246
0.056	0.861	0.083		− 314.5	− 5547
0.008	0.830	0.162		− 49.6	− 6375
0.012	0.827	0.161		− 71.1	− 5945
0.017	0.822	0.160		− 101.6	− 5768
0.026	0.815	0.159	t = 0.1613	− 152.4	− 5542
0.039	0.804	0.157	947 K	− 223.8	− 5557
0.045	0.799	0.156		− 256.9	− 5542
0.050	0.795	0.155		− 281.1	− 5474
0.055	0.791	0.154		− 308.6	− 5449
0.008	0.769	0.223		− 55.8	− 5975
0.015	0.763	0.222		− 94.8	− 5770
0.020	0.760	0.220	t = 0.2250	− 121.3	− 5756
0.025	0.756	0.219	987 K	− 150.2	− 5643
0.031	0.751	0.218		− 185.3	− 5640
0.042	0.742	0.216		− 255.1	− 5640
0.009	0.714	0.277		− 69.4	− 5802
0.016	0.709	0.275		− 113.7	− 6598
0.019	0.707	0.274		− 132.8	− 5768
0.023	0.704	0.273		− 155.8	− 5951
0.028	0.700	0.272	t = 0.2795	− 183.2	− 5677
0.031	0.698	0.271	947 K	− 202.4	− 5594
0.036	0.694	0.269		− 233.8	− 6287
0.041	0.691	0.268		− 260.3	− 5612
0.045	0.687	0.268		− 286.2	− 5708
0.049	0.685	0.266		− 305.7	− 5798

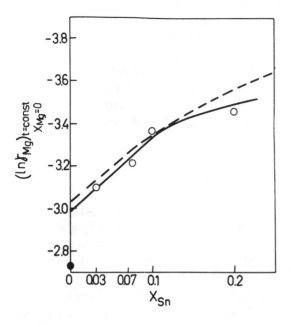

FIG.2. $\cdot(\ln \gamma_{Mg})_{t=const.}$ versus X_{Sn} at $X_{Mg}=0$
923 K for the liquid Mg-In-Sn system,
based on extrapolation of experimental
data (points) and Eq. (2) (broken line).

The first two terms in Eq. (3) (namely $\ln \gamma^0_{Mg(In)} = -2.72$ and $\epsilon^{Mg}_{Mg} = 4.92$) are taken from a previous paper on the Mg-In system [1]. Equation (3) reproduces experimental values of $\ln \gamma_{Mg}$ in dilute magnesium solutions in the ternary system at t = 0.03 to 0.2 with an accuracy of a few per cent. Error analysis performed for $\Delta\bar{G}^*_{Mg}$ and $\Delta\bar{H}_{Mg}$ in the ternary system by linear regression analysis using variances of average estimated values (as in Ref. [7] for the Zn-Ga system) has shown that errors in $\Delta\bar{G}^*_{Mg}$ are of the order of ± 2% to ± 5%, while errors in $\Delta\bar{H}_{Mg}$ calculated from temperature coefficients can even exceed ± 50% in some alloys. It seems, therefore, that in the thermodynamics of alloys, enthalpies (integral and partial) ought to be obtained from a calorimetric method. For certain systems, such as in the previously tested Mg-In system and, in this work, the Mg-In-Sn system, it is not possible to obtain $\Delta\bar{H}_{Mg}$ directly from calorimetric measurements. This is due to oxygen in the liquid indium (or tin) and the reaction of the oxygen with the added magnesium. Therefore, partial enthalpies of magnesium were obtained from integral enthalpies.

Integral enthalpies in dilute magnesium solutions at t = 0.0881, 0.1613, 0.2250 and 0.2795 were determined using a high-temperature Calvet calorimeter. The resulting data are summarized in Table II. They were fitted by a polynomial equation and used for computations of the partial enthalpy of magnesium, $\Delta\bar{H}_{Mg}$, using the Gibbs-Duhem equation. Values of $\Delta\bar{H}_{Mg}$ for various values of t were described by linear equations and extrapolated to $X_{Mg} = 0$. $(\Delta\bar{H}_{Mg})_{t=const.}$ data $X_{Mg}=0$

were used for calculations of the ternary interaction parameter of enthalpy ρ_{Mg}^{Sn}, as in the case of ϵ_{Mg}^{Sn}. In this manner the limiting partial enthalpy, $\Delta\bar{H}_{Mg}^{0} = -5432$ cal/mol, and $\rho_{Mg}^{Sn} = -2668$ were obtained. For comparison, the result for the binary system, $\Delta\bar{H}_{Mg(In)}^{0} = -5303$ cal/mol. Accepting the latter value and a binary interaction enthalpy parameter of $\rho_{Mg}^{Mg} = 11\,502$ from previous calorimetric studies on the Mg-In system [1], Wagner's equation for the partial enthalpy of magnesium in dilute solutions of magnesium in indium for the Mg-In-Sn ternary system takes the form:

$$\Delta H_{Mg(Mg-In-Sn)}(cal/mol) = -5303 + 11502\,X_{Mg} - 2668\,X_{Sn} \qquad (4)$$

In dilute magnesium solutions, values of $\Delta\bar{H}_{Mg}$ calculated from Eq. (4) and compared with the same experimental data recalculated from the integral enthalpy show differences of the order of $\pm 7\%$. As in the case of the Mg-In system, calorimetric values of $\Delta\bar{H}_{Mg}$ are from 400–600 cal more exothermic than those obtained from EMF temperature coefficients. In the case of calorimetric measurements, integral enthalpies were obtained with an accuracy of $\pm 3\%$. Recalculation to $\Delta\bar{H}_{Mg}$ introduces an error of about $\pm 7\%$ and, taking into account the extrapolation to obtain ρ_{Mg}^{Sn}, it seems that values of $\Delta\bar{H}_{Mg}$ calculated from Eq. (4) in dilute magnesium solutions are estimated with an error of $\pm 15\%$. It is, however, about three times as precise as the data for $\Delta\bar{H}_{Mg}$ obtained from EMF studies and, therefore, the calorimetric technique is recommended for enthalpy measurements.

Calorimetric studies with higher magnesium concentrations for $t > 0.3$ will be continued to obtain further comparisons of results with those from EMF studies and further thermodynamic calculations.

REFERENCES

[1]   MOSER, Z., CASTANET, R., paper submitted to Metall. Trans., 1978.
[2]   WAGNER, C., Thermodynamics of Alloys, Addison-Wesley Publ., Cambridge, Massachusetts (1952).
[3]   MOSER, Z., Metall. Trans. **5** (1974) 1445.
[4]   MOSER, Z., FITZNER, K., Thermodynamics of Nuclear Materials 1974 (Proc. Symp. Vienna, 1974) Vol. 2, IAEA, Vienna (1975) 379.
[5]   SHARMA, R.A., J. Chem. Thermodyn. **2** (1970) 373.
[6]   PELTON, A.D., FLENGAS, S.N., Can. J. Chem. **47** (1969) 2283.
[7]   MOSER, Z., Metall. Trans. **4** (1973) 2399.

DISCUSSION

V.V. AKHACHINSKIJ: What is the reason for the comparatively high error of the calorimetric measurements? The Calvet method gives results of higher accuracy (the error is usually a good deal less than ±3%).

Z. MOSER: A high-temperature Calvet calorimeter (T < 1400 K) usually gives an accuracy better than ±3%. However, in the case of the Mg-In-Sn alloys such an error in integral enthalpies is expected because of side reactions between the added magnesium and the oxygen in liquid indium. This was verified when we attempted calorimetric determinations of partial magnesium enthalpies in our previous investigation of Mg-In liquid alloys (paper submitted for publication in Metallurgical Transactions).

A. NAOUMIDIS: Could you please comment briefly on the background of the investigation of the Mg-In-Sn system?

Z. MOSER: The Mg-In-Sn system was investigated in order:

(a) To test some methods of interpreting experimental results on the basis only of the binary system (Krupkowski's formalism); our previous results were obtained from the binary systems Mg-In, Mg-Sn and In-Sn;

(b) To accumulate selected thermodynamic data, i.e. $\Delta\bar{G}^*_{Mg}$ from EMF and $\Delta\bar{H}_{Mg}$ from calorimetric methods, for application of Wagner's linear equations in dilute magnesium solutions and for determination of binary and ternary interaction parameters of both free energy and enthalpy, and also for precise excess entropy calculations;

(c) To use the selected $\Delta\bar{G}^*_{Mg}$ and $\Delta\bar{H}_{Mg}$ data in our future phase-diagram calculations;

(d) To observe the deviation between extrapolated ternary and experimental binary end points on plots of $\Delta\bar{G}^*_{Mg}$ versus t and $(\ln \gamma_{Mg})_{X_{Mg}=0, \; t=\text{const.}}$ versus X_{Sn}. These deviations were also observed in our previous extensive studies on ternary dilute zinc solutions.

INVESTIGATION OF REACTION EQUILIBRIUM IN REACTOR MATERIALS BY EMF METHODS

H. ULLMANN, K. TESKE, T. REETZ, D. RETTIG
Central Institute for Nuclear Research,
Academy of Sciences of the GDR,
Rossendorf, Dresden

F.A. KOZLOV, E.K. KUZNETSOV
Institute of Physics and Power Engineering,
Obninsk,
Union of Soviet Socialist Republics

Abstract

INVESTIGATION OF REACTION EQUILIBRIUM IN REACTOR MATERIALS BY
EMF METHODS.
　　Measurements were made of the chemical activities of oxygen and hydrogen in a
sodium test loop by means of electrochemical cells with solid electrolytes. The reaction
equilibrium of oxygen and hydrogen in dilute solution in sodium was investigated. The
activities of both oxygen and hydrogen decrease with increasing concentration of the
partner in the reaction. From the relation between the activity of one component and the
analytical concentration of the partner in the reaction, the equilibrium constant of the
reaction $O + H = OH$ was determined to be:

$$\log K_{diss} = -(1.502 \pm 0.216) - (1356 \pm 140)/T$$

A discussion of an electrochemical cell using a solid electrolyte and an iron membrane which
is being used to measure the activity of carbon in liquid sodium under carburizing conditions
follows.

1. INTRODUCTION

　　At elevated temperatures, such as exist in coolant circuits of a LMFBR,
chemical reactions and mass transfer processes within materials and between
materials in contact proceed at an increased rate. Thus, states of chemical
equilibrium are obtained locally, and steady-state conditions of mass transfer
are observed in larger systems. The type of chemical reactions and the state of
chemical equilibrium can only be estimated using high-temperature methods of
measurement, such as the solid electrolyte methods. In this paper examples
are presented of work undertaken to determine chemical equilibria in liquid
sodium.

2.   METHODS OF MEASUREMENT

The construction of the solid-electrolyte cells has been described elsewhere. The cell for the determination of the oxygen activity consists of a thoria-yttria electrolyte (Fig.1a) [1]. The oxygen activity in sodium, related to the state of saturation, is given by:

$$a_O = \frac{x_O^0}{x_{O(sat)}} \cdot f_O^H \tag{1}$$

The coefficient of interaction, f_O^H, expresses the change of oxygen activity in solution by interaction with further dissolved components x_i, in our example with hydrogen. The interaction coefficient derives from the change in potential of the solid-electrolyte cell, measured in the presence of the hydrogen in the solution, in comparison with the cell potential measured in a binary Na-O solution:

$$\log f_O^H = -\Delta U \cdot \frac{2F}{R} = -10.08 \; \Delta U \tag{2}$$

where F is the Faraday constant and R is the universal gas constant.

The activity of hydrogen in sodium is:

$$a_H = \frac{x_H^0}{x_{H(sat)}} \cdot f_H^O \tag{3}$$

The hydrogen activity in sodium is in equilibrium with the partial pressure of the hydrogen in the carrier gas flowing over a nickel diffusion membrane, in accordance with a relation analogous to Henry's law, expressed by:

$$x_{H_2} = k \cdot a_H^2 \tag{4'}$$

and

$$x_{H_2} = k' \cdot (x_H^0 \cdot f_H^O)^2 \tag{4''}$$

The content of hydrogen in the steady stream of carrier gas is titrated coulometrically with oxygen by means of a zirconia solid-electrolyte cell (Fig.1b) [2].

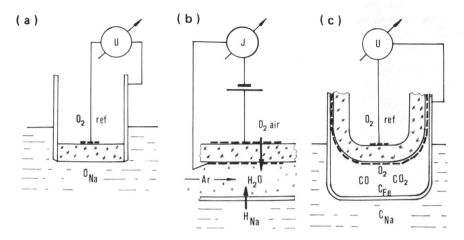

FIG.1. *Schematic view of the cell construction for the detection of: (a) oxygen activity,*
(b) hydrogen activity and (c) carbon activity in sodium.

The activity of carbon in the fluid is in equilibrium with the gaseous
$CO/CO_2/O_2$ phase (Fig.1c) [3] via an iron membrane. The oxygen potential
of the equilibrium is measured potentiometrically by a zirconia solid-electrolyte
cell (T = const):

$$U = k_1 + k_2 \log a_C \qquad\qquad (5)$$

3.   INVESTIGATION OF THE O + H = OH INTERACTION
     IN LIQUID SODIUM

3.1. **Fundamental considerations**

In our example, the reaction between oxygen and hydrogen in dilute solution
in sodium was investigated by determining the relation between the interaction
coefficient and the concentration of the second partner in the reaction. The
differential quotients

$$\frac{\partial \ln f_H^O}{\partial x_O^0} \qquad \text{and} \qquad \frac{\partial \ln f_O^H}{\partial x_H^0}$$

correspond to the interaction parameter ϵ defined by Wagner [4]. At low concentrations the linear relationships, to an approximation, are:

$$\ln f_H^O = \epsilon_H^O \cdot x_O^0 \tag{6'}$$

and

$$\ln f_O^H = \epsilon_O^H \cdot x_H^0 \tag{6''}$$

With $\epsilon \cdot x \ll 1$, the simplification $e^{\epsilon \cdot x} \simeq 1 + x$ then applies for the low concentration range. Hence, approximating, we obtain for the f-values:

$$f_H^O \simeq 1 + \epsilon_H^O \cdot x_O^0 \tag{7'}$$

and

$$f_O^H \simeq 1 + \epsilon_O^H \cdot x_H^0 \tag{7''}$$

The chemical interaction between oxygen and hydrogen in sodium is assumed to be the association reaction $O + H = OH$. The equilibrium concentrations of the reaction partners are then:

$$\left.
\begin{aligned}
x_{OH} &= x_H^0 (1 - f_H^O) = x_O^0 (1 - f_O^H) = - x_H^0 \epsilon_H^O x_O^0 = - x_O^0 \epsilon_O^H x_H^0 \\[2mm]
x_O &= x_O^0 - x_{OH} = x_O^0 (1 - \epsilon_O^H x_H^0) \simeq x_O^0 \\[2mm]
x_H &= x_H^0 - x_{OH} = x_H^0 (1 - \epsilon_H^O x_O^0) \simeq x_H^0
\end{aligned}
\right\} \tag{8}$$

The relation for the concentration of the hydroxide specimen, x_{OH}, shows, that $\epsilon_H^O = \epsilon_O^H = \epsilon$ is valid at constant temperature. The equilibrium constant of the reaction at low concentrations is then given by:

$$K_{diss} = \frac{x_O \cdot x_H}{x_{OH}} \simeq - \frac{1}{\epsilon} \tag{9}$$

3.2. Experimental arrangement

The oxygen activity was measured with solid-electrolyte cells at 350 and 400°C in an 80 ltr sodium loop [5]. The hydrogen concentration of the gaseous

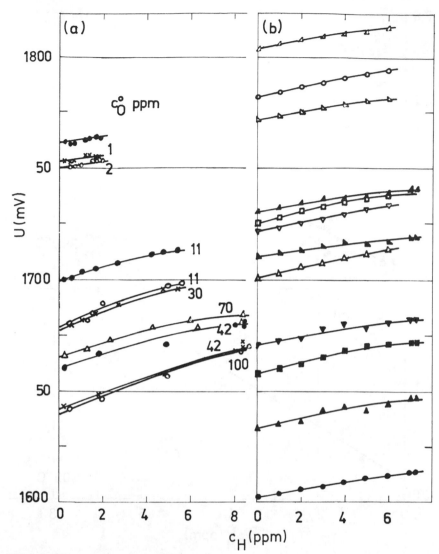

FIG.2. Change of the cell potential of the oxygen-activity measuring cell after injecting
hydrogen into sodium at various oxygen concentrations (differently coded points represent
different cells). (a) 350°C; (b) 400°C (open points: 16 ppm O; full points: 115 ppm O).

phase was determined coulometrically in the carrier gas stream after passing
through a nickel diffusion gauge, operating in sodium at 500°C. The oxygen
and hydrogen contents in the sodium were varied by changing the quantity of
the gases introduced. The concentration of oxygen was analysed by the vacuum
distillation method. The hydrogen concentration in the sodium was found by
computation, knowing the volume of gas introduced.

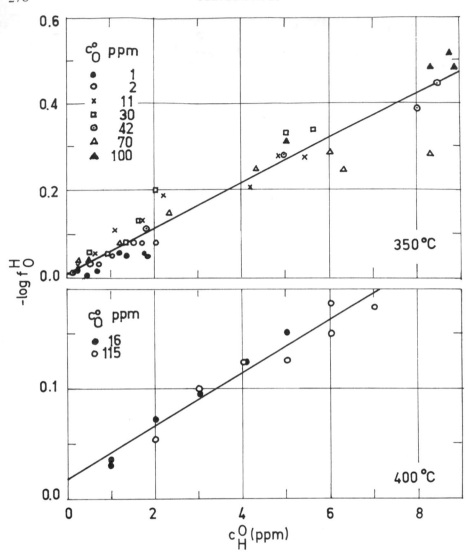

FIG.3.   *Logarithm of the interaction coefficients, f_O^H, as a function of hydrogen concentration in sodium.*

To determine the change in the oxygen activity as a function of hydrogen concentration, the sodium was purified by cold trapping. After shutting down the cold trap, a chosen concentration of oxygen was obtained by introducing a controlled amount of oxygen. Hydrogen under pressure was then continuously fed into the Na-O solution via a nickel membrane ($\sim$ 10 ppm over a period of 2 to 3 hours). The dependence of the hydrogen activity on the oxygen concentration was determined by two different tests. In one, the dependence of

TABLE I. TEST VALUES FOR THE CONCENTRATION OF HYDROGEN, x_{H_2}, IN THE CARRIER GAS AND OF HYDROGEN, c_H^0, AND OXYGEN, c_O^0, IN SODIUM

c_O^0 (ppm)	Mole fraction, x_O^0	Mole fraction, x_{H_2}	c_H^0 (ppm)	Mole fraction, x_H^0
MODE 1				
1.8	0.26×10^{-5}	1.50×10^{-5}	0.51	1.17×10^{-5}
		4.25×10^{-5}	0.82	1.89×10^{-5}
		5.85×10^{-5}	1.05	2.42×10^{-5}
		8.70×10^{-5}	1.23	2.83×10^{-5}
		11.5×10^{-5}	1.46	3.36×10^{-5}
		1.01×10^{-5}	0.42	0.97×10^{-5}
		3.17×10^{-5}	0.72	1.66×10^{-5}
		5.10×10^{-5}	0.90	2.07×10^{-5}
		7.52×10^{-5}	1.12	2.58×10^{-5}
56	8.06×10^{-5}	3.28×10^{-5}	0.85	1.96×10^{-5}
		5.82×10^{-5}	1.17	2.69×10^{-5}
		8.60×10^{-5}	1.46	3.36×10^{-5}
MODE 2				
20	2.88×10^{-5}	4.30×10^{-5}	0.90	2.07×10^{-5}
32	4.60×10^{-5}	3.70×10^{-5}	0.88	2.00×10^{-5}

the hydrogen content in the argon carrier gas on the amount of hydrogen introduced into the sodium at various fixed concentrations of oxygen was measured. In the other test, oxygen was injected into the sodium having a constant concentration of hydrogen; it was again the resulting hydrogen content in the argon carrier gas that was measured.

3.3. Results and discussion

At all concentrations of oxygen an increase of the cell voltage with increasing hydrogen content in the sodium was observed (Figs 2a and b). The interaction coefficients, f_O^H, calculated from Eq.(2), are shown in Fig.3 as function of the hydrogen concentration in the sodium. The dependence of the

TABLE II. COEFFICIENTS AND PARAMETERS OF INTERACTION, CONSTANTS OF EQUILIBRIUM AND FREE ENERGIES OF FORMATION FOR THE REACTION O + H = OH IN DILUTE SOLUTION IN LIQUID SODIUM (c in ppm)

	623 K	673 K	773 K
$\log f_H^O$	$-(0.051 \pm 0.002)\, c_H^0$	$-(0.024 \pm 0.004)\, c_H^0$	$-(0.0013 \pm 0.00015)\, c_O^0$
ε_H^O (mole fraction)$^{-1}$	$-(0.012 \pm 0.009)$	$-(0.018 \pm 0.016)$	
$\log f_O^H$			
ε_O^H (mole fraction)$^{-1}$	$-(5110 \pm 220)$	$-(2394 \pm 354)$	$-(2040 \pm 235)$
K_{diss} (mole fraction)	$(1.96 \pm 0.09) \times 10^{-4}$	$(4.17 \pm 0.70) \times 10^{-4}$	$(4.90 \pm 0.60) \times 10^{-4}$
$\Delta G^0_{OH(Na)}$ (kJ/mol)	$-(44.2 \pm 0.3)$	$-(43.6 \pm 0.7)$	$-(49.0 \pm 1.0)$

hydrogen content in argon on the amount of hydrogen in the sodium measured at oxygen concentrations of 1.8 (after cold trapping) and 56 ppm oxygen (after oxygen injection) are given in Table I (first mode of testing). Further, this table also contains the values of hydrogen contents in the carrier gas measured after injection of oxygen into the sodium containing defined hydrogen concentrations (second mode of testing). A graph of $\ln (x_{H_2}/(x_H^0)^2)$ plotted against the oxygen concentration yields a straight line. By correlation (confidence interval 95%, 14 pairs of values) this becomes:

$$\ln (x_{H_2}/(x_H^0)^2) = -(4079 \pm 470) x_O^0 + (11.62 \pm 0.019) \qquad (10)$$

This results in the relation for $\log f_H^0$ given in Table II. For $\log f_O^H$, the relation was calculated from the pairs of values shown in Fig.3 by correlation (confidence interval 95%; 350°C: 46 pairs of values; 400°C: 13 pairs of values).

Then, the values of ϵ, K_{diss}, and the free energy of formation ΔG_{OH}^0 may be obtained from the relationship for the interaction coefficients. The values are only valid at low concentrations, up to 5×10^{-5} mole fractions of hydrogen and 1×10^{-4} mole fractions of oxygen (corresponding to ~ 2 ppm H and ~ 70 ppm O) in sodium, due to the approximations used above. The temperature-dependent functions of ΔG_{OH}^0 and $\log K_{diss}$ were calculated from the values for the three temperatures investigated[1] (Fig.4):

$$\Delta G_{OH}^0 \text{ (J/mol)} = -(25\,967 \pm 2681) - (28.762 \pm 4.136) \, T \qquad (11)$$

$$\log K_{diss} \quad = -(1.502 \pm 0.216) - (1356 \pm 140)/T \qquad (12)$$

Consequently, the temperature dependence of the reaction in this solution is small. The interaction decreases slightly with increasing temperature. The value of the interaction coefficients is an expression for a negative deviation from Henry's law for solutions of oxygen and hydrogen in sodium.

Measurements of the hydrogen equilibrium pressures after injection of water and hydroxide into sodium yield qualitatively similar results [7]. By calorimetric investigation, Privalov [8] found a nearly temperature-independent value of $K_{diss} \simeq 10^{-4}$ (mole fraction) in the vicinity of a saturated solution. On the other hand, Katsuta [9] reports markedly temperature-dependent values of the free energy of formation $(\Delta G_{OH}^0 \text{ (J/mol)} = -117000 + 180T)$ for saturated solutions between 110 and 180°C. We compared our values of the heat of

---

[1] The temperature-dependent functions given in Ref.[6] were calculated only from values for two temperatures.

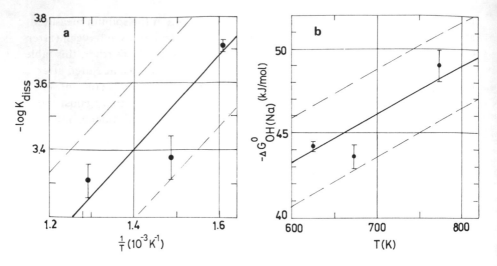

FIG.4. Variation with temperature of (a) the constant of dissociation and (b) the free energy of formation.

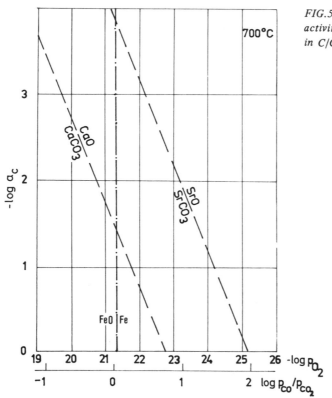

FIG.5. Relation between carbon activity and oxygen partial pressure in C/CO/CO₂ systems (calculated).

TABLE III. HEAT OF FORMATION, ΔH^0, AND OF SOLUTION, L,
IN THE SYSTEM Na-O-H

		ΔH^0 or L (J/mol)	Ref.
Heat of formation of:	Na$_2$O	$-$ 420 600 ± 5024	[10]
	NaH	$-$ 59 870	[11]
	NaOH	$-$ 428 310 ± 5024	[12]
Heat of solution of:	Na$_2$O	$+$ 45 720 ± 676	[13]
	NaH	$+$ 55 266	[14]
	NaOH	$+$ 33 870 ± 2300	[8]
Latent heat of melting of:	Na	$+$ 2 × (2 638 ± 42)	[12]

reaction with the heat of reaction resulting from a Born-Haber cycle as
expressed by:

$$L_{Na_2O} + L_{NaH} + \Delta H^0_{OH(Na)}$$

$$= \Delta H^0_{NaOH} - (\Delta H^0_{Na_2O} + \Delta H^0_{NaH}) + L_{NaOH} + 2\,L_{Na} \qquad (13)$$

A value of $-$ 9682 J/mol was obtained, using literature data for the heat values
of the elements and compounds (Table III).

4. EXPERIMENTS FOR THE DETERMINATION OF THE CARBON ACTIVITY

Efforts aimed at making an electrochemical determination of the thermodynamic activity of carbon in liquid sodium by means of molten salt electrolyte
cells have not yet resulted in a suitable device [3, 15]. Particular problems are
the slower diffusion of carbon in the metallic membranes (iron, nickel) and the
interaction between the membrane and the molten salt. Solid-electrolyte cells,
using a coupling of the carbon activity to the oxygen activity by observing a
carbon monoxide/carbon dioxide equilibrium might yield a useful result.

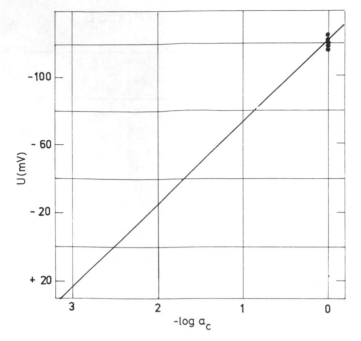

FIG.6.   *Voltage of the carbon cell*
(——— calculated from thermodynamic data;
● measured values).

Variants of such experiments have been described, for example using an encapsulated CO/CO_2 gas mixture [16] or an alkaline-earth-metal carbonate system [3, 15]. We have preferred the latter, encapsulated in an iron membrane.

The method must be capable of measuring the carbon activity over the range 1 to 10^{-3}. To avoid oxidation of the iron membrane owing to reaction with carbon dioxide, we adjusted the CO_2 pressure with the aid of a strontium oxide/strontium carbonate mixture (Fig.5). In addition, the oxygen pressure of the reference electrode was kept low to avoid disturbing the sensitive $C/CO/CO_2/O_2$ equilibrium by permeation of oxygen through the solid electrolyte. A calcium oxide/calcium carbonate/carbon reference system meets this condition.

Up to now we have measured cell potentials over a range of values calculated from thermodynamic data in gases or solid mixtures having a carbon activity of unity (Fig.6). Constant cell potential levels were obtained 8 hours after heating the cell to a temperature of 700°C. The cell output has been stable over a period exceeding 1000 hours. We are now starting experiments to determine carbon activities in decarburizing media.

5.   CONCLUSIONS

Electrochemical cells with solid electrolytes are suited for investigations of chemical equilibria in liquid sodium. Data for the reaction equilibrium of oxygen and hydrogen in liquid sodium over a concentration range of interest for normal operating conditions of LMFBR circuits were obtained. This investigation suggests that it might be possible to investigate other processes in liquid sodium that cause mass transfer in sodium circuits. The extension of the electrochemical methods to determine the carbon activity in liquid sodium is being investigated.

REFERENCES

[1]   ULLMANN, H., TESKE, K., REETZ, T., Kernenergie 16 (1973) 291.

[2]   TESKE, K., WITKE, W., KOZLOV, F.A., KUZNECOV, E.K., Kernenergie 20 (1977) 69.

[3]   RETTIG, D., IRMISCH, R., Kernenergie 20 (1977) 76.

[4]   WAGNER, C., Thermodynamics of Alloys, Addison-Wesley Publ., Cambridge, Massachusetts (1952) 51.

[5]   ULLMANN, H., REETZ, T., TESKE, K., KOZLOV, F.A., KUZNECOV, E.K., KOZUB, P.S., Kernenergie 18 (1975) 221.

[6]   TESKE, K., ULLMANN, H., KOZLOV, F.A., KUZNECOV, E.K., Kernenergie 21 (1978).

[7]   KOZLOV, F.A., KUZNETSOV, E.K., SERGEEV, P.G., VOLKOV, L.G., SOTOV, V.V., Tr. Fiz.-Ehnerg. Inst. Energeticeskovo Instituta (KUZNETSOV, W.A., Ed.), Atomizdat Moscow (1974) 290.

[8]   PRIVALOV, Yu.V., paper presented at the 'Rat für Gegenseitige Wirtschaftshilfe' specialist meeting on sodium technology, Central Institute for Nuclear Research, Rep. ZfK-337 (1977) 9.

[9]   KATSUTA, H., FURUKAWA, K., Nucl. Technol. 31 (1976) 219.

[10] ALCOCK, C.B., STAVROPOULOS, G.P., Can. Metall. Q. 10 (1971) 257 (*NB* calculated from values of Brewer and Kelley).

[11] HOBDELL, M.R., WHITTINGHAM, A.C., Central Electricity Generating Board, UK, Rep. CEGB-RD/B/N-2545 (1973).

[12] KUBASCHEWSKI, O., EVANS, E.Ll., ALCOCK, C.B., Metallurgical Thermochemistry, 4th ed., Pergamon Press, Oxford (1967).

[13] NODEN, J.D., Central Electricity Generating Board, UK, Rep. CEGB-RD/B/R-2146 (1972).

[14] VISSERS, R.D., HOLMES, J.T., BARTHOLME, L.G., NELSON, P.A., Nucl. Technol. 21 (1974) 235.

[15] ROY, P., Int. Conf. Liquid Metal Technology in Energy Production (Seven Springs, USA, 1976), American Nuclear Society, CONF-760 503-Summ (1976) VI B-8.

[16] RUTHER, W.E., SKLADZIEN, S.B., ROCHE, M.F., ALLEN, J.W., Nucl. Technol. 21 (1974) 75.

DISCUSSION

J.-P. MARCON: Could your experimental method be applied to the study of the oxygen/tritium chemical interaction in sodium? What can you say a priori about this interaction, as compared to that between oxygen and hydrogen?

H. ULLMANN: I cannot see any additional difficulty in applying our method to determination of the chemical activity of tritium or to investigation of the oxygen/tritium chemical interaction in liquid metals. No distinction between hydrogen and tritium will be possible by our method.

I suppose that the behaviour of tritium in the reaction with oxygen in liquid metals is very similar to that of hydrogen, assuming a weak isotopic effect in condensed phases.

A. NAOUMIDIS: Is it possible, from the reaction-rate point of view, to use this method at low pressures of CO and CO_2 for the measurement of the O_2 potential in an inert gas with CO/CO_2 impurities?

H. ULLMANN: Yes, oxygen equilibrium pressures can be measured in the range from 1 to about 10^{-20} atm in such gas mixtures with an arrangement such as shown in Fig.1b. The cell operates at a temperature of 700°C.

M. TETENBAUM: Low carbon activity values can be established for a given oxygen potential (defined by the CO/CO_2 ratio) by decreasing the CO pressure (the CO/CO_2 ratio remaining fixed). Have you considered this approach in your work?

H. ULLMANN: Yes, it is one of the methods of establishing certain defined low carbon activities in calibration experiments.

Section E

DIFFUSION STUDIES

INTERDIFFUSION OF KRYPTON AND XENON IN HIGH-PRESSURE HELIUM*

R.J. CAMPANA, D.D. JENSEN, B.D. EPSTEIN,
R.G. HUDSON, N.L. BALDWIN
General Atomic Company,
San Diego, California,
United States of America

Abstract

INTERDIFFUSION OF KRYPTON AND XENON IN HIGH-PRESSURE HELIUM.
The interdiffusion of gaseous fission products in high-pressure helium is an important
factor in the control of radioactivity in gas-cooled fast breeder reactors (GCFRs). As presently
conceived, GCFRs use pressure-equalized and vented fuel in which fission gases released from
the solid matrix of mixed oxide fuel are transported through the fuel rod interstices and internal
fission product traps to the fuel assembly vents, where they are swept away to external traps
and storage. Since the predominant transport process under steady-state operating conditions
is interdiffusion of gaseous fission products in helium, the diffusion properties of krypton-
helium and xenon-helium couples have been measured over the range of GCFR temperature
and pressure conditions (< 675 K and $\leqslant 8.7$ MPa). A high-pressure, high-temperature diffusion
apparatus incorporating scintillation detection of tagged krypton or xenon in helium was used
to measure interdiffusion rates. Interdiffusion values for krypton-helium and xenon-helium
couples were derived by fitting the resulting diffusion plots using the fission product diffusion
code SLIDER. Analysis of the data confirmed the theoretical inverse-pressure dependence
(P^{-1}) and expected temperature dependence to the 1.66 power ($T^{1.66}$) at lower pressures and
temperatures. Additional work is in progress to measure the behaviour of the krypton-helium
and xenon-helium couples in GCFR fuel rod charcoal delay traps.

INTRODUCTION

The interdiffusion of gaseous fission products in high pressure helium is an important factor in the control of radioactivity
in gas-cooled fast reactors. As presently designed, GCFRs use
pressure-equalized and vented fuel in which fission gases released
from the solid-matrix of mixed oxide fuel are transported through
the fuel rod interstices and internal fission product traps to the
fuel assembly vents, where they are swept away to external traps
and storage. The predominant transport process at steady-state

---

* Prepared under United States Department of Energy Contract EY–76–C–03–0167,
Project Agreement No. 23.

operating conditions is interdiffusion of gaseous fission products
in helium. Thus it is important that the gaseous diffusion properties of krypton-helium and xenon-helium couples be known over the
range of GCFR temperature and pressure conditions of interest
(<675 K and ≤8.7 MPa).

The interdiffusion coefficient developed from Chapman-Enskog
dilute gas theory is given by the following equation:

$$D_{AB} = 0.1883 \; \frac{T^{3/2} \left(\frac{1}{M_A} + \frac{1}{M_B} \right)^{1/2}}{P\sigma_{AB}^2 \; \Omega_{D,AB}} \tag{1}$$

where

$\quad D_{AB}$ $\quad$ = $\quad$ interdiffusion coefficient of gases A and B (cm^2/s)

$\quad T$ $\quad$ = $\quad$ temperature (K)

$\quad P$ $\quad$ = $\quad$ pressure (kPa)

$\quad M_A, M_B$ $\quad$ = $\quad$ molecular weight of components A and B

$\quad \sigma_{AB}$ $\quad$ = $\quad$ collision diameter (Å)

$\quad \Omega_{D,AB}$ $\quad$ = $\quad$ collision integral.

The collision integral $\Omega_{D,AB}$ is a function of both the temperature
and the intermolecular potential field ε_A and ε_B for molecules A
and B, respectively. The values of ε_{AB}, σ_{AB}, and Ω_{AB} were derived
using the usual combining laws, i.e.,

$$\sigma_{AB} = 1/2(\sigma_A + \sigma_B), \tag{2}$$

$$\varepsilon_{AB} = (\varepsilon_A \varepsilon_B)^{1/2}, \tag{3}$$

and the Lennard-Jones molecular parameters and collision integrals
presented in Tables B-1 and B-2 of Bird, Stewart and Lightfoot [1].
The values of D_{Kr-He} and D_{Xe-He} at standard conditions derived
from the above treatment are 0.566 and 0.470 cm^2/s, respectively.

To compare measurements under varying temperature and pressure conditions, Eq. 1 may be simplified to

$$D_{AB} = D_{AB,o} \left(\frac{T}{T_o} \right)^x \left(\frac{P_o}{P} \right)^y \tag{4}$$

where

$\quad D_{AB,o}$ $\quad$ = $\quad D_{AB}$ at T_o and P_o

$\quad T_o$ $\quad$ = $\quad$ standard temperature (273 K)

$\quad P_o$ $\quad$ = $\quad$ standard pressure (101 kPa)

The exponent x of the ratio of absolute temperatures has been found to have values ranging from 1.68 to 1.76 in experiments carried out at temperatures up to ~400 K [2-6]. Although use of Eq. 1 indicates only a small change in D_{AB} at higher temperatures, no data were available to confirm these calculations. In addition, the exponent y has been found to deviate from an inverse pressure dependence [7]. This unexpected result needed confirmation in the pressure range of interest for GCFR operation. It should be noted that although the uncertainties in the existing data appear to be small, they have an important impact on the predicted release of short-lived radioactive species undergoing simultaneous diffusion and decay in the narrow annuli of GCFR fuel rods.

A number of researchers have studied the interdiffusion behavior of krypton-helium and xenon-helium binary mixtures. Srivastava [2] determined xenon-helium interdiffusion coefficients as a function of temperature by allowing diffusion to occur between two bulbs separated by a precision capillary tube, deriving experimental values ~7% larger than those predicted by Eq. 1. Srivastava and Barua [3] and Srivastava and Paul [8] extended this work to the krypton-helium system and obtained D_{Kr-He} values which were in close agreement with predictions. The behavior of Kr-85 in helium at pressures up to 41.9 MPa was investigated by Durbin and Kobayaski [7] using ionization and scintillation detection techniques. Analysis of their data revealed that D_{Kr-He} is not inversely proportional to pressure, as predicted by Eq. 1, but instead is proportional to pressure raised to the -0.92 or -0.96 power. Weissman and Mason [4] used viscosity measurements to determine krypton-helium and xenon-helium diffusion coefficients and the temperature dependence of D_{Xe-He}. Malinauskus [5,6] used diffusion measurements to determine the temperature dependence of both diffusion couples up to 394 K. Van Heijningen, et al.,[9], investigated the effect of concentration and temperature and found that the diffusion coefficients exhibited only a small dependence on concentration over the entire spectrum of mole fractions of krypton or xenon in helium. Later work [10-12] confirmed the small concentration dependence.

Gaseous diffusion within fuel rod charcoal traps is an additional barrier to fission gas venting. The extent of holdup and hence radioactive decay that takes place in the traps is dependent upon charcoal type and grain size, trap operating temperature, trap geometry, and other factors. Underhill [13] and Bolch [14] have treated this problem theoretically and experimentally for ambient conditions. However, because of the high-pressure, high-temperature operating conditions of interest in this study, an extension of existing experimental data on diffusion in charcoal traps is required for accurate GCFR fission gas transport predictions.

EXPERIMENTAL

The method used to derive krypton-helium and xenon-helium interdiffusion coefficients was based upon measurements of the

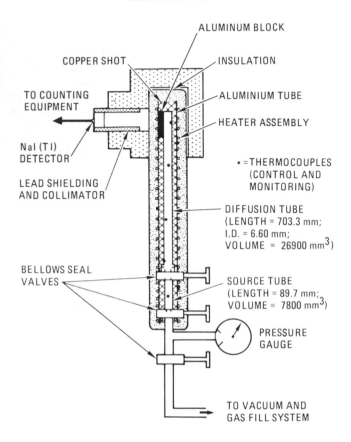

FIG.1. Diffusion apparatus incorporating gamma spectroscopy.

rate of krypton or xenon migration through a helium-filled diffusion tube. Three slightly different experimental arrangements were
utilized, each incorporating a krypton-helium or xenon-helium filled
source region, a helium-filled diffusion tube region, and a pressure-
monitoring region. Early experiments were carried out at room
temperature using a source at one end of the diffusion tube and an
ionization chamber connected to a Cary model 32 vibrating reed electro-
meter at the opposite end. The source and diffusion tube regions were
separated by a bellows-sealed valve capable of withstanding high
pressures and high temperatures. The diameter of the diffusion tube
was equal to that of a GCFR fuel rod. The advantage of this system
was that low concentrations of krypton in helium could be detected by
the decay of Kr-85. However, this system could not be welded, and it
was therefore prone to leakage at high pressure.

To circumvent this problem, a welded system incorporating bellows-sealed valves and gamma scintillation detection of Kr-85 or Xe-127 was used (Fig. 1). The source and diffusion regions were resistively heated, and temperatures were regulated with three-Chromel-Alumel thermocouples connected to Barber-Coleman model 520 solid-state controllers. Temperatures were monitored with five Chromel-Alumel thermocouples coupled to a Fluke model 2100A digital thermometer. The upper end of the diffusion tube was maintained 5 to 6 K higher than the source to minimize convective transport. The upper third of the diffusion tube assembly was surrounded by lead bricks providing shielding and acting as a collimator for a Canberra 50-mm-diameter NaI (Tℓ) scintillation detector. An aluminum block positioned at the top of the tube served as an additional collimator. The output of the detector was fed to a Canberra model 816 amplifier, model 830A single-channel analyzer, model 1481L log/lin rate meter, and model 1772 counter/timer; gross activity was recorded with a Heath model EU-20B recorder.

A third experimental design, assembled without bellows-sealed valves, was used to gather data at ambient conditions. This design was adopted since the dead space and flow path through the bellows-sealed valve of the high-pressure apparatus introduces uncertainties in modelling with the one-dimensional SLIDER code. To provide data more readily modeled by SLIDER, a ball valve was positioned between the source and the diffusion tube regions. When this valve was in the open position, it allowed diffusion of gases in a straight-through passage of constant cross section between the two regions. In addition, there was negligible dead space in the valve. This design was accurately modeled by SLIDER since the apparatus had a uniform cross sectional area along the entire length of the tube.

The helium used in this analysis was provided by Liquid Carbonic Company and had a stated purity of 99.995%. Kr-85 (half-life = 10.72 y) in helium and Kr-85 in stable krypton were obtained from Oak Ridge National Laboratory and Matheson Gas Products and were used at concentrations ranging from 1.25 to 5.25 μCi/cm^3(NTP). Xe-127 (half-life = 36.41 days) in helium was obtained from Isotope Products at a concentration of 16.5 μCi/cm^3(NTP) on the date of vendor analysis. The stated purity levels for all gas mixtures was at least 99.99%.

Experimental runs were initiated by evacuating the heated diffusion tube and source regions, pressurizing the diffusion region with helium to the desired test pressure,and closing the valve between the source and diffusion regions. The source region was then re-evacuated and back-filled with either krypton-helium or xenon-helium mixtures to a predetermined activity level, followed by pressurization with helium to the test pressure. The runs were commenced by opening the valve between the source and diffusion tube regions and then monitoring activity at the top of the diffusion tube region until equilibrium was attained.

DATA REDUCTION

The data from each experiment were analyzed using the one-dimensional fission product diffusion code SLIDER [15]. This code

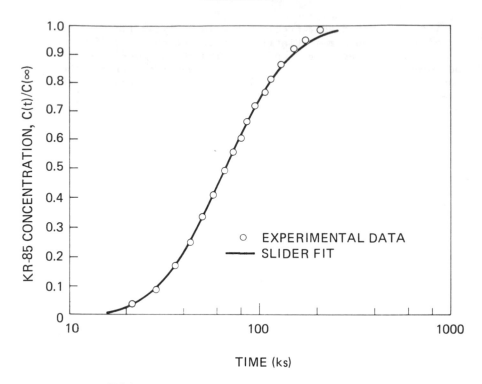

FIG.2.   *Representative diffusion curve and SLIDER fit.*

accounts for the simultaneous birth, release, and decay of radionuclides in solving the following one-dimensional gaseous diffusion
equation using finite difference numerical methods:

$$\frac{\partial C_i(\vec{r},t)}{\partial t} = S_i(\vec{r},t) - \lambda_i C_i(\vec{r},t) - \vec{\nabla}\cdot\vec{J}_i(\vec{r},t) \qquad (5)$$

where

$C_i(\vec{r},t)$ = volume concentration of diffusion species i (mol/cm^3),

$S_i(\vec{r},t)$ = volumetric birth rate of diffusion species i $(mol/cm^3\text{-s})$ (set equal to zero in the current analysis),

λ_i = radioactive decay constant for species i (s^{-1}).

The absolute count rates recorded during each experimental run were
reduced to relative values by dividing the absolute count rate, or
activity, C(t) on the strip chart at time t by $C(\infty)$, the activity
at t→∞. A representative example of experimentally measured diffusion curves and the associated SLIDER fit is shown in Fig. 2.

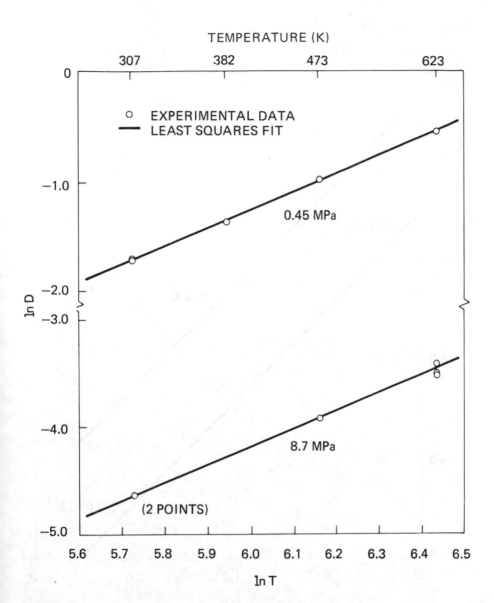

FIG.3. Plot of ln $D_{Kr\text{-}He}$ versus ln T at 0.45 and 8.7 MPa. (T is measured in kelvin, D in cm^2/s.)

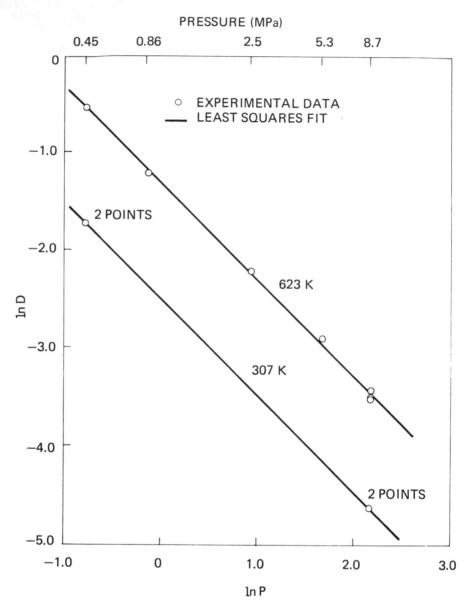

FIG.4. *Plot of ln D_{Kr-He} versus ln P at 307 and 623 K. (P is measured in kelvin, D in cm²/s.)*

TABLE I. EXPERIMENTAL CONDITIONS AND MEASURED DIFFUSION COEFFICIENTS FOR Kr-He AND Xe-He INTERDIFFUSION TESTS

Gas Couples	Apparatus[a]	Temperature (K)	Pressure (kPa)	Kr or Xe Conc. (atom/cm^3)	D_{AB} (cm^2/s)	$D_{AB,o}$[b] (cm^2/s)
Kr-He	LP	289	205	1.9×10^{14}	0.260	0.480
	LP	294	205	1.8×10^{14}	0.308	0.539
	LP	293	205	1.8×10^{14}	0.275	0.497
	LP	293	205	1.8×10^{14}	0.300	0.542
	LP	293	205	1.8×10^{14}	0.310	0.560
	LP	293	205	1.8×10^{14}	0.320	0.578
	LP	293	205	1.8×10^{14}	0.305	0.551
	HP	293	101	5.2×10^{14}	0.680	0.605
	HP	293	101	5.2×10^{14}	0.680	0.605
	HP	293	101	2.5×10^{19}	0.800	0.712
	HP	293	101	2.5×10^{19}	0.800	0.712
	HP	293	1100	1.9×10^{20}	0.0634	0.614
	HP	293	1140	5.2×10^{14}	0.0652	0.655
	HP	293	1140	1.9×10^{20}	0.0621	0.624
	HP	293	1140	2.5×10^{19}	0.0699	0.702
	HP	293	3520	5.2×10^{14}	0.0216	0.670
	HP	293	3520	1.9×10^{20}	0.0200	0.620
	HP	293	3520	1.9×10^{20}	0.0200	0.620
	HP	293	3520	2.5×10^{19}	0.0220	0.682
	HP	293	6900	2.5×10^{19}	0.0111	0.675
	HP	293	6930	1.9×10^{20}	0.0103	0.629
	HP	293	6930	1.9×10^{20}	0.0109	0.666
	HP	293	7140	5.2×10^{14}	0.0116	0.730
	HP	293	7140	5.2×10^{14}	0.0118	0.742
	HP	307	446	3.8×10^{14}	0.179	0.651
	HP	307	446	3.8×10^{14}	0.180	0.655
	HP	307	8714	4.6×10^{14}	0.00963	0.685
	HP	307	8728	4.6×10^{14}	0.00962	0.685
	HP	383	446	3.1×10^{14}	0.252	0.637
	HP	475	446	3.0×10^{14}	0.370	0.655
	HP	475	8679	2.5×10^{14}	0.0197	0.679
	HP	623	446	1.9×10^{14}	0.576	0.652
	HP	623	859	1.9×10^{14}	0.310	0.676
	HP	623	2513	1.9×10^{14}	0.110	0.701
	HP	623	5269	1.9×10^{14}	0.0551	0.737
	HP	623	8714	2.2×10^{14}	0.0300	0.663
	HP	623	8714	2.2×10^{14}	0.0296	0.655
	HP	623	8714	1.9×10^{14}	0.0320	0.708
Xe-He	LP	294	101	6.6×10^{11}	0.500	0.442
	LP	295	101	5.9×10^{11}	0.500	0.440
	LP	295	101	5.8×10^{11}	0.495	0.435
	LP	295	101	5.8×10^{11}	0.490	0.431
	LP	295	101	5.7×10^{11}	0.495	0.435
	HP	298	101	1.3×10^{12}	0.600	0.519
	HP	298	101	1.2×10^{12}	0.600	0.519
	HP	307	446	3.0×10^{12}	0.142	0.516
	HP	473	446	2.0×10^{12}	0.308	0.546
	HP	623	446	4.4×10^{11}	0.460	0.516
	HP	623	790	9.6×10^{11}	0.308	0.612
	HP	623	1479	9.1×10^{11}	0.163	0.607
	HP	623	2513	9.6×10^{11}	0.100	0.632

a. LP = Low Pressure, Low Temperature; HP = High Pressure, High Temperature
b. T = 273K, P = 101 kPa

TABLE II. SUMMARY OF EXPERIMENTAL RESULTS ON INTERDIFFUSION OF KRYPTON AND XENON IN HELIUM

Authors	Ref.	Experimental Method	Temperature (K)	Pressure (kPa)	Kr − He			Xe − He	
					$D_{Kr-He,o}$ (cm²/s) [a]	x [b]	y [b]	$D_{Xe-He,o}$ (cm²/s) [a]	x [b]
Srivastava (1959)	[2]	Capillary diffusion	273–318	10.4–10.9	0.556			0.501	1.76[d]
Srivastava and Barau (1959)	[3]	Capillary diffusion	273–318	10.3–12.4	0.550	1.69[c]			
Srivastava and Paul (1962)	[8]	Capillary diffusion	304	6.41–9.23	0.512[d]				
Durbin & Kobayashi (1962)	[7]	Porous plug-diffusion	308	376–42100	0.554		0.92–0.96[d]		
Weissman and Mason (1962)	[4]	Viscosity	291–550	101	0.526[d]			0.459[d]	1.68[d]
Watts (1964)	[10]	Capillary-diffusion	303	6.70–22.1	0.535				
Malinauskas (1964)	[5]	Capillary diffusion	273–393	101	0.542[d]	1.69		0.463	1.72
Malinauskas (1966)	[6]	Capillary diffusion	273–393	101	0.542[d]				
Van Heijningen et al. (1968)	[9]	Capillary diffusion	112–400	0.2–1.0	0.535	1.65[d]			
Carson and Dunlop (1972)	[11]	Cell diffusion	300	84.3–89.0					
Staker et al. (1974)	[12]	Cell diffusion	300	55.5–65.7					
This work		Low pressure tube diffusion	298	101				0.436	1.69[d]
		High pressure tube diffusion	293–623	101–8710	0.66	1.65	0.98–0.99	0.56	1.66

a. T = 273K, P = 101 kPa; D_o (theoretical) = 0.566 cm²/s

b. Exponents in equation: $D_{AB} = D_{AB,o} \left(\dfrac{T}{T_o}\right)^x \left(\dfrac{P_o}{P}\right)^y$

c. T = 273K, P = 101 kPa; D_o (theoretical) = 0.470 cm²/s

d. Derived from reported data

SLIDER modeling was initiated by converting the volume of the
apparatus to one dimension. This was done by dividing the volume
of each region by the cross-sectional area of the diffusion tube.
Experimental measurements of the tube volumes were derived using a
gas displacement technique. The diffusion apparatus was evacuated
to ∿10 Pa, the valves closed (Fig. 1), and the system connected to
the open end of a partially water-filled burette, which was connected
by means of rubber tubing to a second burette. The volume of each
region was determined six times by sequentially opening the bottom,
middle, and top valves and noting the level of the water in the
burettes, after bring the two menisci to the same height by raising or
lowering one of the burettes. The volumes of the source and diffusion regions of the high-pressure apparatus were 7.80±0.02 and
26.90±0.02 cm^3, respectively.

RESULTS AND DISCUSSION

Studies of krypton-helium and xenon-helium interdiffusion
couples were carried out over a wide range of experimental conditions. Table 1 shows the temperature, pressure, and krypton or xenon
concentration for each test and the corresponding diffusion coefficient, D_{Kr-He} or D_{Xe-He}, derived from SLIDER fits of the measured
results. To derive the temperature dependence of D_{Kr-He}, lnD was
plotted as a function of lnT for pressures of 0.45 and 8.7 MPa (Fig. 3).
The slopes of these lines provided values of x in Eq. 4 of 1.65±0.04
and 1.66±0.06 (80% confidence), for pressures of 0.45 and 8.7 MPa,
respectively, demonstrating that the measured temperature dependence of
D_{Kr-He} is pressure independent over the range studied. These values
are in good agreement with the results reported by Malinauskus
(x = 1.688 273-394 K) [6] and derivable from the work of Srivastava
and Barua (x = 1.69 273-318K) [3] and Van Heijningen et al.
(x = 1.65 169-400 K) [9]. A similar treatment of xenon-helium interdiffusion couples at 0.45 MPa also provided a value of x = 1.66±0.33
(80% confidence) using the three data points available. This is in
accord with the work of Weissman and Mason (x = 1.68 291-550K) [4] and
Van Heijningen, et al. (x = 1.69 231-400 K) [9] and is somewhat smaller
than values obtained from the results of Srivastava (x = 1.76 273-318 K)
[2] and Malinauskas (x = 1.72 273-394 K) [5]. Table 2 summarizes the
results of data on the interdiffusion of krypton and xenon in helium.

A comparable treatment was used to obtain the pressure dependence of krypton diffusion, i.e., y in Eq. 4. Values of the slope
of the plot lnD vs. lnP at 307 and 623 K were plotted (Fig. 4),
and the values y = 0.99±0.03 and 0.98±0.01 (80% confidence),
respectively, were derived. The consistency of the two data sets
confirms that the pressure dependence of the krypton-helium interdiffusion couple is constant over the broad temperature range
studied. These results are in closer agreement with the theoretical inverse pressure dependence than those obtained by Durbin and
Kobayashi [7].

The impact of krypton concentration was examined by using krypton
mole fractions varying from 2.1 x 10^{-8} to 2.5 x 10^{-2}. No measurable

concentration dependence was noted. This is not surprising since the
expected dependence, based on previous work [10-12], is small.

Once the appropriate values of x and y in Eq. 4 were established, $D_{Kr-He,o}$ and $D_{Xe-He,o}$ were determined for each experimental
run, as shown in Table 1. The value of $\overline{D}_{Kr-He,o} = 0.535\pm0.035$
(273K; 101 kPa) obtained from the low-pressure, low temperature
apparatus agrees with the mean value of $D_{Kr-He,o} = 0.540\pm0.015$ derived
from the results of previous authors shown in Table 2. The value of
$D_{Xe-He,o} = 0.436\pm0.038$ is approximately 8% smaller than the mean value
of $D_{Xe-He,o} = 0.474\pm0.023$ of other authors given in Table 2. The explanation for this difference is not apparent, but the values measured
in this work have a high degree of precision ($1\sigma=\pm0.0038$). Values of
$D_{Kr-He,o} = 0.66\pm0.03$ and $D_{Xe-He,o} = 0.56\pm0.05$ were calculated from
results taken from the high-pressure apparatus. These values are
significantly larger than those calculated using Eq. 1 or found by
previously. This difference may be caused in part by the complexity
of the apparatus. The required use of bellows-sealed valves introduces
dead spaces that are significant fractions of the source volume. It is
important to note, however, that this shortcoming does not compromise
the temperature and pressure dependencies derived from this analysis and
the experimental data since the same model was used throughout all SLIDER
analyses, thereby ensuring comparable treatment of all diffusion runs.

Studies of gaseous diffusion in simulated GCFR fuel rods are
continuing with analyses of krypton and xenon diffusion through fuel
rod charcoal delay traps. These results, coupled with the work of
Underhill [13] and Bolch [14], will permit reliable predictions of
fission gas transport by gaseous diffusion in GCFRs under normal and
off-normal operating conditions.

REFERENCES

[1] BIRD, R.B., STEWART, W.E., LIGHTFOOT, E.N., Transport Phenomena, John Wiley
     and Sons, New York (1960) 511.
[2] SRIVASTAVA, K.P., Physica 25 (1959) 571.
[3] SRIVASTAVA, K.P., BARUA, A.K., Indian J. Phys. 33 (1959) 229.
[4] WEISSMAN, S., MASON, E.A., J. Chem. Phys. 37 6 (1962) 1289.
[5] MALINAUSKAS, A.P., J. Chem. Phys. 42 1 (1965) 156.
[6] MALINAUSKAS, A.P., J. Chem. Phys. 45 12 (1966) 4704.
[7] DURBIN, L., KOBAYASHI, R., J. Chem. Phys. 37 8 (1962) 1643.
[8] SRIVASTAVA, B.N., PAUL, R., Physica 28 (1962) 64.
[9] VAN HEIJNINGEN, R.J.J., et al., Physica 38 (1968) 1.
[10] WATTS, H., Trans. Faraday Soc. 60 (1964) 1745.
[11] CARSON, P.J., DUNLOP, P.J., Chem. Phys. Lett. 14 3 (1972) 377.
[12] STAKER, G.R., et al., Trans. Faraday Soc. 70 5 (1974) 825.
[13] UNDERHILL, D.W., Nucl. Appl. Technol. 8 (1970) 255.

[14] BOLCH, W.E., et al., Gas Dispersion in Porous Media Peclet-Reynolds Number Correlations, Sanitary Engineering Research Laboratory, University of California Rep. SERL-6710 (1967).

[15] NORMAN, J.H., et al., "Spheres: diffusion-controlled fission product release and absorption", Advances in Chemistry, Vol.93, Radionuclides in the Environment, American Chemical Society, Washington, DC (1970) 27.

ИЗУЧЕНИЕ ЗАКОНОМЕРНОСТЕЙ ДИФФУЗИИ ВОДОРОДА В ГИДРИДАХ МЕТАЛЛОВ IV ГРУППЫ

Е.Б.БОЙКО, А.М.СОЛОДИНИН, Р.А.АНДРИЕВСКИЙ
Институт химической технологии,
Москва,
Союз Советских Социалистических Республик

Представлен А. С. Пановым

Abstract–Аннотация

STUDY OF HYDROGEN DIFFUSION IN GROUP IV METAL HYDRIDES.

An autoradiography method was used to study the self-diffusion of hydrogen (^{3}H) in hydrides of zirconium (ZrH$_{1.5-1.9}$), hafnium (HfH$_{1.2}$) and titanium (TiH$_{1.6}$) over the temperature range 500–900°C. It is shown that the self-diffusion activation energy of hydrogen in zirconium hydride increases with departure from stoichiometry. The addition of oxygen reduces the diffusion mobility of hydrogen in zirconium hydride, while the addition of zirconium carbide increases it. The results are discussed in terms of statistical thermodynamics.

ИЗУЧЕНИЕ ЗАКОНОМЕРНОСТЕЙ ДИФФУЗИИ ВОДОРОДА В ГИДРИДАХ МЕТАЛЛОВ IV ГРУППЫ.

С использованием авторадиографической методики проведено изучение самодиффузии водорода (^{3}H) в гидридах циркония ZrH$_{1,5-1,9}$, гафния HfH$_{1,2}$ и титана TiH$_{1,6}$ в области температур 500-900°C. Показано, что с отклонением от стехиометрии энергия активации самодиффузии водорода в гидриде циркония возрастает. Добавки кислорода способствуют замедлению, а карбида циркония – возрастанию диффузионной подвижности водорода в гидриде циркония. Обсуждение результатов проведено с позиций статистической термодинамики.

Изучению самодиффузии водорода в гидридах IV группы посвящен ряд работ (например, [1-9]), бо́льшая часть из которых проведена на гидриде циркония. Имеющиеся сведения о параметрах самодиффузии существенно разнятся, а в ряде случаев наблюдаются качественные различия в характере их изменения. Такие различия объясняются рядом причин и, в частности, особенностями использованных методик (опыты по насыщению при наличии градиента [1-3, 9] и результаты безградиентного изучения спектров ЯМР [5-8]), а также, вероятно, недостаточной корректностью в соблюдении условий эксперимента.

В этой связи представляло интерес исследовать самодиффузию водорода в гидридах IV группы с помощью трития, как мягкого β-излучателя (что, насколько известно, ранее не предпринималось) и получить информацию об истинных коэффициентах самодиффузии.

303

а)   б)

Рис.1. а) микроструктура образца $ZrH_{1,60}$; *б) типичная авторадиограмма, полученная с образца гидрида циркония состава* $ZrH_{1,60}$ *(увеличено в 8 раз).*

МЕТОДИКА ЭКСПЕРИМЕНТА

В работе использовались образцы из переплавленного йодидного циркония, титана и гафния ($6 \times 8 \times 10$мм), которые насыщались водородом при 450-900°C в соответствии с P – C – T-диаграммой [1, 10] до заданного равновесного состава. Для гидрида циркония состава $ZrH_{1,6}$ исследовалось также влияние на диффузионную подвижность водорода добавок ZrC и кислорода. Добавка ZrC, как известно [11], измельчая микроструктуру, способствует повышению механических свойств. Специальными опытами контролировалась равномерность распределения концентрации по сечению и отсутствие микродефектов; водород, используемый для насыщения, подвергался очистке на цеолитовых колонках; образцы дополнительно засыпались циркониевым порошком.

После выдержки насыщенных образцов в равновесных условиях для заданного состава (t = 1-50ч) водород при неизменной температуре ($\pm 5°C$) в течение 2-3с заменялся смесью водород – тритий при том же давлении и начиналась диффузионная выдержка. Выдержка выбиралась в зависимости от температуры и состава и для T = 500°C составляла в среднем 50ч, для T = 900°C – 0,4ч. Давление и активность насыщающей среды в процессе опытов оставались неизменными; после окончания выдержки образцы с помощью специального приспособления закаливались со скоростью 300°C/мин. Подготовленные таким образом образцы взвешивались, разрезались, после чего готовились из них шлифы. Для получения качественных авторадиограмм и уменьшения времени экспонирования применялись сухие съемные слои типа М, которые затем проявлялись и фотометрировались на микрофотометре МФ-4. Использовался линейный участок зависимости плотности почернения от времени, поэтому концентрация трития прямо пропорциональна плотности почернения. Авторадиограммы получались и обрабатывались по методике [12]. Для каждой экспериментальной точки было проведено по 2-3 опыта. Точность указания состава H – Me составляет ±0,02.

На рис. 1 показана микроструктура образца гидрида циркония, а также снятая с него типичная авторадиограмма. Характер отпечатка свидетельствует о том, что влияние граничной диффузии заметным образом не проявляется. Характерные кривые почернения приведены на рис. 2. Обработка диффузионных данных проводилась исходя из предположения о существовании постоянного источника. Из опытных данных $S = f(x)$ по табулированным значениям функции ошибок Гаусса определялась величина $x / 2\sqrt{Dt}$. Суммарная ошибка при определении коэффициентов самодиффузии не превышала ± 10%. Переход от параметров самодиффузии трития к самодиффузии водорода проводился с учетом поправки на массу изотопа. Расчет параметров самодиффузии (Q и D_0) и ошибок производился методом наименьших квадратов.

РЕЗУЛЬТАТЫ И ИХ ОБСУЖДЕНИЕ

На рис. 3 приведены температурные зависимости коэффициентов самодиффузии водорода в гидридах циркония, титана и гафния.

Обращает на себя внимание то, что если для составов $ZrH_{1,6-1,9}$ повышение концентрации структурных дефектов увеличивает коэффициент самодиффузии водорода, то состав $ZrH_{1,5}$ проявляет в этом отношении аномалию. Заметим, что аномальное поведение давления диссоциации $ZrH_{1,5}$ отмечалось также в работе [13] и связывалось с влиянием возможного полиморфного превращения.

Температурные зависимости коэффициентов самодиффузии описываются следующими выражениями:

$$D = 3{,}57 \exp\left(\frac{-31{,}7 \pm 2{,}3}{RT}\right) \qquad \text{для } ZrH_{1,50} \qquad (1)$$

$$D = 2{,}17 \exp\left(\frac{-29{,}1 \pm 2{,}6}{RT}\right) \qquad \text{для } ZrH_{1,60} \qquad (2)$$

$$D = 0{,}47 \exp\left(\frac{-26{,}7 \pm 2{,}0}{RT}\right) \qquad \text{для } ZrH_{1,70} \qquad (3)$$

$$D = 1{,}93 \cdot 10^{-2} \exp\left(\frac{-22 \pm 2{,}2}{RT}\right) \qquad \text{для } ZrH_{1,80} \qquad (4)$$

$$D = 3{,}16 \cdot 10^{-3} \exp\left(\frac{-19{,}3 \pm 5{,}8}{RT}\right) \qquad \text{для } ZrH_{1,90} \qquad (5)$$

$$D = 0{,}172 \exp\left(\frac{-19{,}26 \pm 1{,}07}{RT}\right) \qquad \text{для } TiH_{1,60} \qquad (6)$$

$$D = 2{,}9 \cdot 10^{-2} \exp\left(\frac{-10{,}56 \pm 3{,}85}{RT}\right) \qquad \text{для } HfH_{1,20} \qquad (7)$$

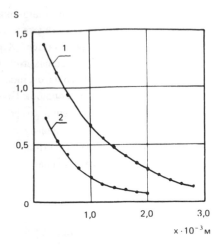

Рис. 2. Характерная кривая почернения, полученная при фотометрировании авторадиограммы, снятой с образца гидрида циркония состава $ZrH_{1,7}$: $1-T=800°\,C$, $t=40\,мин$; $2-T=700°C$, $t=90\,мин$.

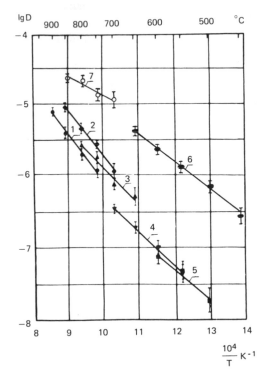

Рис. 3. Температурная зависимость коэффициентов самодиффузии водорода в гидридах циркония, титана и гафния: $1-ZrH_{1,50}$; $2-ZrH_{1,60}$; $3-ZrH_{1,70}$; $4-ZrH_{1,80}$; $5-ZrH_{1,90}$; $6-TiH_{1,60}$; $7-HfH_{1,20}$.

В наиболее широком интервале температур и концентраций исследования проведены для гидрида циркония. Как видно из этих данных, по мере отклонения состава от стехиометрического, энергия активации Q и предэкспоненциальный множитель D_0 возрастают.

Попытаемся обосновать полученные нами данные. Для самодиффузии в фазах внедрения атомов металлов и неметаллов можно получить вытекающее из представлений Верта-Зинера соотношение [14]:

$$\frac{\Delta S_{Me}}{Q_{Me}} \sim \frac{\Delta S_H}{Q_H} \sim \frac{\partial \mu}{\mu_0 \, \partial T} \qquad (6)$$

где $\Delta S_{Me}, \Delta S_H$ — энтропия активации самодиффузии атомов металлов и водорода, соответственно, Q_{Me}, Q_H — энергии активации самодиффузии этих атомов, μ и μ_0 — модули сдвига гидрида при данной температуре и при 0К. Самодиффузия циркония в гидриде циркония исследовалась в работе [15], где были получены для $ZrH_{1,65-1,70}$ следующие параметры: $D_0 = 10^{-4} \div 10^{-6}$ см^2/с, $Q_{Me} = (33,6 \pm 6)$ ккал/г·форм. Используя известные соотношения для оценки ΔS_{Me} и ΔS_H [14,16], а также информацию о частотах колебаний водорода и циркония в гидриде циркония [14], можно получить следующие величины для $ZrH_{1,7}$: $\dfrac{\Delta S_{Me}}{Q_{Me}} \simeq (0,45 \pm 0,2) \cdot 10^{-3}$ (с учетом разброса D_0 и Q),

$\dfrac{\Delta S_H}{Q_H} \simeq 0,33 \cdot 10^{-3}$, а величина $\dfrac{\partial \mu}{\mu_0 \partial T}$ составляет $\sim 0,12 \cdot 10^{-3}$ [5]. Таким образом,

соотношение (6) с учетом его приближенности выполняется, как и в случае карбидов [14], вполне удовлетворительно, что свидетельствует о разумности полученных результатов и их "совместимости" с данными [15] и с температурной зависимостью упругих свойств.

Реальность концентрационной зависимости энергии активации самодиффузии водорода в гидриде циркония можно усмотреть, анализируя рис. 4.

Зависимость $Q_H \sim C_H (1-C_H) E_{см}$, где C_H — концентрация водородных вакансий в гидридах, $E_{см}$ — энергия смешения в водородной подрешетке, была теоретически обоснована в работе [17]. Она хорошо оправдывается для карбидов и некоторых гидридов, что наблюдается и в нашем случае. Энергия смешения, оцененная из этих данных, составила 4 ккал/г·форм, что по порядку величины совпадает с энергией смешения для углеродной подрешетки в карбидах [14, 17]. Знак энергии смешения ($E_{см} < 0$) свидетельствует о том, что содержание структурных вакансий вокруг заданной ниже, чем при статистическом распределении, и это определяет рост энергии активации при отклонении от стехиометрии. Следует, однако, обратить внимание на то, что полученные нами значения Q_H значительно выше по абсолютной величине, чем величины $\overline{Q}_H$, полученные в опытах по изучению спектров ЯМР при темепературах до 200-400°C [5, 7]. Обработка данных [7] показала, что они также следуют зависимости $Q_H \sim C_y (1-C_y) \cdot E_{см}$, но обнаруживают небольшой положительный ближний порядок [17].

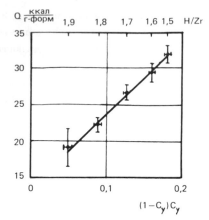

Рис. 4. Зависимость энергии активации самодиффузии от дефектности в металлоидной подрешетке.

ТАБЛИЦА I. ОТНОСИТЕЛЬНОЕ ИЗМЕНЕНИЕ КОЭФФИЦИЕНТОВ САМОДИФФУЗИИ ВОДОРОДА В ГИДРИДЕ ЦИРКОНИЯ $ZrH_{1,60}$, ЛЕГИРОВАННОМ КИСЛОРОДОМ

$T(^\circ C)$	При содержании кислорода (вес %)		
	0,05	0,25	0,57
700	0,85	0,65	0,6
750	0,75	0,6	0,55
800	0,6	0,55	0,55

Выяснение различий между результатами изучения диффузии при низких (ЯМР) и высоких температурах (тритиевая методика) должно стать предметом отдельного исследования. Возможно это различие связано с изменением механизма диффузионной подвижности при переходе от низких температур к высоким.

На рис. 3 представлены также температурные зависимости коэффициентов диффузии водорода в гидриде титана состава $TiH_{1,60}$ и гидриде гафния состава $HfH_{1,20}$. Как видно, энергия активации самодиффузии в гидриде титана ($Q = 19{,}26$ ккал/г-форм) выше значения Q для гидрида циркония того же состава и выше той величины, которая была получена в работе [18]. Абсолютные значения коэффициентов самодиффузии для $TiH_{1,60}$ в исследованной области температур примерно на порядок превышают D для гидрида циркония. Сравнительно низкая энергия активации самодиффузии водорода получена для гидрида гафния ($Q = 10{,}56$); этот результат представляется странным, учитывая высокую концентрацию структурных дефектов в $HfH_{1,2}$. Однако, детальное сопоставление диффузионной подвижности водорода в гидридах переходных металлов IV группы может быть проведено лишь после изучения самодиффузии в области гомогенности гидридов титана и гафния, ибо не исключено, что знак энергии смешения для TiH_x и HfH_x может быть иным, нежели в случае ZrH_x.

Легирование гидрида циркония карбидной добавкой приводит к некоторому (не более 30%) возрастанию коэффициента самодиффузии, что может быть связано, учитывая измельченность структуры, с дополнительным вкладом граничной диффузии.

Влияние кислорода проявляется в снижении диффузионной подвижности водорода. В табл. I представлено относительное изменение коэффициентов самодиффузии $\frac{D^1}{D}$ (D^1 — коэффициент самодиффузии водорода в легированном гидриде) в гидриде циркония состава $ZrH_{1,60}$ при различных температурах в зависимости от содержания кислорода.

Как видно из таблицы, снижение диффузионной подвижности в бо́льшей мере наблюдается при переходе от сплава с 0,05%O к сплаву с 0,25%O. Если учесть, что растворимость кислорода в гидриде циркония при температурах 700-800^{0}C не превышает 0,2 вес%, то тормозящее воздействие кислорода в пределах твердого раствора можно объяснить блокированием тетраэдрических междоузлий атомами кислорода, размещающихся в октапорах. Как отмечается в работе [13], каждый атом кислорода может блокировать две ближайшие тетрапоры.

ЛИТЕРАТУРА

[1] МЮЛЛЕР, В., БЛЕКЛЕДФ, Д., ЛИБОВИЦ, Дж., Гидриды Металлов, М., Атомиздат, 1973.

[2] ALBRECHT, W., GOOD, W., Battelle Memorial Institute Rep. BMI-1426 (1960).

[3] HARKNESS, F., Atomics International Rep. NAA-SR-9297 (1964).

[4] STALINSKI, B., COOGAN, C., GUTOWSKY, H., J. Chem. Phys. 34 (1961) 1191.

[5] ХОДОСОВ, Е.Ф., АНДРИЕВСКИЙ, Р.А., Порошк. Металл. №8 (1967) 65.

[6] ХОДОСОВ, Е.Ф., ПРОКОПЕНКО, А.К., ЛИННИК, А.И., Физ. Мет. Металловед. 38 (1974) 746.

[7] ХОДОСОВ, Е.Ф., АНДРИЕВСКИЙ, Р.А., Физ. Мет. Металловед. 29 (1970) 415.

[8] ХОДОСОВ, Е.Ф., ШЕПИЛОВ, Н.А., САВИН, В.И., Физ. Мет. Металловед. 32 (1971) 189.

[9] PAETZ, P., LÜCKE, K., Z. Metallkd. 62 (1971) 657.

[10] ЛЕВИНСКИЙ, Ю.В., АНДРИЕВСКИЙ, Р.А., БОЙКО, Е.Б., Изв. АН СССР, сер. металлы 3 (1975) 185.

[11] АНДРИЕВСКИЙ, Р.А. и др. Peaceful Uses of Atomic Energy (Proc. 4th Int. Conf. Geneva, 1971) Vol.10, UN, New York, and IAEA, Vienna (1972) 383.

[12] ФРИЦ, М.Е., СВИДЕРСКАЯ, З.А., КАДАНЕР, Э.С., Авторадиография в Металловедении М., Металлургиздат, 1961.

[13] SIMNAD, M., DEE, J., Thermodynamics of Nuclear Materials, 1967 (Proc. Symp. Vienna, 1967), IAEA, Vienna (1968) 513.

[14] АНДРИЕВСКИЙ, Р.А., УМАНСКИЙ, Я.С., Фазы Внедрения, Л., Физматгиз, 1977.

[15] BENTLE, G., J. Am. Ceram. Soc. 50 (1967) 166.

[16] ШЬЮМОН, А., Диффузия в Твердых Телах, М., Металлургия, 1966.

[17] АНДРИЕВСКИЙ, Р.А., ГУРОВ, К.П., Физ. Мет. Металловед. 39 (1975) 57.

[18] KORN, C., ZAMIR, D., J. Phys. Chem. Solids 31 (1970) 489.

ACTINIDE DIFFUSION IN WASTE GLASSES

Hj. MATZKE
Commission of the European Communities,
European Institute for Transuranium Elements,
Karlsruhe

Abstract

ACTINIDE DIFFUSION IN WASTE GLASSES.

The diffusion of ^{233}U and of ^{238}Pu and ^{239}Pu in phosphate glass, in different borosilicate glasses and in glass ceramics was investigated. Glasses with and without added simulated fission products were used. Some glasses also contained Gd_2O_3 as a neutron poison. The method of thin tracer layers and high resolution alpha spectroscopy were used to measure both depth and time dependence of diffusion. Diffusion rates were immeasurably small for $T < 390°C$ (diffusion coefficient, $D < 5 \times 10^{-18}$ cm$^2 \cdot$s^{-1}). Since the softening points of some glasses are as low as 500°C, only a limited temperature range was available for making accurate measurements. Above the softening points, the diffusion profiles became distorted due to plastic flow of the glass. Further complications arose due to nucleation of crystallites and subsequent crystallization of part of the glass for long annealing times and at high diffusion temperatures. The kinetics of the crystallization process, verified by optical microscopy and X-ray diffractometry, could be obtained from the time dependence of tracer penetration since differently crystallized glasses show different diffusion behaviour. In the crystallized glasses (glass ceramics), composite tracer penetration profiles were frequently observed. These could be unfolded to give a set of up to three Gaussians yielding different diffusion coefficients for different components of the glass ceramics. The effects of α-radiation damage and radiation enhancement of diffusion were also studied using curium and americium-containing glasses. In all cases, the measured D-values were small, indicating that, from the diffusion point of view, the glasses are a safe medium in which to store actinides. Observed plutonium gradients in leaching experiments are therefore stable until the gel-like surface layers get mechanically 'pealed off'.

1. INTRODUCTION

The immobilization of radioactive waste in appropriate glasses or glass ceramic products is under consideration as a means of providing safe long-term storage. The high-level radioactive waste (HLW) to be incorporated into the glass is a complex mixture of fission products and actinides, and results from the reprocessing of spent power-reactor fuel elements. Due to the high self-heating capacity of the waste, the glass will experience elevated temperatures for much of its expected long period of storage. Temperatures in excess of 400°C are being considered for the beginning of storage.

In this context, the problem of actinide diffusion is important as it affects a number of factors, such as homogeneity, second-phase formation, crystallization

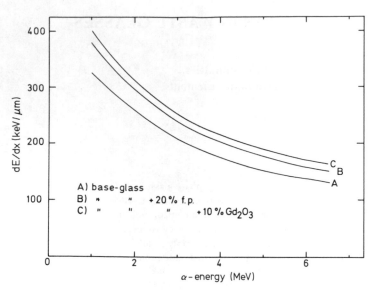

FIG.1. *Differential energy loss, dE/dx, as a function of the energy of the α-particles for the three glasses (see Table II).*

TABLE I.   DEPTH/ENERGY RELATION FOR α-PARTICLES IN THE THREE TYPES OF GLASS[a]

Glass	dE/dx (keV/μm)	Depth, x (μm)
A. Base-glass (fundamental starting material)	$\dfrac{dE}{dx} = \dfrac{1}{0.873\,E + 2.12}$	$x = 0.436\,(E_\alpha^2 - E^2) + 2.12\,(E_\alpha - E)$
B. Glass + f.p.	$\dfrac{dE}{dx} = \dfrac{1}{0.765\,E + 1.82}$	$x = 0.382\,(E_\alpha^2 - E^2) + 1.82\,(E_\alpha - E)$
C. Glass + f.p. + Gd$_2$O$_3$	$\dfrac{dE}{dx} = \dfrac{1}{0.712\,E + 1.76}$	$x = 0.356\,(E_\alpha^2 - E^2) + 1.76\,(E_\alpha - E)$

[a]   x is the depth of the α-emitting atom beneath the surface, or the thickness of the absorbing layer; E_α is the original (or decay) energy, and E is the energy remaining following passage through the absorbing layer; f.p. = fission products (for the calculations on dE/dx, assumed to be ZrO_2).

and, in particular, concentration gradient formation at surfaces if water attacks
the glass and leaches out components thereof. If actinides have a different
leaching behaviour to the rest of the glass, gradients will form which will change the
leaching rate. Such gradients would be rendered less steep by diffusion.

Previous diffusion studies in relevant glasses[1] were mainly concerned with
the highly mobile elements[2] helium and sodium or with the other alkaline metals,
rubidium and caesium, that also occur as fission products. Only a few studies
have been published on the diffusion of other elements (e.g. silver, Ref. [4];
strontium, caesium and cerium, Refs [5—7]). Nothing could be found on the
diffusion of actinides. Therefore, an extensive study of the diffusion of uranium
and plutonium in different glasses and glass ceramics was initiated.

2. EXPERIMENTAL

The non-destructive method of α-energy degradation [8,9] was chosen to
measure diffusion profiles of ^{233}U and ^{238}Pu in the glasses. The tracers were
applied as thin layers ($\sim$ 10 nm for ^{233}U and $\sim$ 1 nm for ^{238}Pu) by flash evaporation.
High-resolution α-spectroscopy was used to follow tracer penetration at different
periods of time [10] at each temperature. Annealing was done in air. In this way,
both the temperature and time dependence [10, 11] of actinide diffusion could
be measured. The latter turned out to be important when following the kinetics
of nucleation and recrystallization.

To convert the measured energy profiles into the desired depth profiles of
the tracer, the energy loss, dE/dx, of α-particles in the glasses must be known.
Reliable experimental data for stopping powers in low atomic number elements
are available and have been tabulated [12, 13]. Though fewer results for compounds
have been reported, some data for nuclear materials are available [14, 15] which
support the applicability of the rule:

$$\epsilon (X_n Y_m)_{mol} = n\epsilon (X)_{atom} + m\epsilon (Y)_{atom} \qquad (1)$$

for calculating energy-loss values ϵ (eV/cm^2) for compounds (e.g. SiO_2, Na_2O,
ZrO_2, Gd_2O_3 etc.) from the values for the elements Na, Si, O, Zr, Gd etc. The
stopping powers, dE/dx (MeV/(mg·cm^{-2}) or keV/μm) can then be obtained via
the relation:

$$\epsilon \ (eV/(10^{15} \ at \cdot cm^{-2})) = \left(\frac{0.6023}{at.wt} \right) \times \left(\frac{dE}{dx} \right) \ (MeV/(mg \cdot cm^{-2})) \qquad (2)$$

---

[1] See Ref. [1] for a review.
[2] See, for example, Refs [2, 3].

TABLE II. CHEMICAL COMPOSITIONS OF GLASSES* A TO D AND APPROXIMATIONS USED TO CALCULATE dE/dx

Glass[a]	Components (wt %)							Fission products	Actinides	Corrosion products	Poison, Gd_2O_3
	Base-glass										
	SiO_2	P_2O_5	TiO_2	Al_2O_3	B_2O_3	CaO	Na_2O				
Actual compositions											
A	50.5	–	4.2	1.4	13.6	2.8	27.5	–	–	–	–
B	40.6	–	3.4	1.1	10.9	2.2	24.3	15[a]	0.5[b]	2.1[c]	–
C	35.4	–	3.0	1.0	9.6	2.0	21.5	15[a]	0.5[b]	2.1[c]	10
Approximate compositions used for calculation											
A	50	–	–	–	–	–	50	–	–	–	–
B	40	–	–	–	–	–	40	20[a]	–	–	–
C	35	–	–	–	–	–	35	20[a]	–	–	10
D	–	58.0	–	3.5	–	0.8	9.0	27[d]	1.2[e]	0.5[f]	–

* Borosilicate glasses except for D, where substitution of P_2O_5 for SiO_2 represents a phosphate glass.

[a] Fission products added were: Nd_2O_3 (2.43 wt %), Ce_2O_3 (1.26 wt %), La_2O_3 (0.66 wt %), Pr_2O_3 (0.60 wt %), ZrO_2 (2.53 wt %), MoO_3 (2.37 wt %), Cs_2O (1.71 wt %), Ru (0.70 wt %), BaO (0.69 wt %), SrO (0.60 wt %), Y_2O_3 (0.34 wt %), Mn_3O_4 (for TcO_2) (0.31 wt%), Te (0.21 wt%), Rh (0.20 wt %), Rb_2O (0.16 wt %), Pd (0.10 wt %), CdO (0.004 wt %).

[b] Actinides were replaced by PbO.

[c] Corrosion products added were: Fe_2O_3 (1.47 wt %), Cr_2O_3 (0.42 wt %), NiO (0.23 wt %).

[d] Fission products added were: Cs_2O (3.0 wt %), Rb_2O (0.3 wt %), BaO (1.3 wt %), SrO (1.1 wt %), Ce_2O_3 (8.6 wt %), Y_2O_3 (0.5 wt %), ZrO_2 (4.3 wt %), MoO_3 (5.2 wt %), Ru (1.8 wt %), Rh (0.3 wt %), Pd (0.1 wt %), TeO_2 (0.5 wt %).

[e] UO_2.

[f] Fe_2O_3.

The resulting dE/dx values as a function of α-energy are shown in Fig.1 and the corresponding functional dependencies of the type [14]:

$$\frac{dE}{dx} = \frac{1}{AE + B} \tag{3}$$

are shown in Table I. Data are given for three typical types of glass (see Table II).

Because of the decrease in dE/dx with increasing energy, a homogeneous glass shows an α-spectrum which is a continuum with decreasing counting rate at decreasing energy (hence increasing depth). Comparison of calculated spectra (as in Fig.2) with measured spectra allows of an experimental verification of the calculated energy-loss data. Any disagreement between calculation and experiment was always $\leqslant 10\%$. The spectra of glasses containing different Pu, Am and Cm isotopes were accordingly more complex but could also be unfolded into their different contributions. With the above information, all experimental energy spectra can be converted to concentration/depth plots. Integration of the relations of Table I yields the thickness of the absorbing layer, x, or the depth of the α-emitting atom for any remaining energy, E:

$$x = \int_{E}^{E_\alpha} \left(\frac{dE}{dx}\right)^{-1} dE \tag{4}$$

Finally, putting E equal to zero gives the range of the α-particles in the glass, and hence the limit of applicability of the present method. As an example, for α-particles of $E_\alpha = 5.50$ MeV ([238]Pu), the ranges in the three glasses A, B and C are 24, 21 and 20 μm, respectively.

The glasses used in this study were either phosphate glasses [16, 17], obtained from W. Heimerl and H. Lahr[3], or borosilicate glasses [18, 19] with and without additions of simulated (i.e. not radioactive) fission products, obtained from W. Guber and U. Riege of the Kernforschungszentrum Karlsruhe. Some of the borosilicate glasses contained Gd_2O_3 as a neutron poison, and some contained Pu, Am and/or Cm [20]. Furthermore, some of the glasses were crystallized to a glass-ceramic by a heat treatment. Table II summarizes the composition of typical glasses.

---

[3] Gelsenberg AG, Essen, Federal Republic of Germany.

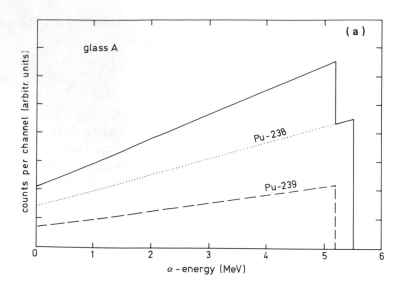

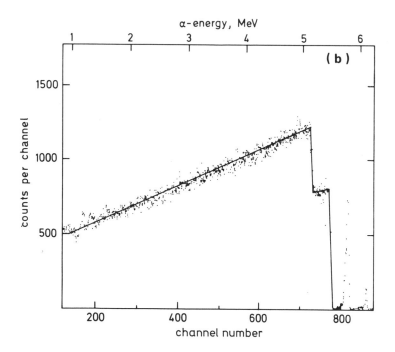

FIG.2. (a) Calculated energy spectrum for glass A containing ^{239}Pu and ^{238}Pu in homogeneous distribution at a ratio of specific activities of 1 : 2, respectively. The solid line shows the spectrum resulting from the overlapping of the two contributions. (b) Measured energy spectrum of a typical glass.

TABLE III. PHYSICAL PROPERTIES OF THE MATERIALS INVESTIGATED

The material designations (VG 98, etc.) are those used by the Kernforschungszentrum, Karlsruhe. Material II was a VC 15 ceramic with Gd_2O_3. FP = fission product; P = Gd_2O_3 poison.

Material	No.	Property			
		Density, ρ (g/cm³)	Coef. of thermal expansion, λ (10^{-6} °C^{-1})	Viscosity, η (poise)	Dilatometric softening temp., T_s (°C)
VC 15[a]					
glass	I	2.54	9.9	95	500
ceramic	III	2.55	8.9	95	715
VC 15 + FP[b]					
glass	II	2.81	8.9	122	500
ceramic	IV	2.85	8.9	122	690
181/1 + FP	V	2.77	15.6	78	575
VG 98 + FP + P[c]	VI	3.00	18.0	58	500
Phosphate glass	VII				450

[a] Type A of Table II.
[b] Type B.
[c] Type C of Table II.

3. RESULTS

3.1. Uranium as tracer

A number of experiments were performed using all the glasses and ^{233}U as tracer at eight temperatures between 180 and 350°C for annealing times of up to 800 h. No reliable indication of tracer penetration was ever measured, thus yielding an upper limit of the diffusion coefficient, D, of $\leqslant 5 \times 10^{-18}$ cm²·s^{-1}, the limit of detection of the present method.[4]

---

[4] Note that the conventional sectioning techniques yield a much higher limit of detection, usually 10^{-14} cm²·s^{-1}. A diffusion coefficient of 5×10^{-18} cm²·s^{-1} corresponds to a very low atomic mobility, e.g. for an annealing time of 1 year, a root-mean-square tracer displacement of less than about 0.2 μm only.

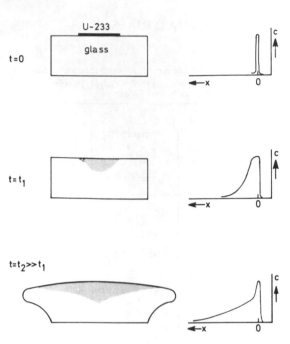

FIG.3. *Development of a distorted tracer profile (right) due to softening of the glass and plastic flow of the specimens.*

At temperatures near to the softening point or recrystallization point (see Table III), strongly distorted non-Gaussian profiles were observed. This was due either to plasticity of the surface, causing the originally Gaussian profile to spread due to an increase in the total surface area (as schematically indicated in Fig.3), or to the formation of differently crystallized phases with different diffusion properties. For the latter case, a computer program was written to fit the measured profile with a calculated curve assuming superposition of up to three different Gaussian profiles, corresponding to three different diffusion coefficients.

Figures 4 and 5 and Table IV show typical examples of the measured time dependence of ^{233}U diffusion, plotted as 4Dt, the 1/e variance of the Gaussian, versus time. Similar data were obtained for the remaining substances.

Two types of experiments were performed. Samples were either annealed at one temperature only for different periods of time, up to ~ 1000 h, or else the temperature was increased stepwise, being held constant for ~ 1000 h at each temperature step. In the latter case, nucleation for recrystallization was facilitated by the long pre-anneals at lower temperatures.

In all cases, negligible diffusion was found in *borosilicate glasses* and ceramics at temperatures below about 500°C, with D $\leqslant$ 2 X 10^{-17} cm$^2 \cdot$s^{-1} (except with glasses V and VI, where D was 4 X 10^{-17} and 6 X 10^{-17} cm$^2 \cdot$s^{-1}, respectively).

Over the temperature range from 500 to 550°C, apparently reliable D-values could be measured which were constant with time and were of the order of 1×10^{-16} to 5×10^{-16} cm²·s⁻¹ for the different compositions. Over the temperature range 550 to 600°C, recrystallization phenomena occurred, causing diffusion coefficients to change with time. Both decreases and increases were observed, depending upon the kinetics of recrystallization. The latter was usually verified by ceramography and X-ray diffractometry. Thus, the time dependence of the uranium tracer penetration gave indications regarding the kinetics of crystallization, as well as data on the differences in D-values in the different crystallization products. These often differed by factors of ten or more.

Over the range of temperatures around 600 to 620°C, good straight lines were usually again observed in plots of 4Dt versus t, indicating no further change in the morphology of the specimens. The resultant D-values were quite low ($\leqslant 2 \times 10^{-16}$ cm²·s⁻¹), as would be expected for largely crystalline specimens.

Above 650°C, and up to 740°C, all samples showed softening of the (remaining) glass matrix, with greater tracer mobility yielding *apparent* high D-values and distorted depth profiles. Such apparent values were in the range of 10^{-15} to 10^{-13} cm²·s⁻¹.

For these various reasons, construction of an Arrhenius diagram is not very meaningful, and data must be treated with care. A typical plot looks as shown in Fig.6, which gives the temperature dependence of the deduced D-values for the phosphate glass. Though an attribution of an Arrhenius function is of limited use, a least-squares fit was performed, yielding:

$$D \ (\text{cm}^2 \cdot \text{s}^{-1}) = 2 \times 10^5 \exp(-70\,000/RT) \qquad (5)$$

(However, care should be taken in applying such equations to extrapolate data to other temperatures since the physical properties — shape, degree of crystallization, etc. — of the specimens change with temperature.)

3.2. Plutonium as tracer

A number of experiments were also performed with ²³⁸Pu and ²³⁹Pu as tracers. These reacted in a similar way to structural changes in the specimens. However, due to handling problems in the glove box, the lower limit of detection of diffusion coefficients was higher ($\sim 5 \times 10^{-17}$ cm²·s⁻¹) than with the uranium tracer. Apart from this restriction, no apparent difference was observed between uranium and plutonium diffusion.

3.3. Effect of α-radiation on the diffusion

Curium and americium-containing glasses were used to study the effect of α-radiation on diffusion. For these highly radioactive specimens, the limit of detection was

MATZKE

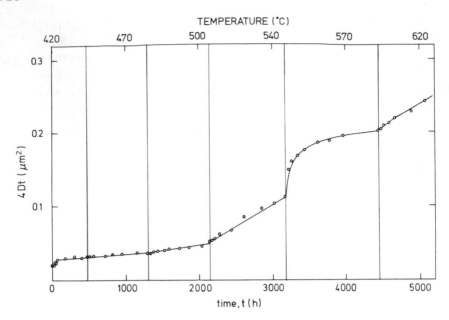

FIG.4. *Typical time dependence of diffusion for a borosilicate glass (No. I).*

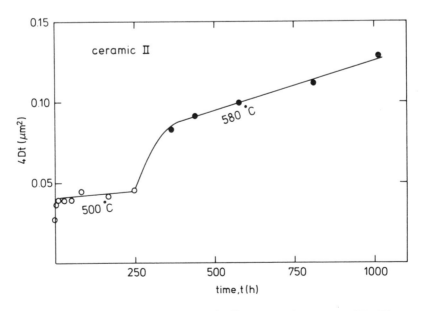

FIG.5. *Typical time dependence of diffusion in a glass ceramic (No. II).*

TABLE IV.   APPARENT DIFFUSION COEFFICIENTS, D, FOR ^{233}U IN
GLASS I AND CERAMIC III DEDUCED FROM LONG-TERM ANNEALS

T (°C)	D for glass I (10^{17} cm^2/s)	D for ceramic III (10^{17} cm^2/s)
420	0.5	1.0
470	0.5	1.2
500	0.9	1.4
540	4.1	5 → 1.4[a]
570	32 → 1.4[a]	14 → 2.4[a]
620	5.0	7.0
650	79 → 1.3[a]	2· → 22[b]
690	15 → 0[a]	31

[a]   Decrease in apparent D with time, for example from 32 × 10^{-17} to 1.4 × 10^{-17} cm^2/s, etc.
[b]   Increase in apparent D with time from 2 × 10^{-17} to 22 × 10^{-17} cm^2/s.

$D \simeq 1 \times 10^{-16}$ cm$^2 \cdot$s^{-1}. An americium content of up to 4.7 wt% alone did not
change the diffusion properties significantly. Curium-containing glasses [20, 21]
that had accumulated about 2.8 × 10^{18} α-decays/g (corresponding to a simulated
storage time of high-level-waste (HLW) glasses of > 10 000 years) showed enhanced,
'stepwise'[5] tracer penetration at the temperatures at which the stored energy is
released [21]; this indicates an accelerated diffusion during energy release. The
amount of tracer penetration was proportional to the amount of stored energy
and the maximum was ~ 3 μm (temperature range 380 to 500°C).

4.   CONCLUSIONS AND SUMMARY

The present study shows that actinide diffusion in phosphate and borosilicate
glasses is very slow indeed up to the softening point. This is even more true in
glass ceramics. Tracer penetration gave good indications of the kinetics of
crystallization, which was also followed ceramographically and by X-ray
diffractometry. No major effort was made to attribute the large number of X-ray

---

[5] Change to a different temperature results in a new step in tracer penetration.

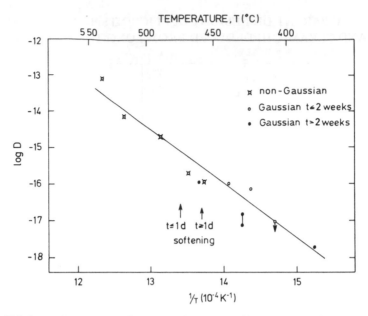

FIG.6.   *Arrhenius type of diagram of uranium diffusion in phosphate glass.*

peaks observed in recrystallized specimens (in particular in the fission-product-containing materials) to specific compositions or minerals (see, for example Ref. [22]), but the shape of the penetration profiles indicated different — but always small — diffusion rates in the different crystallization products.

These experiments showed two additional features. Firstly, the differences in behaviour between glasses and glass ceramics were less than expected, indicating that a certain percentage of glass phase remains in the ceramics. Secondly, the behaviour depended on the thermal pre-treatment.

No significant difference was observed between uranium and plutonium diffusion. Alpha radiation in an americium-loaded glass did not increase D-values above the 10^{-16} $cm^2 \cdot s^{-1}$ level, hence the radiation enhanced diffusion coefficients, D^*, are essentially smaller. This is in accord with theoretical considerations and experiments on nuclear fuel materials which yield D^*-values of the order of 10^{-18} $cm^2 \cdot s^{-1}$. In contrast, enhanced tracer penetration was noted in curium-containing glasses with a simulated storage time of more than 10 000 years. Even here, however, tracer penetration was limited to a few micrometres.

It is difficult to attribute Arrhenius equations to the observed data, since these are limited to a narrow temperature range between the softening point and $\sim 400°C$, where diffusion becomes negligible ($D < 5 \times 10^{-18}$ $cm^2 \cdot s^{-1}$). Apparent activation enthalpies range between 1.6 and 3.0 eV for the different materials.

For these reasons, it is impossible to determine the lattice locations and diffusion mechanism of the actinides. Applying the compensation law [22–24] and comparing the present results with the data by Kahl [5] for caesium and cerium diffusion shows that the measured D-values can be expected to be small for the highly charged ions. However, the measured values are yet smaller than expected, indicating that the constant used to calculate the 'effective ionic radius[6], ρ,' might possibly be larger than 1.59 for 4 and 6-valent ions.

A computer program was written to estimate the relative importance of the present diffusion data in the leaching of waste glasses. Typical leach rates of these glasses at ambient temperature are of the order of 10^{-6} g/cm^2·d [19] and they increase with temperature. For diffusion to eliminate the resultant concentration gradients, D-values of the order of 10^{-13} and 10^{-14} cm^2·s^{-1} would be needed.[7] Such values are a factor of 10^3 to 10^4 higher than the present results, even at 400°C.

From the diffusion point of view, the glasses and glass ceramics are thus a safe medium to store actinides. Observed plutonium gradients in leaching experiments [19] are therefore stable until the surface layers get mechanically 'peeled-off'.

ACKNOWLEDGEMENTS

The author would like to thank Drs W. Guber, L. Kahl, U. Riege and K. Scheffler of the Kernforschungszentrum Karlsruhe and Drs W. Heimerl and H. Lahr of Gelsenberg AG, Essen, for providing the glass samples and for discussions. He is also grateful to G. Guardalben for providing the computer programs and for experimental help. Valuable assistance in performing the experiments, which extended over very long periods of time, was also given by V. Meyritz and G. Zeibig.

REFERENCES

[1] FRISCHAT, G.H., Ionic Diffusion in Oxide Glasses, Diffusion and Defect Monograph Series No. 3, 4, Trans. Tech. Publications.
[2] WILSON, C.G., CARTER, A.C., Phys. Chem. Glasses 5 (1964) 111.
[3] TURCOTTE, R.P., Battelle Pacific Northwest Laboratories Rep. BNWL-2051, UC-70 (1976).

---

[6] The effective ion radius, ρ, is usually taken to be 1.59 r, where r is the ionic radius of a 3-valent ion.

[7] Depending on time, since leaching is a linear function of time, t, whereas diffusion is a linear function of $\sqrt{t}$.

[4] WILLIAMS, J.A., BRUNGS, M.P., McCARTNEY, E.R., Phys. Chem. Glasses **16** (1975) 53.

[5] KAHL, L., PhD Thesis, D 83, Technische Universität Berlin (1974).

[6] KAHL, L., SCHIEWER, E., Reaktortagung Berlin, Deutsches Atomforum eV, Bonn (1974) 301.

[7] RALKOVA, J., SAIDL, J., Kernenergie **10** (1967) 161.

[8] SCHMITZ, F., LINDNER, R., J. Nucl. Mater. **17** (1965) 259.

[9] HÖH, A., MATZKE, Hj., Nucl. Instrum. Methods **114** (1974) 459.

[10] MATZKE, Hj., J. Phys. (Paris) **37** (1976) C7-452.

[11] MATZKE, Hj., Plutonium and Other Actinides, North Holland Publishing Co., Amsterdam (1976) 801.

[12] ZIEGLER, J.F., CHU, W.K., IBM (USA) Rep. RC-4288 (1973).

[13] WARD, D., FORSTER, J.S., ANDREWS, H.R., MITCHELL, I.V., BALL, G.C., DAVIES, W.G., COSTA, G.J., Atomic Energy of Canada Ltd., Rep. AECL-4914 (1975).

[14] HIRSCH, H.J., MATZKE, Hj., J. Nucl. Mater. **45** (1972/73) 29.

[15] NITZKI, V., MATZKE, Hj., Phys. Rev., B **8** (1973) 1894.

[16] HEIMERL, W., Reaktortagung Nürnberg, Deutsches Atomforum eV, Bonn (1975) 695.

[17] Van GEEL,J., ESCHRICH, H., HEIMERL, H., GRZIWA, P., Management of Radioactive Wastes from the Nuclear Fuel Cycle (Proc. Symp. Vienna, 1976) Vol. 1, IAEA, Vienna (1976) 341.

[18] GUBER, W., Kernforschungszentrum Karlsruhe, personal communication (1976).

[19] SCHEFFLER, K., RIEGE, U., LOUWRIER, K., MATZKE, Hj., RAY, I., THIELE, H., Kernforschungszentrum Karlsruhe Rep. KFK-2456, *and* Euratom Rep. EUR-5509e (1977).

[20] SCHEFFLER, K., RIEGE, U., Kernforschungszentrum Karlsruhe Rep. KFK-2422 (1977).

[21] SCHEFFLER, K., RIEGE, U., HILD, W., Kernforschungszentrum Karlsruhe Rep. KFK-2333 (1976).

[22] MALOW, G., SCHIEWER, E., Hahn-Meitner-Inst. für Kernforschung Berlin GmbH Rep. HMI-B-217 (1977).

[23] RUETSCHI, P., Z.Phys.Chem. N.F. **14**, Akademische Verlagsanstalt, Frankfurt/Main (1958) 27

[24] WINCHELL, P., High Temp. Sci. **1** (1969) 200.

DISCUSSION

R. ODOJ: The leaching factors for a ceramic product and a glass are of the same order of magnitude, although on physico-chemical grounds there should be a difference. At all events this seems to depend on the amount of glass phase which is still in the ceramic. Do you think that it is the same with the diffusion coefficient, and would you expect other values in pure crystalline phases which have a different leachability?

Hj. MATZKE: I agree with your first statement. As to diffusion, the *effective* diffusion coefficient is certainly determined mostly by the glass phase remaining. Diffusion coefficients in the pure crystalline phases were not measured, but we used a computer program to unfold the non-Gaussian diffusion profiles which we measured for recrystallized glasses. In this way we obtained D-values which were usually somewhat smaller than those for the glass phase, and attributed these to the newly formed products of crystallization. Since we are dealing with a very complex system yielding a vast number of different crystalline phases, measuring the diffusion properties of each of these separately would be too big a task.

P.R. TREMAINE: In water, at temperatures above 150°C, glasses recrystallize very quickly owing to hydrothermal devitrification. We have been concerned that low-temperature leach tests are not a good criterion of the ability of a glass to retain waste products. Would you care to comment?

Hj. MATZKE: In the paper I have dealt mainly with the question of whether actinide diffusion contributes to the loss of actinide activity from glasses and, if so, to what extent. What I have shown is that even in the glass phase, the contribution of diffusion is small. Diffusion in crystallized glasses is either similar to or slower than that in the parent glasses. Only during crystallization is there actinide mobility (limited in space and time). I fully agree that high-temperature leach tests are important. We plan to do these tests as well; so far the highest temperature in our leach tests was 100°C. Though leach rates will change with temperature, I do not think that this will affect my argument that solid-state diffusion of actinides is too slow to explain their leaching behaviour and that some other effect similar to the one described (mechanical peeling) must still be responsible for actinide loss.

A. NAOUMIDIS: Since glass recrystallizes fast, thanks to the α-emission of the actinides, should we speak about diffusion of actinides in 'glass' or in a 'two-phase matrix'?

Hj. MATZKE: The question of α-decay and α-radiation enhancing crystallization of glasses has, to my knowledge, still not been fully clarified. However, experiments with curium-containing glasses simulating waste glasses older than 10 000 years in age reveal no structural change when stored at ambient temperature.[8] The stability of glasses vis-à-vis α-radiation may be different at higher storage temperatures. The experiments reported here have shown, however, that actinides move faster during either recrystallization or release of stored energy. This mobility is nonetheless limited − to < 4 μm penetration distance in our experiments − and the subsequent diffusion rates in the crystallized products tend to be smaller than those in the previous glass phase, as has already been stated.

R. ODOJ: The concept for final storage of radioactive fission products favoured by the Federal Republic of Germany involves reprocessing of the fuel, and fixing the resulting high-level radioactive waste in a solid matrix. At the "Entsorgungszentrum" at Gorleben, a vitrification process providing a borosilicate glass matrix is envisaged.

Radiation waste can enter the biocycle at two stages of this process: during fabrication, in the off-gas stream, and during storage, owing to leaching by groundwater.

---

[8] See Ref. [20] of the paper.

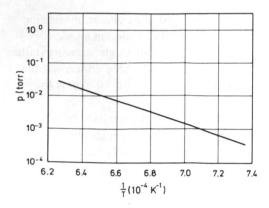

FIG.A. Data on sodium vapour pressure in the Na_2O-SiO_2 binary system obtained by the Knudsen effusion method.

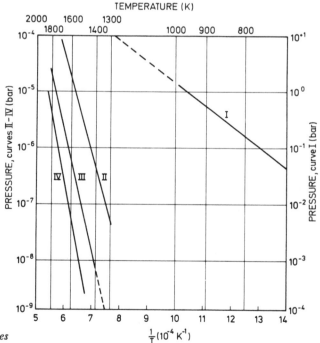

FIG.B. Caesium vapour pressures above pure caesium and the $Cs_2O-Al_2O_3-SiO_2$ ternary system. I pure Cs; II $CsAlSiO_4$; III $CsAlSi_2O_6$; IV $CsAlSi_5O_{12}$.

I   Cs-metal
II $Cs\,Al\,SiO_4$
III $Cs\,Al\,Si_2O_6$
IV $Cs\,Al\,Si_5O_{12}$

The fixation of the waste in the glass involves melting temperatures of ~1200°C. Volatilization is the predominant mechanism of fission product release during the solidification process. The pressure and composition of the atmosphere above the samples is important, since CO_2 and H_2O can increase the vaporization rate by orders of magnitude.

In particular the degree of evaporation of alkali metals such as sodium and caesium is high, because their oxides begin to react with SiO_2 at temperatures above about 600°C, i.e. after the $\alpha \rightarrow \beta$ quartz transformation.

During storage, mobile chemical species can migrate under the influence of thermal gradients and condense in voids and cracks, forming highly leachable phases, e.g. Cs_2MoO_4.

The non-radioactive element, sodium, is an integral part of the glass matrix. High sodium content provides good homogeneity and a low melting point. As the alkali metal content increases, and the composition becomes more complex, the mechanical stability of the glass decreases. Thus the vaporization of non-radioactive elements is of importance, because it affects the long-term storage properties of the glass.

The vapour pressure of sodium over the temperature range 800–1200°C at pressures between 10^{-4} and 10^{-1} torr has been measured by the Knudsen effusion method (Fig.A). The vapour pressure of sodium does not depend on the amount of SiO_2. The incongruent evaporation of sodium from silica glasses occurs by dissociation:

$$Na_2O \cdot nSiO_2 \rightarrow \underline{Na_2O} \cdot SiO_2 + (n-1)SiO_2$$
$$\longrightarrow 2\,Na + \tfrac{1}{2}O_2 \rightarrow Na_2O(g)$$

The vapour pressure depends on the dissociation and, hence, the amount of sodium metasilicate present in the glass.[9]

It should be noted that the vapour pressure above the ternary system Na_2O–CaO–SiO_2 increases with calcium content up to 10% Ca, after which it decreases. The sodium vapour pressure also increases if water vapour is present above the glass.

Caesium, a fission product, is fixed in the glass during melting in compounds such as Cs_2O or $CsNO_3$, and the physical chemistry of these compounds determines the caesium volatility. Mass spectrometric studies have shown that the species resulting from decomposition of Cs_2O are Cs^+, O_2^+ and Cs_2O^+. The last two have vapour pressures one order of magnitude lower than Cs^+.

---

[9] KRÖGER, C., SÖRGSTRÖM, L., Glastechn. Ber. 38 8 (1965) 42.

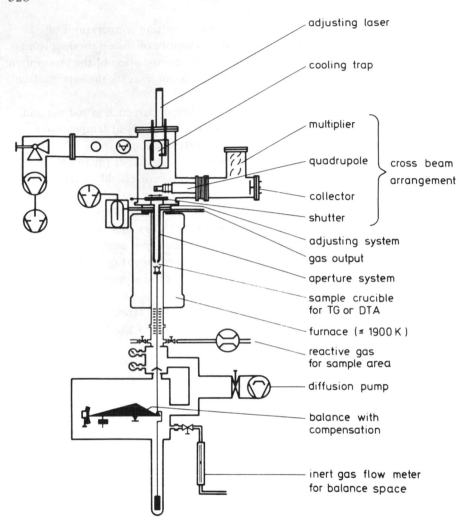

adjusting laser

cooling trap

multiplier

quadrupole   } cross beam
                      arrangement

collector

shutter

adjusting system
gas output

aperture system

sample crucible
for TG or DTA

furnace (≤ 1900 K)

reactive gas
for sample area

diffusion pump

balance with
compensation

inert gas flow meter
for balance space

FIG.C. High-temperature apparatus for alkali-metal volatilization experiments.

Our experiments have shown that there is no significant difference between vaporization of pure $CsNO_3$ and that of a mixture of $CsNO_3$ and sodium borosilicate glass. For the most part, caesium evaporates below the melting temperature of the glass frit.

We have tried to find more stable systems involving caesium and silica. The results of intensive experimental studies on the vaporization behaviour of caesium from the $Cs_2O-Al_2O_3-SiO_2$ ternary system[10] are shown in Fig.B. The vapour pressure of caesium above the ternary system is some 10 orders of magnitude lower than that above pure caesium.[11]

These results suggest that crystalline or ceramic matrices may be advantageous alternatives to glass, being products of greater thermodynamic stability, i.e. having lower free energies than glass.

The final storage matrix must, however, be a satisfactory storage medium for some 40 elements deriving from high-level waste. A matrix which might be advantageous from the point of view of one element may be disadvantageous from that of another element.

To study the volatility of fission products in a more realistic fashion, a new apparatus has been developed (Fig.C). It makes possible thermogravimetric analysis (TG) and differential thermal analysis (DTA) at atmospheric pressure under various atmospheres. Differential pumping provides the high vacuum needed in the mass spectrometer measuring space.

We are using this equipment to obtain detailed information on the volatility of the alkali metals during the solidification process.

[10] ODOJ, R., HILPERT, K., Kernforschungsanlage Jülich GmbH Rep. 1460.
[11] ODOJ, R., HILPERT, K., Materials Research Society Proceedings (Boston, 1978).

Section F

OXIDE FUELS

THERMODYNAMIC AND CHEMICAL PROPERTIES OF THORIA-URANIA AND THORIA-PLUTONIA SOLID SOLUTION FAST BREEDER REACTOR FUELS

R.E. WOODLEY
Hanford Engineering Development Laboratory,*
Westinghouse Hanford Company,
Richland, Washington

M.G. ADAMSON
General Electric Company,
Vallecitos Nuclear Center,
Pleasanton, California,
United States of America

Abstract

THERMODYNAMIC AND CHEMICAL PROPERTIES OF THORIA-URANIA AND THORIA-PLUTONIA SOLID SOLUTION FAST BREEDER REACTOR FUELS.

The paper reviews the available thermodynamic and chemical properties data for thoria-urania and thoria-plutonia solid solution fast breeder reactor fuels. Recent measurements are emphasized and important gaps in the available data are indicated. Certain properties of the thoria-based fuels are compared with those of the better known urania-plutonia solid solutions.

1. INTRODUCTION

A variety of thorium-base fuel types (oxide, carbide, nitride and metal) are currently being considered as candidate alternative fuels in the United States of America's LMFBR development programme. Recently, the existing data for these various types of fuel were critically reviewed to provide a basis for recommending values for the various properties that are required in the design and analysis of fuel pin irradiation tests and core design studies. In carrying out this review, it quickly became apparent that there are many gaps in the existing data and, as a result, experimental programmes were initiated in the USA to provide the data most critically required.

In this paper, we review the available data for key chemical and thermodynamic properties of thoria-urania ($Th_{1-y}U_yO_{2+x}$) and thoria-plutonia ($Th_{1-y}Pu_yO_{2-x}$) solid solution, emphasizing recent measurements of thermodynamic and chemical properties performed in our respective laboratories and indicating

---

* Operated for the United States Department of Energy.

TABLE I. $Th_{1-y}U_yO_{2+x}$ AND $Th_{1-y}Pu_yO_{2-x}$: PROPERTIES AND PHENOMENA OF INTEREST, AND IN WHICH AREA OF NUCLEAR ENGINEERING

Property/phenomenon	Principal use[a]	Importance
Phase diagrams (limits of composition)	D, A, F	● ●
Phase equilibra:		
Solidus-liquidus	D, A, S	● ● ˄
Oxygen potentials	D, A, F, S	● ● ˄
Vapour pressures	D, A, F, S	● ● ˄
Ideality of solid solutions	A, F	●
Thermodynamic functions:		
Heat capacity and enthalpy	D, A, S	● ●
Entropy, enthalpy and free energy of formation	A, S	●
Fuel/cladding chemical interaction	D, A, S	● ● ●
Fuel/coolant compatibility	D, A, S	● ● ˄
Reactivity with fission products	A, S	● ●

[a] Abbreviations: D — design; A — analysis; F — fabrication; S — safety. For a general ranking of importance: ● useful; ● ● important; ● ● ● critical.

important gaps in the data available. The comparison of certain properties of the thoria-based fuels with those of the better known U-Pu-O system afford a heightened perspective to our review and discussion. Our attention is restricted to oxides, because the bulk of extent data refers to this type of fuel. The composition range of interest, $0.1 \leqslant y \leqslant 0.4$, is that pertinent to fast breeder reactors, for which the fissile component of thoria-urania fuels would be ^{233}U or ^{235}U.

Our literature review revealed the existence of a reasonable body of data for $Th_{1-y}U_yO_{2+x}$, but the values of y were generally somewhat lower than required ($y < 0.2$) and the stoichiometry, x, was often imprecisely defined. In contrast, there is an extreme paucity of data for $Th_{1-y}Pu_yO_{2-x}$. The chemical and thermodynamic properties and phenomena to be reviewed are listed in Table I, together with an indication of the principal areas for which these data are of interest.

In the following sections of this paper, the properties tabulated above will be discussed separately for the Th-U-O and Th-Pu-O systems only when more than superficial information is available on a given property. Often this is not the case and the two systems will be discussed together.

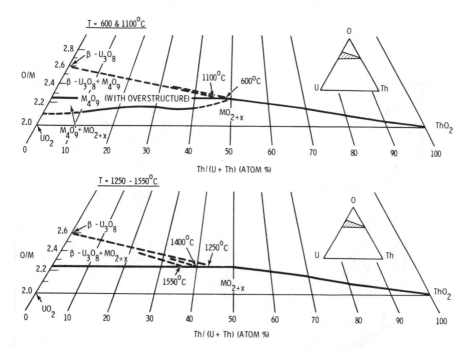

FIG.1. *Sections of the phase diagram for the ternary system* Th-U-O *from 600 to 1550°C* [1].

2. PHASE DIAGRAMS AND PHASE EQUILIBRIA

2.1. Phase diagrams

Although the phase diagrams for the Th-U-O and Th-Pu-O systems are incomplete, ThO_2 is known to be completely miscible with both UO_2 and PuO_2, forming complete series of fluorite cubic solid solutions. Thoria-urania solid solutions exhibit considerable ranges of hyperstoichiometry upon heating to high temperatures in oxidizing atmospheres but remain stable only as long as the ThO_2 content exceeds 40 to 50%. Phase diagrams for the Th-U-O system at temperatures from 600 to 1500°C [1] are shown in Fig.1. Within the range of application of these data to nuclear fuels, the Th-U-O system exhibits only a single cubic solid solution. In Fig.2, the limiting compositions given by Gilpatrick et al. [2] for the precipitation of a second phase of, or similar to, U_3O_8 from uranium-rich solid solutions are illustrated.

At sufficiently high temperatures ($> 1500°C$, approximately) and low oxygen activities, ThO_2-UO_2 solid solutions are capable of becoming hypostoichiometric, but this region of the Th-U-O phase diagram has yet to be explored. In contrast,

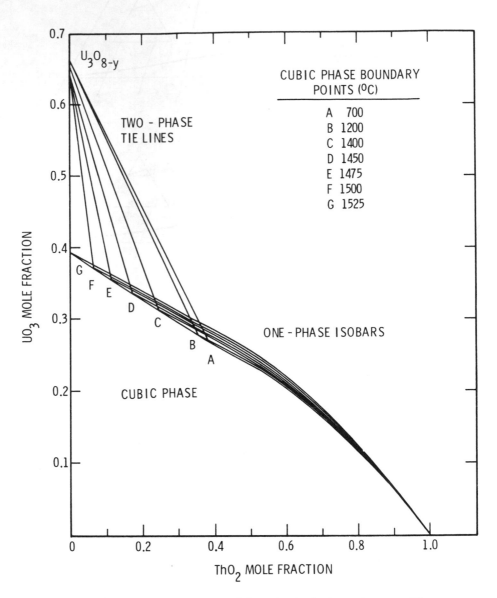

FIG.2. *Limiting composition of phases in the thoria-urania system* [2].

ThO_2-PuO_2 solid solutions cannot become hyperstoichiometric, but instead exhibit a considerable range of hypostoichiometry, as will be noted in the section on oxygen potentials. From this brief discussion, it is apparent that, although only a minimal amount of data exists for either the Th-U-O or Th-Pu-O phase diagram, the available data suggest that only single solid solutions occur within the regions of interest for potential fast reactor fuels.

2.2. Phase equilibria — solidus/liquidus

Solidus/liquidus temperatures have been reported for $Th_{1-y}U_yO_2$ only in the composition range 0–17 mol% ThO_2 and thus beyond the range of primary interest. These data [3], which were obtained by a thermal-arrest procedure on totally encapsulated samples, are illustrated in Fig.3. Encapsulation of the fuel samples was employed to prevent changes in stoichiometry as well as the preferential vaporization of UO_2 at the high melting temperatures. Unfortunately, the other melting-point studies [4, 5], which covered wider ranges of y, involved fuel specimens exposed to the furnace environment. Although compositional uncertainties are consequently implied, the results of these studies are shown in Fig.4, because they do indicate approximate temperature intervals over which melting may be expected.

Some melting points have been reported for $Th_{1-y}Pu_yO_2$ based on measurements employing a tungsten ribbon filament [6]. However, the melting points of both plutonia and thoria (by extrapolation) are low by ca. 120–130°C, indicating that the entire curve should be displaced upward by this amount. Additionally, the bulk of these data were obtained on fuel specimens with y > 0.5, as may be seen in Fig.5. Thus, the majority of melting point measurements for the Th-U-O and Th-Pu-O systems refer to compositions beyond the range of interest for fast breeder reactors.

2.3. Phase equilibria — oxygen potentials

2.3.1. $Th_{1-y}U_yO_{2\pm x}$

Oxygen potentials[1] ($\Delta \bar{G}_{O_2}$) of single-phase $Th_{1-y}U_yO_{2+x}$ have been measured in three separate studies over a restricted temperature range (730–1080°C) but reasonably broad compositional intervals ($0.063 < y < 0.9$; $0.001 < x < 0.157$)[7–9]. However, for compositions where data from the different investigations overlap, agreement is imperfect. This may be seen in Fig.6, in which the data of Aronson and Clayton [8] are compared with those of Tanaka et al. [9] for a temperature of

---

[1] $\Delta \bar{G}_{O_2}$ = RT ln p_{O_2} where p_{O_2} is the oxygen pressure/activity in atmospheres (1 atm = 101 325 Pa).

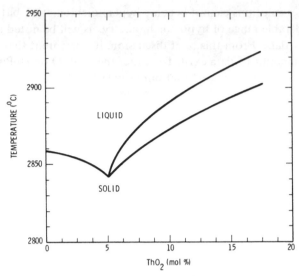

FIG.3. *Partial liquidus and solidus for the thoria-urania system (adapted from Ref.[3]).*

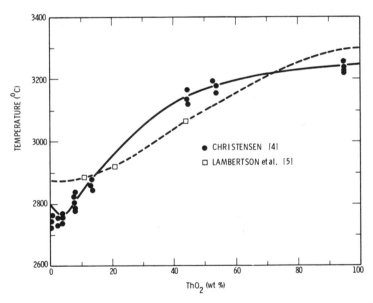

FIG.4. *Melting points of thoria-urania solid solutions.(Data from Refs [4, 5].)*

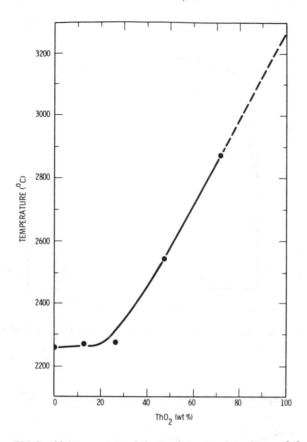

FIG.5. Melting points of thoria-plutonia solid solutions [6].

1250°K. Also shown are data for UO_{2+x} [10, 11] and $U_{0.75}Pu_{0.25}O_{2+x}$ [12, 13]. To a fair approximation the oxygen potentials of thoria-urania solid solutions are independent of the uranium content and appear to be determined primarily by the uranium valence. Because of the scatter displayed by these data and the limited temperature range over which they have been obtained, additional measurements have been initiated at the General Electric Company.

2.3.2. $Th_{1-y}Pu_yO_{2-x}$

Because of a complete lack of oxygen potential data on solid solutions of thoria and plutonia, an experimental programme was undertaken at the Hanford Engineering Development Laboratory to obtain representative data on fuel compositions in the range of particular interest for the fast breeder reactor programme.

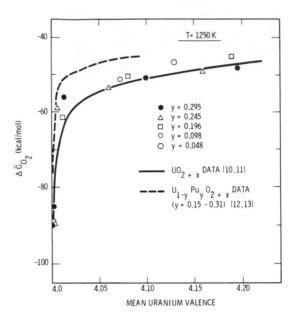

FIG.6. Oxygen potentials for $Th_{1-y}U_yO_{2+x}$, $U_{1-y}Pu_yO_{2+x}$ *and* UO_{2+x} *at 1250 K. (Data from Refs [10–13].)*

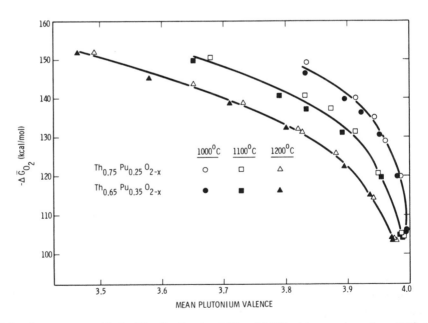

FIG.7. Oxygen potentials for $Th_{1-y}Pu_yO_{2-x}$ *(y=0.25 and 0.35) at temperatures from 1273 to 1473 K.*

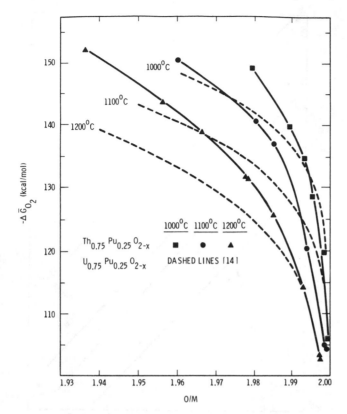

FIG.8.   *Comparison of oxygen potentials for* $Th_{0.75}Pu_{0.25}O_{2-x}$ *and* $U_{0.75}Pu_{0.25}O_{2-x}$ *at temperatures from 1273 to 1473 K.*

A thermogravimetric procedure, which was employed in earlier studies of urania-plutonia fuels [14, 15], was used to determine the oxygen potential/composition relations for thoria-plutonia solid solutions containing 25 and 35 mol% plutonia at temperatures from 1000 to 1200°C. The fuel sample, consisting of three one-gram sintered pellets, was contained in a molybdenum pan and suspended from a recording balance. The sample was exposed to a recirculating H_2 atmosphere to which various partial pressures of water vapour were added to set the oxygen potentials. The sample was initially equilibrated to an O/M ratio of 2, after which subsequent weight changes could be related to its new oxygen stoichiometry.

The results of these measurements are presented in Fig.7, where it is evident that, to a good approximation (at least at 1000 and 1200°C), the oxygen potentials of the two fuels are functions solely of the plutonium valence but not of the plutonium concentration. For a reason unknown at present, the oxygen potentials

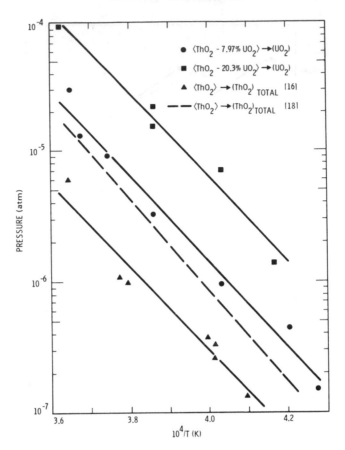

FIG.9. *Partial pressures of urania over thoria-urania solid solutions compared with total pressures of thoria at temperatures from* 2340 *to* 2760 K. *(Data from Refs* [16, 18].*)*

for $Th_{0.75}Pu_{0.25}O_{2-x}$ at 1100°C, particularly at plutonium valences < 3.95, are somewhat more reducing than anticipated based on the values at the other two measurement temperatures. Therefore, the data for the two solid solutions do not coincide as closely at 1100°C as at the other two temperatures.

Thoria-plutonia fuel exhibits oxygen potentials more negative than its urania-based counterpart, which suggests that, if equilibrated to a specified O/M ratio, it may be more compatible with stainless steel cladding than 'standard' urania-plutonia fuel. Figure 8 compares the oxygen potentials of fuels containing 25 mol% plutonia [14]. It appears that in these solid solutions thoria suppresses plutonia reduction relative to comparative plutonia behaviour in $U_{1-y}Pu_yO_{2-x}$ (and PuO_2).

TABLE II. VOLATILITY OF UO_3 AT 1200 TO 1600°C AND UO_2 ACTIVITY AT 1300°C FOR $Th_{1-y}U_yO_{2+x}$ [17]

Composition				
2+x	2.24	2.13	2.10	2.03
y	0.5	0.25	0.20	0.063
Partial pressure (atm)				
$lnP = A - B/T$, A	15.20	13.02	11.43	8.32
$lnP - A - B/T$, B	42 568	39 808	37 313	33 198
Heat of sublimation				
Value (cal/mol)	84 580	79 100	74 143	65 960
Standard deviation	1 400	800	1 500	1 900
Value (kJ/mol)	353.9	331.0	310.2	276.0
Standard deviation	5.8	3.3	6.3	7.9
Entropy of sublimation				
Value (cal/mol·K)	30.20	25.87	22.71	16.53
Standard deviation	0.85	0.45	0.95	1.2
Value (J/mol·K)	126.4	108.2	95.0	69.2
Standard deviation	3.6	1.9	4.0	5.0
UO_2 activity in 0.2 atm O_2 at 1300°C				
For given composition	0.091	0.061	0.059	0.036
Corrected to x = 0	0.5	0.24	0.20	0.05
UO_2 activity coefficient for x = 0	1.0	0.96	1.0	0.83

2.4. Phase equilibria — vapour pressures

Total vapour pressures of metal-bearing species over single-phase $Th_{1-y}U_yO_{2+x}$ have been measured by means of transpiration techniques by Alexander et al. [16] and by Aitken et al. [17]. The former investigators studied solid solutions with y = 0.08 and 0.20 over the temperature range 2100 to 2500°C. Because H_2 or an Ar-H_2 mixture was used as the carrier gas, the thoria-urania solid solutions were presumed to be nearly stoichiometric. The total UO_2 and ThO_2 vapour pressures determined by Alexander et al. are presented in Fig.9, in which they are compared to the total pressure of ThO_2 based on the measurements of Ackermann and Rauh [18]. Although the absolute magnitudes of the partial pressures measured by Alexander et al. are in question, it is interesting that, even when y is as low as 0.08, the urania partial pressure is significantly higher than that of thoria.

TABLE III. COMPARISON OF ACTIVITY COEFFICIENTS FOR PuO_{2-z} IN $Th_{1-y}Pu_yO_{2-x}$ AND IN $U_{0.75}Pu_{0.25}O_{2-x}$ AT 1400 K

Pu Valence	$\gamma_{PuO_{2-z}}$ $(Th_{1-y}Pu_yO_{2-x})$ $y = 0.25$	$y = 0.35$	$\gamma_{PuO_{2-z}}(U_{0.75}Pu_{0.25}O_{2-x})$
3.96	1.03	1.03	1.05
3.92	1.08	1.07	1.05
3.88	1.14	1.11	1.08
3.84	1.20	1.15	1.10
3.80	1.28	1.21	1.13
3.60	1.65	1.58	1.40

Aitken et al. [17] investigated solid solutions with y ranging from 0.063 to 0.5 at temperatures from 1200 to 1600°C. Dry air was employed as the carrier gas and consequently the solid solutions were grossly hyperstoichiometric (x = 0.03 to 0.24). By assuming that the principal uranium-bearing species was UO_3, the results presented in Table II were derived for the four solid solutions studied.

Analogous vapour pressure measurements have not been reported for $Th_{1-y}Pu_yO_{2-x}$.

2.5. Phase equilibria — ideality of solid solutions

2.5.1. $Th_{1-y}U_yO_{2+x}$

By utilizing the results of their transpiration measurements of the UO_3 partial pressure over $Th_{1-y}U_yO_{2+x}$ solid solutions, Aitken et al. were able to derive activities of UO_2 in these fuel materials [17]. The results thus obtained indicate that stoichiometric solid solutions are nearly ideal, but, with the addition of interstitial oxygen to the lattice, the activity coefficients for UO_2 decrease, presumably as a consequence of strong interactions between uranium ions and the excess oxygen. Urania activities at 1300°C for various values of x and y are also shown in Table II.

TABLE IV. SPECIFIC HEAT VALUES FOR ThO_2-10 wt% UO_2 AND ThO_2-20 wt% UO_2 [22]

Temperature (°C)	Specific Heat (cal/g·°C)	
	ThO_2-10 wt% UO_2	ThO_2-20 wt% UO_2
0	0.056_5	0.056_1
100	0.061_1	0.061_5
200	0.063_5	0.064_2
400	0.066_8	0.067_2
600	0.068_0	0.069_1
800	0.069_6	0.070_7
1000	0.071_1	0.072_2
1200	0.072_5	0.073_5
1400	0.073_8	0.074_9
1600	0.075_2	0.076_2
1800	0.076_5	0.077_5
2000	0.077_8	0.078_7

2.5.2. $Th_{1-y}Pu_yO_{2-x}$

Analysis of $\Delta\bar{G}_{O_2}$ measurements on thoria-plutonia solid solutions using the computational technique of Rand and Markin [19] has shown that, although the stoichiometric solutions may be ideal, the activity coefficients for PuO_{2-z} dissolved in ThO_2 increase with decreasing plutonium valence, i.e. increasing values of z. Furthermore, these solid solutions show greater non-ideality than their urania-plutonia counterparts. Activity coefficients for $Th_{1-y}Pu_yO_{2-x}$ and $U_{1-y}Pu_yO_{2-x}$ are compared in Table III. Unfortunately, the plutonia/oxygen potential data presently available in the literature [20, 21] are not consistent, and the calculated activity coefficients, while indicating trends correctly, can at best be considered only semiquantitative.

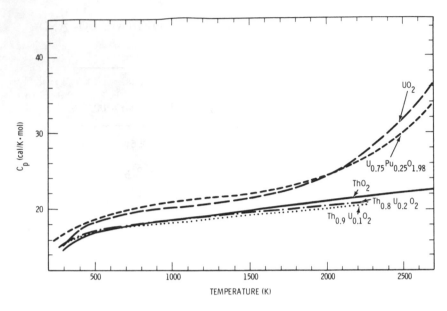

FIG.10. *Comparison of heat capacities of thoria-based and urania-based mixed-oxide fuels. (Data from Refs [22–25].)*

3. THERMODYNAMIC FUNCTIONS

3.1. Heat capacity and enthalpy

A Bunsen ice calorimeter was employed by Springer et al. [22] to determine the enthalpy and heat capacity of thoria and thoria-urania solid solutions containing up to 20 wt% urania over the temperature interval 0 to 2000°C. These data are presented in Table IV as specific heats whose limits of error amount to about ±3%. From a comparison of the values for the different compositions, it is apparent that the specific heat varies only slightly with the solution of up to 20 wt% UO_2 in ThO_2. A further comparison of the heat capacities of the two thoria-urania solid solutions is shown in Fig.10 where they are plotted with recently-assessed data for UO_2 [23], ThO_2 [24] and $U_{0.75}Pu_{0.25}O_{1.98}$[25]. Over the composition and temperature ranges illustrated, the thoria-based materials do not exhibit the characteristic upswing displayed by urania-based fuels above about 1700°C.

No heat capacity measurements have been reported for thoria-plutonia solid solutions.

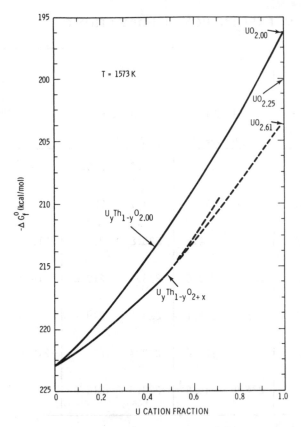

FIG.11.   *Free energy of formation of* $Th_{1-y}U_yO_{2+x}$ *at 1573 K* [17].

3.2. Entropy, enthalpy and free energy of formation

Free energies of formation for $Th_{1-y}U_yO_{2+x}$ at $1300°C$ were derived by
Aitken et al. [17] from their measurements of the volatilization of UO_3 from
thoria-urania solid solutions containing up to 50 mol% urania. These results are
illustrated in Fig.11 for both stoichiometric and hyperstoichiometric solid solutions.
Because the stoichiometric materials are essentially ideal, thermodynamic functions
can be readily derived from the corresponding functions for thoria and urania. In
the hyperstoichiometric case, on the other hand, the solid solutions are not ideal
and a knowledge of the component activity coefficients is required for any
realistic derivation of thermodynamic functions for the solid solutions.

Free energies of formation for $Th_{1-y}Pu_yO_{2-x}(y = 0.25$ and $0.35)$ at temperatures from 1000 to $1200°C$ have been calculated from the recently-measured

TABLE V. FREE ENERGY OF FORMATION OF $Th_{1-y}Pu_yO_{2-x}$

	ΔG_f^o (kcal/mol)					
	1273 K		1373 K		1473 K	
Plutonium Valence	y=0.25	y=0.35	y=0.25	y=0.35	y=0.25	y=0.35
4.0	228.6	225.0	224.3	220.8	220.1	216.7
3.96	228.2	224.6	224.1	220.4	219.9	216.3
3.92	227.9	224.1	223.8	220.0	219.6	215.9
3.88	227.5	223.6	223.4	219.6	219.3	215.5
3.84	227.2	223.1	223.1	219.1	219.0	215.0
3.80	226.8	222.6	222.7	218.6	218.6	214.6
3.60	224.9	219.9	220.8	216.0	216.9	212.2

oxygen potential data and use of the fact that the stoichiometric solid solutions are ideal. The desired free-energy changes were obtained by subtracting the free-energy change accompanying the removal of x/2 moles of oxygen from the free energy of formation of the stoichiometric solid solution, viz. :

$$\Delta G_f^0(x=x) = \Delta G_f^0(x=0) - \frac{1}{2} \int_2^{2-x} \Delta G_{O_2} \, dx \qquad (1)$$

The free energies derived in this manner have been listed in Table V and are illustrated in part in Fig.12. No other thermodynamic data are presently available for either the Th-U-O or Th-Pu-O system.

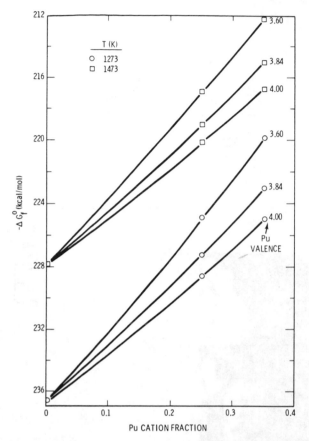

FIG. 12. Free energy of formation of $Th_{1-y}Pu_yO_{2-x}$ at 1273 and 1473 K.

4. FUEL/CLADDING CHEMICAL INTERACTION

Fuel/cladding chemical interaction (FCCI) describes the fission-product-accelerated oxidative attack of the cladding that is frequently observed in stainless steel clad mixed-oxide fast reactor fuel pins. Several FCCI mechanisms are believed to be operative in such pins, involving reactive fission products such as caesium, tellurium and iodine [26]. Caesium and tellurium fission products have been specifically implicated in the most aggressive attack modes (intergranular cladding component transport) undergone by austenitic stainless steel cladding [26, 27].

Although fast-neutron-irradiated $Th_{1-y}U_yO_{2+x}$ and $Th_{1-y}Pu_yO_{2-x}$ fuel pins have yet to be examined, a limited amount of data is available for thermal neutron irradiations of stainless steel and zircaloy-clad elements. Zircaloy-clad elements

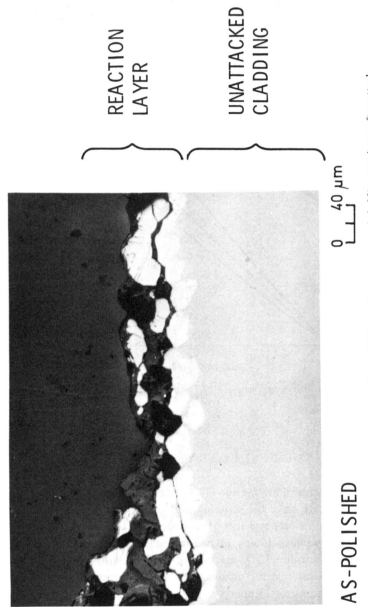

AS-POLISHED

FIG. 13.　Photomicrograph of 20% CW Type 316 stainless steel cladding specimen after attack by 1:1 Cs:Te mixture buffered by $Th_{0.75}U_{0.25}O_{2+x}$ (700°C for 200 h).

irradiated to moderately high burn-ups with cladding inner-surface temperatures as high as 705°C exhibit growth of thin adherent ZrO_2 layers on the cladding inner surface but no severe FCCI effects. The few stainless steel clad elements that have been irradiated for significant periods of time had cladding inner surface temperatures well below the known threshold temperature for FCCI ($\sim$ 525°C) [26] and thus, not surprisingly, did not exhibit significant cladding inner surface attack.

A number of out-of-pile FCCI simulation experiments in Type-316 stainless steel capsules, the methodology of which has been described previously [28], have been performed at General Electric to evaluate the comparative influences of $Th_{1-y}U_yO_{2+x}$ and $Th_{1-y}Pu_yO_{2-x}$ fuels on caesium and tellurium-induced cladding attack. Preliminary assessments had indicated (a) the existence of an initially low oxygen activity (O/M) at the fuel/cladding interface in a $Th_{1-y}Pu_yO_{2-x}$ fuel pin was coupled with a rapid increase of O/M ratio with burn-up, and (b) the existence of an initially high fuel-surface oxygen activity in a $Th_{1-y}U_yO_{2+x}$ fuel pin was coupled with a slow rate of O/M ratio increase with burn-up. Results of experiments simulating conditions at the interface of a pin loaded with $Th_{0.75}U_{0.25}O_{2+x}$ fuel (x = 0.0001 to 0.0005) suggest that rather extensive combined intergranular/matrix attack could occur in pins with a high cladding temperature at moderate to high burn-ups (see, for example, Fig. 13). Results from previous experiments performed at the (low) oxygen activity level characteristic of mixed (U,Pu) oxide fuels [28, 29] are also applicable to $Th_{1-y}Pu_yO_{2-x}$ fuel pins, hence similar FCCI behaviour is expected for these two fuel types.

5. FUEL/ALKALI-METAL COOLANT COMPATIBILITY

Very little data exist on the reactivity or compatibility of dense compacts of $Th_{1-y}U_yO_{2+x}$ and $Th_{1-y}Pu_yO_{2-x}$ with liquid alkali metals (Na, NaK and Cs). Pure thoria shows good compatibility with liquid sodium up to a temperature near its boiling point (1154 K) (and with caesium vapour up to $\sim$ 1873 K), and urania shows similar behaviour provided it is exactly stoichiometric. Cylindrical fuel compacts of $Th_{0.975}U_{0.025}O_2$, irradiated in liquid $Na_{0.33}K_{0.67}$ with outer surface temperatures of 160−275°C, underwent no reactions or dimensional changes. Based on the known behaviour of cylindrical UO_{2+x} and $U_{0.75}Pu_{0.25}O_{2\pm x}$ pellets with liquid sodium, liquid potassium, $Na_{0.33}K_{0.67}$ and liquid caesium [30], hyperstoichiometric thoria-urania pellets are also expected to undergo volumetric expansion as a result of reactions occurring at temperatures exceeding $\sim$ 500°C. As illustrated by the stoichiometry of reaction I:

$$\frac{3x}{2}\left|Na\right| + \left|UO_{2+x}\right|_{ThO_2} = \frac{x}{2}\left\langle Na_3UO_4\right\rangle + \frac{2-x}{2}\left|UO_2\right|_{ThO_2} \quad (I)$$

the extent of reaction swelling (which is determined by the final volume fraction of Na_3UO_4) will be proportional to both x and y. Some exploratory experiments have been performed at General Electric to evaluate the potential for sodium/fuel reaction swelling in $Th_{1-y}U_yO_{2+x}$ pellets. Pellets with $y = 0.30$ and 85% and 94% of theoretical density were adjusted to stoichiometric ($x = 0$) and hyperstoichiometric ($x \simeq 0.01$) compositions by equilibration in flowing mixtures of dry H_2 in argon (24 h at 1200°C) and CO_2/CO in argon (24 h at 1400°C). The compatibility tests consisted of heating the pre-measured pellets in clean liquid sodium for 200 hours at 800°C inside nickel capsules, and then re-measuring the pellets after cleaning off excess sodium. No diameter changes occurred in the stoichiometric pellets ($\Delta d/d < 0.2\%$), but small diameter increases were measured in both hyperstoichiometric specimens (($\Delta d/d$) $\simeq 0.9\%$ for 94% TD and $\simeq 1.5\%$ for 85% TD).

The compatibility of $Th_{1-y}Pu_yO_{2-x}$ pellets with liquid sodium has yet to be evaluated. It is an open question whether PuO_{2-x} will react with liquid sodium to form "Na_3PuO_4" in the absence of a stabilizing matrix of Na_3UO_4[27].

6. CHEMICAL REACTIONS WITH FISSION PRODUCTS

The chemical states of fission products in irradiated ThO_2-based fuels have been discussed extensively by Lindemer [31] and also by others [32, 33]. Lindemer pointed out that the oxygen potential of irradiated $Th_{1-y}U_yO_2$ fuels for high-temperature reactors (i.e. $T_{fuel} > 1000°C$) is determined principally by solution of $PaO_{2.5-x}$, CeO_{2-x}, Y_2O_3 and rare earth oxides $(RE)O_{1.5}$ in the UO_{2+x}-ThO_2 matrix and their concomitant influence on the uranium valence. In fast breeder reactor fuel pins, somewhat lower fuel temperatures are possible; thus, we must consider reactions between caesium (and rubidium) fission products and the fuel that may control the oxygen activity in the fuel/cladding gap (these reactions can, in turn, influence caesium axial transport behaviour, fuel/cladding mechanical interactions, and FCCI).

Although caesium does not react with ThO_2, reactions analogous to those undergone with UO_{2+x} and $U_{1-y}Pu_yO_{2-x}$ [34–36] must be expected for $Th_{1-y}U_yO_{2+x}$ and, possibly, $Th_{1-y}Pu_yO_{2-x}$. With UO_{2+x}, small quantities of caesium (vapour or liquid) react to form the low-density compound Cs_2UO_4, viz.:

$$x Cs + \left\langle UO_{2+x} \right\rangle = \frac{x}{2} \left\langle Cs_2UO_4 \right\rangle + \frac{2-x}{2} \left\langle UO_2 \right\rangle \qquad (II)$$

when $450 \leqslant T \leqslant 1100°C$. An analogous uranoplutonate $Cs_2(U_{1-y}Pu_y)O_4$ is believed to form when caesium reacts with $U_{1-y}Pu_yO_{2-x}$ ($x \leqslant 0.015$ for $y = 0.25$) under equivalent conditions. Under caesium-rich conditions ($a_{Cs} \simeq 1$) both UO_2-based oxides appear to react to form a caesium/fuel compound containing U(V) ions ($Cs_{2-x}MO_3$ or Cs_2MO_{4-x}, where $M = U$ or $U_{1-y}Pu_y$) [35, 36]. The reactions of caesium with ThO_2-based oxides are expected to be analogous to the corresponding reactions of sodium (see §5), the extent of reaction in each case being determined by the values of x and y. (However, the threshold values of x will be somewhat smaller for caesium reactions than for sodium reactions.) Qualitative confirmation of these predictions for $Th_{1-y}U_yO_{2+x}$/Cs chemical interactions is being obtained from the results of caesium thermomigration experiments currently underway at General Electric. The behaviour of caesium over columns of $Th_{0.70}U_{0.30}O_{2+x}$ pellets contained in axial-temperature-gradient capsules (600 to 1400°C) is found to closely resemble its behaviour over slightly hyperstoichiometric UO_{2+x}[36].

REFERENCES

[1] PAUL, R., KELLER, C., J. Nucl. Mater. **41** (1971) 133.

[2] GILPATRICK, L.O., STONE, H.H., SECOY, C.H., Reactor Chemistry Division Annual Progress Report, Oak Ridge National Laboratory ORNL-3591 (1964) 160.

[3] LATTA, R.E., DUDERSTADT, E.C., FRYXELL, R.E.,, J. Nucl. Mater. **35** (1970) 347.

[4] CHRISTENSEN, J.A., Quarterly Progress Report Research and Development Programs Executed for the Division of Reactor Development, October, November, December, 1962, General Electric Co., Richland Rep. HW-76559 (1963) 11.5.

[5] LAMBERTSON, W.A., MUELLER, M.H., GUNZEL Jr., F.H., J. Am. Ceram. Soc. **36** (1953) 397.

[6] FRESHLEY, M.D., MATTYS, H.M., Proc. 2nd Int. Thorium Fuel Cycle Symp., Oak Ridge, Tennessee, May 1966, USAEC (1968) 463.

[7] ANDERSON, J.S., EDGINGTON, D.N., ROBERTS, L.E.J., WAIT, E., J. Chem. Soc. (1954) 3324.

[8] ARONSON, S., CLAYTON, J.C., J. Chem. Phys. **32** (1960).

[9] TANAKA, H., KIMURA, E., YAMAGUCHI, A., MORIYAMA, J., Nippon Kinzoku Gakkai-Shi *(J. Japan Inst. Metals)* **36** (1972) 633.

[10] MARKIN, T.L., BONES, R.J., UKAEA Reports AERE-R4042 and AERE-R4178 (1962).

[11] ARONSON, S., BELLE, J., J. Chem. Phys. **29** (1958) 151.

[12] WOODLEY, R.E., ADAMSON, M.G., J. Nucl. Mater. (to be published).

[13] CHILTON, G.R., KIRKHAM, I.A., Plutonium 1975 and Other Actinides (Proc. 5th Int. Conf., Baden-Baden, 1975: BLANK, H., LINDNER, R., Eds), Elsevier, New York, and North Holland Publ. Co., Amsterdam (1976) 171.

[14] WOODLEY, R.E., J. Am. Ceram. Soc. **56** (1973) 116.

[15] WOODLEY, R.E., J. Nucl. Mater. **74** (1978) 290.

[16] ALEXANDER, C.A., OGDEN, J.S., CUNNINGHAM, G.W., Battelle Memorial Institute Rep. BMI-1789 (1967).

[17] AITKEN, E.A., EDWARDS, J.A., JOSEPH, R.A., J. Phys. Chem. **70** (1966) 1084.

[18] ACKERMANN, R.J., RAUH, E.G., High Temp. Sci. (1973) 463.

[19] RAND, M.H., MARKIN, T.L., Thermodynamics of Nuclear Materials, 1967 (Proc. Symp. Vienna, 1967), IAEA, Vienna (1968) 637.

[20] MARKIN, T.L., RAND, M.H., Thermodynamics (Proc. Symp. Vienna, 1965) Vol. 1, IAEA, Vienna (1966) 145.

[21] ATLAS, L.M., SCHLEHMAN, G.J., in Plutonium 1965 (Proc. 3rd Int. Conf. London, 1965:
      KAY, A.E., WALDRON, M.B., Eds), Barnes and Noble, New York, and Chapman and Hall,
      London (1967) 838.
[22] SPRINGER, J.R., ELDRIDGE, E.A., GOODYEAR, M.U., WRIGHT, T.R.,
      LAGEDROST, J.F., Battelle Memorial Institute Report BMI-X-10210 (1967).
[23] KERRISK, J.F., CLIFTON, D.G., Nucl. Technol. 16 (1972) 531.
[24] GODFREY, T.G., WOOLEY, J.A., LEITNAKER, J.M., Oak Ridge National Laboratory
      Rep. ORNL-TM-1596 (Rev.) (1966).
[25] GIBBY, R.L., LEIBOWITZ, L., KERRISK, J.F., CLIFTON, D.G., J. Nucl. Mater. 50
      (1974) 155.
[26] ADAMSON, M.G., IAEA/IWGFR Technical Committee Meeting on Fuel-Cladding
      Interactions (Tokyo, Feb. 1977), Rep. IAEA-IWGFR-16, IAEA, Vienna (1977) 170.
[27] ADAMSON, M.G., et al., this symposium, Paper IAEA-SM-236/63.
[28] AITKEN, E.A., ADAMSON, M.G., DUTINA, D., EVANS, S.K., Thermodynamics of
      Nuclear Materials 1974 (Proc. Symp. Vienna, 1974) Vol.1, IAEA, Vienna (1974) 187.
[29] ADAMSON, M.G., IAEA/IWGFR Technical Committee Meeting on Fuel-Cladding
      Interactions (Tokyo, Feb. 1977), Rep. IAEA-IWGFR-16, IAEA, Vienna (1977) 108.
[30] ADAMSON, M.G., General Electric Company (USA) Rep. GEAP-14093 (1976).
[31] LINDEMER, T.B., J. Am. Ceram. Soc. 60 (1977) 409.
[32] FÖRTHMANN, R., GRÜBMEIER, H., KLEYKAMP, H., NAOUMIDIS, A., Thermo-
      dynamics of Nuclear Materials 1974 (Proc. Symp. Vienna, 1974) Vol.1, IAEA, Vienna
      (1975) 147.
[33] LEIGH, H.D., McCARTNEY, E.R., J. Aust. Ceram. Soc. 13 (1977) 12.
[34] FEE, D.C., JOHNSON, I., DAVIS, S.A., SHINN, W.A., STAAHL, G.E., JOHNSON, C.E.,
      Argonne National Laboratory Rep. ANL-76-126 (1977).
[35] CORDFUNKE, E.H.P., Thermodynamics of Nuclear Materials 1974 (Proc. Symp. Vienna,
      1974) Vol.2, IAEA, Vienna (1975) 185.
[36] ADAMSON, M.G., AITKEN, E.A., Trans. Am. Nucl. Soc. 17 (1973) 195.

DISCUSSION

A.S. PANOV: When you sought the dependence of the chemical potential of oxygen on the plutonium valence in the $Th_{1-y}Pu_yO_{2+x}$ system, did you consider a possible change in the valence of thorium?

M.G. ADAMSON: The calculations to which you refer were based on the assumption that only the plutonium valence changes.

A.S. PANOV: Did you compare the rates of corrosion of UO_{2+x} and $U_{1-y}Th_yO_{2+x}$ for the same value of x? What part does thorium play in the processes of interaction with caesium and other fission products?

M.G. ADAMSON: Yes, we compared the corrosion behaviour of stainless steel in the presence of caesium and tellurium fission products for UO_{2+x} and $Th_{1-y}U_yO_{2+x}$ in isothermal capsule experiments. However, these experiments were performed at a common uranium valence (about 4.003) rather than at a common value of x. At these oxygen activities, in the presence of high concentrations of caesium and tellurium, quite a severe intergranular attack of the cladding alloy was observed. At a common value of x (say 0.002) I would imagine that attack in the presence of fission products would be considerably more severe in the case of mixed thoria-urania.

Hj. MATZKE: I wonder about the technical potential of the urania-thoria fuel for fast breeder application. Since you have to start with stoichiometric compositions, corrosion is likely to cause problems. How does the O/M ratio change with burn-up if you allow for some plutonium being formed from ^{238}U and for most fissions being due to ^{233}U?

M.G. ADAMSON: Of necessity thoria-urania fuels would establish a somewhat higher oxygen activity than thoria-plutonia fuels at the start of irradiation. However, owing partly to the different fission spectra of ^{233}U and ^{239}Pu, the rate of increase in oxygen activity (O/M) with burn-up will be considerably greater in thoria-plutonia than in thoria-urania. I do not think that breed-in of plutonium in thoria-urania fuel will contribute significantly to the (low) oxygen production rate in this fuel. However, I agree with you that its corrosion potential vis-à-vis the cladding may be high. This aspect of thoria-based fuel behaviour needs further evaluation.

OXYGEN POTENTIALS OF $U_{0.77}Pu_{0.23}O_{2\pm x}$
IN THE TEMPERATURE RANGE 1250–1550°C

G.R. CHILTON, J. EDWARDS
Windscale Nuclear Power Development Laboratories,
United Kingdom Atomic Energy Authority,
Windscale Works,
Sellafield, Seascale, Cumbria,
United Kingdom

Abstract

OXYGEN POTENTIALS OF $U_{0.77}Pu_{0.23}O_{2\pm x}$ IN THE TEMPERATURE RANGE 1250–1550°C.
    Oxygen potentials for $U_{0.77}Pu_{0.23}O_{2\pm x}$ have been measured at four temperatures
between 1250 and 1550°C using a thermogravimetric technique. Additional measurements
made on UO_{2+x} and $U_{0.69}Pu_{0.31}O_{2\pm x}$ are compared with previously published data. Anomalies
in some of the low temperature measurements may be attributable to small-scale variations in
the Pu/(U + Pu) ratio.

1. INTRODUCTION

Uranium plutonium oxide solid solutions can exist over a wide range of
oxygen compositions both above and below the stoichiometric value. The majority
of the oxides used in fast reactors will have oxygen to metal (O/M) ratios in the
range 1.95 to 2.00. Near to stoichiometric composition, large changes in the
oxygen potential of the oxide occur for small changes in the oxygen concentration.
Accurate knowledge of these oxygen potentials is important both in fuel manu-
facture and interpretation of constitutional changes during irradiation.

Low-temperature measurements (up to 1100°C) have been made of the
oxygen potentials of both hyper- and hypo-stoichiometric oxides using an EMF
technique [1, 2]. Work at higher temperatures using various techniques has been
concentrated on either the hypostoichiometric [3–6] or hyperstoichiometric
compositions [7, 8]. Oxygen potential data on the oxides near to stoichiometric
compositions are sparse and there is some scatter in the results [9]. For these
compositions reliance must be placed on the extrapolation of the low temperature
EMF results or mathematical models [10–12] fitted to the non-stoichiometric
higher temperature experimental data.

In the present work the oxygen potentials of $U_{0.77}Pu_{0.23}$ oxide were measured
for near stoichiometric compositions at four temperatures between 1250 and

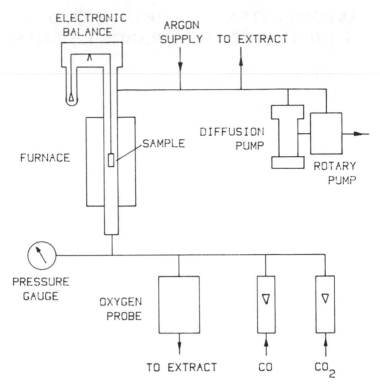

FIG.1. Schematic layout of thermobalance equipment

TABLE I. ANALYSIS OF THE OXIDES

		UO_2	$U_{0.77}Pu_{0.23}O_2$	$U_{0.69}Pu_{0.31}O_2$
Pu/U + Pu		–	23.16	31.59
Americium	(ppm)	–	350	480
Iron	(ppm)	450	270	150
Nickel	(ppm)	15	20	$\leqslant 4$
Chromium	(ppm)	5	30	35
Aluminium	(ppm)	30	30	40
Calcium	(ppm)	10	35	30
Fluorine	(ppm)	$\leqslant 10$	$\leqslant 10$	$\leqslant 10$
Chlorine	(ppm)	$\leqslant 10$	$\leqslant 10$	$\leqslant 10$
Carbon	(ppm)	50	50	95

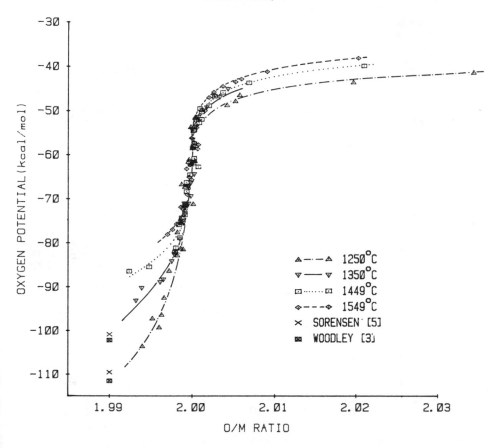

FIG.2. Oxygen potentials for $U_{0.77}Pu_{0.23}O_{2\pm x}$

1550°C using a thermogravimetric technique. Additional measurements were made using UO_{2+x} and $U_{0.69}Pu_{0.31}$ oxide to establish the validity of the technique by comparison of results with previously published data [8, 13, 14].

2. EXPERIMENTAL

The oxide specimens used in this work were in the form of sintered annular pellets (5 mm OD, 1.5 mm ID, 5 mm long). The pellets were produced by granulation, pelleting and sintering of oxide powders. Both the $U_{0.77}Pu_{0.23}O_2$ and the $U_{0.69}Pu_{0.31}O_2$ powders were produced from ammonium diuranate and

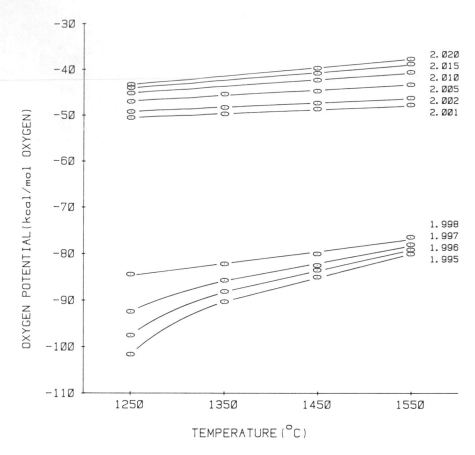

FIG.3.   Change of oxygen potential with temperature for $U_{0.77}Pu_{0.23}O_{2\pm x}$

plutonium hydroxide co-precipitated from nitrate solutions by ammonia. The UO_2 pellets were produced from ceramic-grade UO_2 powder derived from ammonium diuranate. The analysis of the sintered pellets is shown in Table I. The mixed oxide pellets were examined by colour autoradiography and electron probe microanalysis (EPMA) to assess the homogeneity of the uranium and plutonium.

The equipment used for the oxygen potential measurements was similar to that reported previously [8]. The schematic layout is shown in Fig.1. The sample, suspended from a Cahn model RH microbalance, was heated in a high-temperature furnace. The furnace had an alumina work tube and could be used under vacuum

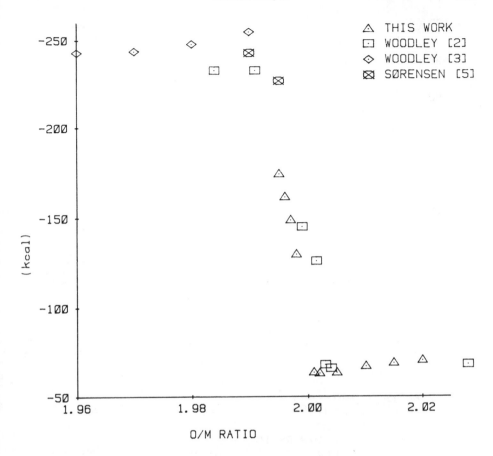

FIG.4. *Partial molar enthalpies of UPuO$_{2\pm x}$*

or flowing gas conditions. The sample weight and temperature were continuously recorded. The nominal precision of the electrobalance is better than 10 μg. However, practical limitations imposed by the glove box installation have reduced this to a usable value of 100 μg, equivalent to an error in O/M ratio measurement of less than 0.0005 O/M units.

The oxygen potentials required for this work were produced by blending CO and CO$_2$. To maintain a constant gas flow at the extremes of the mixing range a total flow of 1000 cm^3/min was required. Of the total flow, 750 cm^3/min were diverted to an oxygen probe housed in a separate furnace at 670°C. The remaining 250 cm^3/min of CO/CO$_2$ flowed over the oxide specimens suspended in the thermobalance furnace.

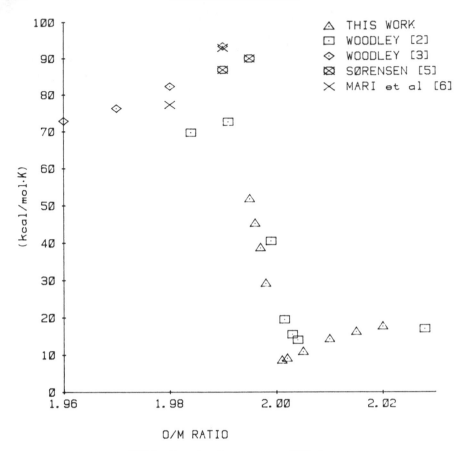

FIG.5. *Partial molar entropies of* $UPuO_{2 \pm x}$

The oxygen potential of the gas mixture at the temperature of the thermo-balance furnace, T_2 (K), was calculated from the equation:

$$RT_2 \ \ln p_{(O_2)} \ (kcal/mol) = \frac{T_2}{T_1} \ (-134108-4FE_0)-134108-3.109 \ T_2 \quad (1)$$

where R is the gas constant, F the Faraday constant, p the pressure (atm), and T_1 the absolute temperature (K) and E_0 the EMF (volts) of the oxygen probe.

For the determinations, four pellets (of approximately 1 g each) were suspended from the balance. The composition of the pellets was initially adjusted to a reference state. For this work the reference was taken as −61.5 kcal at

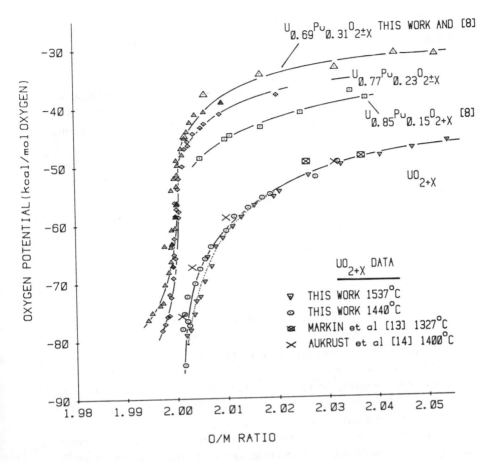

FIG.6. *Oxygen potentials for UPuO$_{2\pm x}$ and UO$_{2}$+x*

1250°C. Equilibrium was judged to have been achieved when identical specimen weights were recorded after two consecutive weighings taken at half hourly intervals. All the weighings were made in static CO/CO$_2$.

The CO/CO$_2$ composition was adjusted to give the required oxygen potential at the particular temperature of the measurement. Once stable conditions were achieved at the oxygen probe, the required proportion of the CO/CO$_2$ mixture was diverted over the oxide specimens until equilibrium was reached.

The experiments were carried out in such a sequence that equilibrium was obtained both by oxidation and reduction; thus gains and losses in weight were recorded. The small scatter in the results show that equilibrium was reached irrespective of the direction of the reaction.

The oxygen potentials of the $U_{0.77}Pu_{0.23}$ oxide were obtained at 1250, 1350, 1449 and 1549°C. Between 25 and 30 determinations were made at each temperature to cover the full range of oxygen ptoentials possible with the CO/CO_2 mixing equipment.

At the higher temperatures and under the more oxidizing conditions, significant evaporation of the oxide occurred. Under these circumstances the oxide was returned to the reference state after every fourth determination and corrections made to compensate for the losses due to evaporation.

In addition to the determinations on $U_{0.77}Pu_{0.23}$ oxide, 33 measurements were made using $U_{0.69}Pu_{0.31}$ oxide at 1542°C to check the agreement with previous work and to extend the data to the hypostoichiometric region. As a further check on the validity of the technique, two series of measurements were made at 1440 and 1537°C using UO_2 specimens to obtain a comparison with established data.

3.   RESULTS

The large number of individual determinations made prevent the inclusion of all the tabulated data in this report. These data will be reported elsewhere [15]. The O/M ratios of the $U_{0.77}Pu_{0.23}$ oxide are plotted against oxygen potential in Fig.2. The O/M ratios of the oxide were calculated from the weight changes and the specimen weights. In addition to the correction applied for evaporation loss, each individual weight change was corrected for the buoyancy effect of the CO/CO_2 mixture and temperature.

The results of this work suggest that the precision of weight measurements and evaporation and buoyancy corrections could produce a maximum error of up to 200 μg. This is equivalent to an error in the O/M ratio of less than 0.001.

The values of the oxygen potentials shown in Fig.2 were calculated according to Eq.(1). The errors in this determination are likely to result from inaccuracies in the measurement of the temperature of the oxygen probe furnace and the thermobalance furnace. A five-degree error in both temperatures would result in a maximum error in the calculated oxygen potential of ~1.7 kcal. The results in Fig.2 suggest that this is a pessimistic estimate.

An overall bias of the results is possible if the thermodynamic data on the CO/CO_2 system incorporated in Eq.(1) are incorrect. Previous work [8] suggests that, based on results obtained with the Ni/NiO system, any overall bias is likely to be less than 1 kcal at 1250°C.

The change of oxygen potential with temperature for a range of O/M ratios is shown in Fig.3. The partial molar enthalpy and entropy data calculated from this information are shown together with those of other workers in Figs 4 and 5.

The additional data obtained on uranium oxide and $U_{0.69}Pu_{0.31}$ oxide are shown together with other relevant work in Fig.6.

4.   DISCUSSION

The agreement between the UO_{2+x} data obtained here and those of other workers demonstrates the validity of this experimental technique. In Fig.6 the results obtained for UO_{2+x} at 1440 and 1537°C are plotted together with the results of Markin et al. [13] at 1327°C and Aukrust et al. [14] at 1400°C. The agreement between the various sets of data is better than 3 kcal.

Similar agreement is shown between the data obtained on $U_{0.69}Pu_{0.31}$ oxide and that published previously [8] for the same oxide composition (Fig.6). The other data for $U_{0.77}Pu_{0.23}O_{2\pm x}$ and $U_{0.85}Pu_{0.15}O_{2+x}$ clearly show the systematic decrease in oxygen potential with decreasing plutonium content.

In the case of the results on $U_{0.77}Pu_{0.23}O_{2\pm x}$, limitations in the oxygen potential possible with the CO/CO_2 mixtures precludes overlap with the data of other workers. The data of Sørensen [5] for $U_{0.80}Pu_{0.20}O_{2-x}$ and Woodley [3] for $U_{0.75}Pu_{0.25}O_{2-x}$ adjusted to 1250 and 1350°C are plotted in Fig.2 together with values from the present work. Taking into account the small differences in plutonium content, there is good agreement in this region.

Assuming stoichiometric compositions occur over the steepest section of the oxygen potential versus O/M ratio plot, the data in Fig.2 suggest that stoichiometry for $U_{0.77}Pu_{0.23}$ oxide exists over a small range (−58 to −62 kcal) between 1250 and 1550°C. At lower oxygen potentials an almost linear change in composition with oxygen potential occurs down to 1.999 at 1550°C and 1.997 at 1250°C. At lower O/M ratios, the change in composition with oxygen potential becomes greater.

The oxygen potentials for these oxides are all less negative than the extrapolation of the early low-temperature EMF measurements [1] would suggest. This is in agreement with the more recent published oxygen potential data on the uranium plutonium oxide [2−8] which show similar, less-negative results.

The results plotted in Fig.3 show that, in the hyperstoichiometric region, the oxygen potential varies linearly with temperature. In the hypostoichiometric region the data at 1250°C for O/M ratios 1.997, 1.996 and 1.995 do not coincide with the linear extrapolation of the three sets of data at higher temperatures.

The partial molar enthalpy and entropy were calculated for this oxide by a least-squares fit to the data in Fig.3. In the hypostoichiometric region, only the data at 1350, 1449 and 1549°C were used for the calculation. The data obtained for $U_{0.77}Pu_{0.23}O_{2\pm x}$ are compared with those of Woodley for $U_{0.75}Pu_{0.25}O_{2\pm x}$ [2, 3], and Sørensen [5] and Mari [6] for $U_{0.80}Pu_{0.20}O_{2-x}$ in Figs 4 and 5. In the region covered by the present work (O/M ratios 1.995 to 2.02), the results are consistent and agree well with the lower temperature EMF measurements of Woodley [2]. The agreement is encouraging since it gives confidence in oxygen potential data over the temperature range 800°C to 1550°C for these Pu/(U + Pu) compositions. The agreement further supports the rejection of the anomalous results obtained in the present work at 1250°C.

It is possible that the anomalous data may have been due to heterogeneity of the uranium and plutonium distribution in the pellets.

The uranium plutonium oxide specimens used in this work, and in most oxygen potential measurements [1–3, 5–8], were made from co-precipitated powder. It is generally accepted that this type of powder, rather than powders produced by physically blending UO_2 and PuO_2, produces a homogeneous product. Although X-ray diffraction analysis of the specimens did not show the presence of secondary phases, further examination by autoradiography showed some fine-scale heterogeneity in the uranium and plutonium distribution. Subsequent detailed examination by EPMA quantified the heterogeneity. The $Pu/(U + Pu)$ ratio of a 0.5 μm diameter volume was measured at 50 μm intervals across a pellet diameter. The pellets were examined before and after the oxygen potential measurements. Each examination involved 33 separate determinations of the $Pu/(U + Pu)$ ratio. The mean of each series of determinations did not show any significant difference (0.238 and 0.234). The standard deviation, however, was reduced by a factor of five during the oxygen potential measurements (0.100 and 0.021). The anomalous data at 1250°C were the first to be obtained with these specimens. It is possible that much of the heterogeneity was reduced during low-temperature measurements and the subsequent higher temperature determinations were unaffected.

In the hyperstoichiometric region, the agreement between the data obtained at all four temperatures suggests that the oxygen potentials in this region are not sensitive to variations in the $Pu/(U + Pu)$ ratio.

6. CONCLUSIONS

A new series of oxygen potential data has been produced covering a range of stoichiometries for $U_{0.77}Pu_{0.23}$ oxide at temperatures between 1250 and 1550°C.

The validity of the experimental technique has been demonstrated by comparison of the results on $UO_{2 + x}$ with established published data.

Anomalies in the oxygen potential data for the hypostoichiometric oxide at 1250°C may be attributable to heterogeneity in the $Pu/(U + Pu)$ ratio within the specimen. Further work is required to determine the influence of cation heterogeneity in the oxygen potentials of these uranium-plutonium oxide solid solutions.

REFERENCES

[1] MARKIN, R.L., McIVER, E.J., in Plutonium 1965 (Proc. 3rd Int. Conf. London, 1965: KAY, A.E., WALDRON, M.B., Eds), Barnes and Noble, New York, and Chapman and Hall, London (1967) 845.

[2] WOODLEY, R.E., Hanford Engineering Development Laboratory (USA) Rep. HEDL-SA-1282 (1977).

[3] WOODLEY, R.E., Hanford Engineering Development Laboratory (USA) Rep. HEDL-TME 72-85 (1972).

[4] TETENBAUM, M., Thermodynamics of Nuclear Materials 1974 (Proc. Symp. Vienna, 1974) Vol.2, IAEA, Vienna (1974) 305.

[5] SØRENSEN, O.T., in Plutonium 1975 and Other Actinides (Proc. 5th Int. Conf. Baden-Baden, 1975: BLANK, H., LINDNER, R., Eds), Elsevier, New York, and North Holland Publ. Co., Amsterdam (1976) 123.

[6] MARI, C.M., PIZZINI, S., MANES, L., TOCI, F., J. Electrochem. Soc. 124 (1977) 1831.

[7] ADAMSON, M.G., CARNEY, R.F.A., J. Nucl. Mater. 54 (1974) 121.

[8] CHILTON, G.R., KIRKHAM, I.A., in Plutonium 1975 and Other Actinides (Proc. 5th Int. Conf. Baden-Baden, 1975: BLANK, H., LINDNER, R., Eds), Elsevier, New York, and North Holland Publ. Co., Amsterdam (1976) 171.

[9] SWANSON, G.C., Los Alamos Scientific Laboratory Rep. LA-6083-T (1975).

[10] BLACKBURN, P.E., Argonne National Laboratory Rep. ANL-7977 (1973).

[11] BREITUNG, W., Kernforschungszentrum Karlsruhe Rep. KFK-2363 (1976).

[12] De FRANCO, M., GATESOUPE, J.P., in Plutonium 1975 and Other Actinides (Proc. 5th Int. Conf. Baden-Baden 1975: BLANK, H., LINDNER, R., Eds), Elsevier, New York, and North Holland Publ. Co., Amsterdam (1976) 133.

[13] MARKIN, T.L., WHEELER, V.J., BONES, R.J., J. Inorg. Nucl. Chem. 30 (1968) 807.

[14] AUKRUST, E., FØRLAND, T., HAGEMARK, K., Thermodynamics of Nuclear Materials (Proc. Symp. Vienna, 1962), IAEA, Vienna (1962) 713.

[15] CHILTON, G.R., EDWARDS, J., UKAEA, Risley, unpublished work, 1978.

DISCUSSION

M.G. ADAMSON: Although the temperature ranges of our respective measurements do not coincide, it is apparent that your new measurements on hypostoichiometric mixed oxides give significantly higher oxygen potentials than the recent combined TGA/EMF measurements of Woodley and myself. Do you think that the fact that you have used CO/CO_2 mixtures (as opposed to H_2/H_2O mixtures, for example) could in any way bias your results?

G.R. CHILTON: In the hypostoichiometric region there is good agreement with the extrapolated TGA data of our Ref.[3]. If any significant bias had been produced by using CO/CO_2, we would have expected to have seen it reflected in our UO_2 data. The agreement with previously published UO_2 data suggests that no bias occurs.

T.M. BESMANN: Commenting on Mr. Adamson's suggestion that the CO/CO_2 atmosphere could react with the (U, Pu) oxide to form oxycarbides, I should like to point out that experiments in which I tried to react CO with uranium mono- and dicarbide have been unsuccessful owing to the sluggishness of the reaction. Thus I would not expect oxycarbides to form.

Out of curiosity I attempted to model $(U, Pu)O_{2-x}$ as an ideal solid solution of UO_2, PuO_2 and Pu_2O_3. The resulting oxygen potentials did not agree with reported observations, and this model apparently will not work.

Hj. MATZKE: I consider that Mr. Adamson's remark is most important. We observed differences in plutonium self-diffusion measurements on $(U_{0.8}Pu_{0.2})O_{2 \pm x}$ near the stoichiometric composition between CO/CO_2 and H_2/H_2O atmospheres indicating some — yet unknown — change at the sample surfaces in CO/CO_2 which may also occur in EMF measurements.

I should also underline the significance of Mr. Chilton's obtaining anomalous results from heterogeneous samples. Certainly, homogeneous samples are needed for reliable EMF measurements, and care should be taken to ensure proper characterization of the specimens. May I offer a practical hint in this regard: if evaporative losses occur from a sample which does not have a quasi-congruently evaporating composition (e.g. $MO_{1.97}$ for 20% Pu), we inevitably get surface gradients in the metal-atom concentration at the surface (e.g. Pu enrichment in $MO_{2 + x}$ where the predominant vapour species is UO_3). These gradients are shallow and will not be detected by conventional chemical analysis. Since metal diffusion rates are slow, low temperature (e.g. 1200°C) anneals do not remove them. It is better to remove these surface layers mechanically by polishing.

OXYGEN DISTRIBUTION IN FAST REACTOR OXIDE FUELS

F.T. EWART
Chemical Technology Division,
Atomic Energy Research Establishment,
Harwell, Didcot, Oxfordshire,
United Kingdom

C.M. MARI
Institute of Electrochemistry and Metallurgy,
University of Milan,
Milan,
Italy

S. FOURCAUDOT, Hj. MATZKE, L. MANES, F. TOCI
Commission of the European Communities,
European Institute for Transuranium Elements,
Karlsruhe

Abstract

OXYGEN DISTRIBUTION IN FAST REACTOR OXIDE FUELS.
   The chemistry of irradiated fuel pins has been examined with respect to oxygen distribution. Two hypostoichiometric mixed-oxide fuel pins were irradiated to high burn-up under controlled conditions in a materials testing reactor. The pins were carefully characterized initially and were examined in detail by ceramographic and electron-probe techniques. The radial oxygen profile at the end of irradiation was measured. Small specimens were extracted from different points on a radius of the fuel; their oxygen potential was measured in a micro-galvanic cell. The results have shown that, although the pins had initially widely different oxygen contents (O/M values of 1.948 and 1.976), the final O/M values were virtually identical at 1.992 and 1.997. This is contrary to the behaviour predicted from existing fuel chemistry theory. Oxidation above 1.997 O/M ratio in the higher initial-O/M pin is prevented by the oxidation of the clad. Rapid failure of the thoria electrolyte in the high-temperature galvanic cell occurred under the influence of minute quantities of irradiated fuel. Such an effect imposes limits on the large-scale application of galvanic-cell techniques to irradiated materials.

1.   INTRODUCTION

   One of the most significant parameters believed to influence the internal cladding corrosion of the LMFBR fuel pin is the oxygen potential at the fuel surface. As yet, this has

not been defined experimentally with sufficient precision to
enable unequivocal statements to be made relating to the
corrosion behaviour of a particular fuel composition.

   Some of the uncertainty arises from the relationship
between the average oxygen to metal ratio (O/M) of the fuel and
the oxygen potential at the fuel surface under operating
conditions. This relationship, under isothermal conditions, is
reasonably well known for unirradiated fuel, but, under the
influence of the extreme radial temperature gradient which
exists in the operating fuel, oxygen migrates to form an
appreciable concentration gradient [1]. Furthermore, during
the fission process, the fissile atoms are replaced by a range
of fission products whose mean valency differs from that of
the fissile atoms thus producing both a change of O/M with burn
up [2] and a change in the composition of the cation species
which in turn affects the oxygen potential associated with a
particular O/M [3].

   These effects have been the subject of numerous theoretical
treatments and experimental measurements on simulated fuels but
the number of measurements on irradiated fuels is limited [4,5,6].
In the case of hypostoichiometric plutonium containing fuel
[4,5], an indirect technique has been used in which the oxygen
potential is inferred from the composition of fission product
inclusions in the fuel. Another indirect technique which was
applied previously [7] uses measurements of lattice parameters
following irradiation.

   In this paper we describe direct measurements of the quantity
of interest, the oxygen potential $\Delta \bar{G}(O_2)$ of the irradiated fuel
and its radial variation, on two well-characterised oxide fuel
pins, which have been irradiated in a Materials Testing Reactor.
The pins have also been subjected to a detailed post irradiation
examination; the results can therefore be discussed in the light
of the electron probe and ceramographic data.

2.   EXPERIMENTAL

   The fuel for these irradiations was 25% plutonium, 75%
uranium mixed oxide which had been taken from a single batch of
material prepared by a coprecipitation method. Sintered annular
pellets were used which had been annealed in wet hydrogen to
adjust the stoichiometry to that required for the experiment.
The fuel pins were 5.84 mm outside diameter and contained
ca. 100 mm of fuel between natural uranium oxide insulator
pellets. The pins were enclosed in static sodium capsules and
irradiated in the Harwell test reactor Dido at a rating of
450 $W \cdot cm^{-1}$ to a nominal burn up of 6% FIMA.[1] The irradiation
vehicle was so designed that the fuel could be quenched rapidly

----

[1]   FIMA: fission metal atoms.

TABLE I. FUEL PIN DETAILS

	Pin 1	Pin 2
Experiment No.	1080	1080
Reactor	DIDO	DIDO
Pin No.	006	007
Fuel Composition $-$ ^{235}U	0.005	0.005
^{238}U	0.759	0.759
^{239}Pu	0.184	0.184
^{240}Pu	0.044	0.044
^{241}Pu	0.008	0.008
Fuel Density($g \cdot cm^{-3}$)	10.4	10.4
Fuel OD(mm)	5.00	5.00
ID(mm)	0.15	0.15
Clad OD(mm)	5.84	5.84
ID(mm)	5.08	5.08
Linear Power($W \cdot cm^{-1}$)	450	450
Burn up (measured)(% FIMA)	6.8	6.4
Initial O/M	1.948	1.974
Clad Temp.(K)	870	870
Max. Centre Temp.(K) } calculated	2500	2400
Final O/M }	1.982	2.005

at the end of the irradiation. The temperature at the fuel
centre was reduced to 500 K within 20 seconds of reactor shut
down. The details of the irradiation conditions are given in
Table I.

Samples for the oxygen potential measurements were taken
at different radial positions. Fine holes were drilled under a
high purity inert atmosphere at accurately known radial positions
and the drillings collected constituted the microsample.
These were then transferred to a specially developed micro
galvanic cell which was able to measure the oxygen potential
over small specimens (minimum 1 mg). The method has been
extensively tested for reliability of the oxygen potential data
and for errors introduced by the specimen handling; these aspects
have been described elsewhere[8]. The technique used here
differed only in the problems which arose from the greatly
increased radioactivity from these fuel pins with shorter

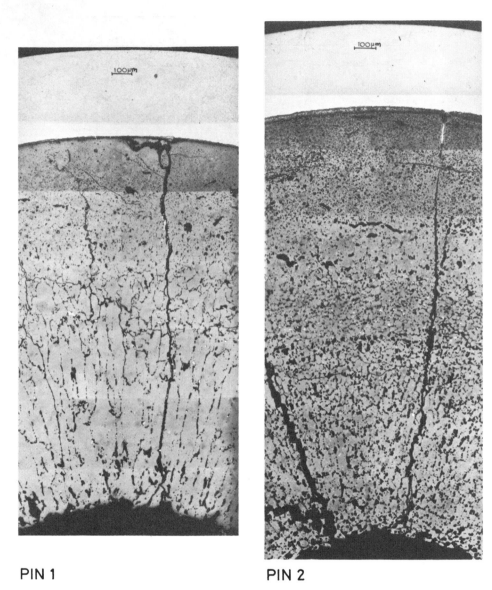

PIN 1 PIN 2

FIG.1.   Radial section of Pins 1 and 2.

cooling times than those used for the earlier experiment. The
previous work had reported that a spurious emf was found within
six hours of the specimen being loaded and was attributed to
radiation induced electronic conductivity. In this experiment
the spurious emf was found within twenty minutes and reliable
results could be obtained only by heating the cell to 1400 K in
order to anneal the 'radiation damage' immediately before each
measurement. This technique was found to give consistent results
which were reproducible over several temperature cycles.

3.   RESULTS

The fuel pins were examined first by neutron radiography
showing that no detectable fuel movement had occurred during
irradiation. Sections were taken therefore over the full fuel
length up to 10 mm from the insulator pellet. Ceramography and
electron probe analyses were made, as far as possible, on fuel
immediately adjacent to that used for oxygen potential
measurements.

3.1 Oxygen potential measurements

Oxygen potential ($\Delta \bar{G}(O_2)$) measurements were made on each
microsample at several temperatures in the temperature range
1000 K to 1273 K. The values measured at (or extrapolated to)
1273 K are quoted in Table II. The conversion of oxygen potential
to O/M was made using the recent data of Woodley [3], for mixed
oxide fuel containing fission products. The conversion was made
at 1273 K because Woodley's data appear to be more complete and
consistent at the higher temperature. No allowance was made for
possible Pu-redistribution and radial profiles in fission product
concentration.

3.2 Ceramographic and electron probe results

Figs. 1 and 2 show views of typical ceramographic sections
from pins 1 and 2. It is clear that the microstructures of the two fuels are quite different; pin 1 shows
clear evidence, from the low porosity and large columnar grains,
of having operated at a higher centre temperature than pin 2.
A number of measurements of clad thickness on two sections from
each pin showed that pin 2 had lost on average a 6 μm clad layer,
using the lower limit in the original clad thickness for this
calculation. Pin 1 showed much less corrosion. Cladding loss
amounted essentially only to the spalled layer (see Fig. 2) and
very few corrosion products were formed in the gap.
Electron probe analysis of material in the original fuel-
clad gap showed, in the case of pin 1, local depositions of
caesium and barium and heavy palladium deposition on the clad

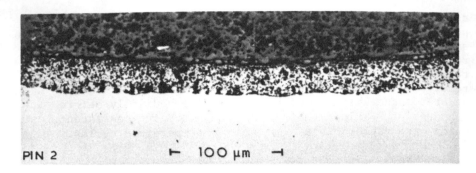

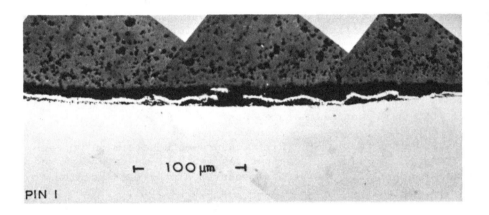

FIG.2. Detail of clad interface, Pins 1 and 2.

inside wall. The 'spalled' layer was heavily depleted in
chromium. In the case of pin 2 the clad was corroded by a
'broad front' type of corrosion and the fuel clad gap was
filled with a corrosion product containing clad constituents
and fission product caesium, tellurium and traces of palladium.
The tellurium-caesium ratio in these phases was significantly
enhanced over the fission yields and varied between 4:1 and 1:4.
No tellurium was found in the fuel-clad gap of pin 1 but, in
one section, tellurium was found in concentrations between one
and five percent in the noble metal inclusions.

3.3 Computer calculations

In order to establish a theoretical background against
which these results may be judged, the behaviour of the fuel
pins has been modelled, on the basis of the design parameters,

using one of the available computer codes for fuel performance,
TPROF [9]. The change in O/M with burn up used for this
calculation was 0.005 O/M units/1% burn up for plutonium fission.
    The oxygen redistribution was calculated using the
expression,

$$\ln x = \frac{-Q*}{RT} + C \qquad \qquad ...(1)$$

where x is the deviation from stoichiometry
    Q* " "   heat of transport for oxygen
    T " "   absolute temperature
    R " "   gas constant
    C " a constant.

Q* is taken to vary with stoichiometry and is here given
tentative values of 10 kcal·mol^{-1} at O/M = 1.97 and 18 kcal·mol^{-1}
at O/M = 2.00.
    The predictions from this model for pins 1 and 2 are shown
in Fig. 3 where the temperature and O/M profiles are shown for
irradiation times of 200 hours and at the end of irradiation
(4600 hours).

4.   DISCUSSION

    Table II shows the experimental values of oxygen potential
and the derived values of O/M for each pin. It will be seen that
the radially corrected average of the O/Ms differs significantly from that calculated on the basis of 0.005 O/M units/
1% burn up. In pin 1 the calculated value (1.982) is less than
that found (1.992) whereas in pin 2 the calculated value (2.005)
is somewhat larger than the experimental (1.997).
    For the case of pin 1, where there is little evidence of
the clad being involved in the fuel chemistry, the change in O/M
with burn up derived from the experimental results and calculated
burn up is 0.0065 O/M units/1% burn up. This is some 30% larger
than the value deduced theoretically from the fission yields and
the predicted chemical state of the irradiated hypostoichiometric
fuel [2]. We are unable to provide an adequate explanation for
this difference which we believe to be real as the experimental
errors in the irradiation experiment necessary to produce an
equal effect would be an error of 0.005 in the O/M determination
together with a 15% error in the burn up. For the former there is
an expected accuracy of ± 0.002 O/M units and for the latter, the
accuracy is better than 0.5%.
    If this new value for the change in O/M with burn up is
then applied to pin 2, then the calculated final O/M is 2.016
which exceeds the experimental value by 0.019 O/M units. This
difference is attributed to the formation of some oxide phase

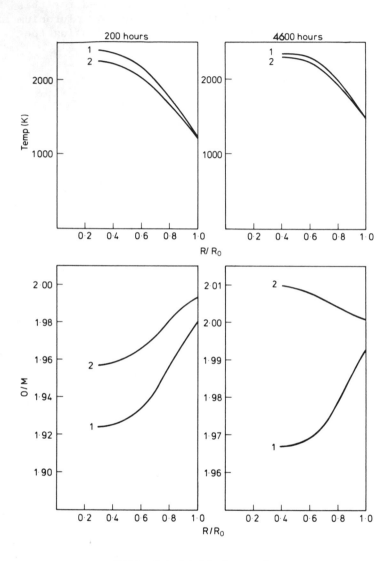

FIG.3.   Calculated fuel parameters.

TABLE II. RESULTS OF $\Delta\bar{G}(O_2)$ MEASUREMENTS

Pin	R/R_o	$\Delta\bar{G}(O_2)$ (kcal.mol^{-1}) 1273 K	Equivalent O/M	Mean O/M in fuel pin
Pin 1	0.95	−119	1.995	
	0.89	−131	1.989	
	0.81	−125	1.993	
	0.78	−127	1.992	1.992
	0.72	−125	1.993	
	0.69	−130	1.990	
	0.40		1.985 (extrapolated)	
Pin 2	0.90		2.000 (extrapolated)	
	0.76	−106.5	1.999	
	0.69	−120	1.996	
	0.65	−119	1.996	1.997
	0.63	−116	1.997	
	0.52	−119	1.996	
	0.29	−124	1.993	

(a) $\Delta\bar{G}(O_2)$ values are converted to O/M using Woodley's data [3]
    for (U,Pu)O$_2$.

(b) Inclusion of extrapolated data points (1) to centre and (2)
    to edge changes the mean O/M value in case (1) not at all,
    due to the $(R/R_o)^2$ weighting, and in case (2) by + 0.001 O/M
    units (1.998).

acting as a 'buffer' preventing an increase in the mean O/M
beyond 1.997. A detailed search has been made for phases such
as Cs$_2$MoO$_4$ or Cs$_2$UO$_4$ but without success; the clad corrosion
and resultant loss of at least 6 µm of clad of wall thickness,
however, provides an adequate buffer. If 6 µm of clad would be
converted to M$_2$O$_3$, this would be equivalent to a change of the
fuel in O/M by nearly 0.04. The required buffering could also
be accounted for by assuming a $\sim$ 14 µm layer (taking the nominal
original clad thickness) to convert all its Cr into Cr$_2$O$_3$. This
would also be compatible with the observation of a porous layer
of metallic appearance in the ceramograph. Moreover, the oxygen

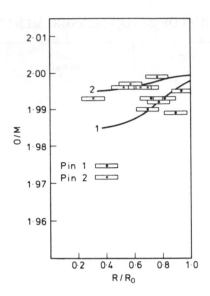

FIG.4. *Experimental results.*

potential near the surface of the fuel, -119 kcal·mol^{-1} at
1273 K corresponds well with the oxygen potential of -115 kcal·
mol^{-1} found by McIver and Teale [10] for the onset of oxidation
in stainless steel (at 1273 K). The values of oxygen potentials
derived are thus consistent with the experimental observations.

This interpretation of the experimental results has been
used to modify the calculations of O/M change in the computer
code. The O/M profiles obtained from such calculations are
shown as solid lines in Fig. 4. Shown also on this figure are
the experimental results. The experimental points are given
error boxes which represent both sample 'diameter' and precision.
It can be seen that, at the final O/M found in these fuel pins,
there is only a shallow oxygen profile to be expected. It is
unfortunate that those shallow profiles, combined with the
scatter in the data which is inevitable in measurements made
close to O/M = 2 where the oxygen potential changes rapidly with
composition, preclude any accurate interpretation with respect to
the equation describing oxygen distribution. It is clear,
however, that there is an upward trend in O/M with increasing
radius and we would expect to obtain better results from fuel
pins which have a lower initial O/M or a lower burn up. The
final O/M would not then be in the region close to 2.00 and
theoretically, steeper profiles would be predicted. The rapid
cooling of the irradiation vehicle gives reasonable confidence
that the oxygen profile which existed during irradiation will be

retained. Using the most recent diffusion data [11], Norris [12] has calculated that a radial oxygen profile will take 15 minutes to decay if the gradient is removed and the fuel is held iso-thermally at 1200 K.

In addition to the results on irradiated fuel, the experiments have provided some new and unexpected results on the behaviour of thoria-yttria electrolytes under irradiation. The radiation damage to the electrolyte, which was caused by a layer of fuel, one year cooled, equivalent to about 150 $mg \cdot cm^{-2}$, was substantial. Attempts to measure oxygen potential by a galvanic cell technique with larger quantities of fuel such as that in a whole fuel pellet are unlikely to succeed. The amount of fuel necessary to cause the enhanced electronic conductivity will be reduced even further by shorter cooling times.

5. CONCLUSION

Two fuel pins have been irradiated under similar conditions and with widely different starting stoichiometries. Measurements of the oxygen potential at a number of points in the fuel show that, contrary to previous fuel chemical assessments, their chemical state at $\sim$ 6.6% burn up is virtually identical. On the basis of detailed ceramography and electron probe analysis this is attributed to the oxidation of the clad which acts as an oxygen absorber thus inhibiting the predicted increase in the fuel O/M above 1.997. The rate of rise of O/M with burn up was found in the case of pin 1 to be larger than previously pre-dicted (0.0065 O/M units/1% burn up c.f. 0.005)[2].

No conclusions could be drawn on the mechanisms for radial redistribution of oxygen because of the unexpected high final O/M of one pin, and the scatter in the results. The derivation of radial O/M profiles is difficult close to the stoichiometric composition due to the shallow slope. It is expected to obtain results more meaningful to this problem from subsequent experiments at lower initial O/M or lower burn ups, which are expected to cause steeper profiles.

Tellurium-caesium ratios significantly greater than the fission yield ratio were found in the fuel-clad gap phase of pin 2. Based on out-of-pile studies, such compositions should, at the oxygen potential found, promote an intergranular corrosion mechanism; the mechanism found was, however, of the broad front oxidation type.

6. ACKNOWLEDGEMENTS

Two of the authors, C.M. Mari and F.T. Ewart, acknowledge the support of the Directorate for Science, Education and Research of the Commission of the European Community and of their parent Laboratories, which made this cooperative programme possible. We acknowledge also the assistance of the staff of the Transuranium Institute in making the post irradiation examination, in particular R. Schreiber for obtaining the micro samples, M. Coquerelle for providing the ceramographic data and C.T. Walker for providing the electron probe data.

REFERENCES

[1] BOBER, M., SCHUMACHER, G., Adv. Nucl. Sci. Technol. **7** (1973) 21.
[2] DAVIES, J.H., EWART, F.T., J. Nucl. Mater. **41** (1971) 143; EWART, F.T., et al., J. Nucl. Mater. **61** (1976) 254.
[3] WOODLEY, R.E., J. Nucl. Mater. **74** (1978) 290.
[4] JOHNSON, I., JOHNSON, C.E., CROUTHAMEL, C.E., SEILS, C.A., J. Nucl. Mater. **48** (1973) 21.
[5] KLEYKAMP, H., J. Nucl. Mater. **66** (1977) 292.
[6] ADAMSON, M.G., AITKEN, E.A., EVANS, S.K., DAVIES,J.H., Thermodynamics of Nuclear Materials 1974 (Proc. Symp. Vienna, 1974) Vol.1, IAEA, Vienna (1975) 59.
[7] BENEDICT, U., et al., J. Nucl. Mater. **45** (1972/73) 217.
[8] EWART, F.T., MANES, L., MATZKE, Hj., MARI, C.M., TOCI, F., SCHREIBER, R., *in* Characterization and Quality Control of Nuclear Fuels (Proc. Conf. Karlsruhe, June 1978), J. Nucl. Mater. **81** (1979) 185.
[9] CALIGARA, F., Euratom, Karlsruhe, private communication, 1975.
[10] McIVER, E.J., TEALE, S., UKAEA Rep. AERE-R 4942 (1965).
[11] D'ANNUCI, F., SARI, C., J. Nucl. Mater. **68** (1977) 357.
[12] NORRIS, D., *in* Fast Breeder Reactor Fuel Performance (Proc. Int. Conf. Monterey, California, 1978: NORMAN, E.C., et al., Eds), American Nuclear Society, Illinois (1979) 292.

DISCUSSION

J.-P. MARCON: The average value of the O/M ratio after irradiation is calculated from experimental values available only for the outer part of the oxide. Does not the lack of measurements in the hot part of the oxide give rise to an error in the determination of the average O/M?

F.T. EWART: Note (b) to Table II refers to the extrapolation of the measured data points. The mean value of O/M ratio in Pin 1, to which your question refers, was found to be very insensitive to the extrapolated value because of the $(R/R_0)^2$ weighting, and for any reasonable extrapolation no change in the mean O/M ratio was observed.

G. SCHUMACHER: How did you calculate the oxygen distribution in the fuel pin? Did you fit the experimental curves by variation of the effective heat of transport Q*, or did you make assumptions about Q*?

F.T. EWART: The oxygen distribution in the fuel pin was calculated using the computer code TPROF and assumed values of Q*. Since we could not

satisfy ourselves that the experimental points were adequately described by
$\ln x \propto T^{-1}$, we showed in Fig.4 the curves obtained from the computer code
(at the corrected O/M ratio) and the data points only for illustrative purposes.
No attempt at curve fitting is implied in Fig.4.

M.G. ADAMSON: I find the following aspects of your results on the
MTR-irradiated fuel pins somewhat puzzling: (i) the high rate of O/M increase
with burn-up even as the fuel approaches a stoichiometry of 2 (high oxygen
activity); and (ii) the tendency of all your microsample EMF measurements
on irradiated mixed-oxide fuels to give O/M values close to 2 *and* small radial
O/M gradients. The implication of (i) is that ternary compounds such as
$Cs_{2-x}MO_{4-y}$, Cs_2MoO_4 etc. are not forming in high oxygen activity fuel, and
the implication of (ii) is that perhaps your fuel samples are undergoing oxidation
during drilling and/or measurement. Would you like to comment on both
these points?

F.T. EWART: The high rate of change of O/M ratio with burn-up has
puzzled us also. The fuel pins were prepared for examination by methods known
to preserve ternary caesium compounds. A number of sections were examined
by ceramography and electron probe analysis in great detail to obtain evidence
for the existence of such compounds. None was found.

The second part of your question relates to the philosophy of our method.
One can measure oxygen profiles in situ by galvanic cell techniques and face
the problems of oxygen re-equilibration during measurement (since the
measurement must be made at temperatures at which oxygen diffusion is
appreciable — Refs [11, 12] of the paper) and of radiation damage to the
electrolyte (which we have shown to be a source of serious difficulties in
high burn-up fuels). We have elected to obtain microsamples from the fuel and
measure these in isolation. This also offers the advantage that the measurement
is made on a specimen taken from a finite depth and not on the mechanically
cut face of a section. We thus have to contend only with the problem of
specimen oxidation during sampling and handling.

We have already described in some detail the testing of this method for
errors arising from a number of sources (see Ref. [8] of the paper). We have
confidence in our results.

R. HESKETH: It may be of some help in this field to state an identity for
each of the two quantities *heat of transport* and *reduced heat of transport*.
A physical system is often made up of several components, for example, a solid
which consists of uranium ions, oxygen ions, uranium ion mono-vacancies,
oxygen-ion mono-vacancies, oxygen-ion di-vacancies and clusters of such vacancies.
To each of these several components one may assign a chemical potential, μ,
identifying each component by the subscript i. Then one may write:

$$\mu_i = H_i - T(S_{i,v} + S_{i,c}) \tag{1}$$

where H_i is the enthalpy of solution of the species i, T is the temperature,

$S_{i,v}$ is the entropy of vibration of the species i, and $S_{i,c}$ is the entropy of configuration of the species i.

For example, H_i might be the enthalpy required to create one lattice vacancy in gold, or to dissolve one carbon atom in steel. In these cases, $S_{i,v}$ represents the perturbation of the vibrational states, in the gold or in the steel, which one lattice vacancy or one carbon atom produces. The quantity $(-S_{i,v})$ is a local perturbation in the heat capacity.

The configurational entropy $S_{i,c}$ is:

$$S_{i,c} = - k \ln c \tag{2}$$

where k is Boltzmann's constant and c is the local concentration of the species i.
The following identities hold. The heat of transport is:

$$Q_i^* \equiv H_i - TS_{i,v} \tag{3}$$

The reduced heat of transport is:

$$Q_i^{**} \equiv - TS_{i,v} \tag{4}$$

Equations (3) and (4) are identities. I have used here the notation customary among chemists. (Physicists sometimes use the symbol U* for the left-hand side of Eq.(3) and describe the quantity as the 'energy of transport'. For the left-hand side of Eq.(4), the physicists use the symbol Q* and describe this as the 'heat of transport'. Casual readers can therefore be easily confused by reading two textbooks, one from each of the above two scientific communities, since the symbol Q* and the associated words heat of transport carry different meanings.)

The usefulness of Eqs (3) and (4) lies in the fact that H_i and $S_{i,v}$ may frequently be determined by experiments which are simpler and more accurate than thermomigration experiments. This is especially true of the simpler solids, but it may well be of less assistance to the worker with highly sub-stoichiometric uranium oxides who has a physical system which contains several species of defect, mono-vacancies, di-vacancies and higher clusters, each with a different value of H_i and $S_{i,v}$ and all of which transport oxygen through the system. Equations (3) and (4) may be of more assistance in systems close to stoichiometry, in which H_i and $S_{i,v}$ are well defined.

Three further points are worth mentioning. Firstly, if free electrons are present, a defect such as a lattice vacancy will perturb them, and this perturbation contributes to $S_{i,v}$.

Secondly, it is necessary to distinguish open systems from closed systems. Lattice vacancies in gold are an open system, for they may be created at the cold surface of a specimen, drift up the temperature gradient with a velocity:

$$\frac{dx}{dt} = \frac{D_i \, S_{i,v}(\nabla T)}{kT} \tag{5}$$

and be destroyed at the hot surface. (D_i is the diffusion coefficient of one unit of the species i.) In such open systems, an Arrhenius plot of the concentration of species i has a slope of $(-H_i)$.

By contrast, in a closed system, a species i may not be created at one surface nor destroyed at another. Uranium dioxide is such a system if there is a vacuum at each surface. If a species i cannot be created at a surface, the flux of that species, through the specimen, is necessarily constrained to be zero. This constraint is termed a *thermomolecular pressure* and the constraint perturbs the slope of the Arrhenius plot, which now takes the value $(-H_i - TS_{i,v})$. Notice that H_i and $TS_{i,v}$ have the same sign, unlike the case in Eq.(1), in which they have opposite signs.

Thirdly, it should be noted that interstitial species do not differ from substitutional species in the slope of the Arrhenius plot; for each, in a closed system, the slope is $(-H_i - TS_{i,v})$. Not infrequently, H_i may be an order of magnitude greater than $TS_{i,v}$. It would therefore be a considerable error to assume that H_i is absent from the Arrhenius slope of an interstitial solute.

In conclusion, it should be noted that the reduced heat of transport is a function strongly dependent upon temperature. The Q^{**} of Eq.(4) is a quantity of thermal energy. It is the local perturbation of the thermal energy of an otherwise perfect material which is created by the injection of one unit of the species i into the material. Conceptually, Q^{**} and Q^* are nothing more than synonyms for quantities with which we are already familiar as terms in the chemical potential.

MESURES EN CONTINU DE LA REDISTRIBUTION DE L'OXYGENE SOUS GRADIENT THERMIQUE DANS UO_{2+x}

R. DUCROUX, M. FROMONT, A. PATTORET
CEA, Centre d'études nucléaires
   de Fontenay-aux-Roses,
Fontenay-aux-Roses,
France

Abstract–Résumé

CONTINUOUS MEASUREMENTS OF OXYGEN REDISTRIBUTION UNDER A THERMAL GRADIENT IN UO_{2+x}.
   The purpose of this work is an out-of-pile study of the redistribution of oxygen in UO_{2+x} in order to provide an explanation at a later stage for the O/U+Pu profiles obtained after quenching irradiated fuel. The thermal gradients were obtained by means of an image furnace. The focal spot, of 0.5 cm^2 in area, is brought to the top of the cylindrical oxide sample, the cold part of which is in contact with a molybdenum furnace; in this way it is possible to maintain in isothermal conditions a solid-electrolyte (ThO_2-Y_2O_3) 'minigauge' containing a chemical reference (Fe/FeO). The minigauge shows the oxygen activity in the cold part of the samples at any given time. It has been tested by measuring the oxygen potentials of different chemical systems. The experiments were performed under an atmosphere of argon purified by means of electrochemical pumps. The UO_{2+x} samples are subjected to a moderate thermal gradient of about $300°C/cm$ in order to limit the evaporation phenomena in the hot part. After continuous monitoring of the EMF, the samples are suddenly quenched and then sectioned, an analysis being made of the O/U ratio of each section. The oxygen is seen to have undergone a considerable migration up the thermal gradient even though the annealings are relatively short and the temperatures are low. A comparison of the results obtained in UO_{2+x} with a computer calculation made on the basis of thermal diffusion in the solid phase for oxygen shows very reasonable agreement.

MESURES EN CONTINU DE LA REDISTRIBUTION DE L'OXYGENE SOUS GRADIENT THERMIQUE DANS UO_{2+x}.
   L'objet de ce travail est l'étude hors-pile de la redistribution de l'oxygène dans UO_{2+x}, de façon à expliquer ultérieurement les profils O/U+Pu obtenus après trempe sur combustible irradié. Les gradients thermiques ont été obtenus à l'aide d'un four à image. La tache focale, d'une surface de $0,5 \text{ cm}^2$, est amenée au sommet de l'échantillon cylindrique d'oxyde, dont la partie froide est au contact d'un four de molybdène; celui-ci permet de maintenir en conditions isothermes une minijauge à électrolyte solide (ThO_2-Y_2O_3) contenant une référence chimique Fe/FeO. La minijauge donne à tout instant l'activité de l'oxygène à la partie froide des échantillons. Elle a été testée en mesurant les potentiels d'oxygène de différents systèmes chimiques. Les expériences ont été réalisées sous atmosphère d'argon purifié par des pompes électrochimiques. Les échantillons d'UO_{2+x} sont soumis à un gradient

thermique modeste de 300°C/cm environ pour limiter en partie chaude les phénomènes d'évaporation. Après le suivi en continu de la force électromotrice les échantillons sont trempés brutalement puis tronçonnés, le rapport O/U de chaque tranche étant analysé. L'oxygène remonte le gradient thermique de façon importante bien que la durée des recuits soit relativement courte et les températures basses. La confrontation des résultats obtenus dans UO_{2+x} avec un calcul à l'ordinateur élaboré sur la base de la thermodiffusion en phase solide pour l'oxygène montre un accord très raisonnable.

I - INTRODUCTION

On sait que les propriétés physicochimiques des oxydes $UPuO_{2+x}$ dépendent fortement du potentiel d'oxygène et de la température. A cet égard, il est nécessaire de bien connaître le sens et les mécanismes de la redistribution de l'oxygène dans les oxydes sous gradient thermique. Parmi les propriétés intéressant les combustibles des réacteurs nucléaires actuels, la conductibilité thermique dépend de la stoechiométrie des oxydes et, de ce fait, la distribution des températures est affectée par la migration de l'oxygène. Par ailleurs, la connaissance du potentiel chimique de l'oxygène est utile pour la prévision de l'état thermodynamique des produits de fission et également indispensable à une bonne compréhension des mécanismes de corrosion à l'interface oxyde-gaine.

Durant les dernières années, la redistribution de l'oxygène sous gradient thermique a été étudiée, tant d'un point de vue expérimental que théorique dans UO_{2+x} et $UPuO_{2+x}$ /̄1 à 10 7̄. D'une manière générale, les études concernant la migration de l'oxygène n'ont permis de déterminer les profils de concentration qu'après trempe des échantillons. L'objectif de ce travail était de réaliser un appareillage qui permette de suivre en continu la cinétique de migration de l'oxygène par une technique de pile électrochimique. Cette technique permet de déterminer l'évolution du potentiel d'oxygène en un point donné de l'oxyde et d'évaluer le temps d'établissement de l'état stationnaire. Nous ne présentons dans cette publication que des premiers résultats obtenus sur le système UO_{2+x} sur lequel ont déjà porté plusieurs études concordantes /̄4, 9 7̄. Bien que d'autres mécanismes de transport puissent être proposés, nous avons tenté d'appliquer les équations de thermodiffusion en phase solide décrites par la Thermodynamique des Processus Irréversibles (T.P.I.), d'application plus simple aux régimes transitoires, pour

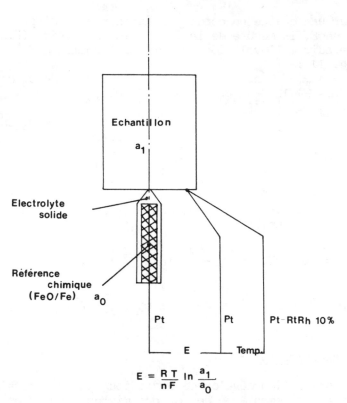

$$E = \frac{RT}{nF} \ln \frac{a_1}{a_0}$$

FIG.1. Schéma de principe de la cellule de mesures.

confronter les cinétiques expérimentales et les profils
de redistribution de l'oxygène avec les résultats déduits
d'un traitement théorique.

II - TECHNIQUES EXPERIMENTALES

- Principe des mesures électrochimiques -

Il consiste à mesurer en continu le potentiel
chimique de l'oxygène à l'extrémité froide d'un échantillon d'oxyde UO_{2+x} soumis à un gradient de température grâce au dispositif décrit ci-après. On réalise la
chaîne électrochimique suivante :

$$\text{Pt, Fe/FeO} \underset{T_1}{//} \underset{ThO_2}{ZrO_2}) \; Y_2O_3 \underset{}{//} UO_{2+x}, \underset{T_1}{Pt}$$

Pour une chaîne en condition isotherme et d'après la loi de Nernst, la mesure de la force électromotrice (F.E.M.) permet de suivre l'évolution du potentiel chimique de l'oxygène (fig. 1) :

$$(1) \qquad E = \frac{RT}{4F} \ln \frac{a_1}{a_2}$$

a_1 activité de O_2 dans UO_{2+x}

a_2 activité de O_2 dans Fe/FeO

$$(2) \qquad \Delta \overline{GO}_2 \ (UO_{2+x}) = EnF + \Delta \overline{GO}_2 \ (Fe/FeO)$$

Dans le cas où les températures de la référence et de l'oxyde ne sont pas identiques, une relation plus générale $\underline{/}\ 11\ \underline{7}$ doit être utilisée :

$$(3) \qquad E = \frac{R}{4F} \ \underline{/}\ T_1 \ln a_1 - T_2 \ln a_2\ \underline{7} + \alpha(T_2 - T_1)$$

α pouvoir thermoélectrique de l'électrolyte $\underline{/}\ 12\ \underline{7}$.

- Aspect expérimental -

L'installation développée avait pour but de travailler dans un premier temps avec des oxydes mixtes $UCeO_{2-x}$ et ultérieurement sur des oxydes $UPuO_{2-x}$. Par suite de l'insuffisance de données sur le système $UCeO_{2-x}$, celui-ci a été provisoirement abandonné au profit de l'oxyde UO_{2+x}. Plusieurs considérations nous ont amenés à définir les caractéristiques du montage :

- Obtention d'un gradient thermique éventuellement élevé. Des essais préliminaires au four solaire d'Odeillo $\underline{/}\ 13\ \underline{7}$ ont permis de réaliser des gradients de température de 5000°C/cm.

- Contrôle et mesure des pressions partielles de l'oxygène des atmosphères.

- Mesure de la cinétique de redistribution de l'oxygène "in situ" à la partie froide de l'échantillon.

Le gradient thermique est réalisé axialement sur des échantillons de 6 mm de diamètre et 9 mm de long. L'extrémité supérieure de l'échantillon est chauffée à l'aide d'un four à image $\underline{/}\ 14\ \underline{7}$, d'une puissance de 6,5 kW (fig. 2). La source de rayonnement, une lampe à

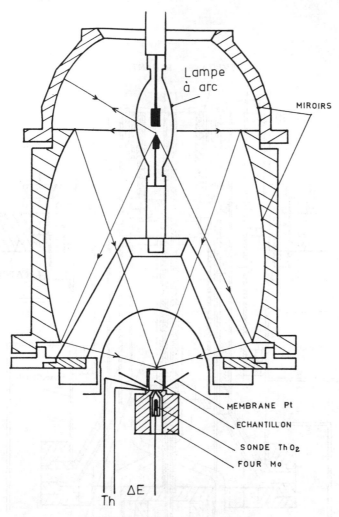

FIG.2.   Schéma de l'appareillage.

arc xénon, est disposée au foyer objet d'un miroir elliptique de révolution. Ce miroir concentre le rayonnement émis sur l'échantillon placé à son foyer image.
La partie du rayonnement de la lampe qui n'est pas reçue
directement par le miroir elliptique est renvoyée sur ce
dernier par un miroir hémisphérique centré sur le foyer
objet du miroir elliptique. Le four à image repose sur une po-
tence à hauteur variable commandée par un système hydraulique
manuel. Une défocalisation de l'échantillon permet une mise
sous gradient thermique progressive , l'allumage de la lampe à

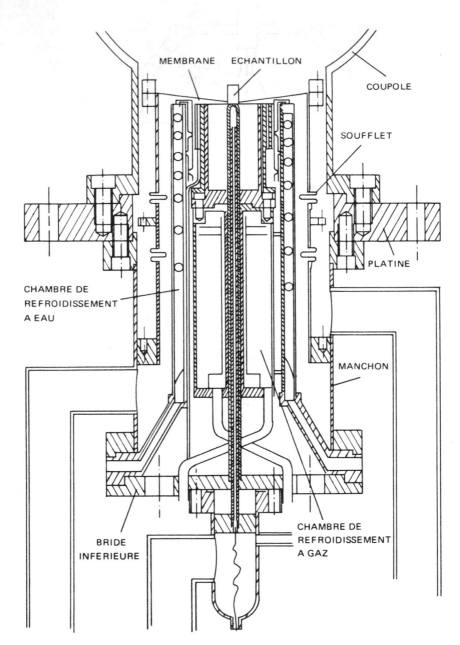

FIG.3. Schéma d'ensemble.

arc étant nécessairement instantané. La face froide des échantillons est maintenue à température fixe par un four en molybdène dont l'élément chauffant est un fil de tungstène isolé
par des baguettes d'alumine. L'alimentation du filament permet
d'atteindre 1100°C à puissance maximale. Cette alimentation
est régulée au degré. Avec ce système de chauffage couplé
four en molybdène-four à image, on réalise des gradients thermiques axiaux de l'ordre de 300°C/cm. Dans l'avenir, sans modification de l'appareillage, des gradients thermiques beaucoup plus élevés pourront être obtenus (jusqu'à 4000°C/cm).

Le profil thermique linéaire a été mesuré à l'aide de
trois thermocouples gainés en chromel-alumel ($\emptyset$ = 0,025 mm)
insérés dans l'échantillon à travers la capsule de platine
iridié. Un thermocouple Pt-PtRh 10 % est soudé au sommet de
la capsule et présent dans toutes les expériences de migration.

De cette mesure ont été déduits les profils thermiques.
La température supérieure de l'échantillon n'a pas dépassé
1100°C pour s'affranchir de l'espèce UO_3 dont la pression
dépend fortement de la température et de l'écart à la stoechiométrie.

La compatibilité chimique de la gaine avec les oxydes
nous a conduits à utiliser,pour les échantillons surstoe-
chiométriques,une capsule en platine iridié dont l'épaisseur
est de 0,2 mm et la hauteur de 9,5 mm. Par l'intermédiaire
de cette capsule, l'échantillon est appliqué fortement sur
une feuille de platine qui l'isole du four en molybdène.

Les éléments principaux de ce montage sont représentés
dans la figure 3 :

- Cellule de mesure électrochimique : la minisonde est en-
  castrée à l'extrémité d'une baguette d'alumine de 6 mm de
diamètre. Cette baguette est maintenue en pression par un
ressort à son extrémité froide pour assurer un bon contact
de la pile sur l'échantillon. Un coaxial prolonge le fil
de mesure au millivoltmètre enregistreur à très haute im-
pédance d'entrée (10^{12} Ω).

III - TECHNIQUES D'ANALYSES DES ECHANTILLONS AVANT ET APRES
      TREMPE

Après trempe au four à image, les échantillons sont
découpés en cinq ou six tranches à l'aide d'une tronçon-
neuse de précision (Macrotom). Le tronçonnage est réalisé

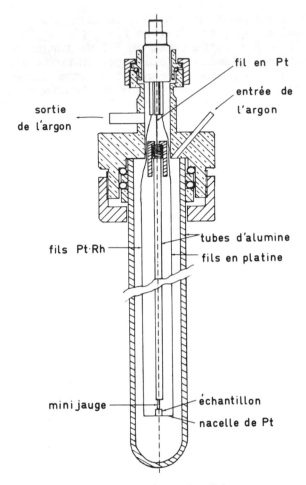

FIG.4. Schéma de la cellule.

avec une lame diamantée de 0,3 mm d'épaisseur. Les échantillons sont dégraissés aux ultra-sons dans un bain de
chlorothène puis nettoyés à l'alccol.

Un banc de mesures électrochimiques a été mis au point
par ailleurs utilisant une minijauge à oxygène identique à
celle utilisée pour l'observation des cinétiques, technique
mise au point par Fabry /‾11‾7. Le schéma de la cellule est
représenté figure 4. Les échantillons à analyser sont placés
sur une petite nacelle de platine qui est accrochée en quatre
points par deux fils de platine et deux fils de platine rhodié
10 %. La mesure de la température est faite entre ces deux

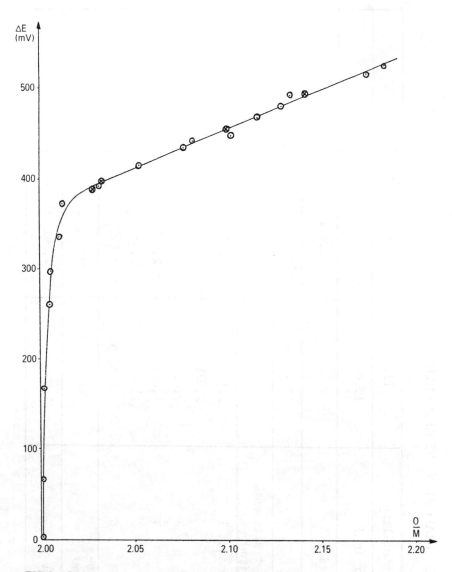

FIG.5. Comparaison de nos résultats avec ceux de Markin et Bones [16] (T = 900°C, réf. Fe, FeO).
⊙ valeurs de Markin et Bones; ⊗ nos valeurs.

TABLEAU I. TRAITEMENT ET RESULTATS

Echantillon	III	IX	X
Longueur (mm)	10,07	9	8,5
O/M initial	2,094	2,048	2,029
$\Delta T(°C)$	1000 – 900	1100 – 900	1120 – 860
durée du traitement (mn)	70	85	110
partie chaude 1	2,104	2,056	2,034
O/M 2	2,091	2,052	2,027
moyen de 3	2,088	2,045	2,024
chaque 4	2,086	2,041	2,023
tranche 5	2,084	2,036	2,021
partie froide 6	n'existe pas	2,033	2,019
Jeu oxyde-gaine (mm)	0,25	0,4	0,0
O/M final	2,090	2,0415	2,027

fils. On a vérifié qu'il n'existait aucun gradient de température sur 4 cm au voisinage de la nacelle, la minijauge travaillant donc en conditions strictement isothermes. Le gaz de balayage est de l'argon purifié par des pompes électrochimiques à oxygène $\underline{/\ 15\ \underline{/}}$, suivies de jauges pour la mesure du PO_2 du gaz.

L'étalonnage (fig. 5) $\Delta\overline{GO}_2 = f(O/M)$ par thermogravimétrie a montré que, dans le domaine monophasé d'UO_{2+x} entre 570°C et 1000°C, les valeurs sont reproductibles avec un intervalle de confiance de $\pm$ 0,5 mV et sont en bon accord avec les résultats de Markin $\underline{/\ 16\ \underline{/}}$.

IV - RESULTATS EXPERIMENTAUX

- Description d'une expérience

L'écart à la stoechiométrie désiré est obtenu par oxydation ménagée en présence d'oxyde de baryum suivant la réaction :

$$BaO_2 \rightarrow BaO + \frac{1}{2} O_2$$

On détermine, au préalable, le rapport O/M à partir d'une mesure de F.E.M. en conditions isothermes. Puis, l'échantillon est placé dans le dispositif de gradient thermique et balayé pendant une nuit sous argon purifié. Le traitement sous gradient thermique proprement dit est précédé d'un chauffage de l'échantillon à l'aide du seul four en molybdène. Dans cette situation, la minisonde est beaucoup plus chaude que la partie inférieure de l'échantillon (chute au contact 100 à 150°C). Lorsque le four à image est allumé et que l'équilibre thermique est réalisé, la base de l'échantillon est sensiblement à la même température que la pile. A ce moment précis, la minisonde fonctionne dans de bonnes conditions de mesures électrochimiques. En fin de traitement, la trempe est réalisée en coupant simultanément le four en molybdène et le four à image. En deux minutes, la température moyenne de l'échantillon chute à 350°C.

Les conditions de traitement et les résultats de ces expériences sont reportés dans le tableau I. Celui-ci montre que, dans toutes les expériences, le bilan en oxygène varie entre 4 et 13,5 % par rapport à l'écart à la stoechiométrie initiale. Cette perte est attribuée à une réduction partielle de UO_{2+x} par les oxydes de tungstène (filament de chauffage du four en molybdène). On ne peut pas attribuer à cette réduction les cinétiques de diffusion de l'oxygène enregistrées par la

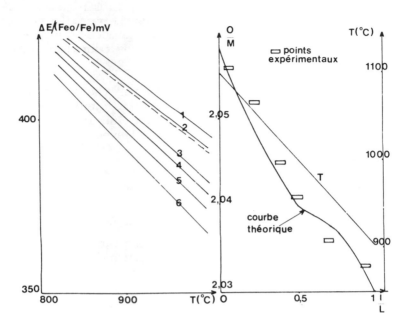

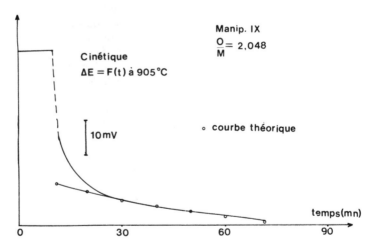

FIG.6.   *Cinétique* $\Delta E = F(t)$ *à 905°C (expérience IX, O/M = 2,048).*

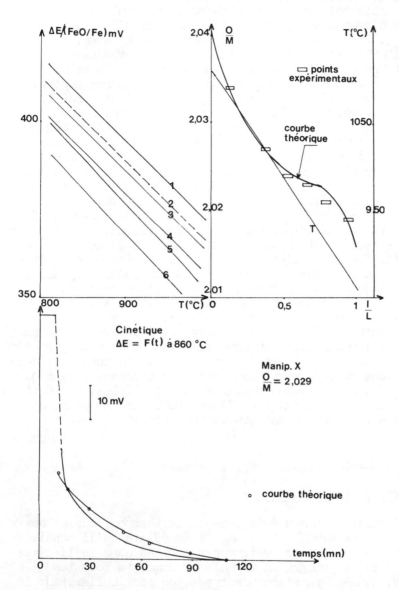

FIG.7. Cinétique $\Delta E = F(t)$ à 860°C (expérience X, O/M = 2,029).

minisonde. En effet, dans une expérience particulière, l'échantillon était totalement gainé de platine, seul un orifice de
1 mm de diamètre permettait à la minisonde d'être au contact
de UO_{2+x}. En 40 minutes de traitement sous gradient thermique
(900-1100°C), l'évolution de la F.E.M. permet d'estimer que le
rapport O/M en partie froide a varié de 0,01. La réhomogéné-
isation de cet échantillon en condition isotherme montre, d'une
part la réoxydation à la partie basse et, d'autre part, que
le bilan en oxygène a été strictement conservé. De ce fait,
nous pensons que la redistribution de l'oxygène est peu affectée par cette réduction. Elle l'est d'autant moins que l'échantillon est fretté dans sa gaine (Expérience X). Deux exemples
de cinétiques sont reportés sur les figures 6 et 7. Dans la
réponse de la pile au cours du temps, il apparaît manifeste-
ment deux stades ; d'abord, une variation brutale de la
F.E.M. de 30-50 mV durant 5 minutes, puis une variation douce
portant sur 21 mV (Expérience X). Cette variation brutale en
début de recuit est imputable au changement de température de
la partie froide de l'échantillon et à la mise en équilibre
thermique du système et thermochimique de la sonde. Dans le
second stade de la cinétique, les variations de la F.E.M.
rendent compte uniquement du départ d'oxygène à la partie
froide. La trempe étant estimée efficace compte tenu des
coefficients de diffusion de l'oxygène, l'analyse de la partie
froide de l'échantillon en condition isotherme nous permet de
calculer exactement la variation de la F.E.M. durant le deuxième stade du traitement sous gradient thermique. Bien que
les gradients thermiques réalisés soient faibles, les profils
de redistribution (fig. 6 et 7) montrent une oxydation appréciable des parties chaudes des échantillons.

V - DISCUSSION

- Théorie phénoménologique -

L'application des équations de la thermodiffusion aux
oxydes du type $UPuO_{2+x}$ n'est pas si immédiate qu'il paraît
d'après les différents traitements publiés dans la littérature. En effet, malgré un formalisme commun à tous les
auteurs, ceux-ci ne s'accordent pas, en particulier, sur la
signification à donner aux coefficients d'activité (défaut
ponctuel ou oxygène). A notre avis ce traitement de la
T.P.I. peut être réalisé sans préjuger de la nature des
défauts, c'est-à-dire en ne raisonnant que sur le potentiel
d'oxygène μO_2. Une justification de ce choix apparaît dans
l'analyse des coefficients de diffusion dans UO_{2+x} faite
récemment par Breitung /¯17¯7. L'application par cet auteur
de la relation de Darken conduit à la relation :

(4) $\dfrac{\tilde{D}_O}{D_O^*}\ \dfrac{2\ RT}{2+x} = \dfrac{d(\Delta\overline{GO_2})}{dx}$ (O indice oxygène
                                                      x écart à la stoechiométrie
                                                      D_O^* coefficient d'autodif-
                                                          fusion de l'oxygène)

Soit encore :

$$\dfrac{\tilde{D}_O}{D_O} = \dfrac{1}{3+x}\ (1 + \dfrac{\partial\log\gamma_O}{\partial\log C_O}\ )$$

avec γ_O coefficient d'activité de l'oxygène dans UO_{2+x}.

Cette relation rend bien compte de l'ensemble des données expérimentales actuellement disponibles.

Etant acquis que le coefficient de diffusion de l'ura- nium est très faible devant celui de l'oxygène, il est légi- time de ne raisonner que sur le sous-réseau anionique.

L'expression du flux d'oxygène d'après de Groots $/\overline{18}\_7$ étant :

(5) $J_O = L_{oo}\ X_O + L_{oq}$ avec $Q = L_{oq}/L_{oo}$

       $X_O = -\ \text{grad}\ (\dfrac{\mu_O}{T})$ μ_O potentiel chimique de
                                              l'oxygène

       $X_q = \text{grad}\ (\dfrac{1}{T})$ T température absolue

en utilisant la relation de Helmholtz :

       $\mu_O = h_O + T\ \dfrac{\partial\mu_O}{\partial T}$ il vient :

(6) $J_O = -\ \tilde{D}_O\ \text{grad}\ C_O - \tilde{D}_O\ \dfrac{Q}{T\ \dfrac{\partial\mu_O}{\partial C_O}}\ \text{grad}\ T$

       $\tilde{D}_O$ coefficient de diffusion chimique
       $C_O = (2+x)\ C_u$ concentration par cm^3
       h_O = enthalpie molaire partielle

                   $Q^* = Q - h_O$

en exprimant $\partial\mu_o/\partial C_o$ en fonction de $\Delta\overline{GO}_2$, et en utilisant la relation (4), on obtient l'expression finale du flux :

$$(7) \quad J_o = - \overset{\vee}{D}_o C_u \underline{/}\,\text{grad } x + \frac{D_o^*}{\overset{\vee}{D}_o}\, \frac{2+x}{RT^2}\, Q^*\, \text{grad } T\underline{\_7}$$

La deuxième loi de Fick permet de calculer la variation de la concentration en fonction du temps :

$$(8) \quad \frac{d_{Co}}{dt} = - \text{div } J$$

Pratiquement les valeurs de D_o^* utilisées sont celles de Marin-Contamin $\underline{/}\,19\,\underline{7}$ et les valeurs de $\overset{\vee}{D}_o$, celles de Lay $\underline{/}\,20\,\underline{7}$. Le paramètre ajustable de ces équations est la chaleur de transport Q^*.

Remarque : L'équation du flux d'oxygène utilisée par les différents auteurs de la littérature est donnée par la relation :

$$(9) \quad J_i = - ND_i \underline{/}\, \nabla x_i + \frac{x_i\, Q_i^*}{1+\dfrac{\partial\ln\gamma_i}{\partial\ln x_i}}\, \frac{\nabla T}{RT^2}\,\underline{\_7}$$

Où i représente l'atome d'oxygène en position interstitielle.

L'intégration de cette relation pour définir l'état stationnaire ($J_i=0$) n'est possible qu'en faisant l'approximation $\gamma_i =$ cste ce qui conduit à l'expression suivante :

$$(10) \quad \ln x \sim \frac{Q^*}{RT} + \text{Cste}$$

A l'état stationnaire, la confrontation des équations (7) et (9) nous a amenés à définir une chaleur de transport apparente :

$$(11) \quad Q_{app} = Q^* \frac{D_o^*}{\overset{\vee}{D}_o}\, \frac{2+x}{x}$$

valeurs que l'on pourra comparer à celles de la littérature.

L'observation des résultats expérimentaux, ainsi que l'examen des cinétiques obtenues par le calcul montrent à

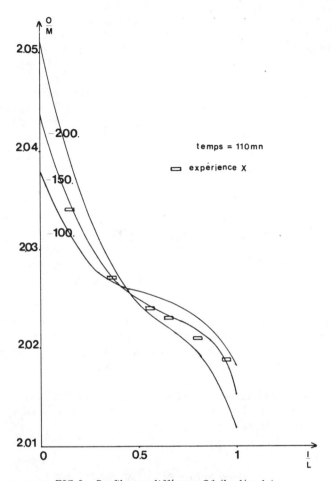

FIG.8.   Profil pour différents Q* (kcal/mole).

l'évidence que l'état stationnaire n'est pas atteint. Un
bon accord existe entre l'expérience et le calcul en
donnant à Q* dans l'équation (7) une valeur d'environ
- 150 kcal/mole (fig. 8). Cet accord semble être d'autant
meilleur que le bilan en oxygène est mieux conservé
(Expérience X). En s'appuyant sur le calcul, on peut estimer le temps au bout duquel l'état stationnaire est
atteint. Pratiquement, après 10 heures, l'évolution du
O/M en partie froide conduirait à une variation de la
F.E.M. non décelable. A l'état stationnaire, d'après (7),
on obtient une valeur du Q_{app} comprise entre -23 et
- 30 kcal/mole, valeurs en accord avec celles données
par Adamson /⎯4⎯7, d'environ - 30 kcal/mole. Pour autant

que l'on puisse se comparer aux résultats de Sari et al
/ 6 / sur $UPuO_{2+x}$, ces valeurs de Q_{app} sont supérieures
à celles déduites de leurs expériences: Le traitement ma-
thématique de la thermodiffusion décrit ici ne faisant
appel qu'à un mécanisme de migration de l'oxygène en phase solide, celui-ci nous semble être, dans nos expériences,
le processus essentiel de la redistribution de l'oxygène.
Dans l'état actuel de nos connaissances, le traitement
purement phénoménologique de ce calcul ne permet pas de
donner une signification physique à Q^* dans l'équation (7).

CONCLUSION

Les expériences de gradients thermiques axiaux ont montré
que la cinétique de diffusion de l'oxygène est très rapide.
Elle intervient dès les premières minutes du traitement. Au
bout d'une heure, la migration de l'oxygène vers les parties
chaudes est appréciable bien que les gradients thermiques
soient faibles.

Un calcul basé sur la T.P.I. rend compte de manière sa-
tisfaisante de la thermomigration de l'oxygène en phase solide.
La chaleur de transport apparente déduite de ce traitement
théorique est comprise entre -23 et -30 kcal/mole.

L'état stationnaire est atteint après une dizaine
d'heures de traitement sous gradient thermique, dans nos
conditions expérimentales.

Remerciements

Nous remercions vivement M. JEAN BAPTISTE pour sa
collaboration à la théorie phénoménologique et pour l'uti-
lisation de son programme de calcul.

Remarques

Un traitement certainement plus précis que celui développé dans le texte peut être
obtenu en exprimant dans l'équation (6) le terme $\partial\mu_0/\partial C_0$ à l'aide des seules données
thermodynamiques.

L'équation (6) devient:

$$(7\text{bis}) \qquad J_0 = -\widetilde{D}_0 C_u \left[\text{grad } x + \frac{Q^*}{\dfrac{1}{2}\dfrac{\partial\Delta\overline{GO_2}}{\partial x}} \text{ grad } T \right]$$

dans laquelle $\dfrac{\partial\Delta\overline{GO_2}}{\partial x}$ est déduite du modèle de DE FRANCO [21].

Dans ce cas, nos résultats expérimentaux s'interprètent avec une chaleur de transport Q^* comprise entre 17 et 26 kcal/mol.

REFERENCES

[1] RAND, M.H., ROBERTS, L.E.J., in Thermodynamics (Proc. Symp. Vienna, 1965) Vol. I, IAEA Vienna (1966) 3.

[2] MARKIN, T.L., RAND, M.H., Ibid., p. 145.

[3] AITKEN, T.L., CRAIG, C.N., EVANS, S.K., J. Nucl. Mater. **30** (1969) 57.

[4] ADAMSON, M.G., CARNEY, R.F.A., AERE-R-6830, UKAEA, Harwell (Oct. 1972).

[5] OLANDER, D.R., J. Nucl. Mater. **44** (1971) 116–120.

[6] ROBERT, M., SARI, C., SCHUMACHER, G., Ibid. **39** (1971) 265.

[7] NORRIS, D.I.R., Ibid. **68** (1977) 13.

[8] BLACKBURN, P.E., Ibid. **46** (1973) 244.

[9] BOWEN, H.K., MARCHANT, D.D., Oxygen redistribution in UO_2 due to a temperature gradient in mass transport phenomena in ceramics, Plenum Press, New York (1974) 97.

[10] FRYXELL, R.E., AITKEN, E.A., J. Nucl. Mater. **30** (1969) 50.

[11] FABRY, P., Thèse de spécialité, Grenoble (1er juillet 1970).

[12] MAITI, H.S., SUBBARO, E.C., J. Electrochem. Soc. (1976) 1057.

[13] COUTURES, J.P., Laboratoires des ultra-réfractaires (Odeillo), Font-Romeu, Communication personnelle.

[14] TRAVERSE, J.P., FLAMAND, R., Third Int. Conf. on Chemical Thermodynamics, Baden near Vienna, 3–7 Sept. 1973.

[15] FOULETIER, J., Thèse, Université et INP de Grenoble (juin 1976).

[16] MARKIN, T.L., BONES, B.J., UKAEA, AERE-R-4042 (1962).

[17] BREITUNG, W., J. Nucl. Mater. **74** (1978).

[18] DE GROOTS, S.R., MAZUR, P., Non-Equilibrium Thermodynamics, North-Holland Publ. Co., London (1962).

[19] MARIN, J.F., CONTAMIN, BACKMANN, J.J., J. Nucl. Mater. **42** (1972).

[20] LAY, K.W., J. Am. Ceram. Soc. **53** (1970) 369.

[21] DE FRANCO, M., GATESOUPE, J.P., Fifth Int. Conf. on Plutonium and Other Actinides, Baden-Baden, 10–13 Sept. 1975, North-Holland Publ. Co., London (1976).

DISCUSSION

M.G. ADAMSON: Your experimental results showed extensive migration of oxygen up the applied temperature gradient in UO_{2+x}. This direction of migration is generally considered to be consistent with a CO/CO_2 gas-phase transport path (whereas the opposite direction is expected for solid-state thermal diffusion). How did you demonstrate that oxygen redistribution was not dominated by gas-phase (CO/CO_2) transport in your experiments?

M. FROMONT: Personally, I have never understood on what basis the direction of thermomigration of oxygen in the solid phase could be predicted, and so I do not think that the direction of oxygen redistribution would lead us to choose one mechanism rather than another. In our experiments there are several arguments in favour of rejecting the CO/CO_2 mechanism:

(1)   The response of the minigauge in the coldest part of samples is quasi-instantaneous;
(2)   Our experiments were performed in an 'open system' and this should be conducive to carbon escaping from the sample;
(3)   The redistribution amplitude and the kinetics are little affected by the presence of an oxide-cladding gap. Even when there is no gap between cladding and sample, oxygen redistribution is observed.

A.S. PANOV: Did you check the content of carbon in the uranium dioxide in your experiments? How do you believe carbon could influence the transfer of oxygen in the temperature gradient?

M. FROMONT: We did not determine the carbon contents of our samples. However, during preparation of UO_{2+x} we avoided any treatment which could lead to a rise in the carbon content. Besides, the good agreement between our experiments and calculation, which only takes into account thermal diffusion of oxygen in the solid phase, justifies a posteriori exclusion of the role of CO and CO_2 species.

MODEL OF THE THERMODYNAMIC PROPERTIES AND STRUCTURE OF THE NON-STOICHIOMETRIC PLUTONIUM AND CERIUM OXIDES

L. MANES, C.M. MARI*, I. RAY
Commission of the European Communities,
European Institute for Transuranium Elements,
Karlsruhe

O. TOFT SØRENSEN
Risø National Laboratory,
Roskilde,
Denmark

Abstract

MODEL OF THE THERMODYNAMIC PROPERTIES AND STRUCTURE OF THE NON-STOICHIOMETRIC PLUTONIUM AND CERIUM OXIDES.

The tetrahedral defect consisting of one oxygen vacancy bonded to two reduced cations, a $2(Me^{3+})'\,V_O^{\bullet\bullet}$ unit, is an important concept, which, as shown in the present work, can explain both the thermodynamic properties and the structures of the phases of the PuO_{2-x} and CeO_{2-x} systems. Based on this concept a statistical thermodynamic model has been developed and this model is described together with some preliminary calculations. A relatively good agreement with experimental thermodynamic data was obtained in this calculation. Using the exclusion principle, defect complexes each containing one tetrahedral defect are derived and it is shown that a systematic packing of these gives a good description both of the non-stoichiometric and the ordered phases observed for these oxide systems.

1. INTRODUCTION

In previous thermodynamic models of non-stoichiometric oxide systems the calculations were based on the assumption that the defects were free and randomly distributed in the lattice. Recent thermodynamic data, however, have given evidence that this is not the case, whereas a realistic thermodynamic model

---

* Work done under a CEC Fellowship. Present address: Institute of Electrochemistry and Metallurgy, University of Milan, Via Venezian 21, I-20133 Milan, Italy.

can be constructed if it is assumed that the defects
are connected together in tetrahedral defects consisting of one oxygen vacancy bonded to two reduced
cations. In this paper we present firstly a general
thermodynamic model based on these tetrahedral defects, and secondly some structural considerations.
These show that the phases derived from thermodynamic
data as well as the Me_nO_{2n-2} phases observed experimentally by X-ray, electron and neutron diffraction
can be described by the packing of defect complexes
based on the tetrahedral defect. Preliminary calculations showing a reasonably good agreement with experimental data are also discussed, but the final
calculations based on the energy terms relevant to
the real structures observed still have to be carried
out.

2.   THE MEANING AND DEFINITION OF THE "TETRAHEDRAL
     DEFECT"

It is generally assumed that the basic point defects in substoichiometric oxides of the fluorite
structure (LnO_{2-x}, with Ln = Pr, Ce, Tb and AnO_{2-x}
with An = Lu, Am, Cm, Bk, Cf) are oxygen vacancies
and reduced cations. For high oxygen deficiencies it
is also reasonable to assume that these point defects
will have a tendency to cluster to form larger ag-
gregates leading to the formation of ordered structures in the parent fluorite lattice at sufficiently
low temperature. For these oxide systems ordered intermediate phases of the Me_nO_{2n-2} series [1] have
been observed between MeO_2 and Me_2O_3.

To link together the high temperature thermodynamics of the disordered MeO_{2-x} system with its
low temperature ordered phases, a basic association of
the point defects must be found which is amenable to
thermodynamic description and at the same time can
act as a building block for describing the low temperature subphases. We define here the "tetrahedral
defect" as a <u>local</u> bond, within an oxygen coordina-
tion tetrahedron in the fluorite structure, between
two reduced cations and one oxygen vacancy. Such a
unit $(2(Me^{3+})'V_O^{\cdot\cdot})$:

(a) is no longer a point defect, but it has its own
    volume in the lattice (the tetrahedral volume);

(b) constitutes a neutral entity with respect to the
    lattice, but it has a dipole moment since the
    centre of positive and negative charge may not
    coincide, especially if a complete charge trans-
    fer has taken place between the vacancy and the
    cations;

(c) represents a local short-range ordering with
    respect to the "free" defects and thus has an
    important effect on the total energy and entropy
    of the crystal;

(d) may be packed in different ways, thus extending
    the range of ordering of the defect population
    as shown in section 3.3.

No special hypothesis is made here on the type of
bond in the tetrahedral defect, but it will depend
on the type of bond existing between the oxygen and
the cation sublattice of the parent fluorite structure.

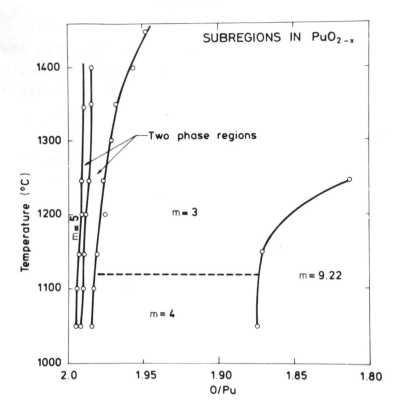

FIG.1. *Pseudo phase diagram for the PuO₂₋ₓ system showing subregions and diphasic regions.*

3.   MODELS

3.1. Non-stoichiometric phases

    In previous publications [2,3,4] the experimental
thermodynamic data for the PuO_{2-x} and CeO_{2-x} systems
were analyzed assuming x (in MO_{2-x}) $\propto p_{O_2}^{-1/m}$, where
the value of m depends upon the type of defect present
- for $V_O^{\cdot\cdot}$ m = 6, for $V_O^{\cdot}$ m = 4 and for V_O^x m = 2, for
instance. Using this proportionality

$$\Delta \bar{G}_{O_2} = RT \ln p_{O_2} \propto -m\, RT \ln x$$

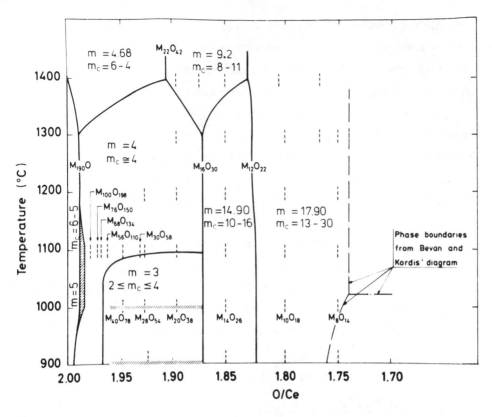

FIG.2.   *Pseudo phase diagram for the* CeO_{2-x} *system showing subregions with possible ordered intermediate phases. The* m_c-*values have been calculated from the thermodynamic model (see text). Key:* ——— *Phase boundaries;* · · · · · · · *Possible intermediate phases;* ////////// *Possible 2-phase regions.*

and a straight line must therefore be expected when $\Delta\overline{G}_{O_2}$ is plotted against ln x, provided that the temperature is constant and that only one type of defect is present.

By plotting the compositions at which the slope changes in isothermal curves of this type in a normal phase diagram, the pseudo-phase diagrams shown in Figs. 1 and 2 are obtained. From these it is clear that both systems can be considered as consisting of subphases, some of which are probably still non-stoichiometric (m ≤ 6) whereas others consist of more ordered phases (m > 6). An interesting feature is that the compositions for the subregion boundaries and the ordered phases observed all follow the general formula Me_nO_{2n-2}, which also describes the compositions of the whole series of low temperature phases observed for the CeO_{2-x} system. Finally it should be pointed out that this treatment is based on the assumption that the defects or defect clusters are independent and non-interacting. This cannot be true especially at larger deviations from the stoichiometric composition, and in the statistical thermodynamic treatment described in the next section defect-defect interactions are also taken into account.

3.2. The thermodynamic model

3.2.1. The basic concepts

The statistical model used to understand the thermodynamics of MeO_{2-x} systems consists of tetrahedral defects (total number; n_ξ) and "free" defects, i.e. oxygen vacancies and reduced cations (total number n_o and $2n_o$, respectively) which are distributed in the fluorite lattice [5,6,7].

The formation energy of one isolated tetrahedral defect, which must contain a term describing the bonding of the three defects together in a tetrahedron and a term describing the perturbation introduced in the fluorite lattice by this unit, can be expressed as:

$$E_{fl} = E_f + E_b + \Delta = E_f + \delta \qquad (\delta = E_b + \Delta) \qquad (1)$$

where E_f is the formation energy of one oxygen vacancy and two reduced cations when they are not bound together, E_b is a bonding energy (E_b,< 0), Δ is a local strain energy introduced in the lattice when a normal oxygen-containing tetrahedron is substituted for a tetrahedral defect (Δ > 0) and δ the resulting energy (δ < 0). The term Δ occurs because the tetrahedral defect is considered to have a larger volume than a normal oxygen-containing tetrahedron.

The tetrahedral defects are considered to be isolated. In this context this means that none of the first neighbour tetrahedra to a given defect contains any other tetrahedral defect. There are 22 such first neighbour tetrahedra in the fluorite structure and an envelope C_1 of 23 tetrahedral positions (including the tetrahedral defect) can therefore be identified as an extended object (see Fig. 6) in the fluorite lattice. This envelope, which is very suitable for statistical calculations, is here referred to as a complex of order 1 and later on in this paper as a T_{23} complex.

At a number $N_{S1} = 2N_M/C_1$ of tetrahedral defects — $2N_M$ is the total number of anionic sites — the lattice will be saturated with isolated defects. At any concentration of tetrahedral defects, however, situations might occur in which $2, 3, \ldots, j, \ldots, L$ tetrahedral defects are in the volume C_1. In these situations in which complexes of orders $2, \ldots, j, \ldots, L$ are formed each tetrahedral defect can be considered as belonging to an envelope containing, roughly, $C_1/j = C_j$ tetrahedral positions.

Formation of a complex j causes a greater strain in the lattice than the isolated tetrahedral defect.

The formation energy E_{fj} for complex j therefore can be written as:

$$E_{fj} = E_f + \delta + \Delta_j = E_{f1} + \Delta_j \qquad (\Delta_j > 0) \qquad (2)$$

where Δ_j is a local strain parameter. This introduces a "site exclusion principle" in the statistics, since the "complex 1" situation would be more stable than the "complex 2" situation and so on.

Subject to the restriction:

$$\sum_{i=0}^{L} n_i = N_v \qquad (3)$$

$$\sum_{j=1}^{L} n_j = n_\xi \qquad (4)$$

where N_v and n_ξ are the total numbers of vacancies and tetrahedral defects, respectively, the total degeneracy Ω of the statistical system can be written as [5,6,7]:

$$\Omega = \Omega_v \Omega_e \prod_{j=1}^{L} \Omega_j \qquad (5)$$

Here Ω_v and Ω_c are due to "free" oxygen vacancies and reduced cations, respectively, and the Ω_j's are introduced by complexes of order j. The total lattice energy E due to defects of any type will be:

$$E = N_v E_f + n_\xi \delta + \sum_{j=2}^{L} n_j \Delta_j + E_{int}(n_o, n_j) \qquad (6)$$

where E_{int} is an interaction energy at larger distances than the tetrahedron. In our theory, this energy is considered to consist of two parts:

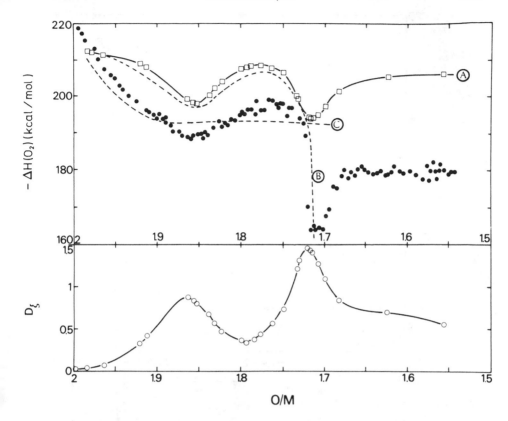

FIG.3.   Calculated $\Delta\bar{H}(O_2)$ and D_ξ (derivative of the concentration of bound vacancies) versus O/M for CeO$_{2-x}$. Experimental data ● from Ref.[9]. Curve A: δ = −4, Δ = 10 and ε = −48 kcal/mol, L = 4. Curve B: same parameters but L = 3. Curve C: fitting of model presented in Ref.[10].

- a charge–charge interaction energy u between "free" point defects, for which a treatment as described in [8] is perfectly suitable;

- a dipolar interaction energy E between tetrahedral defects, which can be described as a Van-der-Waals type, Boltzmann-averaged energy in the fluorite structure [6,7].

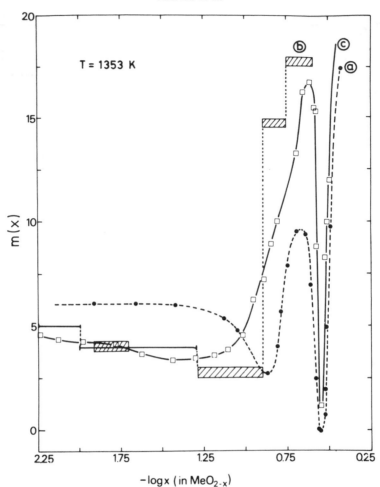

FIG.4.   Comparison between calculated and experimental m-values. Curve a: parameters are the same as those used in Fig.3.   Curve b: m-regions from Fig.2.   Curve c: m-values derived from experimental data of Ref.[10].

3.2.2. The thermodynamic functions

The thermodynamic functions of a statistical system as described depend <u>only</u> on the two concentrations n_ξ and n_o, which represent "order" and "disorder" in the system. These functions have been described elsewhere in detail [5,6,7] and here only two

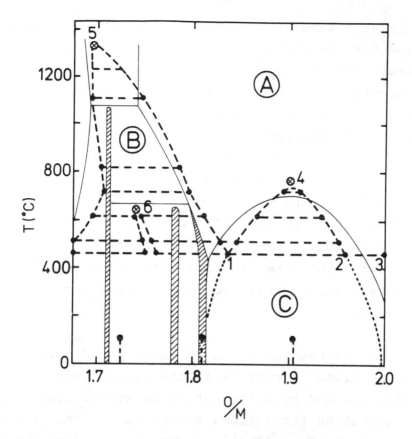

FIG.5. A phase diagram calculated with the present model compared with the Ce-O phase diagram. Parameters are the same as those used in Fig.3. Broken lines: calculated diagram.

features have to be recalled:

(a) The chemical potential of the defective solid μ_s can be conveniently split into three terms:

$$\mu_s = \mu_o(x_o \xi) + \mu_{short\text{-}range}(\xi, \delta, \Delta) +$$

$$\mu_{long\text{-}range}(\xi, x_o, \epsilon, u) \qquad (7)$$

where $x_o = n_o/N_M$ and $\xi = n_\xi/N_M$.

Fig. 3 shows the $\Delta\bar{H}(O_2)$ versus O/M curve calculated for the CeO_{2-x} system. It is shown that the
characteristic minima and maxima which are found experimentally [9] are due to the saturation of the
lattice, at different O/M, with tetrahedral defects
- note the correspondence with the curve for $D_\xi =
\frac{d\xi}{dx}$. It is worth recalling that the presence and the
meaning of these maxima and minima have not been explained previously [10]. The same curve calculated
for the PuO_{2-x} system [11], which shows no experimental minima, is to be related to the lower strain
introduced into the fluorite lattice by tetrahedral
defects [12], perhaps due to a smaller ionicity of
the bond.

(b) When expressing μ_s as in equation (7), the presence of diphasic regions in the phase diagram can be
easily deduced by differentiation with respect to x
and searching for extrema and/or inflection points.
When this is done [7,11], the Sørensen diagrams are
fairly well reproduced at rather high temperatures
as shown in Figs. 2 and 4.

Fig. 5 represents an attempt to calculate the
phase diagram of CeO_{2-x} with the energy parameters
used in the calculation of $\Delta\bar{H}(O_2)$ (Fig. 3). In the
region of validity of the model, which is the fluorite structure region (region A), the diphasic bell
and the 1-5 phase boundary are reproduced fairly
well. In regions B and C the agreement is less good,
but interesting trends are discovered - note, for
instance, the monophasic region terminating at 6.

3.2.3. Further comments on the thermodynamic model

In the thermodynamic study briefly reviewed above, the 23 first-neighbouring tetrahedra of the envelope C_1 are considered equivalent. In reality, when more than one tetrahedral defect is introduced in this envelope, this cannot be true and a careful study of the different possibilities leads to an explanation of the low-temperature structures as described in the next section. The averaging between all non-equivalent positions, as done here, should, however, be sufficiently accurate at higher temperatures.

3.3. Structural considerations of non-stoichiometric oxides

3.3.1. Structure of defect complexes

The fcc structure can be represented as consisting of oxygen cubes alternatively occupied by cations or as cation tetrahedra containing oxygen ions. For purely ionic oxides both descriptions apply equally well, but for the substoichiometric oxide systems considered here in which the cation lattice is quite stable but the oxygen relatively unstable, the second description is preferred. According to Blank [13] this representation is also preferable for the more covalent oxides, such as the PuO_{2-x} system.

In the previous section the exclusion principle was used to define a defect complex consisting of a tetrahedral defect surrounded by 22 tetrahedra each containing one oxygen ion. The structure of this complex - a T_{23} complex - is shown in Fig. 6. Each time

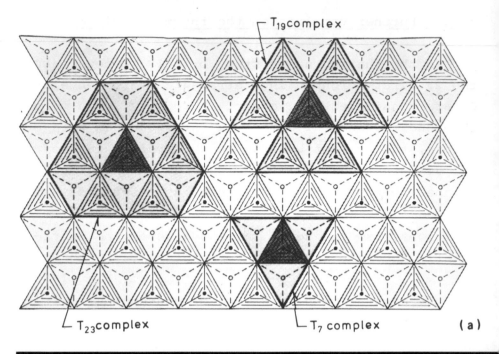

FIG.6. (a) Projection on (111) of a T_{23}, T_{19} and T_7 complex. (b) A perspective view.

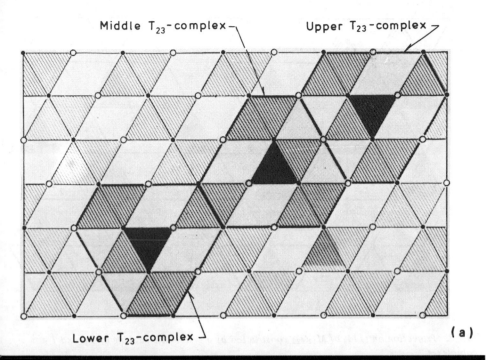

Middle T$_{23}$-complex ⌐   Upper T$_{23}$-complex ⌐

Lower T$_{23}$-complex ⌐

(a)

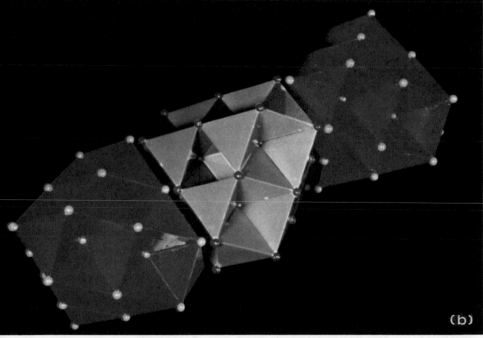

(b)

FIG.7. (a) Projection on (101) of macrocomplex consisting of 3 upper (only one is shown),
1 middle and 3 lower (only one is shown) T$_{23}$ complexes. (b) A perspective view.

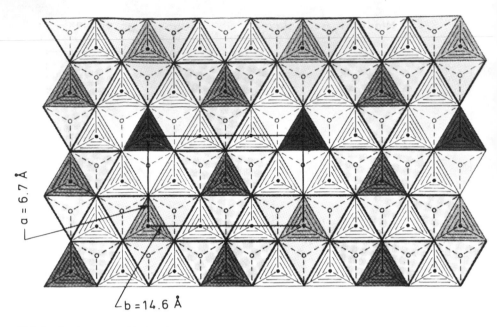

*FIG.8.   Projection on (111) of $M_{12}O_{22}$ constructed by a systematic packing of T_{19} and T_7
complexes.*

a tetrahedral defect is formed the surrounding lattice is strained and the total "lattice energy" is
increased. Considering a T_{23} complex the magnitude of
the local strain created can be assumed to be inversely proportional to the distance from the central
defect. When the whole lattice has been filled by
packing T_{23} complexes in a regular way the smallest
increase of the "lattice energy" by the introduction
of more oxygen vacancies bound into tetrahedral defects will thus be observed when the tetrahedra
already under the smallest strain - those at the long-
est distance from the central defect - becomes a
part of a neighbouring complex. In analogy to section 3.2 this can also be described as the formation

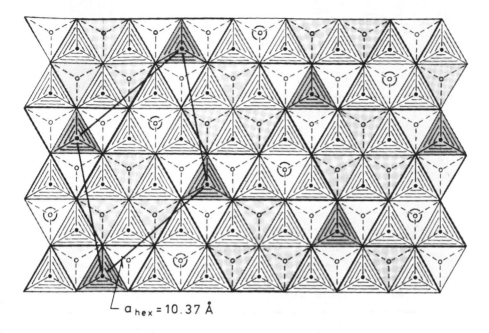

$a_{hex} = 10.37$ Å

FIG.9.   Projection on (111) of M_7O_{12} constructed by a systematic packing of T_7 complexes.

of new defects substituting progressively the most distant positions in the T_{23} complex.

From the T_{23} complex a T_{19} and a T_7 complex, which are energetically different, can thus be derived naturally by removing 4 and 4+4×3 = 16 tetrahedra respectively as shown in Fig. 6.

3.3.2. Interactions between defect complexes

The tetrahedral defect is neutral - it consists of $2(Me^{3+})' V_O^{..}$ - and the coulombic interaction between the defect complexes is probably very small. Experimental $\Delta\overline{H}_{O_2}$ values (see Fig. 3), however, show that a "lattice stabilization" takes place

Table I. Packing of defect complexes in subregions
         observed for CeO_{2-x}

O/M	$\Delta \bar{H}_{O_2}$	Packing
2.00–1.99	constant	Random distribution of T_{23}
1.99–1.96	increasing "lattice stabilization"	Macrocomplexes of 4 T_{23} (1 middle, 3 upper)
1.96–1.92	increasing "lattice stabilization"	Macrocomplexes of 7 T_{23} (1 middle, 3 upper, 3 lower)
1.92–1.88	increasing	T_{19} formed
1.88		close packing T_{19}
1.88–1.83	constant	T_7 formed
1.83 ($M_{12}O_{22}$)		T_{19} and T_7
1.83–1.82	decreasing	T_7 formed
1.82 ($M_{11}O_{20}$)		T_{19} and T_7
1.82–1.80	decreasing	T_7 formed
1.80 ($M_{10}O_{18}$)		T_{19} and T_7
1.80–1.78	decreasing	T_7 formed
1.78 (M_9O_{16})		T_{19} and T_7
1.78–1.71	increasing	T_7 formed
1.71 (M_7O_{12})		T_7

within certain composition ranges both for the
PuO_{2-x} and the CeO_{2-x} systems, indicating that
there must be some attractive interactions between
the complexes at these compositions. In the present
work we consider that these interactions are dipole-
dipole attractions, which, as shown in Fig. 7 could
give rise to the formation of macrocomplexes consisting of, for instance, either 7 T_{23} complexes -
one middle, three upper and three lower. The closer
the tetrahedral defects in the lattice the greater
are the dipole-dipole attractions. This effect,
therefore, must become more pronounced for the composition ranges where the smaller T_{19} or T_7 complexes
are formed. Opposed to this effect, however, is the
larger strain (higher "lattice energy") introduced
when the concentration of tetrahedral defects in-
creases. For the composition ranges, where the $\Delta \overline{H}_{O_2}$
versus x curves show a "lattice destabilization",
this strain effect thus becomes larger than the
dipole-dipole attractive energy.

3.3.3. Packing of defect complexes

The formation of subregions as well as the ordered phases in the PuO_{2-x}- and CeO_{2-x} systems (see
Figs. 1 and 2) can all be explained by a systematic
packing of the T_{23}, T_{19} and T_7 complexes. A close
packing of macrocomplexes comprising 7 T_{23} units,
for instance, corresponds to a composition of $MO_{1.92}$
at which an ordered phase has been observed for the
CeO_{2-x} system. A close packing of T_{19} complexes, on
the other hand, gives $MO_{1.88}$ which also is very close

to subregion boundaries observed for both oxide systems. Furthermore, the ordered phase $M_{12}O_{22}$ can be
constructed by a systematic packing of T_{19} and T_7
complexes as shown in Fig. 8 and, finally, as shown
in Fig. 9, a close packing of T_7 complexes gives a
M_7O_{12} structure with crystallographic parameters
corresponding to those determined by neutron diffraction [14]. In Table I the type of packing giving the
different subregion boundaries and ordered phases observed for the two oxide systems is summarized.

4.   CONCLUSIONS

The present work clearly shows that the tetrahedral defect is a very useful concept in explaining
both the thermodynamic properties and the structures
of the PuO_{2-x} and CeO_{2-x} systems. Structurally, defect complexes consisting of one tetrahedral defect
surrounded by 22, 18 and 6 cation tetrahedra respectively can be derived, and by packing these complexes
in a systematic way the non-stoichiometric and ordered phases observed for the two oxide systems can
be explained. Furthermore, a statistical thermodynamic model based on this concept has also been developed. Preliminary calculations with this model
have shown a rather good agreement with experimental
thermodynamic data although these calculations were
not based on energy terms relevant to the actual
structures observed. With the structural ideas
presented here it should, however, be possible to
get some ideas about the real energies involved and
a more realistic calculation based on such energies

has been planned for the next phase of this collaboration between the Transuranium Institute, Karlsruhe, and Risø National Laboratory, Denmark.

ACKNOWLEDGEMENT

The authors wish to acknowledge the EURATOM grant received by O. Toft Sørensen from the Scientific and Technical Education Service of the Commission of the European Communities.

REFERENCES

[1]   ANDERSON, J.S., Problems of Non-Stoichiometry (Rabenau, A. Ed.), North Holland, Amsterdam, (1970) 1.

[2]   SØRENSEN, O. Toft, J. Sol. State Chem. 18 (1976) 217.

[3]   SØRENSEN, O. T., Risø Report No. 331 (1975).

[4]   SØRENSEN, O. T., Plutonium and other Actinides (Lindner, R., Blank, H. Eds.), Baden-Baden 1975, North Holland, Amsterdam (1976) 123.

[5]   MANES, L., MANES-POZZI, B.M., Plutonium and other Actinides (Lindner, R., Blank, H. Eds.) Baden-Baden 1975, North Holland, Amsterdam (1976) 145.

[6]   MANES, L., PARTELI, E., MARI, C.M., "An order-disorder model theory for the thermodynamic functions of substoichiometric fluorite structure compounds", to be published.

[7]   MANES, L., PARTELI, E., MARI, C.M., "Spinodal points in the G-curves of substoichiometric fluorite structure compounds: Phase diagrams and residual structures at high temperature", to be published.

[8]   ATLAS, L.M., The Chemistry of Extended Defects
      in Non-Metallic Solids (Eyring LeRoy, O'Keeffe,
      M. Eds.), Scottsdale Arizona 1969, North Hol-
      land, Amsterdam (1970) 425.

[9]   CAMPSERVEUX, J., GERDANIAN, P., J. Chem. Thermo-
      dynamics 6 (1974) 795.

[10] CAMPSERVEUX, J., GERDANIAN, P., J. Sol. State
      Chem. 23 (1978) 73.

[11] CHEREAU, P., "Contribution à l'Etude Thermodyna-
      mique des Oxydes de Plutonium, et des Oxydes
      Mixtes Uranium-Plutonium, Thèse de Doctorat,
      Université de Paris-Sud (Orsay) 1972, Report
      CEA-R-4402.

[12] MANES, L., PARTELI, E., MARI, C.M., "Application
      of the model theory based on "tetrahedral" de-
      fects to some substoichiometric fluorite struc-
      ture compounds", to be published.

[13] BLANK, H., Thermodynamics of Nuclear Materials,
      IAEA, Vienna (1974) 45.

[14] RAY, S.P., COX, D.E., J. Sol. State Chem. 15
      (1975) 333.

DISCUSSION

Hj. MATZKE: I would like to add corroborative evidence for your model
of tetrahedral defects which you introduced from a *structural* point of view.
Since you postulate clustering and ordering of point defects, this should show
up in *kinetic* measurements as well. It is promising in this context that the
dependence of the plutonium diffusion coefficient on x in $(U,Pu)O_{2-x}$ at constant
temperatures shows a knee at just the O/M ratio where you predict saturation of
T_{23} complexes. The implied smaller mobility of bigger complexes is indeed
indicated in diffusion rates.

IAEA-SM-236/69

OXYGEN TRANSPORT LIMITATIONS IN USING GETTERS TO CONTROL FUEL/CLADDING CHEMICAL INTERACTION

C.N. WILSON
Hanford Engineering Development Laboratory,*
Westinghouse Hanford Company,
Richland, Washington,
United States of America

Presented by M.G. Adamson

Abstract

OXYGEN TRANSPORT LIMITATIONS IN USING GETTERS TO CONTROL FUEL/CLADDING CHEMICAL INTERACTION.

A series of stainless-steel-clad $Pu_{0.25}U_{0.75}O_{2-x}$ fuel pins was irradiated in EBR-II to assess placement and oxygen-transport limitations using oxygen buffer/getter materials (getters) in LMFBR fuel pins to inhibit fuel/cladding chemical interaction (FCCI). Effective FCCI inhibition required adequate axial dispersal of the getter over the fuel column length. The limiting factor in getter placement was axial oxygen transport from the fuel to axially-isolated getters. The primary axial oxygen transport mode was concluded to be diffusion of oxygen through the fuel. Gas phase axial oxygen transport is most likely not significant at $\Delta\bar{G}_{O_2}$ values less than the threshold for cladding oxidation.

1. INTRODUCTION

Temperature and oxygen potential ($\Delta\bar{G}_{O_2} = RT \ln p_{O_2}$) at the cladding inner surface are generally recognized as being the two most important factors determining the type and rate of cladding corrosion in stainless-steel-clad $(U,Pu)O_{2-x}$ fast breeder reactor fuel pins. Oxygen-to-metal (O/M) ratio of the fuel tends to increase with burn-up, providing a continual source of excess oxygen. In addition, in a hypostoichiometric fuel, oxygen redistributes in the fuel radial temperature gradient, further increasing the O/M ratio at the fuel surface. The thermodynamic threshold for cladding oxidation occurs at the $\Delta\bar{G}_{O_2}$ of the equilibrium between chromium in the stainless steel and Cr_2O_3 (-139 kcal/mol at 873 K). Evans and Aitken [1] have predicted that the cladding oxidation threshold occurs at an approximate fuel-surface O/M ratio of 1.998.

---

* Operated for the United States Department of Energy.

427

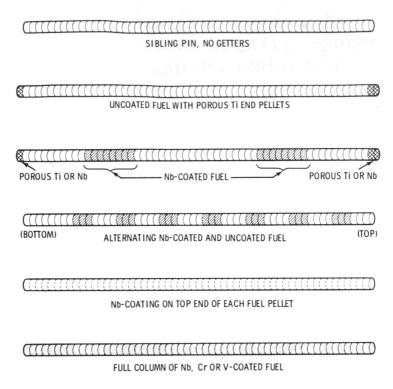

FIG.1. *Fuel column configurations used to assess getter placement requirements.*

Two methods for controlling $\Delta\overline{G}_{O_2}$ at the fuel/cladding interface, thus reducing fuel/cladding chemical interaction (FCCI), are:

- Placement of materials into the fuel pin which getter excess oxygen and maintain the $\Delta\overline{G}_{O_2}$ of the fuel/cladding gap below the threshold for cladding oxidation [2];
- Use of fuels with an initially low O/M ratio [3].

Both methods have been irradiation tested in the United States of America fast breeder reactor 'Reference Fuels Program'. Successful use of getters to control $\Delta\overline{G}_{O_2}$ at the fuel/cladding interface depends upon the ability of oxygen to be transported to, and absorbed by, the getters at $\Delta\overline{G}_{O_2}$ values below which cladding oxidation occurs.

2.   EXPERIMENTAL

In one pin-series of the HEDL P-23 Special Pins Irradiation Test [2], niobium and titanium were used as oxygen getters in various configurations to evaluate oxygen transport and placement requirements for using oxygen getters to inhibit FCCI. Fuel-pin configurations used are shown in Fig.1.

Fuel pellets were sputter coated with 12 μm thick circumferential niobium coatings, or sputter coated with 25 μm thick niobium coatings on one end of the pellets. In addition to niobium-coated fuel pellets, porous titanium or niobium pellets were used at the ends of the fuel column. Sibling pins, fabricated using fuel from the same fabrication lot as the fuel receiving the niobium coatings, were irradiated concurrently to provide experimental control for assessing the effects of getter placement of FCCI. The initial O/M ratio of the fuel lot was 1.985. Burn-up in the getter placement test series was approximately 2.2 at.%, with one pin containing a full column of niobium-coated fuel reaching 5 at.% burn-up.

3.   RESULTS

The results from the HEDL P-23 Special Pins Test indicate that effective FCCI control is dependent upon:

- Adequate getter inventory in the fuel pin;
- Adequate axial dispersal of the getter material over the length of the fuel column.

In the absence of oxygen getters in the fuel pin, cladding matrix attack, characterized by bright granular FCCI reaction product in the fuel/cladding gap, was observed. Matrix attack increased in intensity with increasing cladding temperature. The character of the observed matrix attack in the absence of oxygen getters is shown in Fig.2.

Porous niobium and titanium pellets at the ends of the fuel column had no effect on FCCI. As shown in Fig.3, getter pellets at the top of the fuel column did not even inhibit cladding matrix attack adjacent to the top fuel pellet. After irradiation, the porous titanium pellets were bright in appearance and showed no visible indications of oxidation. The porous niobium pellets darkened in appearance and lost their metallic lustre during irradiation. However, oxygen analysis on irradiated porous niobium pellets indicated that the oxygen content was less than 2 wt%. The post-irradiation oxygen content for the porous titanium pellets was less than 1 wt%.

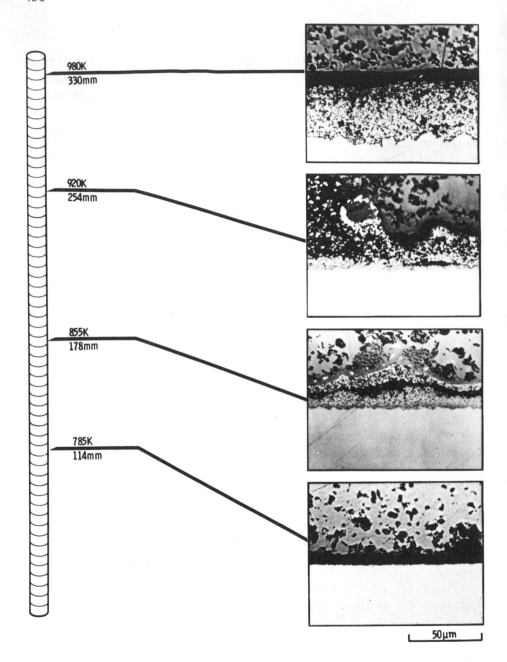

FIG.2.   Fuel/cladding interface micrographs from the 2.2 at.% burn-up sibling pin. Estimated
cladding inner surface temperature (K) and axial position (mm) are given for each micrograph.

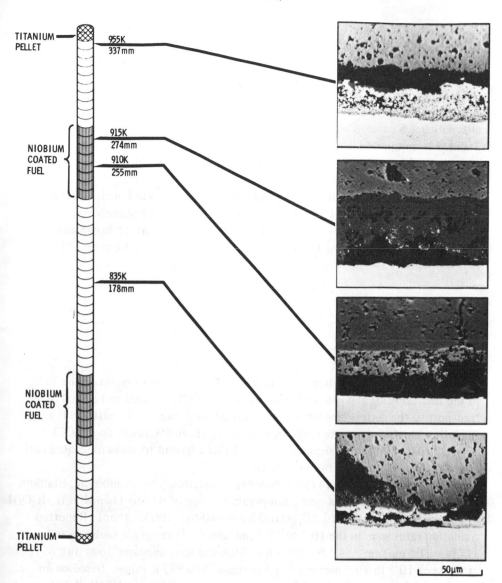

TITANIUM
PELLET

955K
337mm

NIOBIUM
COATED
FUEL

915K
274mm
910K
255mm

835K
178mm

NIOBIUM
COATED
FUEL

TITANIUM
PELLET

50μm

FIG.3.   Fuel/cladding interface micrographs from a pin containing niobium-coated fuel and titanium end pellets.

Cladding matrix attack was eliminated adjacent to the niobium-coated fuel to 5 at.% burn-up, which was the highest burn-up data obtained. In two fuel pins, each containing two 5 cm long axial segments of niobium-coated fuel separated by approximately 13 cm of uncoated fuel, cladding matrix attack equivalent to that observed in pins without getters (at the same cladding temperature and burn-up) was observed adjacent to uncoated fuel. In a pin containing alternating 2.5 cm long axial segments of niobium-coated and uncoated fuel at 2.2 at.% burn-up, no cladding matrix attack was observed. Also, no cladding matrix attack was observed at 2.2 at.% burn-up in the pin which contained niobium at each pellet-to-pellet interface.

Based on the preceding fuel pin examination data, it was concluded that oxygen absorbing (getter) coatings on the fuel are capable of maintaining a fuel surface O/M ratio below the threshold for cladding matrix attack for an axial distance of the order of one to two centimetres from the axial location of the fuel coating.

4.   DISCUSSION

4.1. Getter oxidation rates

The limiting factor when using axially-isolated getters to maintain the $\Delta \bar{G}_{O_2}$ in the fuel/cladding gap below the threshold for FCCI appeared to be axial oxygen transport to the getters and not the in-situ oxidation rates of the getters. If getter oxidation kinetics were rate limiting, axial $\Delta \bar{G}_{O_2}$ gradients, resulting in FCCI adjacent to uncoated fuel segments and no FCCI adjacent to niobium-coated fuel segments, would not have been expected.

Caputi and Adamson [4] have measured oxidation rates of niobium, titanium and 316 stainless steel foils over a temperature range of 873 to 1173 K in $H_2/H_2O/He$ atmospheres equilibrated to $\Delta \bar{G}_{O_2}(1023\,K) = -80$ to -100 kcal/mol. Reported oxidation rates were in the 10^{-4} to 10^{-3} mg·cm^{-2}·s^{-1} range for niobium and 10^{-5} to 10^{-4} mg·cm^{-2}·s^{-1} for titanium. Stainless steel appeared to absorb oxygen at a rate of 10^{-7} to 10^{-6} mg·cm^{-2}·s^{-1} in the 873 to 973 K range. Based on an estimated O/M ratio increase of 0.0034 per at.% burn-up in the HEDL P-23 Special Pins [5], radial oxygen flux across the gap was estimated to be 5.8 × 10^{-8} mg·cm^{-2}·s at steady state.[1] From the above data, it would appear that oxidation rates for the niobium fuel coatings, or for 316 stainless steel above 873 K, were sufficient to absorb excess oxygen as it was released by the fuel.

---

[1] To achieve prototypic linear power in the HEDL P-23 Special Pins, the UO_2 fraction was 65% enriched in ^{235}U. Increase in O/M ratio with burn-up is predicted to be greater in a large breeder reactor which uses natural or depleted UO_2 due to higher noble metal fission-product yields when the primary fissile isotope is ^{239}Pu.

TABLE I. SUMMARY OF ESTIMATED GAS-PHASE TRANSPORT
REQUIREMENTS AND PARTIAL PRESSURES FOR H_2O, CO_2 AND CsOH

- ASSUMED CONDITIONS

    $\Delta \bar{G}_{O_2}$ (873 K) = -139 kcal/mol AT CLADDING

    $\Delta \bar{G}_{O_2}$ (1273 K) = -126 kcal/mol AT FUEL SURFACE

- TRANSPORT SPECIES PRESSURE REQUIREMENT

    RADIAL $\quad 10^{-4}$ TO 10^{-3} Pa

    AXIAL $\quad 10^{2}$ TO 10^{4} Pa

- $H_2O / H_2 \qquad \left\{ P_{H_2} \sim 10^{0} \text{ Pa} \right\}$

    $P_{H_2O} \sim 10^{-4}$ TO 10^{-3} Pa

- $CO_2 / CO \qquad \left\{ a_C \text{ (873 K)} = 0.03 \right\}$

    $P_{CO_2} \sim 10^{-7}$ Pa

- CsOH $\qquad \left\{ P_{H_2} \sim 10^{0} \text{ Pa} \right\}$

    $P_{CsOH} \sim 10^{-2}$ Pa

4.2. Oxygen transport in the fuel/cladding gap

Oxygen transport in the fuel/cladding gap most likely occurred via an oxygen-containing gas phase species. Oxygen transport by solid or liquid fission product phases in the fuel/cladding gap is not considered to be significant at burn-up values represented in the P-23 Special Pins Test.

Calculation of gas-phase oxygen transport fluxes in the fuel/cladding gap is complicated by several factors. The non-equilibrium nature of the fuel/cladding gap (due primarily to the radial temperature gradient) leads to uncertainties in the partial pressures of the transport species calculated from the equilibrium free energy of formation data. Diffusion coefficients in the fuel/cladding gap change with burn-up due to changes in total gas pressure and composition. An additional complicating factor is non-uniform gap width. Assuming a uniform fuel/cladding gap width of 0.001 cm, the diffusion length to cross-sectional area ratio for 1 cm of axial transport would be 10^6 times as great as the same ratio for radial transport

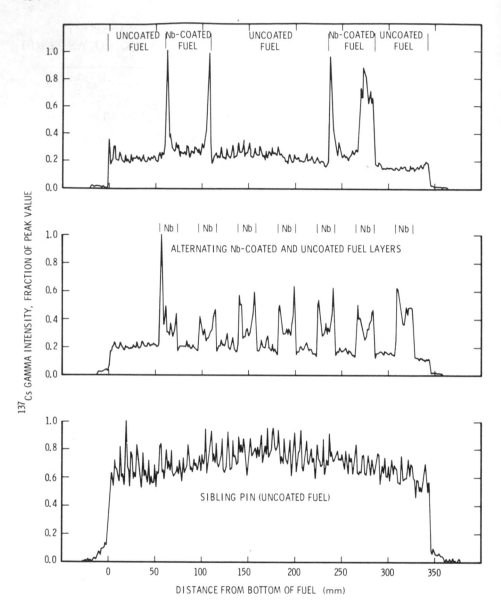

FIG.4.   Caesium-137 axial gamma scan data comparing axial caesium mobility between two pins containing niobium-coated fuel and a sibling pin.

in the fuel/cladding gap. Recognizing the uncertainties involved, order-of-magnitude transport-species partial-pressure requirements were estimated for steady-state radial and axial oxygen transport in the fuel/cladding gap. Results for H_2O, CO_2 and CsOH candidate gas-phase oxygen transport species are summarized in Table I.

4.2.1. Transport by H_2O gas

Bober et al. [6], using Fick's first law, have estimated that H_2O partial pressure (p_{H_2O}) in the fuel/cladding gap would probably be just sufficient to support steady-state radial oxygen transport from the fuel to the cladding. In their analysis, p_{H_2} in the gap was estimated to be in the 2 to 6 Pa range and an average diffusion coefficient of 0.14 cm^2/s was assumed for H_2O vapour in the gap. Using a similar analysis and data particular to the HEDL P-23 Special Pins irradiation, p_{H_2} = 1 Pa was estimated as the minimum p_{H_2} requirement for steady-state radial oxygen transport by H_2O vapour when $\Delta\bar{G}_{O_2}$(1273 K) = −126 kcal/mol at the fuel surface. Minimum p_{H_2O} at the fuel surface was estimated to be in the 10^{-4} to 10^{-3} Pa range for steady-state radial oxygen transport.

4.2.2. Transport by CO_2 gas

Assuming CO/CO_2 equilibrium with carbon in the cladding at an activity of 0.03 and a $\Delta\bar{G}_{O_2}$(873 K) = −139 kcal/mol, p_{CO} is estimated to be 2 × 10^{-3} Pa in the fuel/cladding gap. At the fuel surface, assuming $\Delta\bar{G}_{O_2}$(1273 K) = −126 kcal/mol, p_{CO_2} is then estimated to be of the order of 10^{-7} Pa. Assuming p_{CO_2} = 10^{-3} Pa is required for steady-state radial oxygen transport by CO_2 in the gap, $\Delta\bar{G}_{O_2}$(1273 K) = −85 kcal/mol is calculated as the required oxygen potential at the fuel surface. Under these conditions, p_{H_2O} at the fuel surface should be sufficient to provide radial oxygen transport if p_{H_2} is in the 10^{-3} to 10^{-4} Pa range.

4.2.3. Transport by CsOH gas

If $\Delta\bar{G}_{O_2}$ at the cladding inner surface is maintained below the threshold for FCCI, then p_{Cs} in the fuel/cladding gap may become sufficient with increasing burn-up to cause condensation of fission-product caesium at the bottom of the fuel pin. As $\Delta\bar{G}_{O_2}$ at the fuel surface increases, caesium is expected to react with UO_2, fission product oxides and/or oxidized cladding components, reducing axial caesium mobility. Axial caesium mobility and formation of Cs-Nb-O compounds was observed in fuel pins containing niobium-coated fuel [2] (see Fig.4). No significant axial mobility of fission-product caesium was observed in sibling pins fabricated with fuel having an O/M ratio of 1.985 and no getters.

When $\Delta\bar{G}_{O_2}$ in the fuel/cladding gap is maintained below the $\Delta\bar{G}_{O_2}$ threshold for FCCI, the maximum value of p_{Cs} should be approximately the saturation pressure at the sodium inlet temperature. Assuming a sodium inlet temperature of 660 K, at saturation p_{Cs} is estimated to be approximately 400 Pa. Using reported ΔG_f^0 values for CsOH [7] and H_2O, p_{CsOH} may be estimated from the reaction described by Eqs I and II:

$$Cs(g) + H_2O(g) = CsOH(g) + \tfrac{1}{2}H_2(g) \tag{I}$$

$$\Delta G^0 = -21\,090 + 7.63\,T \tag{II}$$

Assuming that $\Delta\bar{G}_{O_2}(873\ K) = -139$ kcal/mol at the cladding, and that $\Delta\bar{G}_{O_2}(1273\ K) = -126$ kcal/mol at the fuel surface, $p_{H_2} = 1$ Pa, and $p_{Cs} = 400$ Pa; p_{CsOH} is estimated to be of the order of 10^{-2} Pa in the fuel/cladding gap.

At $\Delta\bar{G}_{O_2}$ values below the threshold for FCCI and high burn-up, CsOH is most likely the predominant gas-phase oxygen transport species in the fuel/cladding gap. Gas-phase oxygen transport mechanisms considered appear capable of sustaining steady-state radial oxygen transport across the fuel/cladding gap when $\Delta\bar{G}_{O_2}$ in the gap is approximately equal to the threshold value for FCCI. However, it appears unlikely that H_2O, CO_2 and CsOH vapours in the fuel/cladding gap can account for the 1 to 2 cm of effective axial oxygen transport to the niobium fuel coatings which was observed in the HEDL P-23 Special Pins.

As fuel surface O/M ratio increases, H_2O and CO_2 gas-phase contributions to axial oxygen transport are expected to increase. In the upper portion of the fuel column, where the cladding temperature is greatest (i.e. greater than 850 K), the fuel surface O/M ratio is most likely buffered by the cladding. At lower axial positions in the fuel pin, $\Delta\bar{G}_{O_2}$ in the fuel/cladding gap may rise significantly above the thermodynamic oxidation threshold for cladding. However, the maximum $\Delta\bar{G}_{O_2}$ value for the fuel surface in the lower fuel column is expected to be limited by oxidation of additional fission products (primarily caesium and molybdenum) and by the formation of CsU_xO_y compounds.

4.3. Oxygen transport by the fuel

Transport of oxygen in the mixed-oxide fuel is, in general, to be expected, accounting for oxygen redistribution in thermal gradients and the ability to adjust the O/M ratio of pellets. During irradiation, fuel pellets sinter together in the central restructured regions, providing solid-state diffusion paths between pellets at temperatures estimated to be in the 1800 to 2800 K range. Oxygen transport through the fuel was likely the predominant mechanism contributing to FCCI control for axial distances of the order of a few centimetres from niobium-coated fuel. The interfaces between getter pellets at the ends of the fuel column and fuel did not sinter or fuse. Therefore, the effects of end-pellet getters on fuel column O/M ratio were minimal.

5.   CONCLUSIONS

It is concluded that at $\Delta \bar{G}_{O_2}$ values less than the $\Delta \bar{G}_{O_2}$ threshold for FCCI, axial oxygen transport in the fuel pin occurs predominantly by transport through the fuel. Axial oxygen transport in a stainless steel clad $(U,Pu)O_{2-x}$ fuel pin is not sufficient for axially-isolated oxygen getters, such as getter pellets at the ends of the fuel column, to be effective FCCI inhibitors. Axial oxygen transport is only sufficient to provide FCCI control for a maximum axial distance of a few centimetres from getters coated onto the fuel.

REFERENCES

[1] EVANS, S.K., AITKEN, E.A., "Oxygen redistribution in LMFBR fuels", Behaviour and Chemical State of Irradiated Ceramic Fuels (Proc. Panel Vienna, 1972), IAEA, Vienna (1974).
[2] WILSON, C.N., Assessment of the HEDL P-23 Special Pins Test: Methods to control fuel-cladding chemical interaction, Hanford Engineering Development Laboratory Rep. HEDL TME-78-52 (1978).
[3] WILSON, C.N., Fabrication and Irradiation Performance of $Pu_{0.25}U_{0.75}O_{1.91}$ Low O/M Fuel, Hanford Engineering Development Laboratory Rep. HEDL S/A-1553 (Oct. 1978).
[4] CAPUTI, R.W., ADAMSON, M.G., Experimental Evaluation of the Oxidation Behaviour of Nb, V, and Ti as Candidate FCCI Inhibitors, General Electric Co. (USA) Rep. GEFR-00370 (May 1978).
[5] WOODLEY, R.E., Hanford Engineering Development Laboratory, personal communication.
[6] BOBER, M., DORNER, S., SCHUMACHER, G., "Kinetics of oxygen transport from mixed-oxide fuel to the clad", Fuel and Fuel Elements for Fast Reactors (Proc. Symp. Brussels, 1973) Vol.1, IAEA, Vienna (1974) 221.
[7] JANAF Thermochemical Tables, 1974 Supplement, Dow Chemical Co., Golden (1974).

DISCUSSION

U. BENEDICT: We are not going to reach a conclusion at this session about the relative importance of solid-state diffusion and gas-phase diffusion for the redistribution of oxygen in oxide fuels. However, two papers presented here indicate that solid-state diffusion may be much more important than a gas-phase mechanism: Mr. Fromont has found a satisfactory explanation for his results taking into account solid-state diffusion only, and in his paper, Mr. Wilson concludes that axial transport occurred only by solid-state diffusion in his experiments.

M.G. ADAMSON: Although Mr. Fromont explained his results in terms of a purely solid-state diffusion mechanism, I do not think he unequivocally demonstrated that gas-phase transport was absent. Personally, I feel that gas-phase transport, presumably via CO/CO_2, dominated in his experiments on UO_{2+x}. In Mr. Wilson's experiments solid-state diffusion clearly controlled axial oxygen transport; however, it should be recognized that the presence of reactive metal in his fuel pins also reduces the partial pressures of oxygen-containing gas-phase species — hence biassing oxygen transport towards solid-phase paths.

D.D. SOOD: Mr. Wilson's investigations have shown that the fuel/cladding interaction can be controlled by suitably locating niobium-coated fuel pellets in the fuel pin. This would, however, require coating of a large number of fuel pellets with niobium. Do you think that such a solution is technologically acceptable?

M.G. ADAMSON: I don't think that there is any question about the *technical feasibility* of buffer/getter applications such as the one I have described; however, cost is obviously an important factor in assessment of its technological viability.

SEEBECK EFFECT IN (U, Pu)O$_{2\pm y}$ AND ITS INFLUENCE ON OXYGEN MIGRATION

F. D'ANNUCCI, C. SARI
Commission of the European Communities,
European Institute for Transuranium Elements,
Karlsruhe

G. SCHUMACHER
Institut für Neutronenphysik und Reaktortechnik,
Kernforschungszentrum Karlsruhe,
Karlsruhe,
Federal Republic of Germany

Abstract

SEEBECK EFFECT IN (U,Pu)O$_{2\pm y}$ AND ITS INFLUENCE ON OXYGEN MIGRATION.
Thermoelectric potentials have been investigated in hypo- and hyperstoichiometric uranium-plutonium mixed oxides up to a temperature of 1800 K. Sintered pellets were exposed to a temperature gradient in an axial direction; this was produced by having an induction-heated tungsten cylinder at the top of the pellets and a cooled specimen carrier at the bottom. The thermoelectric potential difference was measured in two holes in the pellet which had a separation of 1 mm in the direction of the temperature gradient and a temperature difference of 80 K. At low temperatures, below 1200 K, the Seebeck coefficient was found to be positive in hypo- and in hyperstoichiometric mixed oxides. It vanished at temperatures between 1200 K and 1800 K, and it changed sign with increasing temperature. This behaviour is explained by a change of the conductivity mechanism from p-type to n-type. The absolute value of the Seebeck coefficient depends strongly on the O/M ratio of the mixed oxide. It is high for the stoichiometric region and low for greater deviations from stoichiometry. Thermo-electric power provides one of the forces which drive ions along temperature gradients. Another force originates in the differences of the lattice forces of the various ions, atoms and vacancies. Both forces can be described by the heat of transport of oxygen ions, which contains a thermo-electric and a thermic part. Values for the parts are calculated using the known value of the overall heat of transport and that of the measured Seebeck coefficient.

1. INTRODUCTION

Oxygen redistribution is an important process in reactor oxide fuels under irradiation. It greatly influences the behaviour of the fuel pin, especially the fuel/cladding reactions and the migration of fuel and fission-product atoms. Redistribution of oxygen can take place via the solid fuel matrix by thermo-migration, and via the gas phase in carrier gases like CO_2 and H_2O or by evaporation and condensation of oxygen-rich species like UO_3 or fission-product oxides like Cs_2O and MoO_2.

The crystal lattice of uranium oxides and uranium-plutonium mixed oxides are of a fluorite type. The binding is mainly ionic, with oxygen vacancies in hypostoichiometric oxides and oxygen interstitial atoms in hyperstoichiometric oxides. Because of the temperature-dependent concentrations of electrons in the conduction band and of holes in the valence band, an electrical potential gradient is built up in oxides subjected to a temperature gradient. This potential gradient results in a force that acts on the oxygen ions and influences their migration via the solid phase in a temperature gradient. A number of redistribution experiments have been conducted with oxide fuels [1–5]. The mechanisms proposed for the oxygen redistribution processes show more or less agreement with the observed processes. Models have been developed which are based on an effective heat of transport for solid-state migration [6], on a combined migration via the solid and gas phases [7], and on the effective enthalpy of vacancy-formation in $(U,Pu)O_{2-y}$ [8]. In other models, the driving force is assumed to be proportional to the gradients in oxygen activity and vacancy concentration in the solid phase [9], or to be dependent on the differences between the activities of the carrier gases and the solid matrix along the temperature gradient [10]. Consideration of the possible flows of oxygen in the solid matrix and via the gas phase in interconnected porosity channels and cracks shows that the dominating process in hypostoichiometric and/or dense oxide fuels (~95% of theoretical density) is migration through the solid matrix. The aim of this paper is to report on the investigations of thermoelectric potentials in uranium and uranium-plutonium mixed oxides for various O/M ratios up to a temperature of 1800 K in experiments involving a controlled temperature gradient and to show the influence of potential differences on redistribution by solid-state thermomigration. The results are compared with other investigations of the Seebeck effect, which have mainly been conducted with uranium dioxide [11–15]: three measurements had been extended to temperatures up to 1000 K with $(U,Pu)O_2$ [15] and to 1400 K with UO_2 [12, 14].

2.   SEEBECK EFFECT IN $(U,Pu)O_{2\pm x}$

2.1. Seebeck coefficient

The Seebeck effect is caused by the differences in the concentrations of ions, electrons and holes that exist in a temperature gradient. The absolute Seebeck coefficient, α, of a material at the temperature T is defined as [16]:

$$\alpha(T) = \frac{dV}{dT} \tag{1}$$

where dV is the potential difference.

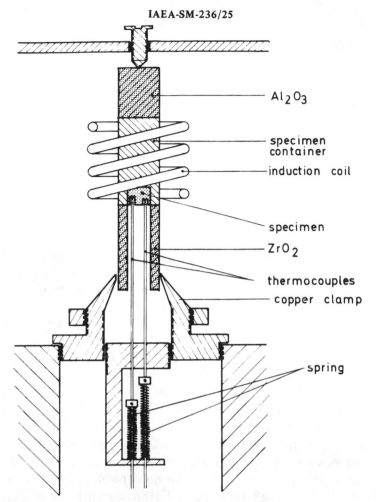

FIG.1. Sketch of the experimental arrangement.

In most cases, it is the relative Seebeck coefficient, α_{ir}, that is measured. This is related to the coefficient of a reference material:

$$\alpha_{ir} = \alpha_i - \alpha_r \qquad (2)$$

where α_i and α_r are the absolute Seebeck coefficients of the materials involved.

The absolute Seebeck coefficient in a semiconductor crystal is given as a function of the temperature by [16]:

$$\alpha(T) = \frac{k}{e} \frac{E_s}{kT} \qquad (3)$$

where k is the Boltzmann constant, e is the electronic charge and E_s (= $E_t - E_f$) is the thermoelectric energy of activation (E_t is the mean energy of the charge carrier, E_f the Fermi energy).

2.2. Experimental

The experiments were carried out in a vacuum vessel containing an induction furnace. Figure 1 shows the principle of the experimental arrangement. The specimen container is of tungsten for hypostoichiometric oxides and of iridium for hyperstoichiometric oxides. The lower end of the specimen container is in contact with a zirconia tube that is cooled by a copper clamp. Thus a slight temperature gradient exists along the specimen during heating. In the lower part of the specimen there are two holes parallel to the axis which contain the thermocouples. The depths differ by 1 mm. Good contact between the thermocouples and the oxide is provided by two springs which press the thermocouples against the bottoms of the holes. The EMF is measured with Pt/6% Rh-Pt thermocouple wires.

The cylindrical specimens consisted of UO_2 and $(U_{0.8}Pu_{0.2})O_{2\pm y}$, having polished surfaces to improve the thermal contact with the container. Before the experiment the specimens were heated in a H_2/H_2O or CO/CO_2 atmosphere in order to establish the desired O/M ratio.

During the experiments a temperature difference of 80 K was maintained between the two measuring points. The temperature was increased by steps of ~100 K and the EMF was measured after each step when steady-state conditions existed. The atmosphere inside the vessel was chosen according to the stoichiometry of the specimens. Hypostoichiometric specimens were annealed in high vacuum. Stoichiometric analyses after the experiments showed agreement with the initial O/M ratios. (With regard to the stoichiometric specimen, it should be noted that the O/M ratio deviated slightly to the hypostoichiometric side, down to ~1.9999 (calculated from the CO/CO_2 ratio)).

2.3. Results

The relative Seebeck coefficients have been calculated from the EMF measurements. These values together with the absolute Seebeck coefficients of Pt and 6%Rh-Pt allow us to determine the absolute Seebeck coefficients of the uranium oxide and uranium-plutonium mixed oxide specimens. Measurements with UO_2 have been carried out in order to provide more values for comparison with the experimental work of other authors.

Figure 2 shows the absolute Seebeck coefficients of UO_2 as a function of temperature as calculated from the EMF measurements. These results are in good agreement with those reported in the literature [11–14] and reveal that UO_2 has

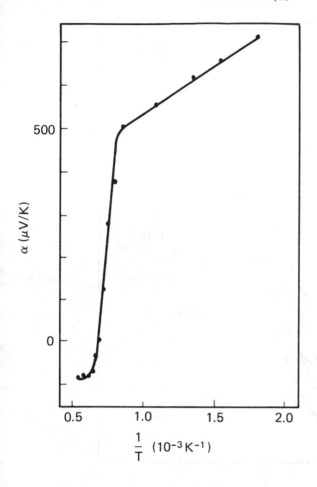

FIG.2. *Absolute Seebeck coefficient of UO₂.*

the typical properties of a classical semiconductor. In the range of extrinsic conductivity (up to 1200 K) the Seebeck coefficient can be described by:

$$\alpha = \frac{k}{e}\left(3.71 + \frac{0.22}{kT}\right) \tag{4}$$

i.e. the energy of activation, E_s, is 0.22 eV.

The results of the experiments with $(U_{0.8}Pu_{0.2})O_{2\pm y}$ are given in Fig.3, which shows the absolute Seebeck coefficient as a function of the reciprocal temperature. The conductivity is of the p-type at low temperatures. The range of this type of conductivity depends on the O/M ratio. The Seebeck coefficient decreases with increasing temperature, linearly with 1/T. The highest values are obtained for

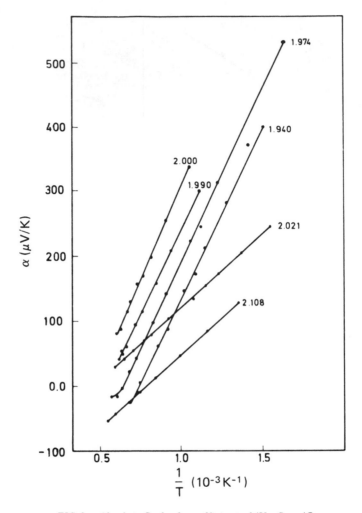

FIG.3. *Absolute Seebeck coefficient of* $(U_{0.8}Pu_{0.2})O_{2\pm y}$.

(O/M) = 2.00. A relation, similar to that used for UO_2, allows of the description of α as a function of temperature:

$$\alpha = \frac{k}{e} \left(-A + \frac{E_s}{kT}\right) \tag{5}$$

The values of the constant, A, and the energy of activation, E_s, for specimens with different O/M ratios are given in Table I. If we express A as a function of y

TABLE I. VALUES OF A AND E_s FOR $(U_{0.8}Pu_{0.2})O_{2\pm y}$

O/M ratio	2.00	1.990	1.974	1.940	2.021	2.108
E_s(eV)	0.57	0.53	0.53	0.53	0.23	0.23
A	3.08	3.33	3.88	4.58	1.32	2.08

(the deviation from stoichiometry), the Seebeck coefficient in hypostoichiometric mixed oxides can be described by:

$$\alpha = \frac{k}{e}\left[-6.16 + 1.68 \log\left(\frac{1-2y}{2y}\right) + \frac{0.53}{kT}\right] \tag{6}$$

and in hyperstoichiometric mixed oxides by:

$$\alpha = \frac{k}{e}\left[-2.61 + 0.95 \log\left(\frac{1-2y}{2y}\right) + \frac{0.23}{kT}\right] \tag{7}$$

Masayoshi and Kurihara [15] carried out measurements of the EMF of $(U_x,Pu_{1-x})O_{2-y}$. The stoichiometry of the pellets varied during the experiments and the calculated values of the Seebeck coefficient cannot be related to a defined O/M ratio. Therefore no comparison with our results is possible.

3. THERMOMIGRATION OF OXYGEN IN $(U,Pu)O_{2\pm y}$

Oxygen migrates rapidly in $(U,Pu)O_{2\pm y}$ at the temperatures existing in fuel pins under irradiation. Typical diffusion coefficients at 1400 K are higher than 10^{-5} cm^2/s for hypo- and hyperstoichiometric $(U_{0.8}Pu_{0.2})O_2$ [17]. Thus, steady-state conditions can be attained in a very short time (a few minutes to a few hours, according to the fuel temperature). The oxygen distribution in a fuel pin and in out-of-pile test specimens heated in a thermal gradient can be calculated from the Soret coefficient, which gives:

$$\ln\left(\frac{y_2}{y_1}\right) = \frac{Q^*}{R}\left(\frac{1}{T_2} - \frac{1}{T_1}\right) \tag{8}$$

where y_1, y_2 are the appropriate deviations from stoichiometry in $(U,Pu)O_{2\pm y}$ at the absolute temperatures T_1, T_2, and Q^* is the effective heat of transport.

Equation (8) describes the steady-state distribution of a solute in a dilute solution. Thus, in hypostoichiometric oxide, Q^* is related to the oxygen vacancies (molar fraction $x_v = y/2$) and in hyperstoichiometric oxides to the interstitial oxygen atoms (molar fraction $x_i = y$). From the phenomenological description of thermal diffusion, a thermoelectric force can also be taken into account and, therefore, an electrical term must be added to the heat of transport, Q_g^*, originating from the lattice forces. Thus the effective heat of transport can be written as:

$$Q^* = Q_g^* - ZT\frac{dV}{dT} \qquad (9)$$

where Z is the electric charge transported by a number of vacancies or interstitials that corresponds to one mole of oxygen.

4.   INFLUENCE OF THE SEEBECK EFFECT ON THERMOMIGRATION OF OXYGEN

The thermoelectric part, Q_s^*, of the effective heat of transport can be calculated with the known Seebeck coefficient, $\alpha = dV/dT$, by using Eq.(9).

A simple experiment was carried out to check Eq.(9). A cylindrical specimen of $UO_{2.030}$ was heated isothermally to 1160 K in a furnace containing an argon atmosphere. During the experiment a potential difference of 50 mV was applied to the specimen in an axial direction. The resulting force caused the oxygen interstitials to migrate along the thermoelectric gradient towards the positive pole of the specimen. After a steady state had been attained (the potential difference stabilized at 45.3 mV after 2 h), the specimen was quenched to room temperature. The oxygen gradient observed is depicted in Fig.4.

There are two possibilities of calculating Q_s^*, using either Eq.(8) or Eq.(9). For this purpose it should be noted that the potential difference of 45.3 mV corresponds to that which is generated in $UO_{2.030}$ heated in a temperature gradient of 210 K/cm between a T_1 of 1260 K and a T_2 of 1060 K [11]. It is, therefore, possible to calculate the value of Q_s^* by introducing into Eq.(8) values of the temperature and oxygen gradient, i.e.:

$$Q_s^* = 7.4 \pm 1.7 \text{ kcal/mol} \qquad (10)$$

For the calculation using Eq.(9), one has to determine the appropriate value of Z, which is given by:

$$Z = 6.02 \times 10^{23} \, q_0 \qquad (11)$$

where q_0 is the electric charge of an oxygen ion. It has been shown that the

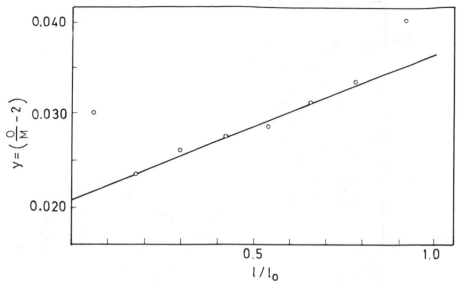

FIG.4. *Oxygen profile caused by a potential difference of 50 mV applied at the extremities of a* $UO_{2.030}$ *pellet heated isothermally at 1160 K (negative pole at l = 0 and positive pole at l = l₀.*

effective charge of oxygen ions is less than 2, because a part of the binding is covalent. For PuO_2, a value, q_0, of 1.15e has been calculated [19]; for UO_2 values of 1.15e [20], 1.64e [21], 1.76e [22] and 1.0e [23] have been found. It is obvious that $q_0 < 2e$, but one cannot decide which is the best approximation. From Eq.(9) a rough estimation of Q_s^* can be made using a mean value of q_0 ($q_0 = 1.4e$) and assuming $Q_g^* = 0$, i.e.:

$$Q_s^* = 7.9 \text{ kcal/mol} \tag{12}$$

which is in good agreement with the value given in Eq.(10). Consequently Eq.(9) can be used for the calculation of Q_s^*.

Calculation of values of Q_s^* in $(U_{0.8}Pu_{0.2})O_{2\pm y}$ were carried out using a value for q_0 of 1.4e, and the results are presented in Fig.5. The absolute value of Q_s^* is higher in near-stoichiometric oxides. It decreases with increasing temperature. At temperatures above 1300 K, Q_s^* changes sign and the temperature at which the sign of Q_s^* changes is higher for near-stoichiometric oxides.

5.   DISCUSSION

Uranium-plutonium mixed oxides have the typical properties of a classical semiconductor. The conductivity is of the p-type in stoichiometric $(U,Pu)O_2$,

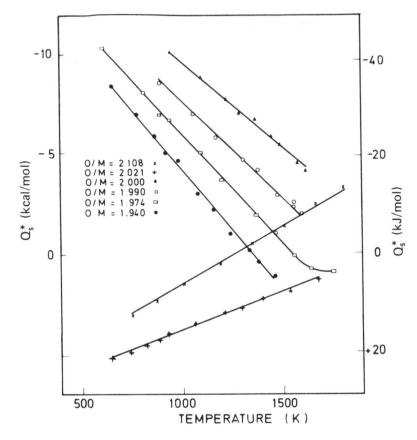

FIG.5.   *Thermoelectric heat of transport, Q_s^*, in $(U_{0.8}Pu_{0.2})O_{2 \pm y}$ as a function of temperature.*

i.e. the electrical conductivity is caused by the movement of the excess of holes
in the valence band with respect to the electrons in the conduction band. In
hypostoichiometric mixed oxide, the plutonium ions change their valency
from 4+ to 3+ by formation of donor levels. The more donor levels that exist,
the greater the number of electrons which can be excited in the conduction band;
these electrons neutralize an equal number of holes in the valence band and
cause a decrease in the absolute value of α. The defect structure in hyperstoichiometric mixed oxide is not well known. The values of α in U_3O_8 [14]
and in $(U,Pu)O_{2 \pm y}$ (see Fig.3) show that the absolute value of α decreases with
increasing O/M ratio in the hyperstoichiometric oxide. Thus, it can be assumed
that, with increasing O/M ratios, the concentration in the donor level increases.

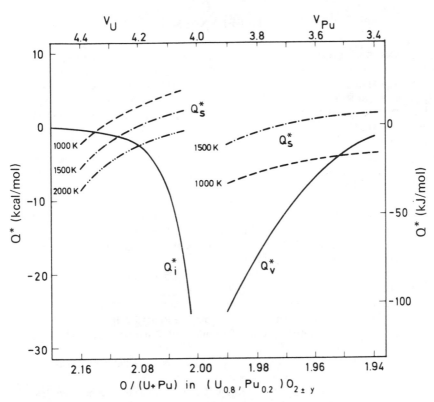

FIG.6. Effective heat of transport of oxygen vacancies, Q_v^*, and interstitials, Q_i^*, as a function of stoichiometry compared with the thermoelectric share, Q_s^*.

It is of interest to discuss the influence of the Seebeck effect on the effective heats of transport Q_i^* (interstitials) and Q_v^* (vacancies). The values of Q_s^* (taken from Fig.3) are compared in Fig.6 with those of Q_i^* and Q_v^* determined in experiments carried out in temperature gradients [3]. At low temperatures, Q_s^* has the same sign as that of the effective heat of transport of vacancies Q_v^*. In contradistinction, at high temperatures and with low stoichiometry, the sign of Q_s^* is opposite to that of Q_v^*, but the value of Q_s^* is very small. In hyperstoichiometric oxides the thermoelectric forces act in a direction opposite to the direction of the force caused by Q_i^* at low temperatures. Both forces have the same sign at high temperatures (T > 1500 K).

On the basis of these considerations, one can tentatively conclude that the contribution of Q_s^* to the total measured heat of transport is generally small in the range of stoichiometry around 2.00. However, in the region of high deviation from stoichiometry, the thermoelectric force could play an important role in the process of oxygen redistribution.

REFERENCES

[1] EVANS, S.K., AITKEN, E.A., CRAIG, C.N., J. Nucl. Mater. **30** (1969) 62.

[2] ADAMSON, M.G., AITKEN, E.A., EVANS, S.K., DAVIES, J.H., Thermodynamics of Nuclear Materials 1974 (Proc. Symp. Vienna, 1974) Vol.1, IAEA, Vienna (1975) 59.

[3] SARI, C., SCHUMACHER, G., J. Nucl. Mater. **61** (1976) 192.

[4] JOHNSON, J., JOHNSON, C.E., CROUTHAMEL, C.E., SEILS, C.A., J. Nucl. Mater. **48** (1973) 21.

[5] KLEYKAMP, H., J. Nucl. Mater. **66** (1977) 292.

[6] BOBER, M., SCHUMACHER, G., Material Transport in the Temperature Gradient of Fast Reactor Fuels, Adv. Nucl. Sci. Technol. 7, Academic Press, New York (1973).

[7] NORRIS, D.I.R., J. Nucl. Mater. **68** (1977) 13.

[8] AITKEN, E.A., J. Nucl. Mater. **30** (1969) 62.

[9] BLACKBURN, P.E., JOHNSON, C.E., Thermodynamics of Nuclear Materials 1974 (Proc. Symp. Vienna, 1974) Vol.1, IAEA, Vienna (1975) 17.

[10] RAND, M.H., MARKIN, T.L., Thermodynamics of Nuclear Materials, 1967 (Proc. Symp. Vienna, 1967), IAEA, Vienna (1967) 637.

[11] ARONSON, S., RULLI, J.E., SCHANER, B.E., J. Chem. Phys. **35** 4 (1961) 1382.

[12] WOLFE, R.A., USAEC Rep. WAPD-270 (1963).

[13] DEVRESSE, J., de CONNINCK, R., Phys. Status Solidi **17** 1966) 825.

[14] SORRIAUX, A., Commisariat à l'énergie atomique Rep. CEA-R-4321 (1972).

[15] MASAYOSHI, n.i., KURIHARA, n.i., Plutonium 1970 and Other Actinides (Proc. 4th Int. Conf. Santa Fé, 1970), (1970) 84.

[16] HEIKES, R.R., URE, R.W., Thermoelectricity, Interscience Pub., New York (1961).

[17] SARI, C., J. Nucl. Mater. **78** (1978) 425.

[18] LEVINE, B.F., SYRKIN, R., DYATKINA, T., Electrodynamical band-charge calculation of nonlinear optical susceptibilities, Russ. Chem. Rev. **38** (1969) 95.

[19] MANES, L., BARISICH, A., Phys. Status Solidi, A **3** (1970) 971.

[20] AXE, J.D., PETTIT, G.D., Phys. Rev. **151** (1968) 676.

[21] NAEGELE, J., Thesis, Fakultät für Physik der Universität Karlsruhe (1975).

[22] BLANK, H., Thermodynamics of Nuclear Materials 1974 (Proc. Symp. Vienna, 1974) Vol.2, IAEA, Vienna (1975) 45.

[23] BROOKS, M., KELLY, P., to be published in J. Phys. (Paris) in 1979.

DISCUSSION

R. DUCROUX: Could I ask you to give a technical explanation about the measurement of the Seebeck coefficient? Can you indicate what is the time needed for measurement of EMF at a given temperature? Also, have you observed a drift of EMF due to oxygen redistribution because of the temperature gradient between the electrodes?

G. SCHUMACHER: The measurements were carried out when the steady state was reached. This occurred in the time range extending from a few minutes to half an hour, depending on the temperature. No influence of oxygen redistribution could be observed.

Section G

FISSION PRODUCTS

SOLID SOLUBILITY OF FISSION-PRODUCT AND OTHER TRANSITION ELEMENTS IN CARBIDES AND NITRIDES OF URANIUM AND PLUTONIUM

U. BENEDICT
Commission of the European Communities,
European Institute for Transuranium Elements,
Karlsruhe

Abstract

SOLID SOLUBILITY OF FISSION-PRODUCT AND OTHER TRANSITION ELEMENTS IN CARBIDES AND NITRIDES OF URANIUM AND PLUTONIUM.

Solubility studies were made in some MX-Me systems (M is U or Pu; X is C or N; Me is a fission product or other transition element) by X-ray diffraction analysis and partly by microprobe determination of solute concentrations. Up to 23 mol% ZrC and 17 mol% TaC dissolved in the PuC phases of sintered PuC-ZrC and PuC-TaC samples; the lattice parameter/concentration relationships were derived. 2 wt% WC dissolved in $U_{0.8}Pu_{0.2}C$ above 2000°C. The solubility decreased between 1700–1200°C, orthorhombic $(U,Pu)WC_2$ appearing as a second phase. The solubility of rhenium in UC is less than 0.1 wt%. The solubility of molybdenum in UN is negligible at 1700°C. A lattice parameter decrease of 0.002 Å was observed for addition of molybdenum to PuN, indicating a slight molybdenum solubility ($\leqslant 0.1$ wt%) in PuN. Lattice parameter decreases were measured in high burn-up irradiated and simulated carbide fuels, reflecting mainly dissolution of zirconium. MN fuels irradiated up to 15 at.% burn-up or simulated for high burn-up showed no significant change of the lattice parameter of the matrix phase with respect to the unirradiated condition. The rare earths which increase the lattice parameter are completely soluble in the nitride and are thus able to counterbalance the lattice-parameter decrease due to zirconium. The relative lattice-parameter difference (RLPD) between MX_y and MeX_y (y: ratio $X/(M+Me)$) was used as a solubility criterion. NaCl-type monocarbides with RLPDs from -10.2% to $+7.8\%$ are completely miscible with UC and PuC. NaCl-type mononitrides with RLPDs from -7.5% to $+8.5\%$ are completely miscible with UN and PuN. The solubility in the sesquicarbides increases with decreasing RLPD and becomes complete in Pu_2C_3 for a RLPD of $+4\%$, and in U_2C_3 for a RLPD of approximately $+1.5\%$. Solubilities are predicted on the basis of these rules for the cases where no experimental results are available. A general review of the experimental and predicted solubilities is given.

1. INTRODUCTION

Some of the solid fission-product (FP) elements formed on nuclear fission partly combine with elements from the fuel or with other fission-product elements to form separate phases. But an important proportion of the fission-product atoms remains dissolved in the fuel-matrix phase, whose properties are affected by their presence.

A striking example of an effect caused by dissolved fission products is the enhancement of plutonium diffusion observed in carbides, nitrides and carbo-nitrides of uranium-plutonium [1−3]. The diffusion coefficient is increased by factors of 30 to 100 when FP elements are added. A similar effect was observed when the transition metals nickel, iron, tungsten, tantalum and vanadium were added to UC [4]. These metals can be present in fuels as impurities; they had also been proposed as additives to improve sintering or swelling behaviour. Uranium-233 diffusion was enhanced in UC by factors of up to 500 as compared with UC without additives. Evidence was presented that the enhancement of diffusion is caused by dissolved metals, and not by the formation of additional phases [4]. It was further shown [4] that, in some cases, diffusion was the faster, the higher the concentration of the additives. Carbonitride fuels with chemical compositions simulating 3 and 10 at.% burn-up showed no difference in diffusion rate, but for simulated burnt nitride fuel, diffusion was faster for 3 at.% burn-up than for 10 at.% burn-up [3]. Detailed knowledge of the solubilities of the individual fission products in the different fuel materials could contribute towards explaining this difference in behaviour.

In advanced fuel compounds diffusion of uranium and plutonium is much slower than diffusion of the non-metals carbon, nitrogen and oxygen. Thus uranium and plutonium diffusion will, in general, be the rate-determining factor in all diffusion-controlled processes such as grain growth, sintering and creep, and so will have a pronounced effect on the fuel behaviour during irradiation.

In this context, and in view of the possible effect of dissolved fission products on other properties, the solubilities of fission products in candidate fuels are considered to be data of importance. Work was carried out in the European Institute for Transuranium Elements, Karlsruhe, to improve knowledge of the solubility of some selected metals in the carbides and nitrides of uranium and plutonium. In addition, literature data on solubility of solid fission products and of some other transition metals in these fuel compounds were collected. The work was restricted to fission products which are not volatile at the operational temperatures of reactors. Caesium, for instance, was not considered because it seems to play a role only in the coldest zones of the fuel element, such as the fuel/clad boundary region. Fission products which are only formed in minor amounts, such as indium, were also excluded from the study. On the other hand, the study does include some elements which are not formed as fission products, but belong to a group in the Periodic Table which contains one or more fission-product elements. Due to the chemical similarity, work with such elements allows some conclusions to be drawn regarding the behaviour of fission-product elements for which information is

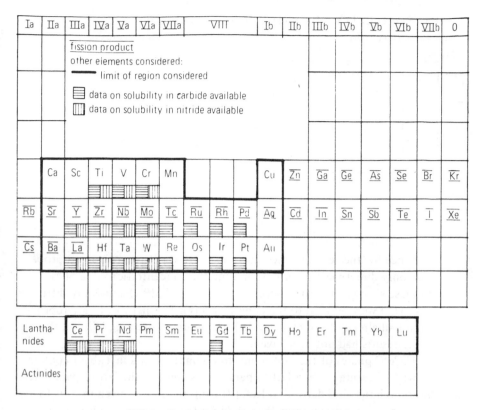

FIG.1.   Region of the Periodic Table considered.

lacking or incomplete. Solubility of uranium and plutonium in the carbides
and nitrides of the transition elements has in general not been considered as it
has no bearing on the properties of the fuels. Figure 1 shows the regions of
the Periodic Table considered in the present work.

2.   STUDY OF SOME SELECTED SOLVENT–SOLUTE SYSTEMS

Some cases for which information was insufficient or uncertain were
checked experimentally.

2.1. Cerium in UC and (U,Pu)C

Cerium solubilities in UC and PuC were derived from lattice-parameter
measurements, using Vegard's law between the solvent and hypothetical
'CeC' [5, 6]. Some uncertainty exists for the estimated lattice parameter of

TABLE I. SOLUBILITY OF CERIUM IN UC AND $U_{0.8}Pu_{0.2}C$

Compacts of cerium hydride with UC or $U_{0.8}Pu_{0.2}C$ were sintered for 4 hours in vacuum at $1700°C$

Sample No.	$\frac{Pu}{U+Pu}$	Content (wt%) C	N	O	ECC[a]	a_{MC} (Å)	Other phases	Cerium in matrix (wt%)	$\frac{Ce}{M+Ce}$
1	0	5.25	0.13	0.17	5.49	4.9681 ± 5	Ce-rich, oxygen-containing precipitates	3.5	0.061
2	0	5.32	0.15	0.16	5.57	4.9670 ± 5		2.9	0.051
3	0.2	4.87	0.25	0.21	5.24	4.9708 ± 5	Ce_2C_3, a = 8.417 ± 4 Å; trace CeC_2	2.9	0.051
4	0.2	5.14	0.20	0.25	5.50	4.9706 ± 5	CeC_2, a = 3.87 Å, c = 6.47 Å	2.9	0.051

[a] Equivalent carbon content.

'CeC' and hence for the cerium concentrations derived [7]. Since a microprobe study is the only direct method of determining the concentrations in a phase of a multiphase sample, electron microprobe analysis (EMPA) and X-ray diffraction analysis (XRDA) were applied to the compositions listed in Table I. Microprobe examination had already been reported [6] for cerium-containing UC, but the details of the results had not been given.

The (U,Ce)C phases of the plutonium-free samples listed in Table I contain 3.5 or 2.9 wt% of cerium. Their lattice parameters confirm the lattice parameter/ concentration relationship derived in Ref. [7] from the results given in Refs [5, 6]. 2.9 wt% cerium was also found in the (U,Pu,Ce)C phases of the plutonium-containing samples with lattice parameters of 4.9706 and 4.9708 Å. The difference of $+ 37 \times 10^{-4}$ Å with respect to the plutonium-free sample constitutes the normal lattice-parameter shift between UC and (U,Pu)C having about 20% plutonium.

2.2. Zirconium and tantalum in PuC

Table II shows that the relative lattice-parameter differences of ZrC and TaC with respect to PuC are practically the same as those with respect to UC. The solubility relationships in the systems PuC-ZrC and PuC-TaC are therefore expected to be similar to those observed for the UC-ZrC and UC-TaC systems, at least over the temperature range where both UC and PuC are stable.

While ZrC and TaC are known to be completely soluble in UC, very limited solubility had been reported for zirconium in PuC [9]. An EMPA and XRDA study [10] has recently shown that up to 23 mol% of ZrC can dissolve in PuC phases and that the lattice parameter/concentration relationship between PuC and ZrC is practically linear (Fig. 2), as is that between UC and ZrC. The fact that solid-solution phases were observed with concentrations spread over

TABLE II. RELATIVE LATTICE-PARAMETER DIFFERENCES (RLPD) OF ZrC AND TaC WITH RESPECT TO UC AND PuC

	UC {a = 4.960 Å [8]}	PuC {a = 4.973 Å (for $PuC_{0.93}$) [8]}
ZrC, a = 4.698 Å	−5.3%	−5.5%
TaC, a = 4.454 Å	−10.2%	−10.4%

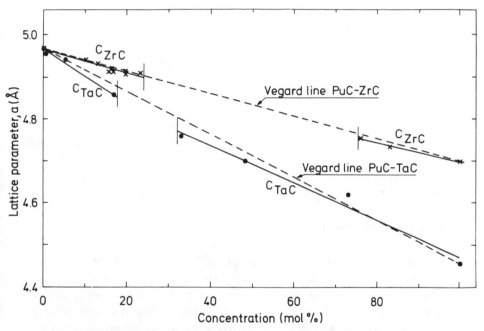

FIG.2. *Lattice parameter versus ZrC concentration, C_{ZrC}, in (Pu,Zr)C (×) and TaC concentration, C_{TaC}, in (Pu,Ta)C (●), as determined by electron microprobe analysis.*

the range between PuC and ZrC indicates that PuC-ZrC solid solutions are stable over large portions of this range, but are not obtained as single-phase samples under the given conditions because diffusion is too slow at the relatively low temperatures used.

A similar study was made for the PuC-TaC system, for which no solubility data had been published before. As in the case of zirconium (and probably for the same reason) single-phase samples were not obtained. The highest solute

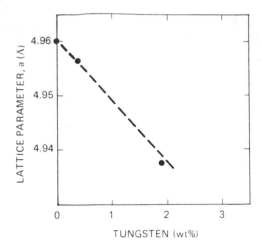

FIG.3. *Lattice parameters of (U,W)C.*

concentrations (about 17 mol% TaC in PuC and 67 mol% PuC in TaC) were obtained after an anneal for 4 h at 1500°C. This leads to the conclusion that, under the given conditions, only a small miscibility gap between 17 and 33 mol% TaC occurs in the PuC-TaC pseudo-binary. As with the corresponding UC system, PuC-TaC seems to present a slight negative deviation from Vegard's law, the maximum deviation existing around 30 mol% TaC (Fig.2).

2.3. Tungsten in UC and (U,Pu)C

The lattice parameters of arc-melted UC and (U,W)C with 0.39 wt% W and 1.90 wt% W are plotted against tungsten content in Fig.3. The value of 0.39 wt% W was derived from an overall tungsten content of 0.35 wt%, considering that approximately 10% of the UC_2 contained in that sample does not dissolve tungsten. The lattice parameter varies linearly with the weight per cent content of tungsten, as in the case of $U_{0.8}Pu_{0.2}C$ [11].

For the U-Pu-W-C system, samples were made from $U_{0.8}Pu_{0.2}C$ ("MC") and W or WC with the aim of preparing a tungsten-modified solid solution (U,Pu,W)C. The terminal solid solubility of tungsten in "MC" had been reported to be (1.3 ± 0.3) wt% at 1700°C and to have a maximum of about 3.2 wt% át 2100°C; the solid solution was in equilibrium with $(U,Pu)WC_{1.75}$ above 2100°C, but with tungsten at lower temperatures when the pseudo-binary section "MC"-W was considered [11]. In the present work, mixtures of "MC" with 1 or 2 wt% W or WC were pressed and sintered at 1700 to 1750°C or arc-melted. The equivalent carbon content (ECC, comprising, in addition to the carbon, the oxygen and nitrogen impurities which fill carbon vacancies up to about the

stoichiometric monocarbide composition) of the sintered samples was in the
range 4.8 to 5.0 wt%, indicating that the compositions were at or slightly above
the "MC"-WC line. Arc melting led to complete solution of 2 wt% WC; some
(U,Pu) sesquicarbide was formed, probably as a consequence of plutonium
evaporation during melting.

The lattice parameter of the solid solution was 4.9391 Å. It increased to
values around 4.944 Å after annealing at 1700, 1360 and 1200°C; at the same
time, a small amount of orthorhombic (U,Pu)WC$_2$ was formed. This confirms
that solubility is less than 2 wt% WC at T ⩽ 1700°C. The fact that lattice
parameters are practically the same at the three temperatures indicates that
diffusion of tungsten is too slow at 1360 and 1200°C to establish the equilibrium
concentrations. The presence of (U,Pu)WC$_2$ indicates that in the "MC"-WC
pseudo-binary, this phase also exists at temperatures below 2100°C, while in
the "MC"-W pseudo-binary [11] only tungsten was found as an equilibrium
phase below that temperature.

The sintered samples also contained the (U,Pu)WC$_2$ phase. If the lattice
parameters of the "MC" phase in the sintered samples are compared to the lattice
parameter/composition curve given in Ref. [11], it must be concluded that,
in most cases, this phase does not approach the 1700°C saturation concentration
of tungsten. This may indicate that the diffusion of uranium and plutonium
into the tungsten or WC particles, forming (U,Pu)WC$_2$, was faster than the
diffusion of tungsten into the "MC" grains.

2.4. Rhenium in UC

Sintered compacts of UC containing zero to 1 wt% Re were heat treated
for 6 h at 1620°C under vacuum. No significant lattice parameter change was
found in this range. It was concluded that rhenium solubility in UC, in the
absence of other rhenium-containing phases, is less than 0.1 wt%. A single
phase was found by XRDA (small amounts of metallic rhenium are below the
detection limit).

Treating the samples at higher temperatures (2000°C) or arc-melting
resulted in a lattice-parameter decrease of 0.005 Å, with respect to samples
treated at 1620°C. A possible explanation for this decrease is oxygen contamination. Also, for the samples treated at higher temperatures, no significant
variation of lattice parameter with rhenium content was observed. The solubility
of rhenium seems thus to be less than 0.1 wt% at all temperatures investigated.

2.5. Molybdenum in PuN and UN

Sintered compacts of PuN and molybdenum were heat treated at 1700°C
for 6 h under nitrogen. With respect to pure PuN treated under the same

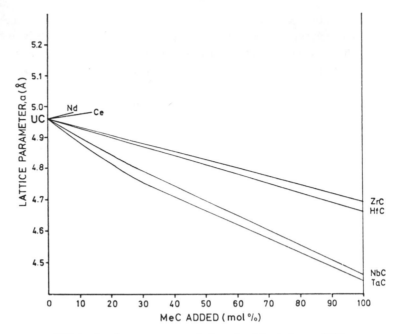

FIG.4. *Lattice-parameter variation of solid solutions of UC.*

TABLE III. MATRIX CONCENTRATION DETERMINED BY EMPA, AND FISSION YIELD OF SOME FISSION PRODUCTS FOR 17 at.% SIMULATED BURN-UP CARBIDE-BASED FUEL [1]

	Matrix concentration (wt%)	Fission yield (wt%)
Lanthanum	0.2	0.57
Cerium	0.1	1.23
Praseodymium	0.1	0.52
Neodymium	0.4	1.58
Samarium	no data	0.37
Zirconium	1.2	1.31
Molybdenum	0.3	1.60

conditions, a lattice-parameter decrease of 0.002 Å was observed for all molybdenum concentrations, the limiting solid solution having a lattice parameter of 4.9054 Å. A single phase was found by XRDA, the molybdenum metal present being below the detection limit.

With the same treatment as for the PuN-base samples, no significant lattice parameter change was observed with any molybdenum addition to UN. Thus, less than 0.1 wt% Mo is soluble in UN at 1700°C.

3.   STUDY OF THE LATTICE PARAMETERS OF IRRADIATED AND 'BURN-UP SIMULATED' FUELS

When a high burn-up carbide fuel was simulated by adding inactive fission product elements to UC or (U,Pu)C, a marked decrease in the lattice parameter of the matrix phase was observed [1, 7, 12]. A decrease of the same order was found for simulated carbonitride fuel with N/(C+N) = 0.2; lattice parameters of 4.953 and 4.949 Å were measured for simulated burn-ups of 3 and 10 at.%, respectively. In nitride fuels simulating 3 and 10 at.% burn-up, the addition of FP elements did not decrease the lattice parameter of the matrix [7]. Lattice-parameter measurements on irradiated fuels, though less accurate due to the high radiation level, confirmed this tendency [7]. They were indicative of local variations in FP concentrations in the matrix.

Amongst the FP elements which are soluble to some extent in the monocarbide phase, zirconium, molybdenum and niobium decrease the lattice parameter, while lanthanum, cerium, praseodymium and neodymium and probably samarium, also, increase it. This is shown in Fig.4 for those elements which have the highest solubilities in UC. Lanthanum and praseodymium, though less soluble, give lattice-parameter increases per unit concentration similar to those of cerium and neodymium [6, 9]. The sum of the fission yields of the lattice-parameter-increasing elements is larger than that of the lattice-parameter-decreasing elements. On the other hand, the fission yields at 10 at.% burn-up are lower than the solubilities in UC, even for the least soluble of the elements mentioned. We would thus expect an overall *increase* of the monocarbide lattice parameter. The *decrease* observed indicates that the solubility of the rare-earth elements in the irradiated carbide fuel is much lower than the sum of the individual solubilities in UC. In fact, only a part of the rare-earth fission yield was found dissolved in the matrix of high burn-up irradiated [13] and simulated [1, 12] carbide fuels; the remaining part formed rare-earth-containing precipitates. It appears that in the monocarbide lattice one has to take into account a *total rare-earth solubility* which limits dissolution of the individual rare-earth elements even if their individual saturation concentrations have not yet been attained.

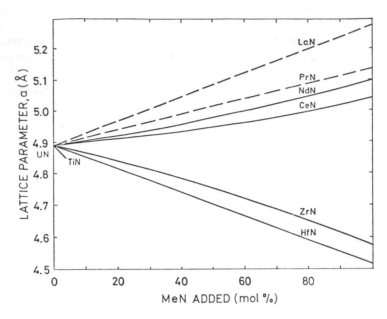

FIG. 5.   Lattice-parameter variation of solid solutions of UN.

Since no such limit exists for zirconium, the total zirconium fission yield can contribute to lattice shrinkage. This is shown in Table III: the zirconium concentration of the monocarbide matrix of a fuel simulating 17 at.% burn-up is approximately equal to the total zirconium concentration which simulates the fission yield. For the rare earths and for the molybdenum, the matrix concentration is much lower than the fission yield. Since part of the molybdenum is dissolved in the matrix, this element also contributes to lowering the lattice parameter. It appears from some measurements made by Lorenzelli [6] that the lattice parameter decrease caused by molybdenum is particularly large (about 0.01 Å for 1 at.% molybdenum dissolved in UC).

The lattice parameter of the mononitride is also increased by the rare earths (except yttrium; YN has a lattice parameter very close to that of UN), and lowered by zirconium and niobium (Fig. 5). Since the rare-earth mononitrides are completely miscible with UN, the total rare-earth fission yield can contribute to expand the lattice. This counteracts the lattice shrinkage caused by zirconium, which is also completely soluble.

The lattice-parameter change caused by dissolved fission products in a nitride fuel at 10 at.% burn-up was estimated from the fission yields of the main solute elements zirconium, lanthanum, cerium, praseodymium and neodymium and from the UN-MeN lattice-parameter curves determined by

Holleck et al. [14–16]. As no direct measurements were made by these authors in the range above 90 mol% UN, the tangents to the curves near UN were used to obtain a linear relationship for that range. An increase of 49×10^{-4} Å was calculated for zirconium, the decrease for (La+Ce+Pr+Nd) was 48×10^{-4} Å. The resulting zero expansion, though in agreement with the above-mentioned experimental results, has to be considered as tentative owing to the imprecisions of the method. Lattice-parameter measurements of UN samples with 0.5 to 3 mol% MeN are required to do a more accurate calculation.

4. DISCUSSION OF THE SOLUBILITY ON THE BASIS OF CRYSTAL STRUCTURE AND LATTICE-PARAMETER DIFFERENCES

Some rules governing the solubility in the carbides and nitrides of uranium and plutonium can be derived from the experimental evidence collected [7] and the complementary studies reported in § 2. To do this, the crystal structure and lattice parameters of the solute compound MeX_y are compared to those of MX_y (M : U or Pu), X being carbon or nitrogen, and y the X/(M+Me) atomic ratio. Where MeX_y does not exist, or only exists with a structure different from that of MX_y, the metal Me is insoluble or shows limited solubility in MX_y because a two-phase field must exist between the terminal single-phase fields. But when the solvent and the solute have the same structure and the same component X, the solubility will depend on the size ratio between the atom M contained in the solvent and the atom Me contained in the solute. This is a frequent case in the solvent-solute systems considered.

There are serious difficulties in assigning atomic or ionic radii to the atoms Me or M in the carbides MeC_y and MC_y, and the nitrides MeN and MN. A rule analogous to Hume-Rothery's size-factor rule using these radii can thus not be derived for the carbides and nitrides considered.

In contrast, lattice parameters are an objectively measurable quantity. They take into account, in an *empirical fashion,* the effect of the particular bonding conditions of the compound concerned on the size of the participating atoms. The difference between the lattice parameters of the solvent compound MX_y and the solute compound MeX_y should thus be a correct measure of the size difference between the Me and M atoms and constitute a criterion for the solubility of MeX_y in MX_y. For this purpose, a *relative lattice-parameter difference* (RLPD) was defined:

$$RLPD = \frac{a(MeX_y) - a(MX_y)}{a(MX_y)}$$

Tables giving the RLPD together with the known solubilities in MC, M_2C_3, and MN have been published elsewhere [17]. The following sections describe the conclusions which can be drawn from these tables.

4.1. Solubility of isostructural compounds in MC

In the four known cases of complete solubility of transition metals in MC, i.e. zirconium, hafnium, niobium and tantalum, the lattice parameters of MeC are 5.3 to 10.2% below that of MC. Where limited solubility in UC was found, the RLPD is −12.8% to −16.0%, thus larger in absolute value than for cases of complete solubility.

From the available evidence, a RLPD between −10.2% and −12.8% may thus be assumed as a lower limit for complete solubility of MeC in MC. Within the region of limited solubility, no systematic variation of the limiting solubility with RLPD can be deduced from the available information.

The only NaCl-type of monocarbides having higher lattice parameters than UC and PuC are the actinide carbides ThC, PaC and NpC. ThC, with RLPDs of +7.8% and +7.5%, is known to be entirely miscible with UC and PuC. The upper limit for complete solubility in UC and PuC is thus higher than + 7.8%.

As a conclusion, a NaCl-type of MeC compound is likely to be completely soluble in MC for values of RLPD between −10.2% and +7.8%.

4.2. Solubility of isostructural compounds in M_2C_3

The values of the corresponding table are plotted in Fig.6. The precision of the graph suffers somewhat from the fact that in two cases (praseodymium and neodymium in Pu_2C_3) only a lower limit has been indicated for the limiting solubility. But on the whole the solubility in Pu_2C_3 seems to be somewhat higher than that in U_2C_3, for the same RLPD. It can be foreseen from the trend of the curves of Fig.6 that terbium, dysprosium and holmium may have a high, if not complete, solubility in U_2C_3 and Pu_2C_3; samarium and gadolinium may be completely soluble in Pu_2C_3, or at least have a high solubility.

Sesquicarbides of the Pu_2C_3 type which have a higher RLPD than +9% with respect to U_2C_3 are not known. But it appears from Fig.6 that for a RLPD of more than +10% there would be only a negligible solubility in U_2C_3.

As a conclusion, it appears that the solubility in M_2C_3 increases with decreasing RLPD, becomes 100% in Pu_2C_3 for a RLPD of about +4%, and becomes 100% in U_2C_3 for a RLPD of (+1.5±1.0)%.

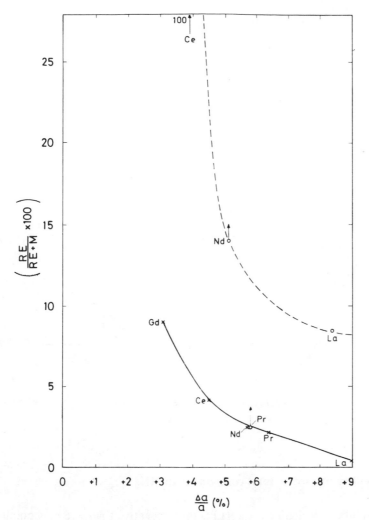

FIG.6. Solubility of cubic rare-earth (RE) sesquicarbides in U_2C_3 ($\times$) and Pu_2C_3 ($\circ$) versus RLPD (δ value is greater than or equal to that indicated by the symbol $\circ$).

4.3. Solubility of isostructural compounds in MC_2

Yttrium and gadolinium are reported to be completely soluble in tetragonal UC_2. The ratios of the unit cell volumes are $V(YC_2)/V(UC_2) = 1.038^3$ and $V(GdC_2)/V(UC_2) = 1.052^3$, which means linear size differences of 3.8% and 5.2%, respectively. It is logical then to assume that the dicarbides of lanthanides higher than gadolinium, which all have lower relative size differences with UC_2 than GdC_2, are completely miscible with tetragonal UC_2 and tetragonal PuC_2.

BENEDICT

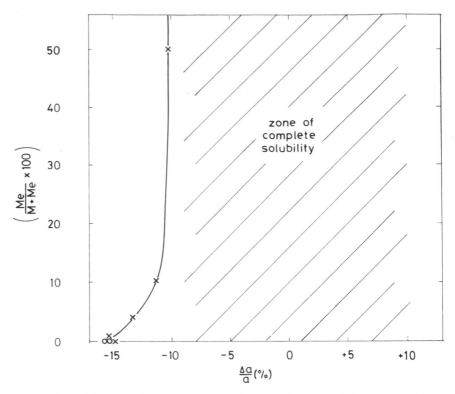

FIG. 7.   Solubility of NaCl-type mononitrides in UN (X) and PuN (○) versus RLPD.

4.4. Solubility of isostructural compounds in MN

Solubility is negligible for a RLPD more negative than −16%. Complete solubility is observed for a RLPD between −7.5% and +8.5%. In the range between −16% and −7.5%, solubility increases sharply with decreasing absolute RLPD (Fig.7). The given solubilities refer to temperatures from 1800°C to 2400°C.

A systematic difference between the solubilities in UN and in PuN, analogous to that observed for M_2C_3 (Fig.6), cannot be discovered from the available evidence. One can expect then that TiN, VN, NbN and WN will also show limited solubility in PuN, and that CrN will show limited solubility in UN.

The data points of Fig.7 refer to the elements which are known to have limited solubility in MN. The remaining rare-earth mononitrides and the known actinide mononitrides have RLPDs within the zone of complete solubility of this figure. They are expected to form complete solid solutions with UN and PuN. ScN, with a RLPD of −9% is at the lower limit of complete solubility;

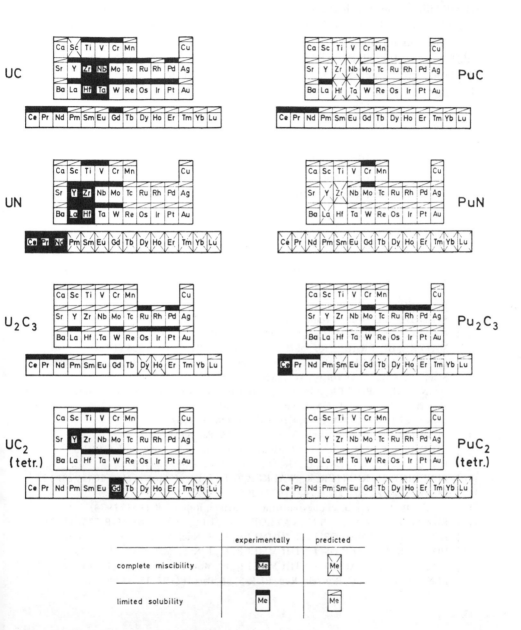

FIG.8.   *General review of experimental and predicted solubilities in uranium and plutonium carbides and nitrides.* [Me] *without any particular indication means that the information available is not sufficient to make a prediction.*

its solubility should be high, maybe complete. TcN, with a RLPD of −18.6% for UN and a RLPD of −18.9% for PuN should have negligible solubility.

As a conclusion, a NaCl-type MeN compound is likely to be completely soluble in MN for values of RLPD between −7.5% and +8.5%. The solubility in MN is negligible when the lattice parameter of MeN is more than 16% smaller than that of MN.

4.5. Prediction for unknown cases and review

The above-mentioned solubility rules were applied to cases where no experimental evidence for solubility is available. Figure 8 shows a review of the determined and predicted solubilities for all the elements and compounds studied.

REFERENCES

[1]   BENEDICT, U., GIACCHETTI, G., MATZKE, Hj., RICHTER, K., SARI, C., SCHMIDT, H.E., Nucl. Technol. **35** (1977) 154.
[2]   MATZKE, Hj., RONCHI, C., USERDA Rep. ERDA-4455 (1977) 218.
[3]   MATZKE, Hj., BRADBURY, M.H., CEC, Brussels, Rep. EUR-5906 (1978).
[4]   MATZKE, Hj., ROUTBORT, J.L., Thermodynamics of Nuclear Materials 1974 (Proc. Symp. Vienna, 1974) Vol.1, IAEA, Vienna (1975) 397.
[5]   HAINES, H.R., POTTER, P.E., Thermodynamics of Nuclear Materials 1974 (Proc. Symp. Vienna, 1974) Vol. 2, IAEA, Vienna (1975) 145.
[6]   LORENZELLI, N., CEA, Paris, Rep. CEA-R-4465 (1973).
[7]   BENEDICT, U., CEC, Brussels, Rep. EUR-5766, Part I (1977).
[8]   STORMS, E.K., The Refractory Carbides, Academic Press, New York (1967).
[9]   HAINES, H.R., POTTER, P.E., UKAEA Rep. AERE-R-6512 (1970).
[10] BENEDICT, U., RICHTER, K., WALKER, C.T., J. Less-Common Met. **60** (1978) 123.
[11] UGAJIN, M., J. Nucl. Mater. **47** (1973) 205.
[12] SMAILOS, E., Kernforschungszentrum, Karlsruhe, Rep. KFK-1953 (1974).
[13] EWART, F.T., SHARPE, B.M., TAYLOR, R.G., UKAEA Rep. AERE-R-7962 (1975).
[14] HOLLECK, H., SMAILOS, E., THÜMMLER, F., J. Nucl. Mater. **28** (1968) 105.
[15] HOLLECK, H., SMAILOS, E., THÜMMLER, F., J. Nucl. Mater. **32** (1969) 281.
[16] HOLLECK, H., SMAILOS, E., THÜMMLER, F., Monatsh. Chem. **99** (1968) 985.
[17] BENEDICT, U., CEC, Brussels, Rep. EUR-5766, Part II (1977).

DISCUSSION

J. FUGER: In the case of the solubility of lanthanides in sesquicarbides, especially U_2C_3, you are predicting a fairly discontinuous pattern of behaviour, from partial solubility to complete solubility and then partial solubility again for the last elements in the series. Would you care to comment on this point and also on the absence of prediction about europium, terbium and ytterbium?

U. BENEDICT: Figure 6 of the paper shows that there is a continuous increase of solubility in U_2C_3 and Pu_2C_3 with decreasing RLPD, attaining 100% for cerium in Pu_2C_3 at a RLPD of +4%. The good fit of cerium in these curves indicates that solubility does not depend primarily on the atomic number of the element in the lanthanide series but on the *actual* atomic size, which is affected by valence peculiarities and is well reflected by the RLPD. From the parallelism between the two curves of Fig.6, one can infer that the U_2C_3 curve should reach 100% solubility at a RLPD of (+1.5±1.0)%. Complete solubility was predicted for dysprosium and holmium because their RLPDs are +1.4% and +1.1% respectively, i.e. lower than the limit quoted. Erbium, thulium and lutetium must have limited solubility because their sesquicarbides do not have the Pu_2C_3-type structure; the lanthanides preceding dysprosium have too high RLPDs to be completely soluble in U_2C_3. No prediction was given for europium or ytterbium as their sesquicarbides are not known, nor for terbium because its RLPD (+2.2%) for U_2C_3 is near the limit for complete solubility given above, and one cannot determine whether its solubility is high (but still limited) or complete.

J.-P. MARCON: Could you please give a value, even if only approximate, for the decrease of parameter observed in a simulated carbide fuel and an irradiated carbide?

U. BENEDICT: The available data are briefly summarized on page 36 of Ref. [7] of the paper. Smailos observed a decrease of about 0.007 Å for 10 at.% burn-up on simulated fuels based on UC. We found $\Delta a \simeq 0.018$ Å in a fuel based on $U_{0.85}Pu_{0.15}C$ but simulating a 17 at.% burn-up, which corresponds to about 0.010 Å for 10 at.% burn-up: in a $U_{0.85}Pu_{0.15}C$ fuel irradiated to 11−12 at.% burn up, we found a Δa, corrected to 10 at.% burn-up, of 0.004 to 0.010 Å, depending on the position on a fuel section of the sampling point.

J.-P. MARCON: Out-of-pile solubility studies are generally carried out at a temperature higher than that existing in an irradiated carbide fuel. Under these conditions, do you have any idea about the variation of the solubility limit as a function of temperature in order to compare the results obtained from simulated and irradiated fuels?

U. BENEDICT: Reference [7] of the paper also gives some indications about the variations of solubility with temperature, e.g. for molybdenum and tungsten in UC, tungsten in (U,Pu)C and titanium in UN, where the solubility increases markedly with temperature. In the case of the lanthanides, however, the available data show little variation between 1000 and 1600°C; there is probably a kinetic obstacle to equilibrium being established at the lower temperatures. The same phenomenon will probably occur in fuels which have never exceeded 1000–1200°C. The diffusion rate of the fission atoms formed at isolated sites in the lattice will often be too low for precipitation to occur. In "cold" fuels we should, therefore, expect to find the diffusion products in supersaturated solution in the matrix. Thus, in some carbide fuels, post-irradiation examination did not reveal precipitates containing fission products in spite of a relatively high burn-up.

SOME PHASE-DIAGRAM STUDIES OF SYSTEMS WITH FISSION-PRODUCT ELEMENTS FOR FAST REACTOR FUELS

H.R. HAINES, P.E. POTTER, M.H. RAND
Atomic Energy Research Establishment,
Harwell, Didcot, Oxfordshire,
United Kingdom

Abstract

SOME PHASE-DIAGRAM STUDIES OF SYSTEMS WITH FISSION-PRODUCT ELEMENTS FOR FAST REACTOR FUELS.

The results of some experimental studies on the uranium-carbon and plutonium-carbon ternary systems with rhenium and technetium are first described. All the systems are characterized by ternary compounds; in particular two new ternary compounds are reported for the U-Tc-C system. Some studies on the Pu-Cr-C system have revealed two ternary compounds whilst there are no such compounds found in the Pu-Ni-C system. In the second part of the paper some calculations of phase diagrams of the binary systems Mo-Tc, Tc-Rh and Tc-Pd together with the ternary systems Mo-Tc-Rh, Mo-Tc-Pd and Mo-Ru-Pd are presented. A regular solution model has been used to describe the thermodynamic properties of the solutions.

1.   INTRODUCTION

In the paper we describe some experimental determinations and also some calculations of a number of phase diagrams of importance to the understanding of any constitutional changes which could occur within fuel pins during irradiation of both oxide and carbide nuclear fuels in fast reactors.

The experimental studies have been constitutional ones for carbide; we have further examined the ternary systems uranium-carbon and plutonium-carbon with rhenium and technetium for which some preliminary data have already been presented [1]. Although rhenium is not a product of fission it has been used frequently to simulate the behaviour of technetium in irradiated fuels and it is thus of relevance to examine the constitution of rhenium-containing alloys. Information on the behaviour of rhenium-containing systems will contribute towards the understanding of the behaviour of the transition elements with these actinide carbides. Some experimental studies on the plutonium-chromium-carbon and plutonium-nickel-carbon ternary systems are also reported.

471

These systems are of importance in both the prediction and
the understanding of any reaction which might occur between
carbide fuels and cladding materials. The results of these
experimental studies are discussed in terms of the present
knowledge of the uranium- and plutonium-transition element-
carbon systems with particular emphasis on the existence of
ternary compounds.

Alloys of the 5-component system, molybdenum-technetium-
ruthenium-rhodium-palladium, are frequently observed in
irradiated oxide fuels for fast reactors [2]; a full understanding and representation of the system could be extremely
useful for the assessment of irradiation conditions in post-
irradiation examinations of fuel. The calculations of phase
equilibria presented in this paper are for three ternary
systems, molybdenum-technetium-rhodium, molybdenum-technetium-
palladium and molybdenum-ruthenium-palladium; calculations of
some relevant binary equilibria for technetium-containing
systems are also given. These calculations will serve as a
basis for the understanding of the quaternary and quinary
systems and they will eventually be extended to describe
those systems.

2.   EXPERIMENTAL STUDIES

2.1 Experimental Techniques

All the alloys were prepared by the arc-melting together
of the elements. The elements were spectrographic graphite
which contained less than 10 ppm of impurities, and uranium
and plutonium each containing less than 2000 ppm of metallic
impurities. Technetium was obtained by the reduction of
ammonium pertechnetate in hydrogen; the finely divided
powder was then melted into a button before use in alloy
preparation. Chromium, rhenium, and nickel were 99.999%
pure. Weight losses during the preparation of alloys were
usually low but nevertheless some checks were made on their
compositions by chemical analyses mainly for carbon and
also for the likely impurities, oxygen and nitrogen [3,4].
In addition to chemical analyses, the alloys were characterized by metallography, X-ray powder photography, and
electron-probe microanalysis. When required the alloys were
annealed in carburised tantalum crucibles within alumina tube
furnaces under vacuum. Annealing temperatures between 800
and 1400°C were employed.

2.2 The Binary Phase Diagrams

The relevant features of the binary systems will be discussed usually with the appropriate ternary systems. However a brief description of the two systems, uranium-carbon and plutonium-carbon, is first given.

2.2.1 The uranium-carbon system

The phase diagram of the uranium-carbon system has been determined by Benz, Hoffmann and Rupert [5]. The system is characterized by three compounds, the monocarbide, the sesquicarbide and the dicarbide. Uranium monocarbide has a face-centred cubic rock-salt structure and exists over a range of composition. Uranium sesquicarbide has a Pu_2C_3-type body-centred cubic structure. The dicarbide exists in two forms; the low temperature α-body-centred tetragonal CaC_2-type structure transforms to the β-face-centred cubic KCN-type structure at higher temperatures. UC and UC_2 melt congruently.

2.2.2 The plutonium-carbon system

The phase diagram for the plutonium-carbon system is essentially based on the work of Mulford et al [6]. There are four compounds in the system, Pu_3C_2, the monocarbide, the sesquicarbide and the dicarbide. Pu_3C_2 is of unknown structure and decomposes peritectoidally at ca. $575^{\circ}C$ into ε-Pu and Pu monocarbide. Pu monocarbide which is isostructural with U monocarbide and also exists over a range of composition, is always hypostoichiometric with respect to carbon. Pu sesqui-carbide is, of course, isomorphous with U_2C_3 but unlike the latter compound exists over a range of composition. Pu dicarbide has the same structure as β-UC_2 and only exists at high temperatures. The form of the phase diagram is very different from that for the U-C system; all the compounds, with the possible exception of the dicarbide, decompose by either peritectoid or peritectic reactions.

2.3 The Ternary Phase Diagrams

2.3.1 The uranium-rhenium-carbon system

Before discussing the experimental data for the ternary system the main aspects of the uranium-rhenium and rhenium-carbon binary phase diagram will be summarized.

2.3.1.1 The uranium-rhenium system

The phase diagram of this system has been determined by Jackson, Williams and Larson [7]. Two compounds have been reported in the system, URe_2 and U_2Re. URe_2 melts congruently at 2200°C forming eutectics with both Re and γ-U. U_2Re forms by a peritectoid reaction below 750°C. URe_2 is reported to occur in two crystallographic modifications; above 180°C URe_2 has the $MgZn_2$-type hexagonal structure with lattice parameters a = 5.433 Å, c = 8.501 Å and below 180°C an orthorhombic structure exists with lattice parameters a = 5.600 Å, b = 9.178 Å, c = 8.463 Å. The structure of U_2Re is monoclinic.

2.3.1.2 The rhenium-carbon system

This system is probably a simple one with an eutectic between the elements, although the microstructures and microhardness data, together with α-ray powder photographs indicated the possible presence of a carbide phase. ReC has been reported to exist at high pressures; Popova and Baiko [8] report a carbide, probably of composition ReC, with an hexagonal structure and lattice parameters, a = 2.840 $\pm$ 0.001 Å and c = 9.85 $\pm$ 0.001 Å at pressures and temperatures above 60 kbar and 800°C. Further work by Popova et al [9] suggested that ReC had a NaCl-type f.c. cubic structure for pressures above 144 kbar, Savitskii et al [10] have also presented evidence for a metastable carbide of composition $Re_{2-4}C$. At ambient pressures, it has been assumed that there are no compounds in the equilibrium diagram; the Re-C eutectic temperature is at ca. 2500°C and the maximum carbon solubility in Re is ca. 11 at.%.

2.3.1.3 An assessment and experimental data on the ternary system

The general features of the phase diagram in this system were determined by Chubb and Keller [11]; the system possessed a ternary compound $UReC_2$, the structure was orthorhombic [11, 12]. The phase fields of the system at 1500°C were UC-$UReC_2$-Re, UC-URe_2-Re, UC-URe_2-liquid, C-$UReC_2$-Re, C-$UReC_2$-U_2C_3, and UC-U_2C_3-$UReC_2$. Preliminary studies confirmed many of the earlier features but a more detailed phase diagram was drawn including the additional features of the system within or on the triangle UC-Re-$UReC_2$ reported by Alexeyeva [13, 14] and Alexeyeva and Ivanov [15]. These studies showed that there are two ternary phases within or on the triangle, $UReC_{2-x}$ (x $\sim$ 0.12) which has a monoclinic structure [13], and $U_5Re_3C_8$ which has a tetragonal structure [14]. One compound

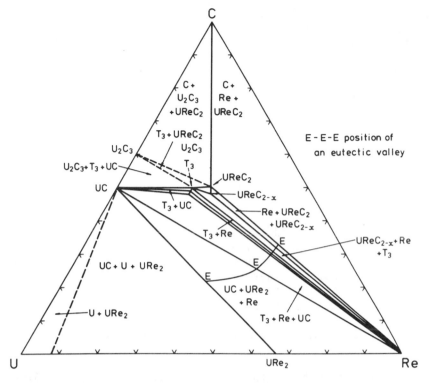

FIG.1. A phase diagram for the U-Re-C system at 800–1100°C.

with the tetragonal structure was designated compound Y by
Alexeyeva and Ivanov [15] on a partial isopleth of the system
for compositions where the Re:U ratio was unity, and by Haines
et al [1] on a phase diagram as compound T_3. T_3 or Y was
positioned with a Re:U ratio of greater than unity because
of the existence of the phase field UC, Y, $UReC_{2-x}$ [15] at Re:U
ratios of unity and with carbon contents between ca. 44 and
49 at.%.

Several alloys have been examined in the arc-melted and
annealed states, and confirm many of the main features and the
existence of the three ternary compounds. Examination of the
structures of the arc-melted alloys indicated the compositions
of any eutectics in the system; the position of the eutectics
is given on a tentative revised phase diagram of the system in
FIG. 1. The composition of the ternary compound T_3 was reported by Alexeyeva to vary from 30 at.% U, 20 at.% Re, 50 at.% C
to U 32 at.%, Re 20 at.%, C 48 at.%; our determinations

also suggest that this compound might exist over a range of
composition, as indicated on the phase diagram. The presence
of this compound does not allow UC to be in equilibrium with
$UReC_2$; further studies to confirm the position of the tie
lines in this region are being carried out.

2.3.2 The plutonium-rhenium-carbon system

The main features of the plutonium-rhenium system will be
discussed before the ternary system.

2.3.2.1 The plutonium-rhenium system

The only compound which has been reported in this system
is $PuRe_2$. This compound is isomorphous with the high temperature form of URe_2; the lattice parameters are a = 5.396 Å and
c = 8.729 Å [16]. Ellinger et al [17] quote some microstructural
evidence for a eutectic between ε-Pu and $PuRe_2$. Bowersox and
Leary [18] have measured the Re solubility in liquid Pu which at
700^oC is 1.57 at.% Re and at 950^oC, 4.05 at.% Re.

2.3.2.2 An assessment and experimental data on the ternary
system

The only reported data on this system are our preliminary
studies [1]. Several alloys were examined in the arc-melted
and annealed states. Alloys were annealed at 1000 and 1250^oC
for ca. 150 hours. No evidence was found for a ternary compound
analogous to $UReC_2$; the constitution of an alloy of this composition in the plutonium system was a three-phase mixture of
C, Pu_2C_3, and $PuRe_2$ and subsequent studies have confirmed this.
There was, however, some evidence of a ternary compound in alloys
with compositions bounded by the phases PuC, Pu_2C_3 and $PuRe_2$, but
further work is required to characterize this phase; the compound may be analogous to that found on a similar region of
the U-Re-C system. The lattice parameters of $PuRe_2$ in an
alloy annealed at 1000^oC of this composition was a = 5.3956
± 0.0007 Å, c = 8.7228 ± 0.0024 Å, and the parameters of
$PuRe_2$ in the presence of carbon-containing phases changed very
little, most probably indicating very little carbon solubility
in this phase.

2.3.3 The uranium-technetium-carbon system

In this section we shall again describe the main features
of the binary systems uranium-technetium and technetium-
carbon.

2.3.3.1 The uranium-technetium system

Darby, Berndt and Downey [19] have reported two compounds
for this system, U_2Tc and UTc_2. U_2Tc is isomorphous with U_2Re;
the lattice parameters of the monoclinic cell are a $_0$ = 13.407
± 0.0003 Å, b = 9.271 ± 0.0001 Å, c = 3.213 ± 0.001 Å,
β = 96°13' ± 1.4'. UTc_2 does not possess the $MgZn_2$ structure;
the X-ray powder pattern could not be indexed. Both YTc_2 [20] and
$ZrTc_2$ [21] are, however, isomorphous with URe_2 and $PuRe_2$.

2.3.3.2 The technetium-carbon system

The technetium-carbon system contains a f.c. cubic
carbide which coexists with Tc. Trzebiatowski and Rudzinski [22]
first reported this compound with a lattice parameter, a =
3.982 Å. Subsequent work reported by Haines et al [1] suggested
that Tc carbide exists over a range of composition; there
was an intragranular precipitate of graphite in the alloy
containing 25 at.% carbon suggesting that the range of composition increases with increase in temperature. It was also
suggested that the range over which the compound
existed was also extended up to a composition corresponding
to Tc_3C.

Further experimental studies on technetium-carbon alloys
(oxygen and nitrogen contents ⩽ 160 ppm and ⩽ 30 ppm respectively) both in the arc-melted and annealed condition now indicate
that the upper limit of carbon content is ca. 20 at.% carbon
(Tc_4C). The highest value of lattice parameter (typical uncertainty ± 0.0005 Å) was 3.9865 Å in an arc-melted alloy
containing 25 at.% carbon; this corresponds to an upper phase
boundary at ca. 20 at.% carbon for the carbide phase. The
highest value of lattice parameter measured in an annealed
alloy, 3.9844 Å, was for one which had been annealed at 1400°C
containing 17.6 at.% carbon. These lattice parameters are in
quite good agreement with the earlier measurements [22] on
alloys with Tc:C ratios of 2:1 and 1:1 which had been annealed
at 700 and 900°C. The lowest value of lattice parameter
measured for the carbide was 3.9560 Å in an arc-melted alloy
which was single-phase; in alloys annealed at 800°C the lowest
value of lattice parameter obtained was 3.968 Å in an alloy
containing 16.2 at.% carbon.

The above results indicate that technetium carbide can
exist over a range of composition from ca. 16.2 at.% carbon
(∿ Tc_5C) to ca. 20 at.% carbon (Tc_4C).

The composition of an eutectic on the carbon-rich side of
Tc_4C is between 25 and 33 at.% carbon. The limit of carbon
solubility in technetium at 800°C was determined to be ca.
10 at.%. The values of lattice parameters for carbon saturated
Tc were a = 2.9848 Å and c_o = 4.4470 Å, compared with values of
a = 2.7375 and c = 4.3950 Å for pure Tc used in these studies.

2.3.3.3 Experimental studies on the ternary system

A tentative phase diagram for this system has been previously presented [1]. The system had many features which
were analogous to those of the U-Re-C ternary system. The
existence of the compound $UTcC_2$ with an orthorhombic structure,
first identified by Farr and Bowman [12] was confirmed. The
values of the lattice parameters of the orthorhombic
cell [1] were very similar to the earlier data [12]. The existence of the UTc_2 phase was confirmed but we have still been
unable to determine the crystallographic structure from the
X-ray powder pattern.

Further experimental studies both on the binary Tc-C
system which were described above and on the ternary system
have enabled us to refine the phase diagram.

A new ternary compound has been identified, the composition of which can be represented by $UTc_3C_{0.55-0.65}$. This
compound has a face-centred cubic structure with lattice parameters which vary from 4.1322 Å to 4.1483 Å with increasing carbon content. It is assumed that this compound is analogous to
the URu_3C and URh_3C compounds which also have a face-centred
cubic-structure [23]. The compound was present in alloys annealed at 800°C for several hundred hours and it was also detected
in arc-melted alloys; the amounts present in arc-melted alloys
indicated that it was most probably formed by a peritectic
reaction. An additional ternary phase (T_3) has been found
with a composition close to that of $UTc_3C_{0.55-0.65}$,
in alloys annealed at 800°C of nominal composition U 20 at.%,
Tc 60 at.%, C 20 at.%; such an alloy contained about 90%
of this phase. The first 26 lines of the X-ray powder
pattern were indexed as a tetragonal structure with lattice
parameters, a = 5.030 + 0.005 Å and c = 12.493 + 0.015 Å.
A tentative phase diagram is shown in FIG. 2; the separate
phase $UTcC_{2-x}$ is included in the diagram by analogy with
the rhenium system although as yet we have obtained little
evidence for its existence.

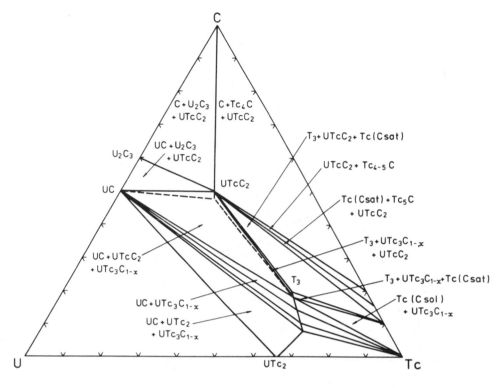

FIG.2. A phase diagram for the U-Tc-C system at 800–1000°C.

2.3.4 The plutonium-technetium-carbon system

The only previous information reported on this system
was that of Haines et al [1]. The compound $PuTc_2$ was identified
in the plutonium-technetium system but, as for UTc_2, the X-ray
powder pattern could not be indexed; the pattern is very
complex and it is not clear that the uranium and plutonium
compounds are isomorphous. There are possibly two
ternary compounds in the system; $PuTcC_{2-x}$ which has a monoclinic structure and is isomorphous with $UReC_{2-x}$, and a
compound, (T), the structure of which has not been determined
but the composition of which is ca. 30 wt.% C, and also
lies close to a line joining $PuTc_2$ with Pu monocarbide. This
compound may be analogous to the compound found in a similar
region of the Pu-Re-C system. In alloys annealed in the
temperature range 1000-1450°C we have found no evidence for
an orthorhombic structured compound analogous to $UTcC_2$. A
tentative phase diagram is shown in FIG. 3.

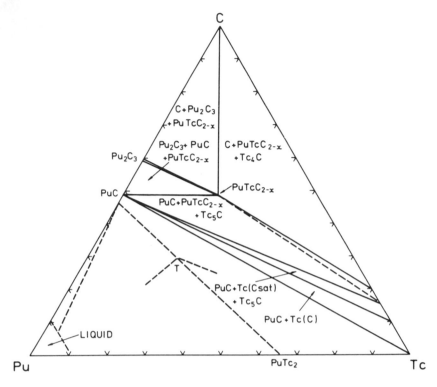

FIG.3. A tentative phase diagram for the Pu-Tc-C system at 1000°C.

2.3.5   The plutonium-chromium-carbon system

The two binary systems plutonium-chromium and chromium-carbon will be briefly discussed before the experimental data on this ternary system.

2.3.5.1   The plutonium-chromium system

The data available for this system have been reviewed by Ellinger et al[17]; there are no compounds formed and there is an eutectic composition very rich in plutonium.

2.3.5.2   The chromium-carbide system [24]

There are three compounds in this system, $Cr_{23}C_6$, Cr_7C_3, and Cr_3C_2 which have face-centred cubic, hexagonal and orthorhombic structures respectively.

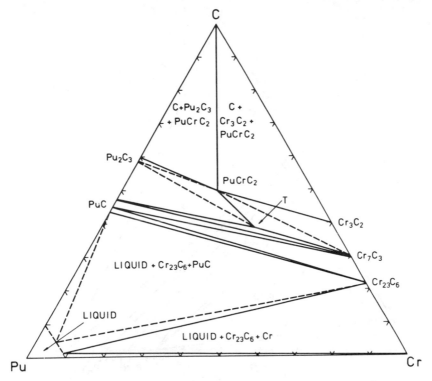

FIG.4. A tentative phase diagram for the Pu-Cr-C system at 1000°C.

2.3.5.3 An assessment and experimental data on the ternary system

We have found two ternary compounds in this system: $PuCrC_2$; this compound is most probably isomorphous with the compound $UCrC_2$ which has an orthorhombic structure [25]. The second compound has been found with a Cr:Pu ratio greater than unity and with carbon contents of ≤ 40 at.% carbon; this compound is labelled T on FIG. 4 and the powder pattern was indexed as a tetragonal structure with lattice parameters a = 3.656 ± 0.001 Å and c = 15.809 ± 0.003 Å. These parameters are very similar to those given by Farr and Bowman [12] for a compound $UCrC_x$; the lattice parameters of this compound were a = 3.636 ± 0.003 Å and c = 15.739 ± 0.008 Å. Alexeyeva and Ivanov [15] presented evidence for a tetragonal structured compound of composition $U_2Cr_9C_9$, but the Cr:Pu ratio is considerably less in the plutonium tetragonal compound than in this uranium compound. A tentative phase diagram is shown in FIG. 4.

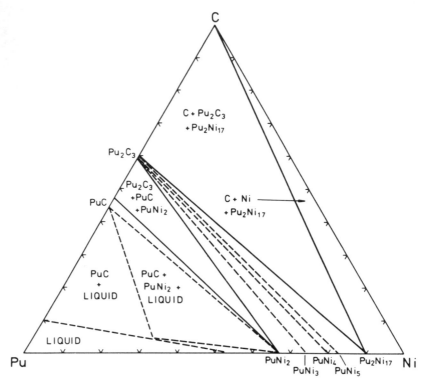

FIG.5. *A tentative phase diagram for the Pu-Ni-C system at 1000°C.*

2.3.6 The plutonium-nickel-carbon system

A brief summary of the binary systems plutonium-nickel and nickel-carbon is first given.

2.3.6.1 The plutonium-nickel system

The data for this system have been reviewed by Ellinger et al [17] ; there are six components in this system, PuNi, $PuNi_2$, $PuNi_3$, $PuNi_4$, $PuNi_5$ and Pu_2Ni_{17}. These compounds have orthorhombic (TlI-type), face-centred cubic (Cu_2Mg type), rhombohedral, monoclinic, hexagonal ($CuZn_6$ and Ti_2Ni_{17} type) structures respectively. Only $PuNi_5$ melts congruently.

2.3.6.2 The nickel-carbon system [26]

The nickel-carbon system is an eutectic system. The hexagonal structured compound Ni_3C is not thermodynamically stable under ambient pressures.

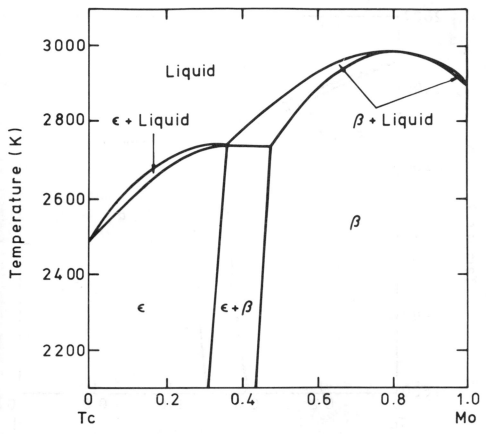

FIG.6.   A calculated Mo-Tc phase diagram.

2.3.6.3   An assessment and experimental data on the ternary system

The limited experimental information obtained from only three alloys suggests that a phase diagram of the form shown in FIG. 5 is appropriate. No evidence was found for a ternary compound analogous to the tetragonal structured compound $UNiC_2$ of the U-Ni-C system. An alloy of composition Pu 25 at.%, Ni 25 at.%, C 50 at.%, showed no evidence for a ternary compound in the arc-melted condition or after annealing at 850°C and 1000°C. The phases present were graphite, Pu_2C_3 and Pu_2Ni_{17}. Similarly, the constitution of alloys of composition Pu 40 at.%, Ni 20 at.%, C 40 at.% and Pu 33.3 at.%, Ni 16.7 at.%, C 50 at.% both in the arc-melted state and after annealing at 1000°C were consistent with the

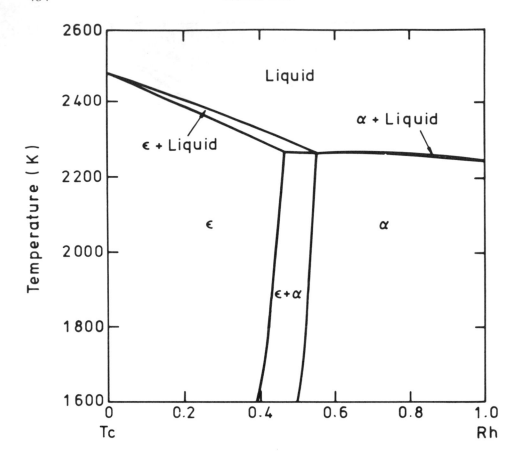

FIG.7. A calculated Tc-Rh phase diagram.

compatibility tie-line diagram shown in FIG. 5. The lattice
parameters of PuC and Pu_2C_3 indicated little or no solubility of Ni (PuC, a = 4.976-4.978 Å; Pu_2C_3, A = 8.128-8.132 Å).
The lattice parameter of $PuNi_2$ was 7.140 Å, in good agreement
with the published data, thus suggesting little carbon
solubility in this phase.

These tie-line data allow the checking of the consistency of the limited thermodynamic data for the compounds of
the Pu-Ni binary system. There is only one measurement
from which thermodynamic data can be obtained, that of
Campbell [27] using a molten salt EMF technique to measure

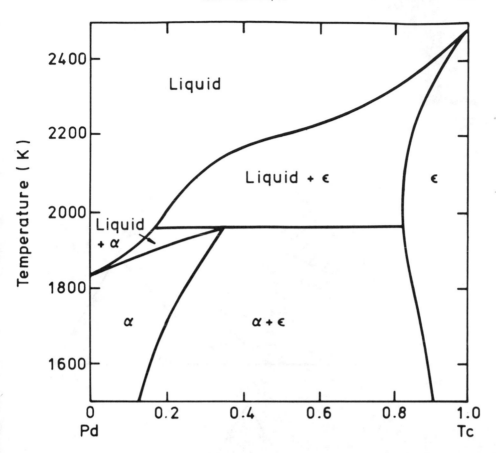

FIG.8. A calculated Tc-Pd phase diagram.

the Pu potential in, what has been assumed to be, the Ni + Pu_2Ni_{17} phase field from 930-1113K. These data give for

$$2Pu(l) + 17Ni(s) \rightarrow Pu_2N_{17}(s)$$

$$\Delta G_f^o = -40100 + 5.9T \; cal \cdot mol^{-1}$$

There are also some estimates of the heats of formation of all the Pu-Ni compounds by Miedema [28].

For the phase diagram of the form shown in FIG. 5, the following conditions must be satisfied.

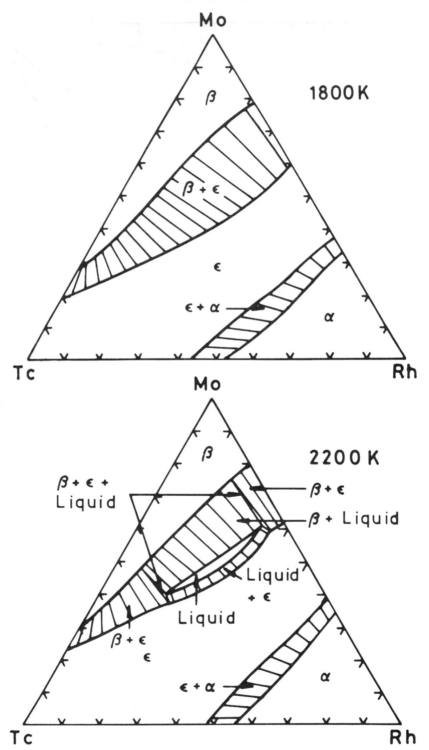

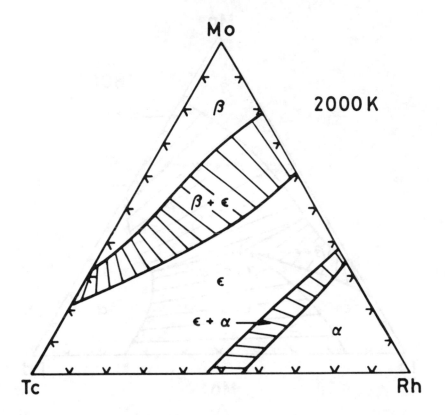

FIG.9.   Some calculated isothermal sections of the Mo-Tc-Rh phase diagram.

$$Pu_2C_3 + 17Ni \rightarrow Pu_2N_{17} + 3C$$

and thus $\Delta G^o_{fPu_2N_{17}} < \Delta G^o_{fPu_2C_3}$

$$\Delta G^o_{fPu_2C_3} = -40830 - 5.04T \text{ cal} \cdot \text{mol}^{-1} \text{ [29]}$$

This condition is never satisfied with the values of $\Delta G^o_{fPu_2N_{17}}$ given by Campbell; a more negative value is required to satisfy the phase diagram.

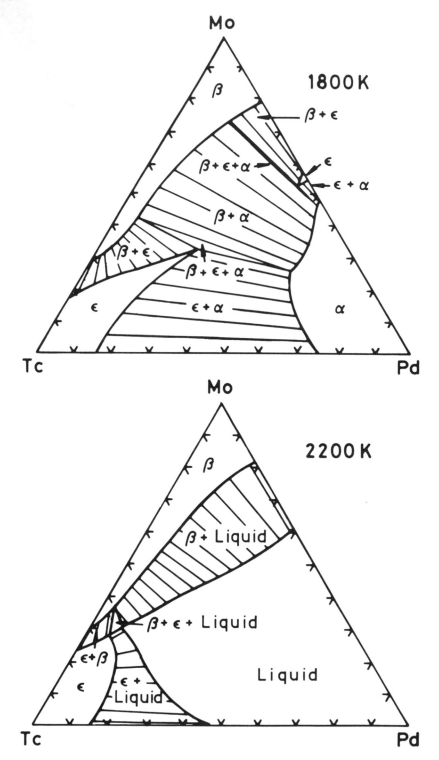

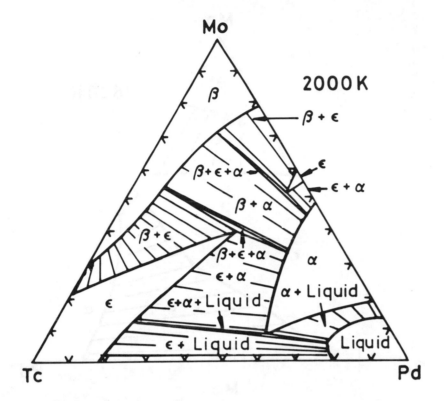

FIG.10. Some calculated isothermal sections of the Mo-Tc-Pd phase diagram.

Another condition is that

$$3.41 \, PuC_{0.88} + 11.99 \, Ni \rightarrow Pu_2C_3 + 0.705 \, Pu_2Ni_{17}$$

and thus

$$0.705\Delta G^o_{fPu_2Ni_{17}} < 3.41\Delta G^o_{fPuC_{0.88}} - \Delta G^o_{fPu_2C_3}$$

$$\Delta G^o_{fPuC_{0.88}} = -12452 - 0.91T \, cal \cdot mol^{-1} \, [29]$$

This condition is satisfied by both the data of Campbell
and Miedema.

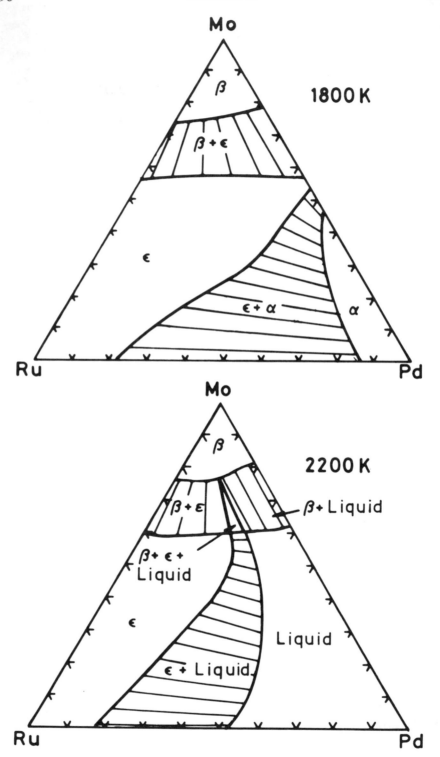

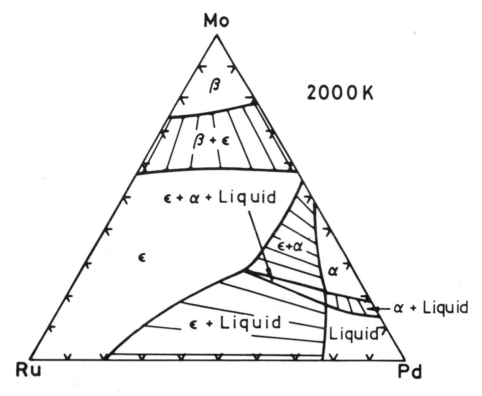

Mo

β

$\beta + \epsilon$

2000 K

$\epsilon + \alpha + $ Liquid

$\epsilon + \alpha$

ϵ

α

$\alpha + $ Liquid

$\epsilon + $ Liquid

Liquid

Ru

Pd

FIG.11.   Some calculated isothermal sections of the Mo-Ru-Pd phase diagram.

In order to estimate the Gibbs energy of formation of $PuNi_2$ the reaction

$$3.41 \; PuC_{0.88} + 2.82 \; Ni \rightarrow Pu_2C_3 + 1.41 \; PuNi_2$$

must be considered and,

$$1.41 \Delta G^o_{fPuNi_2} < 3.41 \Delta G^o_{fPuC_{0.88}} - \Delta G^o_{fPu_2C_3}$$

thus $\Delta G^o_{fPuNi_2} < -2298 + 2.74T \; cal \cdot mol^{-1}$

The heat of formation estimated by Miedema is in fact an order of magnitude greater than the heat of formation term in this two-term expression.

Table I

Compounds of general formula $PuXC_2$ and $PuXC_{2-x}$ where X is a
transition element of groups VIa to VIII of the periodic table

Group VIa	Group VIIa	Group VIII		
$PuCrC_2$	Pu-Mn-C system not examined	$PuFeC_2$	Pu-Co-C system not examined	Pu-Ni-C no ternary compound
$PuMoC_2$ $PuMoC_{2-x}$	$PuTcC_{2-x}$	Pu-Ru-C Pu-Rh-C Pu-Pd-C No ternary compounds		
$PuWC_2$ $PuWC_{2-x}$	Pu-Re-C no ternary compound	Pu-Os-C Pu-Ir-C No ternary compounds		Pu-Pt-C system not examined

2.3.7   A summary of the experimental data on the ternary systems

The uranium-containing systems which we have examined
are characterized by ternary compounds in addition to those
which are isomorphous with the orthorhombic structured compound $UMoC_2$ [30] and the monoclinic structured compound
$UMoC_{2-x}$ [31]. Three additional compounds have been found. One
with a variable composition but with a formula close to
$U_5Re_3C_8$, and which has a tetragonal structure: $UTc_3C_{0.55-0.65}$
which has a face-centred cubic structure and may be similar
in structure to the, now well characterised, compounds
URu_3C and URh_3C [23]. The third compound has a tetragonal
structure and is very close in composition to the previous
compound, but contains slightly more carbon.

From these studies it is possible to add to the summary
of the plutonium-transition element-carbon systems given
earlier by Holleck [23]. The ternary phases which have general formula $PuXC_2$ and $PuXC_{2-x}$ where X is a transition element
of group VIa to group VIII of the periodic table are given
in Table I.

For the second and third row group VIII elements although
there are also no uranium ternary compounds of general formula UXC_2, there are ternary compounds of general formula
U_2XC_2 with tetragonal or orthorhombic structures for all
these elements except palladium. There are no analogous Pu
compounds although the Pu-Pt-C system remains to be
examined.

It is well known that carbon can enter the Cu_3Au-type
lattice of some of the actinide X_3 and lanthanide X_3 compounds [32], where X is Ru or Rh, to give compounds of general
formula UX_3C_{1-x}. In the Pu-containing systems the compounds
$PuRu_3C$ and $PuRh_3C_{1-x}$ have been obtained [33]; the latter is
completely miscible with $PuRh_3$. Preliminary studies have
also indicated that there are probably no analogous ternary
phases in the Pu-Os-C and Pu-Ir-C systems in alloys of
overall composition PuX_3C (X = Os, Ir) which were annealed
at 1250°C for 250 hours; the annealing temperatures, however,
may be somewhat high as the phase $PuRu_3C$ was only readily
detected when the appropriate alloys were annealed at 800°C.

It is probable that there are some other ternary phases
in the Pu-Re-C and Pu-Tc-C systems; the Pu:Re and Pu:Tc
ratios in these compounds are greater than unity and for the
Tc compound the carbon content is ca. 30 at.%. Another
ternary compound for which no analogues have been found is
$Pu_3Fe_4C_5$ [34] which has a body-centred cubic structure.

3. THE CALCULATION OF SOME TRANSITION ELEMENT
TERNARY PHASE DIAGRAMS

3.1 Background of the Calculations

Of the new phases which appear during the irradiation
of uranium-plutonium oxide, a 5-component alloy of Mo-Tc-
Ru-Rh-Pd is the most frequently observed [2,35].   This alloy
can appear as quite large ingots within the hot columnar
grain region of a fuel pin, and has been observed with the
hexagonal ε -Ru structure. Attempts have been made [36]
to use the reaction

Mo (dissolved in Tc-Ru-Rh-Pd) + O_2 (in oxide) $\rightleftharpoons MoO_2$

to determine the thermodynamic oxygen potential of the fuel
in the vicinity of the alloy inclusion by the determination
of the Mo concentration in the alloy and the concentration
of MoO_2 dissolved in the oxide fuel matrix. An increase in
oxygen potential would result in the formation of more MoO_2.

The oxygen potential environment of an operating pin is such that the alloy can be found with varying amounts of Mo. Because of the relatively higher vapour pressure of Pd this element could also be lost from the alloy in the hottest temperature regions of a pin. It would thus be desirable to examine the influence of variations of both Mo and Pd concentrations in the alloy on its constitution and structure. Such changes may affect the dissolution rate of this phase during reprocessing of the fuel and its constitution and structure should also help in the determination of the thermodynamic environment within the fuel in the vicinity of the alloy.

As part of a programme to examine the variation of composition on the constitution and structure of these alloys we have calculated some of the relevant binary and ternary phase diagrams which are constituent systems of the 5-component system. In this paper we have paid particular attention to some of the technetium-containing systems for which there have been practically no experimental studies. The ternary systems are Mo-Tc-Rh and Mo-Tc-Pd. We have also calculated the phase diagrams of the Mo-Rh-Pd system.

3.2   The Method of Calculation

In order to calculate the phase diagram we require the lattice stabilities of all the relevant elements. The values for the lattice stabilities, with the exception of those for Tc, have been taken from the published data of Kaufman and Bernstein [37].

Tc has an hexagonal close packed structure (ε) which is isomorphous with Ru. We have taken the melting point determined by Anderson et al [38] as 2473 K and we have assumed that the value of the entropy of melting is identical to that of Re. For the lattice stabilities of the face-centred cubic phase (α) and the body-centred cubic phase (β) we have used the same entropies as for the analogous structured Re allotropes and interpolated between Mo and Ru so that the Gibbs energies of the lattices bear the same relationship to those for Mo and Ru that Re bears to W and Os.

The lattice stabilities expressed as Gibbs energies for the various transitions are shown in Table II.

In these preliminary calculations we have used regular solution models to describe the four types of binary solution, namely liquid, body-centred cubic (β), hexagonal close-packed (ε), and face-centred cubic (α). This simple model

Table II

The lattice stabilities or Gibbs energies
of transition for the relevant elements

Element	$\Delta G^{\beta \to L}$	$\Delta G^{\epsilon \to L}$ (cal $\cdot$ mol^{-1})	$\Delta G^{\alpha \to L}$
Mo	5800-2.0T	3800-2.0T	3289-2.15T
Tc	4992-2.4T	4946-2.0T	5116-2.3T
Ru	3976-2.8T	5100-2.0T	4984-2.8T
Rh	2837-3.05T	4322-2.15T	4480-2.0T
Pd	2296-2.8T	3381-2.3T	3640-2.0T
W	7300-2.0T	5300-2.0T	4795-2.15T
Re	6504-2.4T	6900-2.0T	6647-2.3T
Os	5488-2.8T	6600-2.0T	6468-2.8T

requires only one parameter, the interaction parameter.
These parameters for the relevant binary systems have been
estimated in the manner described by Kaufman and Bernstein,
and are designated L for the liquid solution and B, E and A
for the respective b.c.c., h.c.p. and f.c.c. solid solutions.
L is the sum of two terms,

$$L = \epsilon_o + \epsilon_p$$

where $\epsilon_o = 2[H^L[(i+j)/2] - 0.5 H_i^L - 0.5 H_j^L]$ cal$\cdot$mol^{-1} and
$H^L[(i+j)/2]$, H_i^L and H_j^L are related to the enthalpies of the
ij, ii, and jj bonds; $^JH_{ij}^L$, H_{ii}^L, H_{jj}^L. For example
$H_i^L = 0.5 NZH_{ii}^L$, where N is Avogadro's number and Z is the
co-ordination number in the i-j system; i and j are the
group number of the periodic table of which a given element is
a member. (The last two columns of group VIII in the periodic
table are designated groups 9 and 10 here.)

$$\varepsilon_p = 0.3 \ (V_i + V_j) \ [(-H_i/V_i)^{\frac{1}{2}} - (-H_j/V_j)^{\frac{1}{2}}]^2 \ \text{cal·mol}^{-1}$$

where H_i and H_j are the enthalpies of vaporization and V_i and V_j the volumes of the binary components, the value of ε_p the internal pressure coefficient is calculated from data for the elements at 298 K.

For the three solid solutions,

$$B = \varepsilon_1 + \varepsilon_2 \ \text{cal · mol}^{-1}$$

where ε_1 is a strain energy term depending on the size or volume difference between i and j, and

$$\varepsilon_1 = -0.5 \ (H_i + H_j) \left[\frac{V_i - V_j}{V_i + V_j}\right]^2 \ \text{cal · mol}^{-1}$$

The electronic component ε_2 for the β-phase is given by

$$\varepsilon_2 = 2\{\Delta H^{L \to \beta}[(i+j)/2] - 0.5 \ \Delta H_i^{L \to \beta} - 0.5 \Delta H_j^{L \to \beta}\} \ \text{cal · mol}^{-1}$$

where $\Delta H_i^{L \to \beta}$, $\Delta H_j^{L \to \beta}$ and $\Delta H_{(i+j)/2}^{L \to \beta}$ are the enthalpy changes for the liquid $\to$ β-phase transition of the elements of group i, group j and intermediate positions $[(i+j)/2]$.

$$E = B + \varepsilon_3 \ \text{cal · mol}^{-1}$$

and $\varepsilon_3 = 2\{\Delta H^{\beta \to \varepsilon}[(i+j)/2] - 0.5 \Delta H_i^{\beta \to \varepsilon} - 0.5 \Delta H_j^{\beta \to \varepsilon}\} \ \text{cal · mol}^{-1}$

and the three quantities $\Delta H^{\beta \to \varepsilon}$, similarly as for ε_2, are the enthalpies for the transitions from the β-phase to the ε-phase.

$$A = E + \varepsilon_4 \ \text{cal · mol}^{-1}$$

and $\varepsilon_4 = 2\{\Delta H^{\varepsilon \to \alpha}[(i+j)/2] - 0.5 \Delta H_i^{\varepsilon \to \alpha} - 0.5 \Delta H_j^{\varepsilon \to \alpha}\} \ \text{cal · mol}^{-1}$

and again as for ε_2 and ε_3, the three quantities $\Delta H^{\varepsilon \to \alpha}$ are the enthalpies of transition from the ε-phase to the α-phase.

All the regular solution interaction parameters have been taken from Kaufman and Bernstein with the exception of those for the Tc systems which have been estimated by us. The values of these parameters are given in Table III.

3.3 The Binary Systems

The phase diagrams of the relevant binary systems except those containing Tc were presented by Kaufman and Bernstein.

Table III

Calculated regular solution interaction
parameters for the relevant binary systems

L, B, E and A are the interaction parameters for the liquid,
b.c. cubic, hexagonal and f.c. cubic solutions respectively.

System	Interaction Parameter			
	L	B	E	A
		(cal)		
Tc-Mo	-10000	-9700	-9300	-9190
Mo-Pd	-1391	-1682	-4822	-4832
Pd-Tc	7625	8548	7548	7408
Ru-Mo	-5662	-5721	-7401	-7521
Pd-Ru	7009	7800	7020	6850
Mo-Rh	-5915	-6267	-8287	-8267
Tc-Rh	3293	3393	3043	3088

The systems of interest here are Mo-Ru, Mo-Rh, Mo-Pd, and Ru-Pd.
The general features of the systems were reproduced quite
satisfactorily with the single parameter but further detailed
representation is required to include much known detail. In
the Mo-Ru system the presence of the Mo_5Ru_3, σ-phase [24] is
neglected. In the Mo-Rh phase diagram the width of the ε-
phase extends to higher Rh concentrations than the calculations
predict. In the Mo-Pd the experimental and calculated positions of the ε-phase differ by 10 at.%, and in the Ru-Pd
system the agreement is good between experimental and calculated phase diagrams providing there are no intermediate
phases [24].

It is recognised that the description of the solution
phases must be refined more but they are capable of giving
some of the essential features of the phase relationships.

In FIGS. 6-8 we present some calculations of possible phase diagrams for the systems Mo-Tc, Tc-Rh and Tc-Pd. The calculations were carried out with a computer programme which finds a particular combination of phases which minimises the Gibbs energies of the system at a given temperature and composition.

There are very little experimental data for the binary systems containing Tc [26], although there is a ε-phase in the Mo-Tc system at 65-75 at.% Tc. There are no compounds in the Tc-Rh and Tc-Pd systems. There is extensive solubility of Rh in Tc (ca. 70 at.% at 1050-1500°C) and the solubility of Tc in Rh is ca. 4.5 at.% at 1500°C. There is also extensive solubility of Pd in Tc (47-50 at.% at 1050-1500°C) and that of Tc in Pd is also quite high (ca. 28 at.% at 1500°C). The calculated phase diagrams for the Tc-Rh and Tc-Pd systems give a much narrower region of ε-phase than the limited experimental data would indicate. The agreement for the extent of the α-phase is somewhat better for the Tc-Pd system.

3.4   The Ternary Systems

In FIGS. 9-11 some calculations of the phase diagrams for three ternary systems, Mo-Tc-Rh, Mo-Tc-Pd and Mo-Ru-Pd are shown. The computer programme we have employed for these calculations finds the combination of phases and their composition which gives a minimum in the total Gibbs energy of the system at a given temperature and overall composition. Isothermal sections of these three ternary systems are given for three temperatures, 1800, 2000 and 2200 K.

3.5   Conclusions Concerning the Calculations of Phase Diagrams

The calculations show that providing there is not extensive melting all the ternary systems possess a region of ε-phase extending from the Tc or Ru corners of the ternary phase diagrams. The replacement of Tc by Ru in the systems Mo-(Tc-Ru)-Pd results in an increase in the area of existence of the ε-phase at a given temperature. The replacement of Pd by Rh in the Mo-Tc-[Rh or Pd] systems also has the same effect in increasing the area of existence of the ε-phase field.

A typical composition of the 5-component alloy which might be encountered in an irradiated oxide fuel would be Mo 31 at.%, Tc + Ru 40 at.%, Rh + Pd 29 at.%. The composition of such an alloy, expressed as a ternary composition, would lie in the ε + α region of the Mo-Tc-Pd system at 1800 K, but very close to the ε-phase boundary, at 2000 K,

the composition of the alloy would be on the $\alpha + \varepsilon$ phase
boundary at the ε-phase side, and in the liquidus at 2000 K.
For the other systems, Mo-Tc-Rh and Mo-Ru-Pd, the composition
of the 5-component alloy expressed as a ternary composition
would lie on the single phase ε region for all cases except
the Mo-Ru-Pd system at 2200 K where it would lie in the liquid
+ ε phase region.

These preliminary calculations would indicate that it
is possible to allow some variation of composition of the 5-
component fission product element alloy and to remain within
the ε phase field.

It should be stressed that these calculations are based
on very simplified models; the solution phases are regular
and for the ternary systems no ternary interaction parameter has been introduced. The calculations will be refined
to describe the binary systems, wherever possible, more
accurately and to check both the predictions for the binary
and ternary systems with some experimental measurements.

REFERENCES

[1]   HAINES, H.R., MARDON, P.G., POTTER, P.E., Plutonium 1975
      and other actinides (Blank, H., Lindner, R., eds.).
      Amsterdam, North Holland-American Elsevier 1976, p. 233.
[2]   BRAMMAN, J.L., SHARPE, R.M., THOM, D., YATES, G.,
      J. Nucl. Mats. 25 (1968), 201.
[3]   MILNER, G.W.C., PHILLIPS, G., JONES, I.G., CROSSLEY, D.,
      ROWE, D.H., Carbides in Nuclear Energy (Russell, L.E.,
      et al eds.), London, Macmillan 1964, Vol. I, p. 447.
[4]   TAYLOR, B.L., MILNER, G.W.C., BIRKS, F.T., PRIOR, H.A.,
      ibid p. 457.
[5]   BENZ, R., HOFFMAN, C.G., RUPERT, G.N., High Temp. Sci.
      1 (1969), 342.
[6]   MULFORD, R.N.R., ELLINGER, F.H., HENDRIX, G.S.,
      ALBRECHT, E.D., Plutonium 1960 (Grison, E., Lord, W.B.H.,
      Fowler, R.D., eds.) London, Cleaver-Hume, London 1961,
      p. 301.
[7]   JACKSON, R.J., WILLIAMS, D.E., LARSON, W.L., J. Less
      Common Mets. 5 (1963) 443.
[8]   POPOVA, S.V., BAIKO, L.G., High Temps. High Press. 3
      (1971) 237.
[9]   POPOVA, S.V., FORNICHEVA, L.N., KHVOSTANSEV, L.G.,
      JETP Letters (Eng. Transl.) 16 (1972) 429.

[10]   SAVITSKII, E.M., TYLKINA, M.$\overline{A}$., KONIEVA, L.Z.,
        KASHIN, V.I., KIEBANOV, Izv. Vyssh. Ucheb. Zaved.
        Tsvet. Metl. No. 5 (1972) 137.
[11]   CHUBB, W., KELLER, D.L., Carbides in Nuclear Energy
        (Russell, L.E. et al eds.), London, Macmillan 1964,
        Vol. 1, p. 208.
[12]   FARR, J.D., BOWMAN, M.G., ibid, p. 184.
[13]   ALEXEYEVA, Z.M., J. Nucl. Mats. 49 (1973/4) 333.
[14]   ALEXEYEVA, Z.M., ibid 64 (1977) 303.
[15]   ALEXEYEVA, Z.M., IVANO$\overline{V}$, O.S., Thermodynamics of
        Nuclear Materials 1974, Vienna 1975, Vol. II, p. 175.
[16]   ELLINGER, F.H., The metal plutonium (Coffinberry, A.S.,
        Miner, W.N., eds.), Univ. of Chicago Press 1961 p.299.
[17]   ELLINGER, F.H., MINER, W.N., O'BOYLE, D.R.,
        SCHONFELD, F.W., Report LA 8870 (1968).
[18]   BOWERSOX, D.F., LEARY, J.A., J. Nucl. Mats. 21
        (1967), 219.
[19]   DARBY, J.B., jr., BERNDT, A.F., DOWNEY, J.W., J. Less
        Common Mets. 9 (1965) 466.
[20]   DARBY, J.B., jr., NORTON, L.J., DOWNEY, J.W., ibid 6
        (1964) 165.
[21]   DARBY, J.B., jr., LAM, D.T., NORTON, L.J., DOWNEY, J.W.,
        ibid 4 (1962), 558.
[22]   TRZEBIATCWSKI W., RUDZINSKI, J., Z. Chem. 2 (1962) 158.
[23]   HOLLECK, H., Thermodynamics of Nuclear Materials 1974,
        I.A.E.A. Vienna 1975, Vol. II, p. 213.
[24]   HANSEN, M., Constitution of binary alloys, New York,
        McGraw-Hill, 1958.
[25]   NOWOTNY, H., KIEFFER, R., BENESOVSKY, F., LAUBE, E.,
        Monatsh. Chem. 89 (1958) 692.
[26]   SHUNK, F.A., Constitution of bindary alloys, second
        supplement. New York, McGraw-Hill 1969, p. 151.
[27]   CAMBELL, G.M., J. Chem. Thermod. 6 (1974), 1110.
[28]   MIEDEMA, A.R., Plutonium 1975 and other actinides,
        (Blank, H., Lindner, R., eds.), Amsterdam, North
        Holland-American Elsevier, 1976, p. 3.
[29]   RAND, M.H., unpublished assessment 1978.
[30]   CROMER, D.T., LARSON, A.C., ROOF, R.B., Acta Cryst. 17
        (1964), 272.
[31]   BOUCHER, R., BARTHELEMY, P., MILET, C., Plutonium 1965
        (Kay, A.E., Waldron, M.B. eds.), London, Chapman and
        Hall (1967), p. 483.
[32]   HOLLECK, H., J. Nucl. Mats. 39 (1971), 226.
[33]   HAINES, H.R., POTTER, P.E., Thermodynamics of Nuclear
        Materials 1974, Vienna 1975, Vol. II, p. 145.
[34]   NICHOLS, J.L., MARPLES, J.A.C., Carbides in Nuclear
        Energy, (Russell, L.E., et al, eds.), London,
        Macmillan, 1964, Vol. I, p. 246.

[35] Behaviour and Chemical State of Irradiated Ceramic
      Fuels, Panel Proceedings Series I.A.E.A. Vienna 1974.
[36] JOHNSON, C.E., JOHNSON, I., BLACKBURN, P., BATTLES, J.E.,
      CROUTHAMEL, C.E., ibid p. I.
[37] KAUFMAN, L., BERNSTEIN, H., Computer Calculation of
      Phase Diagrams, New York and London, Academic Press,
      1970.
[38] ANDERSON, E., BUCKLEY, R.A., HELLAWELL, A., HUME-
      ROTHERY, W., Nature 188 (1960), 48.

DISCUSSION

L. MANES: How was the assessment of pair—pair interaction parameters made?

P.E. POTTER: The method which was used to calculate the regular solution interaction parameters for all the phases is identical with that described by Kaufman and Bernstein (Ref. [37]). The estimation of the liquid-phase interaction parameter was based on the method of Hildenbrand and Scott, which was described in their book 'Solubility of Non-electrolytes, 3rd ed.' [Van Nostrand, Princeton, N.J. (1950)]. As is the case for the solid phases, the interaction parameter of the liquid consists of two terms, one due to internal pressure or strain energy and the other due to electronic effects.

A.S. PANOV: I should like to make a comment on the papers presented by Messrs Benedict and Potter. In discussing the solubility of fission products in carbide and nitride fuels, it is necessary to take into account the non-stoichiometry of the transition-metal nitrides and carbides. This is all the more important because the stoichiometric coefficient of UC_x and UN_x changes with burn-up.

May I ask Mr. Potter whether he took the non-stoichiometry of the carbide fuel into account in his considerations of the U-C-Ni, etc. systems?

P.E. POTTER: No, it was not necessary to include seed considerations in the determination of the constitution of these systems. We were mainly concerned with the formation of ternary compounds.

THERMOCHEMICAL ASPECTS OF FUEL/CLADDING AND FUEL/COOLANT INTERACTIONS IN LMFBR OXIDE FUEL PINS

M.G. ADAMSON, E.A. AITKEN, R.W. CAPUTI
General Electric Company,
Advanced Reactor Systems Department,
Vallecitos Nuclear Center,
Pleasanton, California,
United States of America

P.E. POTTER, M.A. MIGNANELLI
Atomic Energy Research Establishment,
Harwell, Didcot, Oxfordshire,
United Kingdom

Abstract

THERMOCHEMICAL ASPECTS OF FUEL/CLADDING AND FUEL/COOLANT
INTERACTIONS IN LMFBR OXIDE FUEL PINS.

This paper examines several thermochemical aspects of the fuel/cladding, fuel/coolant
and fuel/fission product interactions that occur in LMFBR austenitic stainless steel clad mixed
(U, Pu)-oxide fuel pins during irradiation under normal operating conditions. Results are
reported from a variety of high-temperature EMF cell experiments in which continuous
oxygen activity measurements on reacting and equilibrium mixtures of metal oxides
(UO_{2+x}, $U_{0.75}Pu_{0.25}O_2$, Nb_2O_5, V_2O_3, TiO_2) and (excess) liquid alkali metal (Na, K, Cs)
were performed. Oxygen potential, $\Delta\bar{G}_{O_2}^{eq}$, and O:M thresholds for sodium/fuel reactions are
re-evaluated in the light of new measurements and newly-assessed thermochemical data, and
the influence on $\Delta\bar{G}_{O_2}^{eq}$ of possible U-Pu segregation between oxide and urano-plutonate
(equilibrium) phases has been analysed. Conditions for and likely products of reactions
between oxide fuel and sodium vapour and the reaction swelling behaviour of UO_{2+y} and
$U_{1-x}Ce_xO_{2+y}$ pellets in liquid sodium are also discussed. The roles of important thermochemical parameters in both oxidative and non-oxidative modes of fuel/cladding chemical
interactions are examined, and a chemical transport mechanism based on formation of iron,
chromium and nickel tellurides in a liquid fission-product phase is described that accounts for
the cladding component transport often observed in irradiated fuel pins. Finally, the results
of EMF cell oxygen-activity measurements and a variety of capsule experiments performed on
caesium/oxide mixtures are interpreted in terms of the relative thermodynamic stabilities of
reaction products such as Cs_2UO_4, '$Cs_{2-x}UO_{4-y}$', 'Cs_2NbO_3', 'Cs_2VO_3' and '$CsTiO_2$'.

1.   INTRODUCTION

In this paper thermodynamic and chemical aspects of the fuel-coolant, fuel-cladding, and fission product-fuel interactions that are known to occur in stainless steel-clad LMFBR* oxide fuel pins under normal operating conditions will be examined critically using as a basis recent results from out-of-pile experiments and analytical studies. The chemical interactions of particular interest are sodium-fuel (SFCI), fuel-cladding (FCCI), cesium-fuel, and cesium-buffer/getter oxidation products. Progress on sodium-fuel reactions, which are expected to occur in pins operating with breached cladding, was previously reported at Vienna in 1974 in the form of a detailed evaluation of oxygen potential and oxygen concentration thresholds.[1] Out-of-pile experimental results that helped delineate key thermochemical features of FCCI were also described at the Thermodynamic Symposium in 1974.[2] Cesium-fuel reactions that can lead to fuel-cladding mechanical interaction (FCMI) in intact fuel pins have also been reported previously,[2-4] but this is not the case for Cs-vanadia, Cs-niobia and Cs-titania reactions. These latter reactions may themselves lead to FCMI effects in intact fuel pins containing oxygen buffer/getter metals such as V,Nb or Ti.

2.   THERMOCHEMISTRY OF OXIDE FUEL PINS (Background Review)

The key thermodynamic factor that influences fuel chemistry in general and fuel-coolant, fuel-cladding and fuel-fission product interactions in particular is the oxygen chemical potential $\Delta \bar{G}_{O_2}$ ($=RT \ln p_{O_2}$, where p_{O_2} is the oxygen partial pressure at temperature T/K). In an unirradiated oxide fuel such as $U_{1-x}Pu_xO_{2+y}$, the oxygen potential is simply determined, at a particular temperature, by the ratios O/U+Pu ($=2+y$) and Pu/U+Pu ($=x$). Additional factors that influence the stoichiometry (O/M), and hence $\Delta \bar{G}_{O_2}$, of irradiated oxide fuel are dissolved fission product oxides and the formation of separate ternary oxide phases such as Cs_2MoO_4, $SrZrO_3$ and $Cs_{2-x}(U,Pu)O_{4-y}$. The steep radial temperature gradient that exists in a typical oxide pin during irradiation leads to redistribution of oxygen, the fissile actinide metals, and mobile/reactive fission products such as Cs and Mo. The resulting O/M-, Pu/U+Pu- and Cs concentration gradients determine - in a manner that is not yet fully understood - the radial $\Delta \bar{G}_{O_2}$ profile of the fuel. Because the chemical interactions with which we are presently concerned involve the cool outer region of the fuel, it is the oxygen potential at this location that is of primary interest. Various quantitative models for temperature gradient-induced oxygen redistribution and the effect of burnup on fuel stoichiometry have been proposed since that of Roberts, Rand and Markin,[5,6] and there have been recent advances in knowledge of fuel chemistry and thermodynamics,[7,8] yet our confidence in predictions of $\Delta \bar{G}_{O_2}$ at the outer surface of irradiated oxide fuel is still no better than ±15 kcal/mol (±63 kJ/mol).

---

* Liquid Metal Fast Breeder Reactor

As yet, this uncertainty has not been reduced by direct ΔG_{O_2} measurements such as those performed previously on irradiated urania fuel.[9] Knowledge of the thermodynamic stabilities of fission product-containing ternary oxide phases that form in irradiated mixed oxide fuel has relevance not only for predicting the evolution of oxygen (and cesium) activity during normal irradiation, but also for estimating the availability of oxygen inside breached pins undergoing sodium-fuel reaction. This aspect of irradiated fuel thermochemistry will be discussed further in Section 3.3.2.

3.   COOLANT-FUEL REACTIONS

Overall reactions between (excess) sodium and oxide fuel or breeder materials can be represented by

$$3y \; Na(\ell, \text{ dissolved } O) + 2 \; UO_{2+y}(s) \rightarrow y \; Na_3UO_4(s) + (2-y)UO_2(s) \qquad (1)$$

and

$$3(z-y) \; Na(\ell, \text{ dissolved } O) + (2+z) \; U_{1-x_3}Pu_{x_3}O_{2-y}(s) \rightarrow$$

$$(z-y)Na_3U_{1-x_2}Pu_{x_2}O_4(s) + (2-y)U_{1-x_1}Pu_{x_1}O_{2-z}(s) \; (z>y) \qquad (2)$$

The corresponding equilibrium reactions are

$$3Na(\ell) + UO_2(s) + O_2(\text{dissolved}) = Na_3UO_4(s) \qquad (3)$$

and

$$3Na(\ell) + U_{1-x}Pu_xO_{2-z}(s) \; (\tfrac{2+z}{2}) \; O_2(\text{dissolved}) = Na_3U_{1-x}Pu_xO_4(s) \qquad (4)$$

Formation of the low density Na_3MO_4-type reaction products in breached LMFBR fuel pins is of concern because of possible adverse consequences of the concomitant fuel volumetric expansion.[10] The studies to be described were performed as part of a continuing effort to delineate these reactions and the conditions under which they occur.

3.1.   Oxygen Potential Threshold, $\Delta \bar{G}^{eq}_{O_2}$

The equilibrium oxygen potentials for reactions (3) and (4) define the concentration of oxygen in liquid sodium ($C^{eq}_{O_{Na}}$) above which additional sodium-fuel reaction will take place.   Oxygen concentration thresholds ($C^{eq}_{O_{Na}}$) calculated from $\Delta \bar{G}^{eq}_{O_2}$ are very sensitive to the uncertainty associated with this thermodynamic parameter,[1,11,12] hence it is important to know $\Delta \bar{G}_{O_2}$ as precisely as possible.

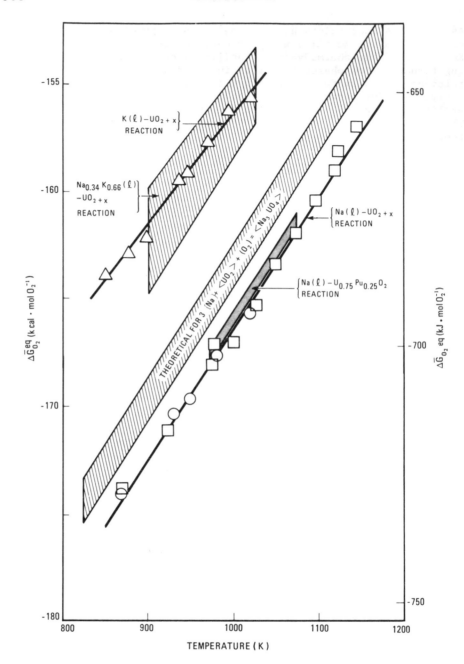

FIG.1. *EMF cell results for liquid alkali metal oxide fuel equilibria expressed as* $\Delta \bar{G}_{O_2}^{eq}$ *versus T plots.*

3.1.1 Recent Measurements of $\Delta\overline{G}_{O_2}^{eq}$

Adamson and Aitken extended their earlier measurements on the Na–U–O system[1,11] by performing two series of experiments with the electrochemical cell

$$In(\ell), In_2O_3, (W)//ThO_2-7\tfrac{1}{2} \text{ w/o } Y_2O_3 // [0]_{Na}, UO_2(s), Na_3UO_4(s), (W) \qquad (5)$$

reference electrode   electrolyte   working electrode

over the temperature range 873–1148K.[13] The oxygen potential-temperature relationship obtained, based on seventeen experimental points, is represented by the equation

$$\Delta\overline{G}_{O_2} (kJ.mol\ O_2^{-1}) = -952.4 + 0.256T(\pm1.8) \qquad (6)$$

Oxygen potential measurements on the systems resulting from $Na(\ell)$ + $U_{0.75}Pu_{0.25}O_2(s), Na_{.34}K_{.66}(\ell)$ + $UO_{2+y}(s), K(\ell)$ + $UO_{2+y}(s)$ and $Cs(\ell)$ + $UO_2(s)$ + $Cs_2UO_4(s)$ reaction mixtures with excess alkali metal were also performed in this latter study; with the exception of the Cs–U–O system, which is discussed in Section 5.1, these results are summarized in Figure 1.

In attempting to obtain reliable measurements for the Na–U,Pu–O system, a variety of cell arrangements were tried. The best results (shown in Figure 1) were obtained from the cell

$$Na(\ell) + UO_{2+y}(s), (W)//ThO_2-7\tfrac{1}{2}w/oY_2O_3// Na(\ell) + U_{0.75}Pu_{0.25}O_2(s), (W) \qquad (7)$$

in which $Na(\ell)$ + $UO_{2+y}(s)$ (or, at equilibrium, $Na(\ell$, dissolved 0) + $UO_2(s)$ + $Na_3UO_4(s)$) served as the reference electrode. Difficulties encountered during these measurements appear to be associated with slower equilibration kinetics for the $Na(\ell)$+polycrystalline $U_{0.75}Pu_{0.25}O_2$ reaction than for the $Na(\ell)$ + polycrystalline UO_{2+y} reaction.

3.1.2 Comparison Between Measured and Calculated/Theoretical $\Delta\overline{G}_{O_2}^{eq}$ Values

For the equilibrium represented by reaction (3) we have

$$\Delta\overline{G}_{O_2} = \Delta G_f^{\circ}(Na_3UO_4,s) - \Delta G_f^{\circ}(UO_2,s) - 3\Delta\overline{G}(Na,\ell) \qquad (8)$$

where $\Delta G_f^{\circ}(Na_3UO_4,s)$ and $\Delta G_f^{\circ}(UO_2,s)$ are Gibbs energies of formation and $\Delta\overline{G}_{Na}$ is the partial molar free energy of liquid sodium (essentially zero because of the exceptionally low concentration of dissolved oxygen). Using the most recently assessed data for the

TABLE 1. VARIATION OF $\Delta \bar{G}_{O_2}$ WITH TEMPERATURE FOR THE
EQUILIBRIUM REACTION (3):

$3\ Na\ (\ell) + UO_2\,(s) + O_2\,(dissolved) = Na_3UO_4(s)$

Temperature (K)	$-\Delta \bar{G}_{O_2}$ (kJ·mol O_2^{-1})		
	Calculated(Eqn.11)†	Calculated (Ref.1)	Measured (Eqn.6)
800	736.2	729.8	747.2
900	710.1	705.9	721.5
1000	684.0	682.0	695.9
1100	657.9	658.1	670.2
1200	631.8	634.2	644.6

†Estimated uncertainty is $4.2kJ \cdot mol\ O_2^{-1}$.

Gibbs energies of formation, viz.

$$\Delta G_f^\circ (Na_3UO_4,s) = -2027.2 + 0.432T\ kJ.mol^{-1} \tag{9}$$

(reference 14)

and

$$\Delta G_f^\circ (UO_2,s) = -1082.2 + 0.171T\ kJ.mol^{-1} \tag{10}$$

(reference 15)

we obtain the following expression for the equilibrium oxygen potential corresponding to reaction (3)

$$\Delta G_{O_2}(3) = -945.0 + 0.261T\ (\pm 4.2)\ kJ.mol\ O_2^{-1} \tag{11}$$

Values of $\Delta \bar{G}_{O_2}^{eq}$ calculated for different temperatures from equation
(11) are listed in Table 1 along with previously calculated (theoretical) values[1,13] and the measured data represented by equation
(6).

Comparison of these data shows that the latest calculated/theo-
retical values are some 2 kJ. mol O_2^{-1} closer to the experimental
values than the previously calculated values; however, their temperature dependences are somewhat different. At 1000K the theoretical
and experimental $\Delta \bar{G}_{O_2}^{eq}$ values fall within 11.9 kJ.mol O_2^{-1} (2.8 kcal.
mol O_2^{-1}; which is only slightly greater than the combined error
limits. Calculation of the equilibrium oxygen potential for reaction (4), which refers to the analogous phase field in the Na–U–Pu–O
system, is discussed below in Section 3.1.3.

3.1.3 Effect of Dissolved Plutonia on $\Delta\overline{G}_{O_2}^{eq}$ for the Liquid Sodium-Urania Reaction

Following reaction between Na(ℓ) and urania-plutonia solid solutions, a phase field is entered that contains Na(ℓ), urania-plutonia solid solution and trisodium uranoplutonate. In general, for a four-component system of this type, the concentrations of U and Pu in the two condensed phases at equilibrium will be different (this is illustrated in Section b of Figure 2). Clearly, this possible segregation of U and Pu, together with possible non-ideality of the mxied oxide and uranoplutonate solid solutions, will have some influence on $\Delta\overline{G}_{O_2}^{eq}$. It is of interest to estimate the magnitude of such effects, especially in view of the existence of some experimental evidence that the reaction product phase in failed fuel pins becomes enriched in uranium.[16]

Let us consider equilibrium reaction (4) in terms of the separate equilibria for the uranium and plutonium components of the system. For the uranium system the equilibrium reaction is

$$3Na(1) + UO_{2-\frac{y}{x_1}}(\text{dissolved in } PuO_2) + O_2 \text{ (dissolved)} \rightleftarrows Na_3UO_4 \text{ (dissolved in } Na_3PuO_4) \qquad (12)$$

and for the plutonium system the equilibrium reaction is

$$3Na(1) + PuO_{2-\frac{y}{x_1}} \text{ (dissolved in } UO_2) + (1 + \frac{y}{2x_1})O_2 \rightleftarrows Na_3PuO_4 \text{ (dissolved in } Na_3UO_4) \qquad (13)$$

The values for the oxygen potentials obtained by considering the uranium and plutonium reactions must obviously be the same.

The partial Gibbs energy of oxygen ($\Delta\overline{G}_{O_2}^{eq}$) is given by the expression

$$\Delta\overline{G}_{O_2} = \Delta G_f^\circ(Na_3UO_4,s) + E_2 x_2^2 + RT\ln(1-x_2) - \Delta G_f^\circ(UO_2,s) - E_1 x_1^2 - RT\ln(1-x_1) \quad (14)$$

where $\Delta G_f^\circ(Na_3UO_4,s)$ and $\Delta G_f^\circ(UO_2,s)$ are the Gibbs energies of formation of the two components Na_3UO_4 and UO_2, E_1 and E_2 are the regular solution interaction parameters for urania-plutonia and for Na_3UO_4 -Na_3PuO_4, and R is the gas constant.

For the ternary uranium system $\Delta\overline{G}_{O_2}^{eq} = \Delta G_f^\circ(Na_3UO_4,s) - \Delta G_f^\circ(UO_2,s)$, thus

$$\Delta\overline{G}_{O_2}^{eq}(\text{U-Pu system}) = \Delta\overline{G}_{O_2}^{eq}(\text{U system}) - E_1 x_1^2 + E_2 x_2^2 - RT\ln\frac{(1-x_1)}{(1-x_2)} \qquad (15)$$

From equation (15) we see that when $x_1 = x_2$ and $E_1 = E_2$,

$$\Delta\overline{G}_{O_2}^{eq}(\text{U-Pu system}) = \Delta\overline{G}_{O_2}^{eq}(\text{U system}).$$

A condition that the quaternary equilibrium oxygen potentials should be identical with those of either ternary system is that there should be no segregation of U and Pu and, for the chosen solution model, namely regular, the interaction parameters must be identical.

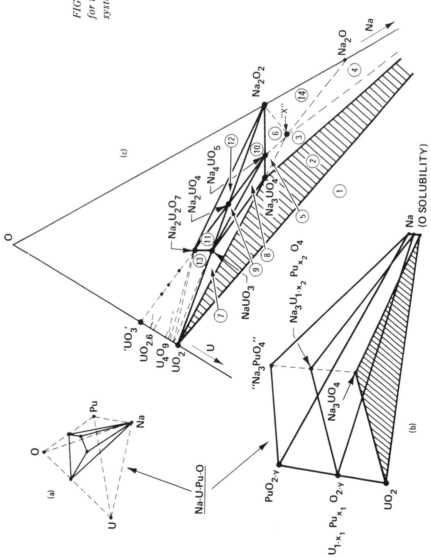

FIG.2. Sections of phase diagrams for the Na-U-Pu-O and Na-U-O systems at 1000 K.

TABLE 2. THE INFLUENCE OF SEGREGATION AND DIFFERENT REGULAR SOLUTION INTERACTION PARAMETERS ON THE OXYGEN POTENTIAL FOR THE PHASE FIELD:

$$Na(\ell) + U_{1-x_1} Pu_{x_1} O_{2-y}(s) + Na_3 U_{1-x_2} Pu_{x_2} O_4 (s)$$

$$\Delta \bar{G}_{O_2}^{eq} \text{ (U-Pu system)} = \Delta \bar{G}_{O_2}^{eq} \text{ (U system)} + \Delta$$

Parameters for solid solutions				Temperature (K)	$\Delta^{\dagger}$ (cal $\cdot$ mol O_2^{-1})
x_1	x_2	E_1	E_2 (calories)†		
0.15	0	0	0	600	194
		4000	*		104
		-4000	*		284
		0	0	1000	323
		4000	*		233
		-4000	*		413
0.4	0	0	0	600	609
		4000	*		-31
		-4000	*		1249
	0	0	0	1000	1015
		4000	*		375
		-4000	*		1655
0.15	0.15	0	0	independent of temperature	0
		0	4000		90
		0	-4000		-90
0.3	0.3	0	0		0
		0	4000		360
		0	-4000		-360
0.15	0.99	0	0	600	-4881
				1000	-8135
0.4	0.99	0	0	600	-5297
				1000	-8827

*Δ is independent of the value of E_2
†1 calorie ≡ 4.184 joules.

The influence of the variation of the U and Pu concentrations and the interaction parameters on the (calculated) threshold oxygen potentials are shown in Table 2.

It will be noted from the values of Δ in Table 2 that changes due to variations in x_1 and x_2, as well as in E_1 and E_2, are generally within the estimated uncertainty of the values of $\Delta \bar{G}_{O_2}^{eq}$ for the uranium system; namely, ± 1 kcal.mol O_2^{-1} (+4.2 kJ.mol^{-1}). Only for

TABLE 3.   EXPERIMENTAL OBSERVATIONS ON REACTIONS BETWEEN
URANIA PELLETS AND LIQUID SODIUM

Composition and density of oxide pellet	Heat Treatment		Observations
	Time (h)	Temp.(°C)	
$UO_{2.00}$ (98% TD)	120	800	Pellet intact. No swelling.
$UO_{2.009}$ (98% TD)	120	450	Pellet intact. No swelling.
$UO_{2.009}$ (98% TD)	120	800	Pellet intact. No swelling.
$UO_{2.015}$ (97% TD)	45	750	Pellet split in half. Slight swelling < 1% vol. increase
$UO_{2.020}$ (98.5% TD)	20	750	Pellet cracked into pieces.
$UO_{2.020}$ "Single crystal" (100% TD)	40	800	Intact
$UO_{2.026}$ (99% TD)	20	750	Pellet cracked into pieces.
$UO_{2.050}$ (98% TD)	120	450	Complete disintegration of pellet.
$UO_{2.073}$ (98% TD)	120	800	Complete disintegration of pellet.
$UO_{2.10}$ (98% TD)	120	450	Complete disintegration of pellet.
$UO_{2.10}$ (98% TD)	120	800	Complete disintegration of pellet.
$UO_{2.10}$ (98% TD)	20	750	Pellet cracked into pieces – Na and urania heated to temperature before reaction.

cases when the interaction parameter $E_1 \leqslant -4$ kcals, or when the plutonium is greatly segregated into the Na urano-plutonate, do the values
of Δ lie outside the band of $\Delta\bar{G}_{O_2}^{eq}$ for the uranium system. It is also
evident that strong segregation of U into the Na_3MO_4 phase – should it
occur – would only have a minor influence on $\Delta\bar{G}_{O_2}^{eq}$. From the experimental result that $\Delta\bar{G}_{O_2}^{eq}(4) = \Delta\bar{G}_{O_2}(3)$, i.e., within 2 kJ.mol O_2^{-1} at ca
1000K, it follows that large deviations from ideality in the
urania-plutonia solid solution and strong segregation of Pu into the
Na_3MO_4 phase are both extremely unlikely.

3.2 O/M Thresholds

3.2.1 Na(ℓ) + Urania-Plutonia Solid Solution

Oxygen-to-metal ratios of the urania-plutonia solid solution in equilibrium with Na(ℓ) and Na_3MO_4 have been measured using a variety of techniques.[1,11-13,17] Expressed as Pu valence, V_{Pu}, these results ranged from ca 3.4^{17} to $3.6_0-3.7_5$[1,11,12] and appeared to be fairly insensitive to temperature. The latest equilibrium oxygen potential data, both experimental (Section 3.1.1) and theoretical (3.1.2 and 3.1.3), can be used to derive V_{Pu} values by comparison with recent $\Delta\bar{G}_{O_2}$-O/M-T data for $U_{1-x}Pu_xO_{2-y}$ solid solutions in the composition range of interest.[18,19] This comparison yields a value of 3.68+.04 for V_{Pu} (x = 0.25), which appears to confirm the O/M threshold measurements of Blackburn[12] and Adamson, et al.[1,11] but not those of Housseau.[17] It is also apparent from this comparison that the earlier $\Delta\bar{G}_{O_2}$-O/M-T data of Markin and McIver[20] for hypostoichiometric compositions is seriously in error.[1,19]

3.2.2 Na (ℓ) + Urania Pellets

The O/U threshold for reaction (3) is exactly 2,[11,12] however, previous work had indicated that the stoichiometry threshold for destructive reaction swelling of dense UO_{2+x} pellets may be considerably higher than 2 - and temperature dependent.[13] To obtain more information, a series of tests were performed in which urania pellets were heated with liquid sodium in sealed nickel capsules at different temperatures. The results from these tests are shown in Table 3.

The results of these experiments indicate that at relatively low values of x (<0.015) the amount of swelling due to reaction is small and confined to the periphery of the pellet. However, for higher values of x considerable swelling occurs, which, in certain cases (x ⩾0.05), results in complete collapse of the pellet with formation of powder at all temperatures. One pellet of $UO_{2.10}$ was mixed with sodium after preheating both reactants at 750°C; the pellet cracked into several pieces, but there was no evidence of formation of powder after 20 hours at temperature. In contrast, the reaction of a pellet of $UO_{2.10}$ in which both reactants were in contact during heating from room temperature to 750°C resulted in complete disintegration of the pellet into powder. A possible interpretation of these observations is that pellet destruction is caused by formation of Na_2O at temperatures below ca 550°C; the density of Na_2O is low (ca 3.7 $g.cm^{-3}$), and the presence of such a voluminous reaction product in the grain boundaries of sintered pellets is likely to cause their destruction. When the reactants are introduced to one another at temperatures above ca 580°C, the first reaction product phase to form is Na_3UO_4, which is denser than Na_2O and capable of epitaxy with the UO_2 matrix.

TABLE 4. EXPERIMENTAL OBSERVATIONS ON REACTIONS BETWEEN URANIA-CERIA PELLETS AND LIQUID SODIUM

Composition and Density of Oxide Pellet	Heat Treatment		Observations
	Time (h)	Temp. (°C)	
$CeO_{2.00}$ (95.5% TD)	120	800	Complete disintegration of pellet.
$CeO_{1.90}$ (95.5% TD)	120	800	Complete disintegration of pellet.
$CeO_{1.80}$ (95.5% TD)	120	800	Partial disintegration of pellet.
$CeO_{1.70}$ (95.5% TD)	120	800	Cracking of pellet.
$CeO_{1.5}$ (98% TD)	120	800	Pellet intact - no swelling.
$U_{0.85}Ce_{0.15}O_{2.00}$ (96% TD)	5	200	Pellet intact - no swelling.
$U_{0.85}Ce_{0.15}O_{2.00}$ (96% TD)	120	800	Complete disintegration of pellet.
$U_{0.97}Ce_{0.03}O_{2.00}$ (96% TD)	5	400	Peripheral cracking only.
$U_{0.97}Ce_{0.03}O_{1.985}$ (96% TD)	5	400	Pellet intact - no swelling.
$U_{0.80}Ce_{0.20}O_{2.00}$ (96% TD)	120	800	Complete disintegration.
$U_{0.80}Ce_{0.20}O_{1.99}$ (96% TD)	120	800	Complete disintegration.
$U_{0.80}Ce_{0.20}O_{1.97}$ (96% TD)	120	800	Cracking of pellet
$U_{0.80}Ce_{0.20}O_{1.90}$ (98% TD)	120	800	Pellet intact - no swelling.

3.3 Effect of Dissolved Fission Product Oxide (Ceria) on the Liquid Sodium-Urania Reaction

3.3.1 Out-of-Pile Results

Additional capsule experiments were performed on ceria (CeO_{2-y}) and urania-ceria ($U_{1-x}Ce_xO_{2-y}$) pellets as part of an effort to determine the influences of ceria additions on pellet reaction swelling behavior and the availability of reactive oxygen (O/M thresholds).

These experiments, like the Na(ℓ) + UO$_{2+x}$ capsule experiments, incorporated attempts to characterize reaction products using x-ray
powder diffraction, ceramography and electron probe microanalysis.
Results are summarized in Table 4.

Observations on the CeO$_{2-y}$ pellet-Na(ℓ) interaction experiments
indicate that the extent of reaction (and concomitant pellet swelling)
is related to the initial O/Ce ratio of the pellets. From the observation that the extent of reaction depends on the relative amounts of
Na(ℓ) and CeO$_{2-y}$, as well as on (O/Ce)$_{initial}$, it appears that Na(ℓ)
removes oxygen from the ceria until the oxygen potentials of the two
phases are equal. Thus far no evidence for ternary Na-Ce-O compounds
has been found either from these experiments or from attempted solid
state reactions between Na$_2$O and CeO$_{2-y}$.

For reactions between Na(ℓ) and pellets of stoichiometric
U$_{1-x}$Ce$_x$O$_{2.00}$ the extent of reaction - as measured by $\Delta V/V$ - was
found to be dependent on x, the Ce concentration. The extent of
reaction increased with increasing x up to ca 0.15 when reaction
caused complete destruction of the pellet. This behavior is consistent with the expected increase in $\Delta \bar{G}_{O_2}$ (and hence in the quantity
of available oxygen) as x increases in mixed oxide at constant O/M.
The extent of reaction was also found to be dependent on the initial
O/M of pellets with fixed x; no reaction occurred with fully reduced
oxide (U$_{1-x}$Ce$_x$O$_{2-0.5x}$), but its extent increased as the value of
O/M$_{initial}$ increased.

3.3.2   Influence of Irradiation on the Availability of Oxygen for Na(ℓ)-Fuel Reaction

Assuming that the oxygen potential thresholds for reactions (3)
and (4) remain unchanged during irradiation, the net result of production of dissolved fission product oxides is an increase in the
quantity of reactive oxygen. As indicated by the results discussed
in the previous section, this is equivalent to a lowering of the
effective threshold O/M. To estimate the quantity of reactive oxygen resulting from fission in a particular oxide fuel or breeder
material, it is also necessary to take account of oxygen combined
in binary/ternary oxides that form as separate phases. Such oxides
include Cs$_2$MoO$_4$, Cs$_2$UO$_4$, Cs$_y$U$_{1-x}$Pu$_x$O$_z$, (Ba,Sr)UO$_3$, (Ba,Sr)ZrO$_3$,
Cr$_2$O$_3$, Cs$_4$CrO$_4$, and also buffer/getter oxidation products. Determination of the thermodynamic stabilities of these phases relative to
Na$_3$UO$_4$/Na$_3$U$_{1-x}$Pu$_x$O$_4$ has been performed with the aid of the Ellingham
diagram shown in Figure 3. The oxygen potential data used to con-
struct this diagram were calculated from the most reliable thermochemical information currently available for the various product
phases. The identities of these products, the most stable oxide
phases existing under typical fuel operating conditions, were established by carefully evaluating available phase equilibrium data or,
where such data were lacking, by constructing plausible, thermo-
dynamically-consistent, constitution diagrams.

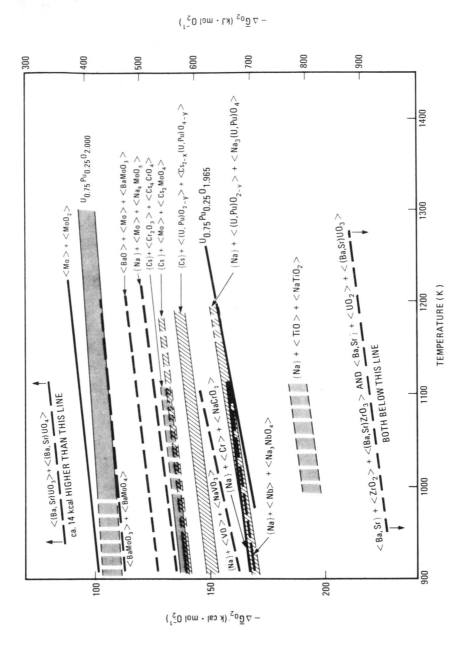

FIG.3. Ellingham diagram illustrating relative thermodynamic stabilities of complex oxide phases.

Using this approach for irradiated $U_{1-x}Pu_xO_{2-y}$ fuel ($x = 0.15$ to 0.30), total available oxygen for Na-fuel reaction, $\Delta O/M$, could be represented as follows:

$$\Delta\, O/M = \frac{1}{2}[\,(1-x)V_U + xV_{Pu}] - (2 - .16x') + A.B - \delta \qquad (16)$$

where V_U and V_{Pu} are the valencies of U and Pu in the original fuel,
  x and x' are the mole fractions of Pu in the fuel at zero and
  B atom % burnup, respectively,
  δ represents the quantity of oxygen combined as cladding and/
  or buffer-getter oxidation products,
  and A is a factor that depends on the degree of ^{235}U enrichment
  (0.005 when % $^{235}U \cong 65$).

It is not expected that oxygen combined with cladding components or buffer-getter metals in the fuel-cladding gap (δ) will make a significant contribution to Na-fuel reaction swelling in a breached pin; however, upon prolonged exposure to sodium at 600°C or higher, these compounds would be converted to stable Na-M-O ternary phases (e.g., $NaCrO_2$).

3.4   Reactions between Urania and Sodium Vapor

In this section of the paper we discuss briefly the conditions for reaction between urania and sodium vapor. The occurrence of such reactions in breached LMFBR fuel or breeder elements is possible because reasonably high Na partial pressures, established by liquid sodium in the relatively cool fuel-cladding gap near the defect, may arise at more remote locations occupied by hot, reactive fuel. These reactions may be important because of the possibility that formation of certain low density products (such as Na_2O) will lead to excessive reaction swelling. Our assessment of the likely behavior of oxide fuel with Na(v) has been performed for the Na-U-O system; although the corresponding phase equilibria in the Na-U-Pu-O system are somewhat more complex, the thermodynamics of the two systems are sufficiently similar that conclusions reached for one apply to both.

A convenient approach is to represent the isothermal section of a ternary phase diagram as a plot of oxygen potential against sodium pressure or sodium potential. A conventional composition plot of the Na-U-O system at 1000K, such as that shown in Figure 2(c), is represented in Figure 4 as a plot of oxygen potential against sodium pressure. The compounds which have been included in the representations are $Na_2U_2O_7$, $NaUO_3$, Na_2UO_4, Na_3UO_4, Na_4UO_5, Na_2O, Na_2O_2 and UO_2. We have also included a compound "X", the composition of which is assumed to be close to the formula Na_6UO_6; some evidence for this compound is the observation that Na_2O and Na_3UO_4 may not coexist. The regions of existence of the various single-phase regions are bounded by lines the loci of which give the relationship between the sodium pressure and oxygen potentials for the

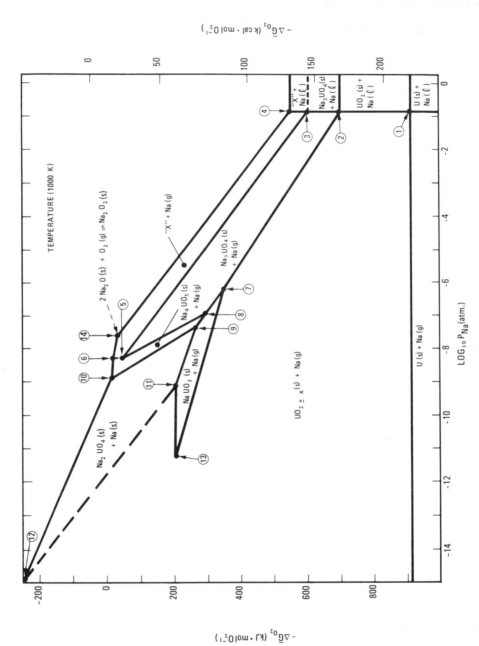

FIG.4.   Oxygen potentials and sodium vapor pressure in the Na-U-O system at 1000 K.

coexistence of two condensed phases. The numbered points give the
sodium pressure and the oxygen potential for the various three-phase
fields. If the sodium pressure is that over liquid at 800K (log
p(atm) $\simeq$ -2.1), the system at 1000K would require urania to have a
substantially higher oxygen potential (ca -620 kJ.mol O_2^{-1}) before reaction occurred to form Na_3UO_4 than with Na at unit activity (ca
-684 kJ.mol O_2^{-1}). At higher temperatures even higher oxygen potentials would be required before reaction could occur (ca -415 kJ.mol O_2^{-1}
at 1500K); under these conditions $NaUO_3$ could then be the first compound to form. The compound "X" is only expected to form at high
oxygen activities and somewhat higher sodium activities than those
required for formation of Na_3UO_4 or $NaUO_3$. Formation of Na_2O as an
initial reaction product would require even higher oxygen potentials
($\geqslant$-550 kJ.mol O_2^{-1} at 1000K) at corresponding sodium activities. It
appears, therefore, that reactions to form compounds such as Na_2O,
$NaUO_3$ and "X" would not take place in fuel pins in which the fuel
was already in oxygen equilibrium with coexistent Na(ℓ) + Na_3UO_4(s)
+ UO_2(s) in the fuel-cladding gap; however, such reactions are possible under special circumstances in freshly breached fuel or breeder
elements.

4. FUEL-CLADDING REACTIONS (FCCI)

Post-irradiation examinations of irradiated stainless steel-
clad oxide fuel pins have shown that fission product-induced fuel-
cladding chemical interaction (FCCI) occurs in three rather distinct
forms at temperatures exceeding ca 500°C. Whereas the first two
types - intergranular attack (IGA) and uniform or matrix attack -
involve oxidation of one or more cladding components, the third form
- CCCT or dissolution attack - is essentially nonoxidative, the
cladding undergoing partial dissolution (and subsequent transport)
in a liquid fission product medium.[8] In addition to the presence
of Cs and Te fission products, each of these identified modes of
FCCI is sensitive to several thermodynamic parameters, as revealed
by results from a variety of carefully controlled out-of-pile experiments and irradiated pin tests. The following sections focus on
results derived from laboratory simulation experiments and thermodynamic analyses of FCCI.[2,8]

4.1 Oxidative Attack Modes (IGA and Uniform/Matrix)

In these 'conventional' forms of FCCI the basic reaction process
is oxidation of the cladding elements. Of the major cladding components (Fe,Cr,Ni), Cr is the least noble with respect to oxidation
and yields Cr_2O_3 (or a complex ternary oxide incorporating the fission product Cs). The other major elements may also oxidize if the
oxygen activity is sufficiently high, although the relative amounts

of the various oxidation products are eventually limited by the sup-
ply of oxygen from the fuel. These net oxidation reactions may be
represented as follows:

1) 'Intermediate' oxygen activity $(-556 < \Delta G_{O_2}^{-1000K} < -418 \text{ kJ.mol}^{-1})$

$$[Cr]_{alloy(or\ carbide)} + O_2 \rightarrow Cr_2O_3 (or\ Cs_xCrO_y) \qquad (17$$

2) 'High' oxygen activity $(\Delta G_{O_2}^{-1000K} > -418 \text{ kJ.mol}^{-1})$

$$[Fe,Ni,Cr]_{alloy} + O_2 \rightarrow FeO,NiO,Cr_2O_3 (spinels) \qquad (18$$

3) Initially 'high' oxygen activity - later stage

$$FeO,NiO + [Cr]_{alloy} \rightarrow Cr_2O_3 + Fe_{1-x}Ni_x \qquad (19$$

Thermochemical factors that determine whether oxidative FCCI is manifest as deep intergranular attack or the relatively shallow matrix/
uniform attack, and the severities of these different forms of attack,
have been delineated from isothermal capsule experiments incorporating
a variety of fission product-oxygen buffer mixtures.[2,8] The most important of these factors are: oxygen activity (and O availability),
Te activity (Cs:Te ratio), availability of Te(+ Cs), carbon activity
(and C availability), and temperature; however, the specific parameter,
or combination of parameters, responsible for the nonuniform distribution of IGA in irradiated fuel pins remains unclear. The influences
of oxygen activity (p_{O_2}), carbon activity (a_C) and tellurium activity
(a_{Te}) are now discussed in turn.

4.1.1   Influence of p_{O_2} (and Temperature)

Under otherwise optimal conditions for fission product-induced
attack of Type-316 stainless steel cladding ('excess' Cs + Te,
1:1⩽Cs:Te⩽3:1, T ≈725°C, C present) the character of FCCI changes from
"deep IGA" to a shallow combination of intergranular and matrix reactions as the oxygen activity increases from 4.6×10^{-30} –56×10^{-30} atm
("low" regime) to ca 54×10^{-27} - ca 3×10^{-18} atm ("high" regime). The
lower and upper bounds of these two regimes correspond to $U_{0.75}Pu_{0.25}O_2$
fuel with O/M = 1.997 ±.001 and 2.000 ±.001, respectively.
At higher activities $(p_{O_2} > 3 \times 10^{-18}$ atm) deep IGA can be caused by Cs
without the aid of Te (i.e., by $Cs_2O(\ell)$ or $CsOH(\ell)$); however, oxygen
activities of this magnitude - which correspond to hyperstoichiometric fuel - are not expected to develop in normal fuel pins. Both
the IGA and matrix forms of Cs,Te-induced attack accelerate as
temperature increases above ca 500°C; however, the apparent activation energies and kinetics are different in each case.

More detailed information about the influence of p_{O_2} on the
compatibility of Type-316 SS with Cs,Te mixtures near to the lower
oxygen activity thershold for FCCI has also been obtained from isothermal capsule experiments. In these experiments, which were
performed at 725°C, coexisting pairs of phases from the Nb-O and

TABLE 5. EQUILIBRIUM $\Delta \bar{G}_{O_2}/p_{O_2}$ VALUES FOR SELECTED BUFFER MIXTURES AT 725°C AND RESULTING EFFECT ON Cs, Te-TYPE 316 SS COMPATIBILITY

(Reaction conditions: Cs:Te = 1:1, 725°C, 100 h)

Buffer Mixture	$-\Delta \bar{G}_{O_2}$ $(kJ \cdot mol_{O_2}^{-1})$	p_{O_2} (atm)	Observation
<Nb>/<NbO>	654	5.95×10^{-35}	No attack
<NbO>/<NbO$_2$>	587	2.00×10^{-31}	No attack
<Cr>/<Cr$_2$O$_3$>	585	2.56×10^{-31}	No attack
316 SS Oxidation	565 ± 8	2.75×10^{-30}	
<VO>/<V$_2$O$_3$>	564	2.90×10^{-30}	Deep IGA; 152 μm
<Cr$_3$C$_2$>/<Cr$_2$O$_3$> /<C>	540	5.74×10^{-29}	Deep IGA; 127 μm
<NbO$_2$>/<Nb$_2$O$_5$>	481	5.38×10^{-26}	IGA; 102 μm

V-O systems were used as buffer mixtures, with and without added carbon. Results from these and previous experiments are summarized in Table 5, which also shows $\Delta \bar{G}_{O_2}^{eq}$ and p_{O_2} values corresponding to the various buffers. Although reaction between Cs and certain oxides in the buffer mixtures is possible (Section 5.2), this type of interaction, which could influence $\Delta \bar{G}_{O_2}^{eq}$, was not significant in the experiments due to preferential reaction of Cs with Te (Section 4.1.3 and 4.2.2). According to the data in Table 5, NbO/NbO$_2$ establishes an oxygen activity just below the oxidation threshold for Type-316 SS, while VO/V$_2$O$_3$ and Cr$_2$O$_3$/Cr$_3$C$_2$ establish activities just above this threshold. Results from these buffers are in such good agreement with the estimated/measured threshold for uncatalyzed (gaseous) oxidation of Type-316 SS that the two thresholds are believed to be coincident.

Intergranular attack also occurred with the NbO$_2$/Nb$_2$O$_5$ buffer, which establishes an oxygen potential ca 60 kJ.mol O$_2^{-1}$ higher than the upper $\Delta \bar{G}_{O_2}$ limit for stability of grain boundary carbides in the cladding (Section 4.1.2). Although the maximum depth of intergranular penetration with this buffer falls between the measured limits for the 'low p_{O_2}/carbide stability' and 'high p_{O_2}/carbide instability' regimes (ca 140μm and ca 40μm, respectively), the response of the attack to added carbon identifies it as a predominantly 'high p_{O_2}' form of FCCI.

4.1.2 Influence of a_C

Carbon, which is an unavoidable impurity in oxide fuels, has been shown to markedly accelerate Cs,Te-induced IGA of Type-316 SS cladding in the 'low P_{O_2}' regime (ca 4.6×10^{-30} to 56×10^{-30} atm); however, in contrast, the same impurity has essentially no influence on the rate of FCCI either at higher O_2 activities ($p_{O_2} > 54 \times 10^{-27}$ atm) or in the absence of Te(Cs alone). Although precise data on carbon activities are lacking, carbon exceeding a concentration of ca 100 ppm in the cool outer regions of oxide fuel is assumed to be at unit activity. A high carbon activity tends to be maintained in an irradiated oxide fuel pin by continuous outward radial transport of carbon.[21] High carbon activities at the cladding inner surface appear to promote progressive conversion of the $M_{23}C_6$-type intergranular carbides to higher carbides (M_7C_3, M_3C_2) which are susceptible to Cs,Te-assisted oxidation, thus resulting in accelerated IGA. The observed decrease in catalytic activity of carbon upon Cs,Te-induced FCCI as p_{O_2} increases is explained on the basis of a parallel decrease in thermodynamic stability of grain boundary carbides. In the presence of C at unit activity, $\Delta\bar{G}_{O_2}/p_{O_2}$ regions of stability exist for different groups of carbides. For example, at 725°C the stable carbides at $\Delta\bar{G}_{O_2} < -556$ kJ.mol O_2^{-1} are $< Cr_{23}C_6 > + < Cr_7C_3 > +$ $< Cr_3C_2 >$. When $-556 < \Delta\bar{G}_{O_2}$ (kJ.mol O_2^{-1}) < -536, only $< Cr_7C_3 >$ and $< Cr_3C_2 >$ are stable. Only iron carbides are stable between -536 and ca -397 kJ.mol^{-1} and, above ca -397 kJ.mol^{-1}, no carbides can exist. Because the presence of Cr-rich carbide phases in cladding grain boundaries appears to be essential for occurrence of fission product-induced IGA, this form of attack, and accompanying C-catalysis, is not expected to be significant at oxygen activities higher than ca 8.9×10^{-29} atm (-536 kJ.mol^{-1} at 725°C). The recent results obtained with NbO_2/Nb_2O_5 buffer mixtures ($p_{O_2} \approx 54 \times 10^{-27}$ atm; see Table 5) show only a small C-catalytic effect but somewhat greater IGA than expected.

4.1.3 Effect of a_{Te}

A critical dependence of the incidence of Cs,Te-induced oxidative attack of Type-316 SS on Cs:Te ratio was recognized early in the GE out-of-pile program. This was presumed to be a a_{Te}-related effect resulting from mass action in the equilibrium

$$2Cs(\ell,melt) + Te(\ell,melt) = Cs_2Te(\ell,melt) \tag{2}$$

viz.

$$a_{Te} = \frac{a_{Cs_2Te}}{a_{Cs}^2 \cdot K} \tag{2}$$

Additional experiments have provided information on the dependence of Type-316 SS attack morphology on Cs:Te (and hence, a_{Te}) in the low oxygen activity regime. Results from these experiments are summarized in Table 6.

TABLE 6. EFFECT OF Cs:Te RATIO ON MORPHOLOGY OF FCCI AT LOW OXYGEN ACTIVITY

Test Conditions — 'excess' Cs + Te
$\quad\quad\quad\quad$ Cr/Cr$_2$O$_3$/C buffer (low p_{O_2})
$\quad\quad\quad\quad$ 100 h at 725°C
$\quad\quad\quad\quad$ Type-316 SS (annealed)

Cs:Te Ratio	Conditions	Result/Observation
0:1	'low' a_{Te} (Cr:Te=4:1)	No attack (<5μm)
0:1	'high' a_{Te} (Cr:Te=1:2)	Combined matrix + IGA (∿50μm)
0.2:1	'low' a_{Te} (Cr:Te=4:1)	IGA[†] (∿100μm)
0.2:1	'high' a_{Te} (Cr:Te=1:2)	IGA[†] (∿100μm)
1:1	*	IGA (∿125μm)
2:1	*	IGA (∿125μm)
3:1	*	Predominantly IGA (∿100μm)
4:1	*	Shallow matrix attack (∿50μm)
8:1	*	No attack
1:0		No attack, shallow carburization

† Deep IGA only observed at location of Cs addition
* a_{Te} determined by Cs:Te rather than Cr:Te

Te alone produced either a characteristic shallow mixed matrix-intergranular attack or essentially no attack, depending on the Te activity. Representative 'high' and 'low' values of a_{Te} were established by varying the Cr:Te in the buffer mixture (this equilibrium is considered further in Section 4.2.2). Absence of attack was associated with 'low' a_{Te}, whereas 'high' a_{Te} produced attack.

The effect of adding a relatively small quantity of Cs to Te was dramatic. Insufficient Cs was added in these tests to significantly change a_{Te} via reaction (20), but the Cs was added as a single drop to the tilted capsules just prior to sealing to ensure that it was not uniformly distributed on the inner surface of the cladding. Deep IGA (∿100μm) only occurred where Cs was present; other locations showed the much shallower attack characteristic of Te alone.

The deep IGA induced by Cs,Te mixtures was shown to occur only when Cs:Te$\leqslant$3:1. Between ratios of 3:1 and 4:1 the character of attack changed from predominantly intergranular to predominantly matrix/uniform. At the high ratio of 8:1 there was no observable attack, which is consistent with the postulated mass action effect of Cs on a_{Te}.

Unfortunately, insufficient data are presently available to establish a quantitative/theoretical basis for the influence of a_{Te} on the rate and character of oxidative FCCI. This situation con-trasts with that for the nonoxidative mode of FCCI (see below) in which the controlling chemical reactions and equilibria appear to be identified. Finally, it is worth noting that available evidence[8] indicates a direct proportionality between Te inventory/availability and the extent of either form of oxidative FCCI.

4.2   Nonoxidative Attack Mode; Cladding Component Chemical Transport (CCCT)[22]

4.2.1   Characteristics of CCCT and Proposed Mechanism(s)

The dissolution or erosive cladding attack that frequently occurs in irradiated mixed-oxide fuel pins at high burnup is gener-ally associated with transport of metallic cladding components across the fuel-cladding gap into or onto the oxide fuel. The cladding appearance in this form of attack – rounded grain surfaces and sub-surface voids/porosity – suggests that cladding components are being selectively dissolved by a liquid medium. Several explanations for cladding transport into the fuel have been offered in the past, the most popular of which has been an iodine-iodide vapor transport mechanism similar to the Van Arkel-De Boer process; however, most of these have been shown to be untenable on thermodynamic grounds.

Recent results from out-of-pile thermal gradient experiments and irradiated miniature fuel pins doped with Cs,Te mixtures indi-cated that $Cs_{1-x}Te_x$ in the approximate composition range $0.6 > x > 0.3$ is an effective (liquid) medium for transport of Fe, Cr and Ni. The existence of such a phase in the fuel-cladding gap of normal (un-doped) fuel pins during irradiation is plausible because most Cs-containing fission product compounds melt between 550 and 940°C (e.g., CsI, Cs_2Te-Te mixtures, Cs_2MoO_4). Along with oxygen, which will vary in concentration according to its activity, Cs, Mo, I and Te are the most likely constituents of a melt that will penetrate the fuel to a depth (temperature) where the stabilizing effect of capillary action is offset by its tendency to decompose or vaporize. Of these fission products Te is the most likely to react with the cladding elements, thereby markedly increasing their solubility in

the melt.   Indeed, Fe, Cr and Ni each form a series of stable tellurides, the most Te-rich of which melt in the temperature range of
interest.

From the foregoing, a cyclic process was proposed[22] for transport
of cladding components (M) that is based on the reversible reaction

$$M(s,alloy) + n\ Te(\ell,melt) = MTe_n(\ell,melt); \qquad T_1(\sim 650°C) \rightarrow T_2\ (T_2 > T_1) \qquad (22)$$

The following mechanistic steps comprise the overall cyclic transport process:

  Step 1 - formation of Fe, Cr and Ni tellurides (MTe_n) at the
           cladding inner surface (T_1);
  Step 2 - dissolution of MTe_n compounds in the Cs-Te melt;
  Step 3 - transport of dissolved MTe_n up the radial temperature
           gradient;
  Step 4 - decomposition of MTe_n, and deposition of M, at some
           high temperature location in the fuel (T_2); and
  Step 5 - return of Te and Cs down the thermal gradient to the
           cladding inner surface.

4.2.2   Thermodynamic Basis for CCCT

Tellurium is a member of the chalcogenide family of elements,
Group VIA of the Periodic Table, and in many of its chemical reactions it resembles other members of the family (O,S,Se). Thus,
with major constituents of austenitic cladding alloys it forms compounds such as Cr_2Te_3, $FeTe_{0.9}$ and $NiTe_{1.1}$ that are analogous to
$Cr_2O_3(Cr_2S_3)$, $Fe_{1-x}O(FeS)$ and $Ni_{1-x}O(NiS)$, respectively. In spite
of these compositional analogies, tellurides and oxides have substantially different thermodynamic stabilities, as illustrated by
the data listed in Table 7.

In an irradiated fuel pin fission product Te also combines with
Cs; however, this interaction is, according to the data in Table 7,
stronger than its interactions with any of the major cladding components. The Te activity in the melt will thus be determined by the
relative concentrations of Cs and Te. Under certain conditions fission product Cs also reacts with oxide fuel to form rather stable
ternary compounds (see 5.1), and, clearly, through its effect on a_{Cs},
such an interaction could influence a_{Te}. These various equilibria
are represented by equations (20), (22) and (23):

$$xCs(\ell,melt) + FO_{2-y}(s) + zO(dissolved\ in\ fuel) = Cs_xFO_{2-y+z}(s) \qquad (23)$$

where F = U, Pu. In the following treatment other possible interactions
between Te, Cs and O (fuel) have been ignored - principally because
they are not expected to exert a significant influence on a_{Te}, but
partly because of incomplete phase diagram and/or thermochemical
data.

TABLE 7. ENTHALPIES, ENTROPIES AND FREE ENERGIES OF FORMATION OF SELECTED TELLURIDES AND OXIDES[a, b, c]

	$-\Delta H^{\circ}_{298}$ (kcal.mol^{-1})	S°_{298} (cal. deg.$^{-1}$mol^{-1})	ΔS°_{298}[d] (cal. deg.$^{-1}$mol^{-1})	Additional Information on Tellurides
$CrTe_{1+x}$	[18.5]	[21]	2.2	x = 0 to 0.2, stable above 1300K; estimated Cp(T) data available[25]
Cr_2Te_3	[72]	49.87	2.8	
Cr_2O_3	270.0	19.4	-65.5	
MnTe	26.6	22.4	2.0	melts at 1438K; Cp(T) data available[25]
$MnTe_2$	[30.0]	34.66	3.3	Cp(T) data available[25]
'MnO'	92.0	14.3	-17.8	
$FeTe_{0.9}$	5.5	19.14	1.9	melts at 1200K; $\Delta \bar{G}_{Fe}$ and $\Delta \bar{G}_{Te}$ values available[25]; Cp(T) data available[25]
$FeTe_2$	17.3	23.94	-6.3_5	melts at 933K; $\Delta \bar{G}_{Fe}$ and $\Delta \bar{G}_{Te}$ values available[25]; Cp(T) data available[25]
'FeO'	63.2	14.05	-16.9	
$NiTe_{1.1}$	13.7	20.09	8.55	Cp(T) data available[25]
$NiTe_2$	[21.0]	28.76	-2.2	
'NiO'	57.5	9.1	-22.5	
Cs_2Te	[70]	[40]	[-14]	melts at 953K
Cs_2O[26]	82.7	35.1	-30.1	melts at 763K

a) Most of the data in this table are taken from compilations by Kubaschewski, Evans and Alcock [24] and K. C. Mills.[25]
b) Data enclosed by square brackets [ ] are estimated by Mills[25] or the present authors.
c) Uncertainties are not listed in the Table; however, they probably range from 1 to 5 kcal mol^{-1} (measured) to ca 10 kcal.mol^{-1} (estimated) for ΔH°_{298} values, and 0.1 to 2 cal. deg^{-1} mol^{-1} (measured) to 5 cal.deg^{-1}mol^{-1} (estimated) for S°_{298} and ΔS°_{298} values.
d) Standard entropies of formation, S°_{298}, were calculated using standard entropy values from Reference 24.

From the generalized, reversible reaction that describes inter-action between Te and cladding components, (22), we obtain the equilibrium relationship

$$\frac{a_{MTe_n}}{a_M \cdot a_{Te}^n} = K_M = \exp^{\frac{\Delta S_M}{R}} \cdot \exp^{-\frac{\Delta H_M}{RT}} \tag{24}$$

where a_i is the activity of species i,
ΔS_M and ΔH_M are the entropy and enthalpy changes corresponding to the reaction $M(s) + nTe(\ell) = MTe_n$ (s or ℓ),
K_M is the mass action equilibrium constant, and
R and T have their usual meaning.

Since we are dealing with dilute solutions, the activity of M in the melt, a_{MTe_n}, should be proportional to its concentration, C_{MTe_n}, viz.

$$a_{MTe_n} = k_M C_{MTe_n} . \tag{25}$$

Substituting (25) into (24), we get

$$C_{MTe_n} = \frac{1}{k_M} \cdot a_{M(alloy)} \cdot a_{Te(melt)}^n \cdot \exp^{\frac{\Delta S_M}{R}} \cdot \exp^{\frac{-\Delta H_M}{RT}} \tag{26}$$

which may be simplified by assuming k_M, $a_{Te(melt)}$ and ΔS_M are each independent of T over the temperature range appropriate to the fuel-cladding gap (600 to 1000°C, approximately). Furthermore, the activities of the major cladding constituents Fe, Ni and Cr in the alloy, a_M, may be approximated by their mole fractions, N_M, so equation (26) becomes

$$C_{MTe_n} = k_M' \cdot N_M \cdot a_{Te}^n \cdot \exp^{-\frac{\Delta H_M}{RT}} \tag{27}$$

Since the melt is bounded by a source (the cladding inner surface) and a sink (the fuel outer surface), each at a different temperature (T_1 and T_2, respectively), the difference in concentration of M in the melt between these two bounds can be estimated using

$$\Delta C_M = C_{M,T_2} - C_{M,T_1} = k_M' \cdot a_{Te}^n \left(N_{M,T_2} \exp^{-\frac{\Delta H_M}{RT_2}} - N_{M,T_1} \exp^{-\frac{\Delta H_M}{RT_1}} \right) \tag{28}$$

where N_{M,T_2} and N_{M,T_1} represent the compositions of the condensed Fe-Cr-Ni alloy phase at the sink and source, respectively. Assuming that $N_{M,T_2} \approx N_{M,T_1}$, equation (28) predicts that the concentration of component M in the melt will be higher at the low

temperature side of the gap than at the high temperature side for
typical (negative) values of ΔH_M (see Table 7). Such a concentration gradient, aided by eventual deposition of M at the high
temperature side, will result in a net drift or flow of M away from
the cladding surface up the temperature gradient. Although the
principal criterion governing the direction of flow of M is the
sign of ΔH_M, the relative values of ΔH_M and ΔS_M must also influence
both the direction and efficiency (rate) of mass transport, as enun-
ciated by Schäfer in a set of rules developed for transport involving
heterogenous equilibria.[23] Thus, as the stability of a dissolved
telluride MTe_n increases beyond some optimum value, the predicted
transport rates (ΔC_M) are expected to decrease. The solubilities
of MTe_n species in the melt

$$c^\circ_{MTe_n} = \frac{1}{k_M}$$

(29)

might also vary amongst the different cladding components M, and,
according to equation (28), such differences would influence predicted values of ΔC_M. In the absence of the required solubility
data we assume that Ni, Cr and Mn tellurides have similar (high)
solubilities to that of Fe, whose quantitative transport behavior
is currently being used to calibrate our transport rate equations.[22]
Although chromium and manganese form the most stable tellurides of
the cladding components under consideration, their oxides are also
considerably more stable than the oxides of nickel and iron. This
proclivity to form stable oxides may actually inhibit dissolution
of Cr and Mn in the melt. However, with low O:M fuel, with oxygen
activities below the threshold values for their oxidation, the predicted flow directions for Cr and Mn are the same as those for Fe
and Ni.

    Equation (27) predicts that the concentration of a particular
cladding component in the melt is dependent, _inter alia_, on a_{Te}, the
activity of tellurium. As described earlier, a_{Te} is itself influenced by the interaction between Te and Cs, which in turn is
influenced by the interaction between Cs and fuel (reactions 20 and
23). Application of the law of mass action to the equilibria represented by these reactions yields equation (21) and

$$\frac{a_{Cs_x FO_{2+z-w}}}{p_{O_2}^{z/2} \cdot a_{FO_{2-w}} \cdot a_{Cs}^x} = K_{23}$$

(30)

which simplify to

$$a_{Te} \cdot a_{Cs}^2 = K' \quad \text{and} \quad p_{O_2}^{z/2} \cdot a_{Cs}^x = K', \text{ respectively,}$$

when $a_{FO_{2-w}} \approx 1$, $a_{Cs_x FO_{2-w+z}} = 1$, and a significant fraction of Te
in the melt is combined as Cs_2Te. Substitu-
tion of these two expressions into (27) yields a relationship that,

qualitatively at least, describes the influence of P_{O_2} on the concentration of cladding component M in the melt:

$$C_{M,melt} = k_M'' \cdot N_M \cdot P_{O_2}^{\frac{zn}{x}} \cdot \exp^{-\frac{\Delta H_M}{RT}} \tag{31}$$

where k_M'' is a composite constant. From this equation it can be seen that the impact of oxygen activity on C_M depends on the relative values of the stoichiometry terms n, z and x. Thus, if $x \gg nz$, the effect of P_{O_2} on C_M will be small; or, if $x \gg nz$, a strong dependence is expected. Using some reasonable numbers for these stoichiometry terms (n = 2, z = 1 to 2, x = 0.5 to 2), we find that the exponent of P_{O_2} is 3 ±1, which suggests a strong dependence of C_M on P_{O_2}.

Due to the onset of competing processes and interactions at 'high' and 'low' oxygen activities inside an operating fuel pin, this predicted dependence of C_M (and hence transport rate) on P_{O_2} O:M can only be expected to hold over a limited range (i.e., 1.97<O:M≥2.00, according to irradiated fuel data).[22] For example, if P_{O_2} is below the threshold value for formation of Cs-fuel compound (10^{-33} to 10^{-29} atm, depending on T_{fuel}), reaction (23) could not occur, and equation (31) would no longer express the effect of P_{O_2} on Cs activity (which would now be close to unity). This, in turn, would maximize the effect on Te activity expressed by equation (21). However, when a_{Cs} is close to unity, cesium is sufficiently volatile that it readily undergoes gaseous migration to lower temperature regions, thus tending to reduce the local ratio of free Cs: combined Cs. In this way, an excessive reduction in oxygen activity could actually limit the extent to which tellurium activity is reduced. In contrast, an increase in oxygen activity/O:M has the effect of raising the tellurium activity in the melt, but if P_{O_2} becomes too high (say, above the oxidation threshold for Fe) the cesium activity is likely to be suppressed below the level required for Cs to participate in the CCCT mechanism. Under these conditions other transport mechanisms may operate and/or the matrix/intergranular modes of FCCI may dominate cladding inner surface wastage. In reality, the predicted strong influence of a_{Te} on the concentration of dissolved cladding components (equation (27)) is expected to dominate CCCT behavior because Te is directly involved in the transport mechanism. Fuel composition effects that have been observed, i.e., a tendency for increased CCCT as O:M increases from 1.97 to 2.00,[22] may simply reflect the combined influences of changes in a_{Cs}' and P_{O_2} on a_{Te}.

The effect of temperature on CCCT in irradiated fuel pins is complex and not fully understood. Equations (27) and (31) indicate that the primary influence of temperature on the concentration of M in the fission product melt is through the sign and magnitude of ΔH. Using a typical value of ca −80 kJ.mol^{-1} (i.e., $FeTe_n$ and $NiTe_n$ as dissolved species; see Table 7) we see that a rise in temperature actually produces a <u>decrease</u> in concentration. However, <u>temperature gradient</u>, as determined by the local pin power, is expected to exert a larger, positive effect on ΔC_M, the concentration gradient across the gap, and concomitant increases in fuel temperature would also tend to increase a_{Te}. Generally, according to observations, the net effect of raising cladding inner surface temperatures is to increase the rate of CCCT.

5. CESIUM FISSION PRODUCT REACTIONS

5.1 $\overline{\Delta G}_{O_2}$ Threshold for Cs-Fuel Reactions

Under certain conditions fission product Cs can react with either the mixed-oxide fuel or the urania axial blanket material in irradiated LMFBR oxide fuel pins. It has been shown recently that both types of chemical interaction can lead to localized cladding strain (diametrical expansion) in stainless steel-clad test pins. In the case of axial blanket swelling the responsible reaction appears to be

$$\frac{2(y-y')}{2-y'} Cs(\ell \text{ or } v) + UO_{2+y}(s) = \frac{y-y'}{2-y'} Cs_2UO_4(s) + \frac{2-y}{2-y'} UO_{2+y'}(s) \tag{32}$$

and it takes place under conditions particularly conducive to formation of the U^{VI} ternary oxide (high p_{O_2}, high UO_{2+y}:Cs ratio). For Cs activities close to unity the value of y' is essentially zero. Although the available thermodynamic and phase diagram data suggest that the only Cs-fuel compounds capable of forming inside irradiated fuel pins are those based on U^{VI} (i.e., Cs_2UO_4 and possibly $Cs_2U_2O_7$),[27] a significant amount of evidence has been obtained that Cs–U–O and Cs–U,Pu–O compounds containing at least some U^V can form when fission product Cs at high activity reacts with oxide fuel at low p_{O_2}.[2,4] These compounds, designated $Cs_{2-x}(U,Pu)O_{4-y}$ to indicate existing uncertainty about their composition(s), resemble the U^V compounds formed in the Na–U–O system (Section 3.4): however, present evidence suggests that exact analogues such as $CsMO_3$ and Cs_3MO_4 do not exist. In view of this uncertainty and to provide quantitative information on the relative thermodynamic stabilities of $Cs_2(U,Pu)O_4$ and $Cs_{2-x}(U,Pu)O_{4-y}$, equilibrium oxygen activity measurements on the Cs–U–O system at 550 to 600°C have been performed using the high temperature galvanic cell technique previously employed in our studies of Na,K(ℓ)-oxide equilibria (Section 3.1.1).

In one of these experiments a mixture of liquid Cs(5g.), stoichiometric UO_2 powder (10g.) and Cs_2UO_4 powder (5g.) was prepared as the working electrode in the electrochemical cell

$$In(\ell), In_2O_3(s), (W) \,/\!/\, YDT \,/\!/\, O(\text{dissolved in } \ell.Cs), UO_2(s), Cs_2UO_4(s)(W)$$

(YDT: solid solution as for the cell in Eq. (7).)

which was then operated at temperatures in the range 500 to 600°C for up to 40 hours. The rationale for starting with $Cs_2UO_4(s)$ in the reaction mixture, suspecting that it is in fact a nonequilibrium phase in the presence of $Cs(\ell)$, was that it would be converted slowly to the more stable $Cs_{2-x}UO_{4-y}$ phase during the high temperature run. Since a_{Cs} was expected to remain constant, any reaction would be manifest as a decrease in p_{O_2} (or increase in EMF cell output). In view of cesium's relatively low boiling point (690°C) and anticipated problems resulting from evaporation of Cs out of the reaction cell, the cell measurements were restricted to the highest practical temperature (600°C).

After ca 1 hour at 600°C the initially rapid rise in cell EMF output diminished ($\sim$530mV), and, for the next 20 hours the EMF slowly increased up to 590mV at which it leveled off and held steady (+4mV) for six hours. Next, the cell temperature was lowered to 550°C, following which the cell EMF finally settled at 590 +4mV. After approximately 12 hours at this temperature the EMF became unstable, and at this point the experiment was terminated. The theoretical (Nernstian) equation for an electrochemical cell utilizing a $In(\ell)$, $In_2O_3(s)$ reference electrode,[13] viz.

$$\Delta\overline{G}_{O_2}^{sple}(kcal.mol\ O_2^{-1}) = -143.7 + 0.0484T(K) - 92.275\ E(v) \tag{33}$$

was used to convert the steady EMF values obtained in this experiment to $\Delta\overline{G}_{O_2}$ values. These results can be summarized as follows:

Temperature (°C)	Cell EMF (mV)	$-\Delta\overline{G}_{O_2}$ exptl. ($kcal.mol\ O_2^{-1}$)	$-\Delta\overline{G}_{O_2}$ exptl. ($kJ.mol\ O_2^{-1}$)	$-\Delta\overline{G}_{O_2}$ Eq.(34) ($kJ.mol\ O_2^{-1}$)
600(after 1 hr)	$\sim$530	$\sim$150	$\sim$628	629+4
600(after 22 hr)	590+4	155.9+.3	652+1	629+4
550	590+4	158.3+.3	622+1	640+4

'Theoretical' oxygen potentials were calculated for the hypothetical equilibrium

$$2Cs(\ell) + UO_{2+y}(s) + \frac{2-y}{2}O_2 = Cs_2UO_4(s) \tag{34}$$

using Gibb's energy of formation data for UO_2[15] and Cs_2UO_4,[14,28] viz.

$$\Delta\overline{G}_{O_2}(34)(kJ.mol^{-1}) = -814.59 + 0.2124T/K. \tag{35}$$

The results (which are also presented diagrammatically in Figure 5) show quite clearly that whereas the oxygen potential of the $Cs(\ell)$ + $UO_2(s)$ + $Cs_2UO_4(s)$ reaction mixture closely corresponds to the theoretical value for reaction (34) after one hour at 600°C, at longer times it falls to a considerably more negative value (lower by ca 5.5 $kcal.mol^{-1}$/22 $kJ.mol^{-1}$). Extrapolation of the present $\Delta\overline{G}_{O_2}^{eq}$ results to 870°C, the temperature at which an O/M threshold value for Cs-mixed (U,Pu) oxide fuel reaction was obtained from Cs thermomigration experiments,[2] yields -143 ± 3 $kcal.mol\ O_2^{-1}$. This value falls well within the $\Delta\overline{G}_O^{eq}$ range estimated from the direct O/M threshold

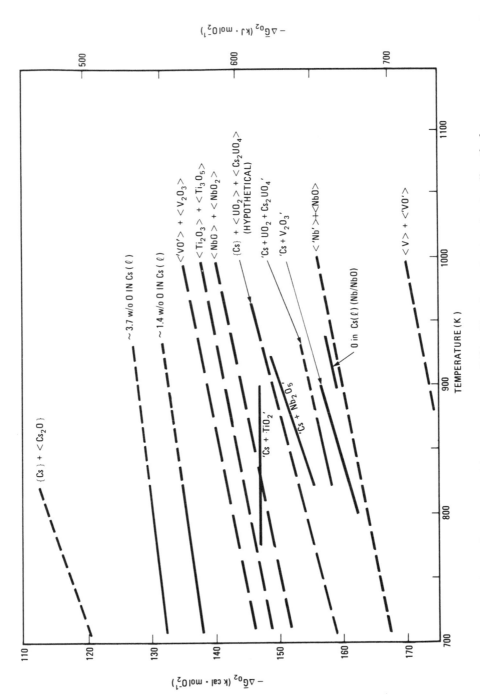

FIG.5. Oxygen potential measurements on Cs(ℓ)-oxide mixtures by the galvanic cell method.

measurement (-135 to -148 kcal.mol O_2^{-1}). Although a confirmatory
phase identification was not accomplished on the reaction product
mixture from the EMF cell experiment, these EMF cell results appear
to constitute strong evidence that Cs_2UO_4, the normal cesium uranate (VI), is not the most stable ternary phase coexisting with $Cs(\ell)$
at low oxygen activity. By inference, we conclude that a $Cs_{2-x}UO_{4-y}$
compound containing some U(V) is more stable than Cs_2UO_4 under high
a_{Cs}/low p_{O_2} conditions.

5.2   Stabilities of Ternary Oxide Phases Formed in $Cs(\ell)-Nb_2O_5(s)$, $Cs(\ell)-V_2O_3(s)$ and $Cs(\ell)-TiO_2(s)$ Reaction Mixtures

A variety of out-of-pile experiments have been performed at
General Electric to evaluate the chemical interactions that occur
between Cs fission products and oxidized buffer-getter metals inside
irradiated fuel pins and to determine the relative thermodynamic
stabilities of the various complex oxide reaction products. The
following brief descriptions summarize key thermodynamic results
obtained from high temperature EMF cells and miscellaneous capsule
experiments.

5.2.1   Galvanic Cell Measurements ($\Delta\bar{G}_{O_2}^{eq}$)

Oxygen activity measurements on $Cs(\ell) + Nb_2O_5(s)$, $Cs(\ell) + V_2O_3(s)$,
and $Cs(\ell) + TiO_2(s)$ reaction mixtures in which liquid Cs is in excess
have been performed at temperatures in the range 550–650°C using the
same EMF cell technique as that successfully applied to the Na-U-O
(Section 3.1.1) and Cs-U-O (Section 5.1) systems. Prior capsule
experiments in which pressed pellets of NbO, NbO_2, 'VO', V_2O_3,
'V_2O_4', and TiO_2 were exposed to Cs(v) and $Cs(\ell)$ at ca 800°C had
shown, by a combination of x-ray powder diffraction analysis and
Cs retention measurements, that the higher oxides of each metal
reacted to form a lower binary oxide plus a complex, presumably
ternary, oxide; e.g.,

$$6Cs(\ell) + 2 Nb_2O_5(s) = NbO(s) + 3'Cs_2NbO_3' \tag{36}$$

(Note: for this reaction 'Cs_2NbO_3' is only a suggested composition
for the stable Cs-Nb-O reaction product; it is based on measured
weight gains and the fact that formation of $Cs_2Nb_2O_6$, the only
known Cs-Nb-O phase, was not suggested by the x-ray diffraction
results.)

None of the Cs-containing reaction product oxides formed in
the present investigation were identified; however, based on analogous reactions undergone by other alkali metals, they are presumed
to be 'Cs_2NbO_3', 'Cs_2VO_3' and '$CsTiO_2$'. The EMF cell results for
the $Cs(\ell) + Nb_2O_5(s)$, $Cs(\ell) + V_2O_3(s)$ and $Cs(\ell) + TiO_2(s)$ reaction

mixtures after 20-30 hours at temperature are illustrated in Figure 5 as $\Delta\bar{G}_O$ -T data. In this figure solid lines represent EMF cell measurements[2] and dashed lines denote calculated or extrapolated data. Steadiness of the EMF readings after ca 20 hours was taken to indicate that equilibrium had been reached in each case. The measured $\Delta\bar{G}_{O_2}$ values thus refer to equilibrium in phase fields such as $Cs(\ell) + NbO(s) + 'Cs_2NbO_3'(s)$, or, expressed as an equilibrium reaction,

$$2Cs(\ell) + NbO(s) + O_2 = 'Cs_2NbO_3'(s). \tag{37}$$

Figure 5 also includes some measured data for dilute solutions of oxygen in $Cs(\ell)$: 1.4 w/o, 3.7 w/o, and the level established by coexistent $Nb(s)/NbO(s)$.

The results at 550-650°C indicate that $'Cs_2NbO_3'$ is only slightly less stable than $Cs_{2-x}UO_{4-y}$ or $'Cs_2VO_3'$, but more stable than $'CsTiO_2'$. Assuming that the same order of stability applies at typical fuel-cladding interface temperatures, these results suggest that fission product Cs will tend to concentrate at the location of (oxidized) Nb and V buffer-getter metals. According to the stoichiometry of reaction (37), the $Cs-NbO_x$ ($Cs-VO_x$) chemical interaction will effectively extend the oxygen buffering capacity of a given quantity of Nb(V) additive.

5.2.2 Isothermal and Temperature Gradient Capsule Experiments

Isothermal heat treatments (775° to 800°C) of a variety of $Cs(\ell$ and v$) - NbO_x$ and $Cs(\ell$ and v$) - VO_x$ mixtures were performed in nickel capsules for periods of 100 hours or more. Following these heat treatments excess Cs was distilled from the solid products in vacuo at temperatures ranging from 300° to 1000°C; weight changes were recorded following each distillation at successively higher temperatures. In some cases attempts were made to analyze the solid reaction products using x-ray powder diffraction. The niobium oxides, NbO_x, were found to react with both $Cs(v)$ and $Cs(\ell)$ at ca 800°C when $x \geqslant 2$, the usual reaction products being NbO plus an unidentifiable complex (ternary) phase. The Cs-containing reaction product(s) exhibited considerable stability at high temperature, up to 30 w/o of Cs being retained following treatment under vacuum at 1000°C. Vanadium oxides, VO_x, behaved similarly, reaction occurring when $x \geqslant 1.5$ with 'VO' as the binary oxide product; however, high temperature vacuum heat treatments of reacted VO_x specimens (800-1000°C) resulted in greater loss of combined Cs than in the analogous $Cs-NbO_x$ experiments.

Additional information about the reaction tendencies of elemental Cs with the various Nb and V oxides at 600-1300°C was obtained from a series of thermomigration experiments. In these experiments, small quantities of ^{137}Cs-tagged cesium metal ($\sim$50 mg) were initially

TABLE 8.   STABILITY DATA FOR $Cs_x MO_y$ (M = Nb, V) COMPOUNDS FROM
CAESIUM THERMOMIGRATION EXPERIMENTS

Oxide Buffer	T(K) for Stability of '$Cs_x MO_y$'	$-\Delta \overline{G}_{O_2}(T)^\dagger$ $(kJ \cdot mol\ O_2^{-1})$	$-\Delta \overline{G}_{O_2}(1273K)^\dagger$ $(kJ \cdot mol\ O_2^{-1})$
'Nb'(s)+NbO(s)	<873	–	611
NbO(s)+NbO$_2$(s)	<1323	533	541
NbO$_2$(s)+Nb$_2$O$_5$(s)	≊1573	415	450
'VO'(s)+V$_2$O$_3$(s)	≊1123	543	518
V$_2$O$_3$(s)+'VO$_2$'(s)	≊1403	233	252

†Calculated from data in Reference 24.

placed at one end of 22cm–long molybdenum capsules containing columns
of coexistent Nb and V oxide pellets (50:50 molar mixtures of 'Nb' +
NbO, NbO + NbO$_2$, NbO$_2$ + Nb$_2$O$_5$, 'VO' + V$_2$O$_3$ and V$_2$O$_3$ + 'VO$_2$'). The
capsules were then heated in a 1300°C-600°C temperature gradient for
two consecutive 100–hour periods with the ends containing the Cs
additive in the hottest zone. Migration of cesium down the axial
temperature gradient was followed by gamma-scanning the capsules
after each 100–hour treatment. The results[29] vividly demonstrated
the increasing reactivity of Cs with Nb oxides and V oxides as oxygen activity (O:M ratio) increases. In the Nb + NbO capsule (O:M =
1:2), cesium had completely migrated to the cool end within 100 hours,
such behavior being typical for elemental Cs. In the NbO + NbO$_2$ capsule (O:M = 1.5:1), all the Cs had concentrated at the midpoint of
the capsule after 200 hours, a position which corresponded to ca
1050°C in the axial temperature gradient. In the NbO$_2$ + Nb$_2$O$_5$ capsule (O:M = 2.25:1), after 200 hours Cs had scarcely moved from the
hot end, being located in a sharp peak at ∿1300°C. After 250–hour
treatments of the VO$_x$ capsules, Cs was distributed as follows: with
VO + V$_2$O$_3$(O:M = 1.25:1) Cs concentrated in a large peak at ca 850°C
and also migrated to low temperatures, with V$_2$O$_3$ + 'VO$_2$' (O:M = 1.75:1)
the Cs concentrated in the two equal-sized peaks at ca 1130°C and ca
680°C. From these results it is obvious that the high temperature
stability of '$Cs_x NbO_y$' is greater than that of '$Cs_x VO_y$' at comparable oxygen activities. This finding is consistent with behavior
inferred from the isothermal capsule results but contrary to the
lower temperature EMF cell results (which refer to 'lower-valent'
Cs-Nb-O and Cs-V-O compounds). The Cs thermomigration/$Cs_x MO_y$ stability results are summarized below in Table 8.

References

1. M. G. Adamson, E. A. Aitken, Thermodynamics of Nuclear Materials 1974 Vol. I, IAEA, Vienna, 1975, p. 233.

2. E. A. Aitken, M. G. Adamson, D. Dutina, S. K. Evans, ibid, p. 187.

3. I. Johnson, C. E. Johnson, ibid, p. 99.

4. E.H.P. Cordfunke, Thermodynamics of Nuclear Materials 1974 Vol. II, IAEA, Vienna, 1975, p. 185.

5. Rand, M. H., Roberts, L.E.J., Thermodynamics 1965 Vol. I, IAEA, Vienna, 1966, p. 3.

6. T. L. Markin, M. H. Rand, Thermodynamics of Nuclear Materials 1968, IAEA, Vienna, 1968, p. 637.

7. P. E. Potter, "Behavior and Chemical State of Irradiated Ceramic Fuels," Proceedings of a Panel, IAEA, Vienna, 1974, p. 115.

8. For recent review of U.S. work see M. G. Adamson, "Fuel and Cladding Interaction," IAEA/IWGFR Technical Committee Meeting, Tokyo, Japan, February 21-25, 1978 (IWGFR/16, p. 108 and 170).

9. M. G. Adamson, E. A. Aitken, S. K.Evans, J. H. Davies, Thermodynamics of Nuclear Materials 1974 Vol. I, IAEA, Vienna, 1975, p. 59.

10. E. A. Aitken, S. K. Evans, G. F. Melde, B. F. Rubin, "Fast Reactor Fuel Element Technology," Proceedings of the ANS Conference at New Orleans, Louisiana, April 13-15, 1971, ed. R. Farmakes (ANS), p. 459.

11. M. G. Adamson, E. A. Aitken, S. K. Evans, "Behavior and Chemical State of Irradiated Ceramic Fuels," Proceedings of a Panel, IAEA, Vienna, 1974, p. 411.

12. P. E. Blackburn, ibid, p. 393.

13. M. G. Adamson, E. A. Aitken, D. W. Jeter, "Sodium-Fuel Reaction Studies: Fuel Pellet Swelling Behavior and Alkali Metal-Oxide Fuel Reaction Thermodynamics," General Electric Co., GEAP-14093 (March 1976); see also paper presented at the International Conference on Liquid Metal Technology in Energy Production, May 3-6, 1976, Champion, Pennsylvania (CONF-760503-P2, p. 866).

14. O'Hare, P.A.G., Cordfunke, E.H.P., "The Chemical Thermodynamics of the Actinide Elements and Compounds, 3: Miscellaneous Actinide Compounds." IAEA, Vienna, 1978.

15. Ackermann, R. J., Grønvold, F., Pattoret, A., Rand, M. H., "The Chemical Thermodynamics of the Actinide Elements and Compounds; The Oxides," IAEA, Vienna, to be published.

16. H. Kleykamp, private communication (1974).

17. M. Housseau, G. Dean, F. Perret, "Behavior and Chemical State of Irradiated Ceramic Fuels," Proceeding of a Panel, IAEA, Vienna, 1974, P. 349.

18. R. E. Woodley, J. Nucl. Matl. 74 (1978) p. 290.

19. R. E. Woodley, M. G. Adamson, J. Nucl. Mat., in press.

20. T. L. Markin, E. J. McIver, Plutonium 1965, ed. A. E. Kay and M. R. Waldron (1967) p. 845.

21. M. G. Adamson, E. A. Aitken, M. H. Rand, J. Nucl. Mat. 50 (1974) p. 217.

22. M. G. Adamson, E. A. Aitken, "Cladding Component Transport in Mixed-Oxide Fast Reactor Fuel Pins: An Additional (High Burnup) Mechanism for FCCI," General Electric Co., GEFR-00404 (September 1978).

23. H. Schäfer, "Chemical Transport Reactions," Academic Press, New York-London (1964).

24. O. Kubaschewski, E. L. Evans, C. B. Alcock, "Metallurgical Thermochemistry," 4th Edition, Pergamon, London (1967).

25. K. C. Mills, "Thermodynamic Data for Inorganic Sulphides, Selenides and Tellurides," Butterworths, London (1974).

26. H. E. Flotow, D. E. Osborne, J. Chem. Thermodynamics 6 (1974) P. 135; J. L. Settle, G. K. Johnson, W. N. Hubbard, ibid, p. 263.

27. D. C. Fee, C. E. Johnson, J. Inorg. Nucl. Chem. 40 (1978) p. 1375.

28. P.A.G. O'Hare, H. R. Hoekstra, J. Chem. Thermodynamics 6 (1974) p. 257; D. W. Osborne, H. E. Flotow, H. R. Hoekstra, ibid, p. 179.

29. See Reference 8, p. 132.

DISCUSSION

Hj. MATZKE: I would like to refer to your deduction from your study of the system $Na-(U, Pu)O_{2-x}$ that the $\Delta\bar{G}_{O_2}-O/M-T$ data of Markin and McIver for $(U, Pu)O_{2-x}$ may be in error. There is further evidence supporting this statement. Using the EMF cell described by Ewart (paper IAEA-SM-236/07, these Proceedings) measurements were made on unirradiated $(U, Pu)O_{2-x}$. These EMF measurements yielded $\Delta\bar{G}_{O_2}$ values in good agreement with the extrapolated results of Woodley's thermogravimetric measurements, provided that *well characterized homogeneous* specimens were used. This confirms that the Markin-McIver data are less recommendable. Further measurements offered a possible reason for the less negative $\Delta\bar{G}$ values and the smaller entropy of the Markin-McIver data. With less homogeneous samples (for example, as obtained by physical mixing of or from samples showing two phases in X-ray diffraction) anomalous results, resembling those of Markin and McIver, were obtained. Successive temperature cycling to 1350 K in the EMF cell gradually decreased ΔG and

increased the entropy until, finally, constant values in agreement with those of
'normal' specimens were obtained. The details of this work are in the press.*

M.G. ADAMSON: Recent $\Delta \bar{G}_{O_2}$ measurements on $U_{0.75}Pu_{0.25}O_{2\pm x}$ by
Woodley and myself using a combined EMF cell (TGA technique) have also
confirmed our conclusion about the earlier Markin-McIver EMF cell data. In this
paper, which will shortly be published in the Journal of Nuclear Materials, we
have examined in some detail the reasons for the discrepancies between these
various sets of EMF cell data. Although I agree that macroscopic heterogeneities
in mixed oxide samples might influence both the measured $\Delta \bar{G}_{O_2}$ and the derived
$\Delta \bar{H}_{O_2}$, $\Delta \bar{S}_{O_2}$ values, I do not think this explanation is valid in the case of the
Markin-McIver experiments. The samples used in their work were almost certainly
homogeneous, having received more than one heat treatment at a temperature in
excess of 1650°C (T.L. Markin, private communication). As for your own work,
1350 K seems too low a temperature to produce the U-Pu ion interdiffusion
required to homogenize two-phase specimens. Perhaps some other changes were
taking place in the mixed oxide during these measurements? My own opinion
is that heterogeneity on the cation sub-lattice may be more important than
macroscopic heterogeneity in bringing about these discrepancies; as several defect
models have assumed, the state of order on the cation sub-lattice will influence
$\Delta \bar{G}_{O_2}$. Since achievement of a given cation distribution requires a temperature at
which the cations become mobile, the temperature history of a given mixed-oxide
specimen could be an important factor.

J.-P. MARCON: Na_3UO_4 and Na_3PuO_4 do not have the same crystallographic structure. Do you think that this fact can be of importance for the
results of your study of the compound $Na_3U_{1-x}Pu_xO_4$?

M.G. ADAMSON: Yes, I think such a difference in structures could
influence the calculated equilibrium oxygen potential for mixed oxide with high
plutonia content. The difference Δ $(= \Delta \bar{G}_{O_2}^{eq}$ (U system) $- \Delta \bar{G}_{O_2}^{eq}$ (U, Pu system))
would become significant if the interaction parameter for $Na_3U_{1-x}Pu_xO_4$ (E_2)
has a high value. Such would be the case if there is a serious mismatch in the
structures of Na_3UO_4 and Na_3PuO_4.

J. FUGER: Is it important to study potassium-uranium and plutonium
ternary oxides in connection with the use of Na-K as a coolant? Our experience
is that conditions for the synthesis of potassium ternary oxides may be very
different from those for the synthesis of sodium compounds.

M.G. ADAMSON: Though NaK alloys have been used as heat transfer fluids
in irradiation test capsules and as a coolant in certain test reactors, they have not,
for a number of reasons, been seriously considered as a coolant in large liquid-metal-
cooled fast breeder reactors. We have measured the oxygen activities established
in $K(\ell) + UO_{2+x}$ and $Na_{0.33}K_{0.67}(\ell) + UO_{2+x}$ mixtures containing excess alkali
metal. From the measurements on the $K(\ell) + UO_{2+x}$ system we were able to
derive the free energy of formation of the K-U-O ternary oxide which coexists
with $K(\ell)$ and $UO_2(\ell)$, namely KUO_3.

---

* TOCI, F., et al., J. Appl. Electrochem. (1979, in press).

ETUDE HORS-PILE DES REACTIONS DE UO_2, PuO_2 ET $(U, Pu)O_2$ AVEC LE CESIUM

R. LORENZELLI, R. LE DUDAL, R. ATABECK
CEA, Centre d'études nucléaires
de Fontenay-aux-Roses,
Fontenay-aux-Roses,
France

Abstract–Résumé

OUT-OF-PILE STUDIES OF REACTIONS OF UO_2, PuO_2 AND $(U,Pu)O_2$ WITH CAESIUM.

In connection with the general studies on oxide-cladding reactions and on the migration of fission products during irradiation, the authors investigated the system M-Cs-O, M being U, Pu or (U,Pu). (a) In the case of the UO_2-Cs system, the existence of three monovariant systems was shown: (1) UO_2-$Cs_2U_4O_{12}$-Cs_2UO_4; (2) UO_2-Cs_2UO_4-X; (3) UO_2-X-Cs. This new X-phase, the exact composition of which has not yet been determined, would appear — on the basis of its X spectrum — to be identical with that observed by Aitken. This finding suggests that it should be assigned a formula similar to Cs_3UO_4. (b) The study of the PuO_2-Cs system established the fact that no compound of the plutonate type is obtained under low partial pressures of oxygen. Conversely, in air and at temperatures close to $400°C$, certain isomorphic compounds of uranates ($Cs_2Pu_4O_{13}$, Cs_2PuO_4) could be identified. (c) In the $(U,Pu)O_2$-Cs system, when liquid caesium is present, the $Cs_2M_4O_{12}$ phase is in equilibrium with MO_2; in this phase the presence of plutonium is probable. The O/M ratio of the MO_2 matrix, deduced from measurements of the crystalline parameter, corresponds to an oxide which is very slightly substoichiometric, i.e. O/M = 1.997 ± 0.001. (d) The out-of-pile results tend to prove that, in irradiated fuels, both in the UO_2 shims and in the $(U,Pu)O_2$ fissile column, the formation of uranate is possible because of the redistribution of oxygen.

ETUDE HORS-PILE DES REACTIONS DE UO_2, PuO_2 ET $(U, Pu)O_2$ AVEC LE CESIUM.

Dans le cadre des études générales concernant la réaction oxyde-gaine et le déplacement des produits de fission en cours d'irradiation, les auteurs ont étudié le système M-Cs-O avec M = U, Pu ou (U,Pu). a) Pour le système UO_2-Cs, trois domaines monovariants ont été mis en évidence: 1) UO_2-$Cs_2U_4O_{12}$-Cs_2UO_4; 2) UO_2-Cs_2UO_4-X; 3) UO_2-X-Cs. Cette nouvelle phase X, dont la composition exacte n'a pu être encore établie, semble être identique, d'après son spectre X, à celle observée par Aitken. Ce résultat suggère de lui attribuer une formule proche de Cs_3UO_4. b) L'étude du système PuO_2-Cs a permis de constater qu'aucun composé de type plutonate n'est obtenu sous des pressions partielles d'oxygène faibles. Par contre, sous air, vers $400°C$, certains composés isomorphes des uranates ont pu être identifiés ($Cs_2Pu_4O_{13}$, Cs_2PuO_4). c) Dans le système $(U,Pu)O_2$-Cs, en présence de césium liquide, la phase $Cs_2M_4O_{12}$ est en équilibre avec MO_2, et dans cette phase, la présence de Pu est probable. Le rapport O/M de la matrice MO_2, déduit des mesures de paramètre cristallin, correspond à un oxyde très légèrement sous-stœchiométrique, soit O/M = 1,997 ± 0,001. d) Les résultats hors-pile tendent à prouver que, dans les combustibles irradiés, tant au niveau des cales en UO_2 qu'au niveau de la colonne fissile en $(U,Pu)O_2$, la formation d'uranate est possible grâce à la redistribution de l'oxygène.

I - INTRODUCTION

Dans le cadre des études générales concernant la
réaction oxyde-gaine et le déplacement des produits de
fission en cours d'irradiation, nous avons effectué des
études hors-pile sur le système $/\!\!-M\text{-}Cs\text{-}O\,\_7$, où M = U,Pu
ou le mixte (U,Pu).

Ces études sont justifiées par le fait que, dans la
littérature $/\!\!-1$, 2, $3\,\_7$, les résultats des études hors-
pile sur le système $U\bar{O}_2$-césium aboutissent à des conclusions contradictoires quant à la nature des uranates
susceptibles de se former en pile et que, par ailleurs,
aucun résultat n'a été publié à ce jour sur le système
$(U,Pu)O_2$-césium. Le but des expériences hors-pile est
de connaître la nature des composés (U-Cs-O) et
$/\!\!-(U,Pu)\text{-}Cs\text{-}O\,\_7$, stables sous faible potentiel d'oxygène,
ainsi que le rapport O/M de la matrice en équilibre avec
ces composés. Pour ce faire, nous avons fait varier la
stoechiométrie initiale de l'oxyde sous forme de pastilles denses et effectué les réactions en système fermé
(tube scellé en nickel) avec ou sans excès de césium
métallique de manière à simuler une aiguille de combustible en cours d'irradiation. Dans les réacteurs rapides,
on constate, en effet, tant au niveau des colonnes fertiles en UO_2 que de la colonne fissile en $(U,Pu)O_2$, que
les concentrations élevées en césium correspondent à
des zones biphasées constituées d'oxyde et d'uranate.

On décrira donc successivement les résultats hors-
pile des systèmes UO_2-Cs, PuO_2-Cs et $(U,Pu)O_2$-Cs, dans
le double but d'établir les diagrammes de phase et de
prédire le comportement du césium en pile.

II - TECHNIQUES EXPERIMENTALES

Les échantillons d'oxydes : $UO_{2,00}$, UO_{2+x}, $PuO_{2,00}$,
$(U,Pu)O_{2+x}$ sont obtenus par frittage et ajustement de
la stoechiométrie, la densité variant de 90 à 95 % de
la densité théorique.

Ils sont scellés dans une capsule en nickel passivé
chimiquement, avec des proportions variables de césium
(Cs/(U+Cs) compris entre 0,1 et 3). Le nickel est choisi
de préférence à l'acier inoxydable, par suite des réactions secondaires pouvant intervenir entre les uranates
formés dans un premier temps et le chrome de l'acier. La
quantité calculée de césium est introduite dans la capsule au moyen d'une micropipette, et connue avec préci-

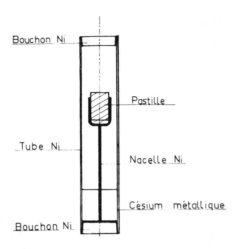

FIG.1. Dispositif à double compartiment.

sion. On utilise soit un type de capsule à deux compar-
timents, identique à celui utilisé dans l'étude des
réactions oxyde-sodium /⁻4⁻7 qui permet des études avec
du césium en phase vapeur (figure 1), soit un type de
capsule à un seul compartiment pour les études en phase
liquide (fig. 2). L'étanchéité de la capsule est vérifiée
par ressuage à l'hélium.

  Les traitements thermiques ont lieu à des températures
variant entre 680°C et 1100°C, pendant des temps compris entre 24 et 300 heures.

II.1 - Méthodes d'examens

Examens non destructifs : la réaction est contrôlée par
radiographie en suivant l'évolution de la morphologie des
pastilles et du niveau de césium en fonction du temps de
traitement (t = 24, 50 et 170 heures). Après chaque traitement, les dimensions des colonnes d'oxyde ainsi que le
niveau du césium sont mesurés sur les films à ± 0,10 mm.
L'action du césium sur MO_2 se traduit si elle a lieu par
une baisse du niveau de césium et par une augmentation du
volume global des pastilles (fig. 2). Par ailleurs, si le
césium n'est pas pollué en oxygène, on peut vérifier par
retournement des tubes qu'il est solide à la température
ambiante et que le ménisque n'a pas changé de forme.

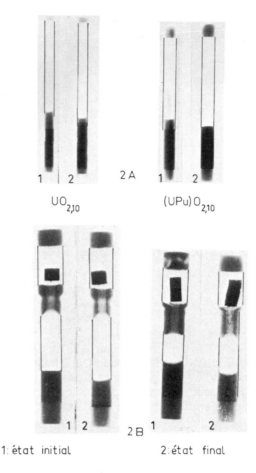

$UO_{2,10}$ $(UPu)O_{2,10}$

1: état initial   2: état final

FIG.2. Contrôle de la réaction oxyde-césium par radiographie X.

Examens destructifs : la détermination des phases après
réaction ainsi que la mesure précise des paramètres
cristallins sont effectuées par diffraction X (diffrac-
tomètre Philipps). Des examens micrographiques permettent
par ailleurs d'observer l'étendue et la morphologie de la
réaction, bien que la présence de césium libre en rende
l'interprétation difficile.

III - RESULTATS - DISCUSSIONS

III.1 - Système U-0-Cs

    Le pseudo-ternaire UO_2-césium a été largement étudié.
Plusieurs uranates ont été caractérisés par diffraction

de RX $\boxed{5}$. Dans la majorité de ceux-ci, l'uranium est
à l'état hexavalent, ils peuvent donc être considérés
comme des composés de Cs_2O et UO_3 (par exemple $Cs_2U_5O_{16}$,
$Cs_2U_4O_{13}$, Cs_2UO_4 ...). Par contre, dans $Cs_2U_4O_{12}$, l'uranium est à une valence inférieure à 6.

Pour étudier ce pseudo-binaire, nous avons procédé
de différentes manières :

- soit en faisant réagir en tube scellé UO_2 de stoechiométrie variable (entre $UO_{2,00}$ et U_3O_8) avec des quantités connues de césium,

- soit en faisant réagir des uranates de césium préparés
  dans des conditions expérimentales semblables à celles
définies par Cordfunke $\boxed{6}$ (tableau I), avec des quantités connues de césium,

- soit en faisant réagir en tube scellé des mélanges
  d'uranates et d'UO_2.

L'identification des produits obtenus est effectuée
à partir des données cristallographiques fournies par
Van Egmond $\boxed{5}$.

Généralement, comme l'indiquent les contrôles radiogra-
phiques, on ne constate plus d'évolution des réactions après
24 heures de traitement.

Dans le cas où il y a réaction avec le césium, on constate que le gonflement de la pastille a comblé tout le jeu
initial entre la pastille et le tube et on vérifie également
que le césium est solide à la température ambiante, ce qui
implique, conformément au diagramme Cs-O, que sa teneur en
oxygène est extrêmement faible (< 0,5 at %).

Les résultats des réactions après refroidissement rapide à l'ambiante sont reportés dans le tableau II. Les
chemins parcourus dans le ternaire U-Cs-O sont représentés
sur le diagramme de phase (fig. 3) par des trajets limités
au départ et à l'arrivée et portant un numéro correspondant à la réaction. Les phases en équilibre dans les différents domaines sont les suivantes:

- Domaine I : dans ce domaine sont à l'équilibre les phases UO_2, $Cs_2U_4O_{12}$, Cs_2UO_4.

TABLEAU I. PREPARATION DES URANATES A PARTIR DES REACTIONS $UO_2 + Cs_2CO_3$

Mélange $UO_2 + Cs_2CO_3$	Réaction	Phases	Structures cristallines paramètres	Observations
Cs/U=0,50 (1)	800°C – 70 h sous air	$Cs_2U_4O_{13}$ traces : $Cs_2U_5O_{16}$	Orthorombique centré Monoclinique centré	Composés stables sous air couleur ocre
Cs/U=0,50 (2)	1000°C – 24 h sous air	$Cs_2U_4O_{13}$ traces : $Cs_2U_5O_{16}$	idem	Composés stables sous air couleur ocre
Cs/U=0,50 (3)	800°C – 12 h sous vide $pO_2 < 10^{-6}$ atm	$Cs_2U_4O_{12}$ (α)	Orthorombique	Composé stable sous air couleur gris foncé

TABLEAU II. RECAPITULATION DES REACTIONS UO_{2+x} — CESIUM

Etat initial compositions exprimées en moles	Conditions tube scellé en nickel	Phases paramètres	O/M de UO_2	Réaction avec le césium d'après la radiographie	Observations
1 $UO_{2,00}+1$ Cs	680°C-48 h	UO_2: a=5,4702 ±0,0008	2,00	non césium résiduel	
2 $UO_{2,06}+1$ Cs	680°C-48 h	uranate : phase X UO_2:a=5,4706 ±0,0008	2,00	oui césium résiduel	présence d'uranate difficile à identifier
3 $UO_{2,10}+0,1$ Cs	680°C-48 h	uranates : (Cs_2UO_4 (($Cs_2U_4O_{12}$ UO_2:a=5,4702 ±0,0008	2,00 2,00	oui pas de césium résiduel	$Cs_2U_4O_{12}$ + Cs_2UO_4 en équilibre avec UO_2
4 $UO_{2,26}+0,4$ Cs	680°C-48 h	uranate : Cs_2UO_4 UO_2:a=5,4703 traces $Cs_2U_4O_{12}$	2,00	oui pas de césium résiduel	Cs_2UO_4 : orange composé instable sous air
5 $U_3O_8+2,5$ Cs	680°C-24 h	uranate : Cs_2UO_4 UO_2:a=5,4670	n.d. 2,03	oui pas de césium résiduel	La limite de réduction de UO_2 n'a pas été atteinte
6 U_3O_8+6 Cs	680°C-48 h	uranate : X + X' UO_2:a=5,4702	2,00	oui césium résiduel	X et X' peuvent s'indexer comme 2 cfc : a = 5,696 a = 5,633 composé instable sous air
7 U_3O_8+6 Cs	800°C-48 h	uranate : X + X' UO_2:a=5,4698	2,00	oui césium résiduel	idem : 2 cfc a = 5,698 a = 5,626 composé instable sous air
8 U_3O_8+6 Cs	1100°C-48 h	uranate X UO_2:a=5,4699	2,00	oui césium résiduel	un seul cfc a = 5,698 composé instable sous air
9 $Cs_2U_4O_{13}$ + 6 Cs	680°C-24 h	uranate : Cs_2UO_4 (pur)		oui pas de césium résiduel	composé orange instable sous air
10 $Cs_2U_4O_{13}$ + 10 Cs	680°C-24 h	uranate : X et X' UO_2:a=5,4702	2,00	oui césium résiduel	2 phases cfc X:a= 5,635 X':a=5,711 instable sous air
11 $Cs_2U_4O_{12}$ + 10 Cs	680°C-24 h	uranate : X et X' UO_2:a=5,4706	2,00	oui césium résiduel	2 phases cfc X:a= 5,6235 X':a=5,711 instable sous air
12 $Cs_2U_4O_{13}$	vide secondaire 800°C	$Cs_2U_4O_{12}$			
13 U_3O_8 + Cs_2CO_3 Cs/U=0,5	vide secondaire 800°C	$Cs_2U_4O_{12}$ Cs_2UO_4 UO_2		pas de césium libre	état d'équilibre final non atteint

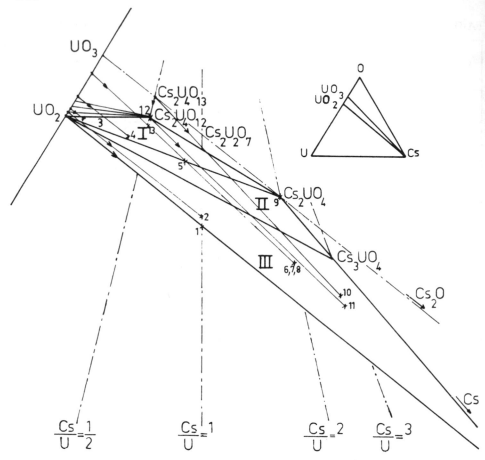

FIG.3. Diagramme U-Cs-O. Section isotherme entre 680°C et 1100°C.

- Domaine II : dans ce domaine sont à l'équilibre les phases UO_2, Cs_2UO_4 et une phase nouvelle, appelée "X", dont la composition et la structure exactes sont en cours d'étude.

- Domaine III : dans ce domaine sont à l'équilibre les phases UO_2, la phase "X" et le césium métallique. La structure cristallographique de cette phase X dépend de la température, probablement par une mise en ordre des atomes de césium et d'uranium. A 1100°C, "X" cristallise apparemment dans un système cubique à faces centrées avec un paramètre a = 5,70 Å. X', la phase stable entre 680° et 900°C, est probablement une surstructure de X ; sur le diagramme de phase X et X' sont donc confondus.

Dans les domaines à trois phases, le paramètre cristallin de UO_2 (tableau II) correspond à celui de $UO_{2,00}$. Dans le domaine bivariant, $Cs_2U_4O_{12}$ - UO_{2+x}, le paramètre de UO_2 varie en fonction de x (fig. 3).

Les phases Cs_2UO_4 et X sont hygroscopiques et disparaissent très rapidement après exposition à l'air à la température ambiante alors que $Cs_2U_4O_{12}$ est stable dans les mêmes conditions.

Par ailleurs, à 800°C sous une pression partielle d'oxygène de l'ordre de 10^{-20} atmosphères (ΔGO_2 = - 100 kcal) imposée par un mélange gazeux ($Ar-H_2$), $Cs_2U_4O_{12}$ et Cs_2UO_4 sont instables en système ouvert conformément aux résultats de Cordfunke [1] et décomposés en UO_2 et césium.

En conclusion, nos résultats sont en désaccord avec ceux de D.C. Fee et Johnson [3, 7]. Ces auteurs proposent, en effet, à partir de considérations thermodynamiques et sans vérification expérimentale, l'existence des domaines à trois phases : [$Cs_2U_4O_{12}$ - $Cs_2U_2O_7$-UO_2] et [Cs_2UO_4-$Cs_2U_2O_7$ - UO_2], totalement incompatibles avec le domaine à trois phases [$Cs_2U_4O_{12}$ - Cs_2UO_4 - UO_2] réellement observé dans nos expériences (fig. 4).

Par ailleurs, on observe un nouveau composé de césium en équilibre avec Cs métal et UO_2, différent de Cs_2UO_4 et dont la composition est en première approche voisine de Cs_3UO_4. L'existence d'un tel composé de césium, dans lequel la valence de l'uranium est inférieure à 6, avait été proposée par Aitken [2]. Nos propres observations (réactions 12 et 13) suggèrent cependant pour ce composé une composition nettement différente de celle de Aitken : $Cs_{1,3}UO_3$, bien que les spectres de diffraction X présentent de fortes analogies [8] et incitent à penser qu'il s'agit du même composé. Le domaine à trois phases Cs_2UO_4-UO_2-césium liquide proposé par Johnson [3] est donc incompatible avec nos résultats et ceux d'Aitken.

III.2 - Système PuO_2-césium

A l'exception de l'étude de M. Pages [9] sur la synthèse et l'identification de composés de plutonium heptavalent (préparés par oxydation

a)

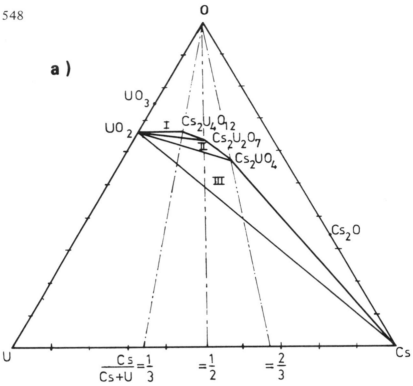

b)

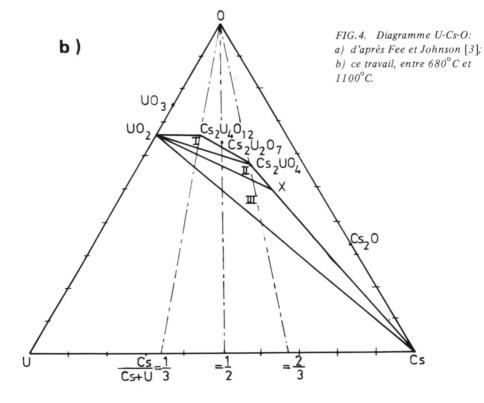

FIG.4. Diagramme U-Cs-O:
a) d'après Fee et Johnson [3];
b) ce travail, entre 680°C et
1100°C.

du plutonium IV par un superoxyde alcalin), aucune étude n'a été consacrée au système Pu-O-Cs. Il nous a donc paru intéressant d'explorer ce système de la même manière que le système U-O-Cs, afin de mettre en évidence l'existence des plutonates.

- Réaction $PuO_2 + CO_3Cs_2$

En faisant réagir sous air à 450°C des mélanges de PuO_2 et de CO_3Cs_2, on a observé la formation de composés isomorphes de certains uranates, tels que $Cs_2Pu_4O_{13}$ et Cs_2PuO_4. A plus haute température (700°C), sous air, ces composés sont instables. Le composé Cs_2PuO_4, isomorphe de Cs_2UO_4, avait été préparé par M. Pages par décomposition thermique de Cs_3PuO_5 [9].

- Réaction $PuO_2 + césium$

En tube scellé en présence de Cs métallique, aucun produit de réaction n'est observé entre PuO_2 et le césium. Ceci distingue fortement le système PuO_2-Cs, du système PuO_2-Na. L'absence de plutonate de césium sous faible pression partielle d'oxygène est donc **probable**. Cependant, une augmentation du paramètre cristallin, correspondant à une légère réduction de $PuO_{2,00}$, est observée (tableau III), en présence de Cs liquide.

III.3 - Système $(U,Pu)O_2$ - césium

III.3.1 - Diagramme de phase

La même technique d'étude est utilisée pour ce système que pour le système UO_2-césium. D'une part, l'oxyde mixte de composition Pu/U+Pu = 0,20 et de rapport O/M compris entre 2,00 et 2,10 est mis à réagir entre 680°C et 900°C dans des tubes de nickel scellés avec des quantités connues de césium : on vérifie avant l'ouverture des tubes qu'il reste bien un excès de métal et que l'oxyde a effectivement réagi avec le césium (fig. 3) ; d'autre part, l'oxyde mixte est mélangé à CO_3Cs_2 et chauffé sous vide secondaire pendant 48 heures avec un $pO_2 \sim 10^{-6}$ atmosphère mesuré à l'aide d'une pompe électrochimique. Les résultats obtenus sont rassemblés dans le tableau IV.

TABLEAU III.　REACTION PuO_2 — CESIUM

Etat initial	Conditions	Phases	Paramètre	O/M
4 PuO_2 + Cs_2CO_3 (1)	450°C sous air	$Cs_2Pu_4O_{13}$ $Cs_2Pu_5O_{16}$ PuO_2	a = 5,3960	2,00
1 PuO_2 + 1Cs_2CO_3 (2)	450°C sous air	Cs_2PuO_4 PuO_2	a = 5,3960	2,00
	700°C sous vide	PuO_2 seul	a = 5,3959	2,00
1 PuO_2 + 1Cs (3)	tube scellé nickel 40 h à 800°C PuO_2 a = 5,3939	PuO_{2-x} césium résiduel	a = 5,3990	$2-x$ $x \sim 0,02$

TABLEAU IV. REACTION $(U,Pu)O_2$ — CESIUM

Etat initial	Conditions	Phases	Paramètres de MO_2	O/M	Réaction avec le césium	Observations
$(U,Pu)O_2+Cs_2M_4O_{13}$ (1)	680°C–48 h tube scellé nickel	$Cs_2M_4O_{12}$ MO_2	$a=5,4540\pm8\ 10^{-4}$	$O/M\simeq2,012$	oui pas de césium résiduel	une seule phase type MO_2 ; pas de phase PuO_2
$(U,Pu)O_2+2\ Cs$ (2)	680°C–48 h tube scellé nickel	MO_2	$a=5,4552\pm3\ 10^{-4}$	$O/M\simeq2,000$	pas de réaction césium résiduel	le paramètre initial de $(U,Pu)O_{2,00}$ est $= 5,4552$
$(U,Pu)O_{2,10}+2\ Cs$ (3)	idem	$Cs_2M_4O_{12}$ MO_2	$a=5,4555_0\pm3\ 10^{-4}$	$O/M\sim1,999$	réaction visible sur les radiographies	déviation faible de O/M
$(U,Pu)O_{2,10}+2\ Cs$ (4)	idem	$Cs_2M_4O_{12}$ MO_2	$a=5,4558_5\pm3\ 10^{-4}$	$O/M=1,997_5$	idem césium résiduel	déviation significative de O/M
$(U,Pu)O_2+Cs_2CO_3$ (Cs/M = 2) (5)	680°C–48 h sous vide secondaire	$Cs_2M_4O_{13}$ $Cs_2M_5O_{16}$ (traces) 2 phases type MO_2	$a=5,4540\pm5\ 10^{-4}$ (majoritaire) $a=5,424\pm1\ 10^{-3}$ (minoritaire)	non défini mais $> 2,00$	oui	enrichissement en PuO_2 de la phase MO_2 minoritaire

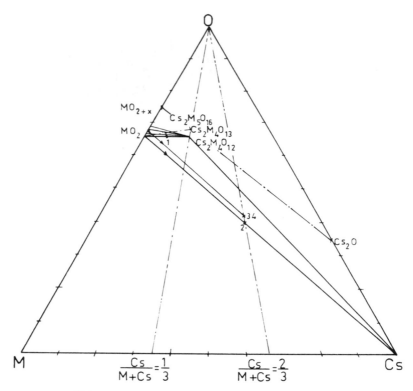

FIG.5. Diagramme MO_2-Cs. Section isotherme à 700°C.

On peut en tirer les conclusions suivantes :

- L'uranate stable formé en présence de césium liquide est, d'après le spectre de RX, identique à $Cs_2U_4O_{12}$ et dans ces conditions, contrairement à ce que laissaient prévoir les observations sur les combustibles irradiés /‾11‾7, la démixtion de PuO_2 n'est pas observée.

- Dans les réactions avec le carbonate de césium (réaction 5), il y a formation d'uranate et on observe une démixtion partielle de PuO_2 sous la forme d'une deuxième phase MO_2 enrichie en PuO_2 ; toutefois, la démixtion péut être attribuéé à la pression d'oxygène élevée lors de la dissociation du carbonate.

Le pseudo-ternaire MO_2-Cs est représenté sur la figure 5. Il met en évidence l'existence d'un

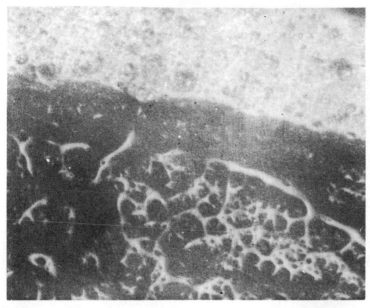

200 μm

FIG.6. Pénétration intergranulaire du césium et formation de composés (U, Pu)-Cs-O.

domaine à trois phases : MO_2, $Cs_2M_4O_{12}$, césium
liquide conformément à la règle des phases ; on
suppose que M est le même pour toutes les phases
en présence.

Par rapport au système UO_2-Cs, on remarquera
l'absence du composé type Cs_2UO_4 qui peut s'expliquer par le fait que, dans Cs_2UO_4, tout l'uranium
est à la valence 6 et qu'il est impossible, dans
ces conditions, d'introduire des ions Pu^{4+} dans
la maille de Cs_2UO_4. Par contre, dans $Cs_2U_4O_{12}$, une
partie des U^{+4} peut être remplacée par du Pu^{+4}.

Le fait que l'on observe la phase $Cs_2U_4O_{12}$ au
lieu de Cs_2UO_4 milite en faveur de l'existence
d'urano-plutonate, bien que la composition exacte
de cet uranate mixte "$Cs_2M_4O_{12}$" n'ait pu être déter-
minée ; l'existence de ce composé a pu être contrôlée micrographiquement, bien que la présence
de césium libre rende très difficile l'observation
(fig. 6).

TABLEAU V.   INFLUENCE DU RUBIDIUM DANS LA REACTION
$(U, Pu)O_2$ − CESIUM

Etat initial	Conditions tube scellé 680°C-24 h	Phases	Paramètre	O/M (déduit)
$(U,Pu)O_{2,10}$+ Cs (avec excès)	"	$Cs_2M_4O_{12}$ MO_2	a=5,4562	$1,997_1$
$(U,Pu)O_{2,10}$+ Rb (avec excès)	"	rubidinate MO_2	a=5,4559	$1,997_9$
$(U,Pu)O_{2,10}$+ (Rb+Cs) $\frac{Rb}{Cs} \sim 0,15$at %	"	$Cs_2M_4O_{12}$ MO_2	a=5,4560	$1,997_6$

III.3.2 - Stoechiométrie de la phase MO_2 dans le domaine MO_2-Cs - $Cs_2MO_4O_{12}$

Par suite de la présence de césium libre en excès, il est impossible de faire des mesures directes du rapport O/M de la phase MO_2.

Pour mesurer cette stoechiométrie, on utilise une relation établie entre la variation du paramètre cristallin et le rapport O/M. Dans le cas des oxydes mixtes lacunaires en oxygène, $(U_{1-y}Pu_y)O_{2-x}$, cette loi est exprimée sous la forme :

$$a = 5,470 - 0,074 \, y + 0,34 \, x$$

La précision sur le paramètre cristallin étant de l'ordre de $3 \cdot 10^{-4}$, une variation de stoechiométrie de $1 \cdot 10^{-3}$ est significative.

Comme on peut le constater sur le tableau IV, le rapport (O/M) de la matrice MO_2, déduit des mesures de paramètre cristallin, tout en étant très voisin de 2,000, correspond à un oxyde très légèrement sous-stoechiométrique, fixé à 1,997±0,001 (réaction 4).

III.3.3 - Etude du système MO_2-césium-rubidium

Afin de mettre en évidence une influence éventuelle de certains produits de fission de même nature

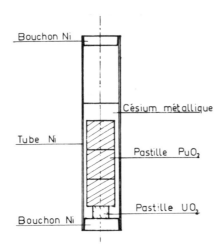

FIG.7. Manipulation type démixtion.

chimique que le césium, sur la nature des composés formés et le rapport O/M de la matrice à l'équilibre, du
rubidium a été ajouté au césium dans des proportions
variables.

Les réactions MO_2, césium, rubidium, ont été
effectuées dans des conditions identiques aux
réactions MO_2-césium. Après analyse des paramètres
cristallins, on constate que l'oxyde initial a
bien été réduit et que la limite de réduction est
sensiblement la même. La présence d'uranate type
$Cs_2M_4O_{12}$ a été identifiée dans tous les cas.

Par rapport aux résultats obtenus avec le
césium pur, on peut donc affirmer que le rubidium
ajouté au césium n'a pas d'effet appréciable sur
la limite de réduction de MO_2 qui reste toujours
très voisine de 2,000 (tableau V).

III.4 - Démixtion de PuO_2

Comme on le verra plus loin, certains examens de combustibles irradiés mettent en évidence
une démixtion de PuO_2 en périphérie du combustible, or, dans les réactions "hors-pile", ce phé-
nomène n'est pas observé. Il nous a donc paru
intéressant de simuler cette démixtion quasiment
totale de PuO_2, afin de connaître les conséquences
éventuelles sur la physicochimie du combustible.

TABLEAU VI. EXPERIENCE DE DEMIXTION

Réaction (exprimée en mole)	Conditions tube scellé	Phases	Paramètre initial	O/M initial	Paramètre final	O/M final
1(UO$_2$)+10 PuO$_2$ + 11 Cs	800°C 20 h 40 h 100 h	uranate formé sur la pastille type Cs$_2$U$_4$O$_{12}$				
		UO$_2$	a=5,4702	2,00	a=5,4708	2,00
		PuO$_2$	a=5,3939	2,00	a=5,4005	1,98

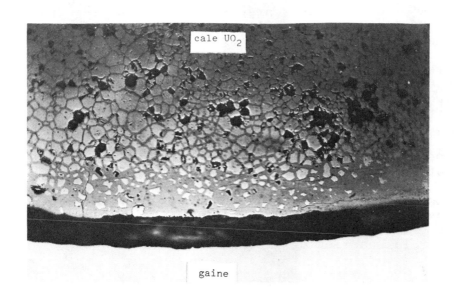

cale UO$_2$

gaine

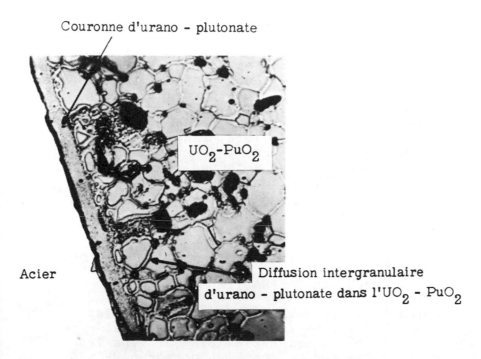

Couronne d'urano - plutonate

UO$_2$-PuO$_2$

Acier

Diffusion intergranulaire
d'urano - plutonate dans l'UO$_2$ - PuO$_2$

FIG.8.   Aspect métallographique du combustible irradié.

L'expérience présentée sur la figure 7 a donc
été réalisée. Une pastille de UO_2, rigoureusement
stoechiométrique, obtenue après traitement sous
argon-hydrogène, est mise en contact avec une petite colonne de PuO_2 stoechiométrique (calciné
sous O_2 à 300°C), dans un tube en nickel en présence d'un excès de césium liquide. Après traitement thermique (tableau VI), on observe la formation "d'uranate" dans la pastille d'UO_2. Cette formation d'uranate, dans des conditions où $UO_{2,00}$
stoechiométrique ne réagit pas avec le césium, ne
peut s'expliquer que par un transfert d'oxygène à
partir de PuO_2, légèrement réduit sur UO_2, par
l'intermédiaire du césium liquide. Le potentiel
d'oxygène de $UO_{2+\epsilon}$ devient alors suffisant pour
permettre la formation d'uranate. On peut schéma-
tiser la réaction de la manière suivante :

$$PuO_2 + Cs \rightarrow PuO_{2-x} + \frac{x}{2} O_2 + Cs$$

$$\frac{x}{2} O_2 + UO_2 \rightarrow UO_{2+x}$$

$$UO_{2+x} + Cs \rightarrow uranate$$

La cinétique de formation de l'uranate dans
ces conditions est probablement plus lente que
dans la réaction UO_2-Cs. De la valeur du paramètre de PuO_2 après réaction, on déduit une
limite de réduction égale à 1,98, correspondant
au potentiel d'oxygène à partir duquel la formation d'uranate est possible. Dans le cas de
l'oxyde mixte à 20 % de PuO_2, pour ce même potentiel, on obtiendrait la composition limite
$(U_{0,80}Pu_{0,20})1,996$, valeur très proche de celle
que l'on peut déduire des mesures de paramètre
(1,997 ± 0,001).

IV - COMPARAISON AVEC LES RESULTATS OBTENUS SUR LES COMBUSTIBLES IRRADIES

On sait que le césium se déplace vers les
parties froides $/\overline{}_{12}\overline{}/$, c'est-à-dire radiale-
ment vers la périphérie et axialement vers les
cales. On pense que ce sont les composés M-Cs-O
qui jouent un rôle important lors de cette redistribution (M = U,Pu,Mo,Cr).

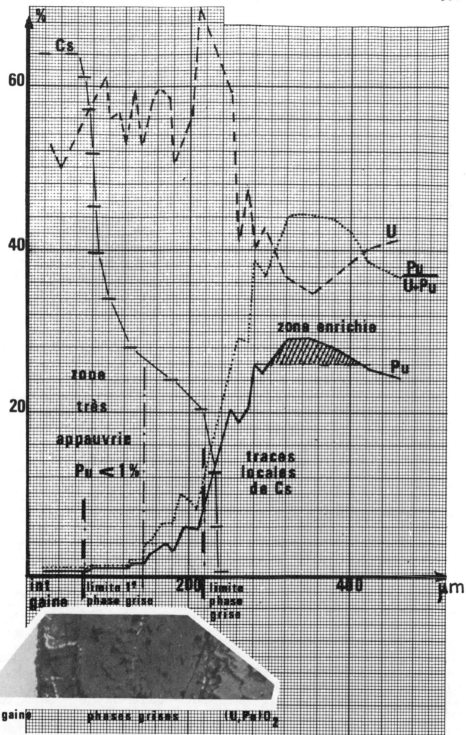

FIG. 9. *Répartition de U, Pu, Cs dans la phase en périphérie de l'oxyde mixte.*

Que devient le césium en périphérie du combustible ? Il semble qu'il puisse se combiner aux
éléments de l'oxyde et aux autres produits de
fission pour former des composés dont l'aspect
micrographique est montré sur la figure 8. Une
étude à la microsonde de Castaing (fig. 9) montre la répartition du Cs, de l'uranium et du
plutonium dans cette zone et révèle que les deux
couches visibles en micrographie sont toutes deux
exemptes de plutonium.

Les résultats hors-pile confirment que la
formation d'uranate est possible au niveau de la
colonne fissile dans la mesure où l'oxyde atteint
la stoechiométrie 1,997 en surface grâce à la redistribution de l'oxygène $\overline{/\ 12\ \overline{/}}$. La démixtion de
PuO_2, ou l'enrichissement en PuO_2 de la matrice,
n'ont pas été observés hors-pile, et pourraient être
plus spécifiques de l'irradiation. Par contre, on a
montré que la démixtion de PuO_2 favorise et entre-
tient la formation des uranates grâce à un transfert de l'oxygène de PuO_2, sur UO_2, ou sur $(U,Pu)O_2$.
Ainsi, sous le double effet de la redistribution dé
l'oxygène et de la démixtion, la couche d'uranate
peut croître à la surface de l'oxyde. Selon toute
vraisemblance, l'uranate formé est du type $Cs_2M_4O_{12}$,
avec un rapport Pu/U+Pu qui peut être différent de
celui de la matrice, mais il n'existe pas, à ce
jour, d'analyses chimique ou cristallographique
qui permettent de l'affirmer. Par contre, la sta-
bilité à l'air de cet uranate semble confirmer
qu'il s'agit bien de $Cs_2M_4O_{12}$.

Que devient le césium aux extrémités de la colonne
fissile ?

Les pics de concentration maximale de
césium semblent se localiser de façon aléatoire,
soit au niveau des interfaces "UO_2-inox, UO_2-
$(UPu)O_2$", soit au niveau de la cale d'UO_2 elle-
même.

Les résultats hors-pile confirment la
possibilité de formation des uranates au niveau
de la cale dont la composition dépend de la stoechiométrie de UO_2. Compte tenu de l'évolution du
potentiel d'oxygène à l'interface cale-oxyde
mixte, il est très probable que seul l'uranate
Cs_2UO_4 puisse exister dans le domaine à deux

phases Cs_2UO_4-$UO_{2\pm x}$. Sur ce point, nos résultats
sont en bon accord avec ceux de D.C. Fee et
C.E. Johnson /‾3‾7. L'extrême sensibilité de
cette phase à l'humidité la rend très instable et
explique que l'on n'ait jamais pu l'isoler ou la
caractériser par diffraction X, après irradiation.

CONCLUSIONS

L'étude expérimentale dans le domaine 600-1100°C
des diagrammes ternaires U-O-Cs, Pu-O-Cs et $(U_{0,8}Pu_{0,2})$
-O-Cs au voisinage des pseudo-binaires oxyde-césium,
réalisée par mise à l'équilibre en tubes de nickel
scellés des réactifs en proportions voulues a montré
que :

- pour le système U, O, Cs, les uranates en équilibre avec le dioxyde UO_2 sont par ordre de pression
partielle d'oxygène décroissante $Cs_2U_4O_{12}$, Cs_2UO_4 et
Cs_3UO_4, la formule de ce dernier composé n'étant pas
définitivement établie. Les domaines monovariants
correspondants sont UO_2+$Cs_2U_4O_{12}$+Cs_2UO_4 (I), UO_2+Cs_2
UO_4+Cs_3UO_4 (II) et UO_2+Cs_3UO_4+Cs (III).

Dans le domaine III, le césium liquide à l'équilibre est pratiquement exempt d'oxygène (<0,5 at %).

Le fait que $Cs_2U_2O_7$, contrairement aux résultats de la littérature, ne puisse être en équilibre
avec UO_2 entraîne une révision des données thermodynamique publiées d'au moins une des trois phases
$Cs_2U_4O_{12}$, Cs_2UO_4 et $Cs_2U_2O_7$.

Dans le système Pu-O-Cs, on a mis en évidence
deux plutonates $Cs_2Pu_4O_{13}$ et Cs_2PuO_4 qui ne peuvent
exister que sous des pressions d'oxygène élevées.
Sous pO_2 faible, RT lg pO_2 < ∿ - 40 kcal, l'existence
des plutonates est peu probable. Le césium liquide réduit
simplement $PuO_{2,00}$.

Dans le système (U,Pu)-O-Cs, un seul urano-
plutonate $Cs_2M_4O_{12}$ peut exister en équilibre avec
l'oxyde mixte MO_2, donnant lieu à un domaine monovariant MO_2, $Cs_2M_4O_{12}$, Cs. A 700°C, la stoechiométrie de l'oxyde mixte à l'équilibre correspond
certainement au domaine sous-stoechiométrique
quoique très proche de la stoechiométrie et peut
être estimée égale à 1,997, correspondant à un
$\overline{\Delta G}O_2$ de -131 kcal à T = 700°C [10]. Aucune formation

d'uranoplutonate n'est donc possible pour les stoechiométries inférieures à cette valeur.

En première approximation, les valeurs de M, dans les phases $Cs_2M_4O_{12}$ et MO_2 sont voisines pour la composition étudiée ($M = U_{0,8}Pu_{0,2}$).

L'application de ces résultats au comportement physico-chimique du césium au cours de l'irradiation du combustible oxyde montre qu'en principe le césium ne devrait pas réagir pour former de l'uranate de césium, ni au niveau de la colonne fertile, ni à celui de la colonne fissile (UO_2, O/M = 2,00 (U,Pu)O_2 O/M $\simeq$ 1,98), tout au moins au début de l'irradiation. Ceci est contraire à l'observation courante, qui montre toujours une concentration locale d'uranate au niveau des premières pastilles fertiles et une répartition régulière d'uranate (env. 80 % du césium formé) tout au long de la colonne fissile /¯11_7.

C'est évidemment la migration de l'oxygène sous gradient thermique vers les parties froides de l'oxyde (fertile et périphérie du combustible) qui permet à l'oxygène d'atteindre un potentiel suffisant pour rendre possible la formation d'uranate. D'après les résultats donnés précédemment, ce potentiel est au moins égal à celui qui correspond à la stoechiométrie O/M = 1,996 pour la composition $U_{0,8}Pu_{0,2}$.

De plus, le composé de césium formé à la périphérie du combustible sous irradiation est non pas un uranoplutonate, comme le montrent les expériences hors-pile présentées ici, mais un uranate accompagné d'un enrichissement concomittant en PuO_2 /¯11_7. Il paraît donc vraisemblable que l'état le plus stable impliquant la démixtion en deux phases distinctes de l'uranium et du plutonium ne puisse être réalisé hors-pile mais seulement sous irradiation par l'effet des fragments de fission.

Remerciements

Les auteurs remercient M. PASCARD pour les nombreuses et fructueuses discussions.

REFERENCES

[1] CORDFUNKE, E.H.P., in Thermodynamics of Nuclear Materials (Proc. Symp. Vienna 1974), Vol. II, IAEA, Vienna (1975) 185.
[2] AITKEN, E.A., ADAMSON, M.G., DUTINA, D., EVANS, S.K., Ibid., Vol. I, p. 187.
[3] FEE, D.C., JOHNSON, C.E., J. Nucl. Mater. 78 (1978) 219.
[4] HOUSSEAU, M., DEAN, G., PERRET, F., in Behaviour and Chemical State of Irradiated Ceramic Fuels (Proc. Panel Vienna, 1972), IAEA, Vienna (1974) 349.
[5] VAN EGMOND, A.B., Reactor Centrum Nederland, Petten, The Netherlands, RCN-246 (May 1976).
[6] CORDFUNKE, E.H.P., VAN EGMOND, A.B., VAN VOORST, G., J. Inorg. Nucl. Chem. 37 (1975) 1433.
[7] FEE, D.C., JOHNSON, I., DAVIES, S.A., SHINN, W.A., STAAHL, G.E., JOHNSON, C.E., ANL-76-126 (1977).
[8] AITKEN, E.A., ADAMSON, M.G., DUTINA, D., EVANS, S.K., LUDLOW, T.E., GEAP-12418 (1973).
[9] PAGES, M., NECTOUX, F., FREUNDLICH, W., Radiochem. Radioanal. Lett. 8 3 (1971) 147.
[10] INTERNATIONAL ATOMIC ENERGY AGENCY, The Plutonium-Oxygen and Uranium-Plutonium-Oxygen Systems: A Thermochemical Assessment, Technical Reports Series No. 79, IAEA, Vienna (1967) 86 p.
[11] CONTE, M., VIGNESOULT, N., Communication présentée à la Réunion de spécialistes sur la chimie interne des barres de combustible de réacteurs à eau, AIEA, TC-221, Erlangen, 22–26 janvier 1979.
[12] MARCON, J.P., Communication présentée à la Réunion de spécialistes sur l'interaction entre le combustible et la gaine, Tokyo, 1977.

DISCUSSION

M.G. ADAMSON: Your conclusion that a Cs-U-O compound of uranium valence less than six is the stable phase coexisting with UO_2 and $Cs(\ell)$, rather than Cs_2UO_4, is essentially the same as the conclusion reached by Aitken and myself from capsule test results. As regards its exact composition, there still appears to be some doubt. We attempted to prepare both $CsUO_3$ and Cs_3UO_4 by reacting appropriate mixtures of Cs, UO_2, UO_3 etc. in nickel capsules, but never succeeded in preparing these phases. We now prefer to designate this phase $Cs_{3-x}UO_{4-y}$ or $Cs_{2-x}UO_{4-y}$ so as to acknowledge that it may have variable stoichiometry or that there may be more than one such phase.

You mentioned that U-Pu segregation between the oxide and uranate phases may be due to oxygen transport from Pu-rich regions to U-rich regions. I think a transport path for U and/or Pu must also be involved, possibly through a liquid (Cs-rich) fission product phase in a steep temperature gradient. Could you please comment both on the uncertainty in composition of the lower-valence Cs fuel compound and on the mechanism of U-Pu segregation?

R. LORENZELLI: In reply to the first part of your question, concerning
the exact chemical formula of the compound in equilibrium with UO_2 and $Cs(\ell)$,
we believe this to be close to Cs_3UO_4 but we have not carried out a sufficiently
accurate chemical analysis to be able to confirm that the composition is indeed
Cs_3UO_4. The possible analogy of the crystal structure with Na_3UO_4 speaks in
favour of a composition close to Cs_3UO_4.

As regards the uranium-plutonium segregation observed in the outer part
of the irradiated $(U, Pu)O_2$ mixed oxides, the results of our out-of-pile experiments
do not enable us for the time being to explain this segregation. It is likely,
however, that the demixing is associated with the formation of a monoplutonate
of caesium with a U/Pu ratio appreciably different from that of the oxide matrix.

E.H.P. CORDFUNKE: Your results on the caesium uranate formed in
liquid caesium agree nicely with ours[1] in that it is a caesium uranate in which
uranium has a lower valence. This phase is described by you as Cs_3UO_4 and by
us as Cs_2UO_{4-x} ($x \cong 0.5$); it thus has the same U^{6+}/U^{4+} ratio, and differs only by
half a mole of Cs_2O. Did you compare the X-ray pattern given in our paper with
yours? In other words, do we have the same uranate phase?

R. LORENZELLI: We think thorough and comparative examination of
the diffraction spectra obtained for this new caesium compound should enable
us to establish its structure.

---

[1] Described in these Proceedings in paper IAEA-SM-236/34.

FISSION-PRODUCT BEHAVIOUR IN IRRADIATED HTR FISSILE PARTICLES AT HIGH TEMPERATURES*

R. BENZ, R. FÖRTHMANN, H. GRÜBMEIER, A. NAOUMIDIS
Kernforschungsanlage Jülich GmbH,
Jülich,
Federal Republic of Germany

Abstract

FISSION-PRODUCT BEHAVIOUR IN IRRADIATED HTR FISSILE PARTICLES AT HIGH TEMPERATURES.

The chemical state and transport behaviour of solid fission products in irradiated fissile coated particles with very high burn-ups ($\sim$70% FIMA) were investigated by means of electron microprobe analysis as a function of the kernel composition. UO_2, UC_2, UO_2+10% UC_2 kernels were used. In UO_2 kernels the fission products molybdenum, technetium, ruthenium, rhodium and palladium form metallic inclusions consisting of two phases which contain very different amounts of molybdenum. In UC_2 kernels, as well as in UO_2-UC_2 kernels, these fission products form complex carbides which also contain uranium. The rare-earth fission products exist in the form of oxides in both UO_2 and UO_2-UC_2 kernels, whereas the alkaline earths form carbides, even in UO_2-UC_2 kernels. Core heat-up simulation experiments using unirradiated particles with and without additions of artificial fission products for simulation of a high burn-up showed that carbon monoxide permeation through the coating controls the failure rate of oxide particles at very high temperatures, and that the migration of metallic fission products causes corrosion of the silicon carbide layer.

1. INTRODUCTION

Planning for electricity and process-heat generating high temperature reactors (HTRs) in the Federal Republic of Germany envisages the use of the uranium-thorium nuclear fuel cycle together with highly enriched uranium (93% ^{235}U). This decision was based upon the unique advantages that the U-Th fuel cycle offers with regard to problems of economy and the use of limited resources. The principle of keeping fissile and fertile materials separate by having them in different, coated particles allows of recycling of the ^{233}U, separated from the other uranium isotopes, and such a system can be considered a high-conversion-ratio system.

* This work was performed in joint partnership with GAG, GHT, HOBEG, HRB, KFA, NUKEM and SIGRI/RW under the 'HTR-Brennstoffkreislauf Project' supported by the Ministry of Research and Technology of the Federal Republic of Germany and by the Federal State of Nord-Rhein Westfalen.

Virtually all the fission products must be retained in both the fertile and the fissile particles under normal reactor conditions. The fuel particles can be designed to satisfy this requirement only if the effects, amongst others, of the fission products on the particle coating, which in an HTR fuel replaces the 'canning', are known. In fuel particles, especially those which attain a burn-up of more than 70% FIMA,[1] the fission-product concentrations are so high as to induce considerable change in the structure and chemical behaviour of the fuel. Various effects can result from an interaction of the fuel and of the fission products with the coating, e.g. the amoeba effect [1, 2] or SiC corrosion [3], all of which lead to failure of the particle.

On the basis of previous experience with the irradiation of fuel particles, the effects leading to failure of the coating may be summarized as follows:

(a)   With uranium carbide kernels, the released fission products — especially the rare earths — can lead to a corrosion of the SiC coating. The process is enhanced by the presence of a small quantity of chlorine. Although chlorine makes this process possible in oxide particles too, oxide particles are more resistent to the SiC corrosion.

(b)   Kernel migration (the amoeba effect) can occur in oxide kernels as a result of an increase in the oxygen potential, caused by burn-up. By contrast, the kernel migration of uranium carbide particles is appreciably slower and is independent of burn-up.

A combination of the advantages of both types of particle may possibly be obtained with a kernel consisting of a combination of uranium oxide with a lesser amount of uranium carbide. Whether or not this combination is effective depends primarily on the chemical binding of the fission products after high burn-ups.

In addition to the effects which the fission products induce under normal operating conditions of an HTR and which can, under some conditions, lead to particle failure, the fission products also determine the overall behaviour of the coated fuel particles during temperature transients. Depending upon the type of system failure, the temperature could rise up to 1700°C for brief periods of time. Or, under hypothetical accident conditions, such as with loss of coolant accompanied by failure of the whole of the emergency cooling system, an uncontrolled heating up of the core could occur, during which the temperature might even rise to 2500°C [4].

The purpose of the work reported here was (a) to investigate the chemical states of the fission products after high burn-up in particles with various kernel compositions (UO_2, UC_2 and UO_2+10% UC_2) with the aim of drawing conclusions regarding the optimization of kernel composition in the fissile particles, and (b) to determine the effect of different fission-product groups on the kernel/coating interaction under simulated core heat-up conditions.

---

[1]   FIMA: fission metal atoms.

TABLE I. KERNEL TYPES AND IRRADIATION CONDITIONS IN
EXPERIMENT BR2-P16

Kernel composition	Burn-up (% FIMA)	Fast neutron fluence (E > 0.1 MeV) (10^{21} cm^{-2})	Temperature (°C)
UO_2	65	8.4	
UC_2	68	9.5	<1000
90% UO_2 + 10% UC_2	69	9.3	

2. FISSION-PRODUCT BEHAVIOUR UNDER HIGH BURN-UP AND NORMAL REACTOR CONDITIONS

2.1. Experimental conditions

The BR2-P16 is an irradiation experiment that allows of a direct comparison
of kernel compositions of a variety of fissile particles under nearly identical
irradiation conditions. The three types of kernels which were examined are listed
in Table I, together with the main irradiation parameters.

Polished sections of the irradiated particles were examined metallographically
and by means of an unshielded electron microprobe. The aim of these measurements was to detect the distribution of various fission products which appeared
during irradiation to high burn-up values. As an example, Fig.1 shows the reflective
bright phases which were detected in oxide kernels with a nominal carbide content
of 10%.

For these purposes the local distributions of fission products were measured
by X-ray scanning image techniques. The atomic ratios of the elements could
also be estimated in favourable cases, i.e. for the larger inclusions, from the results
of semiquantitative point analyses.

2.2. Results of microanalysis

From these results a general picture was obtained of the effect of the initial
kernel composition on the chemical behaviour of important fission-product
groups, such as the rare earths, the alkaline earths and metallic fission products,
and on the distribution of such fission-product groups in the different fission-
product phases. Special attention was given to the latter problem for the case
of carbide-containing oxide kernels.

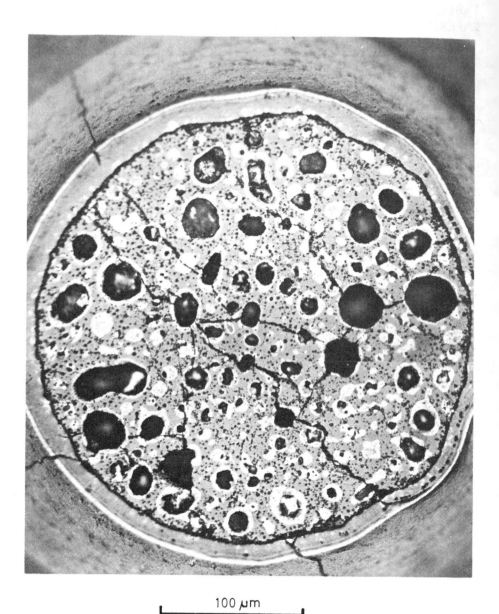

100 μm

FIG.1.   Micrograph of an irradiated kernel with an initial composition of 90% UO₂−10% UC₂ after a burn-up of 69% FIMA.

2.2.1. Particles with oxide kernels (UO₂)

The results are broadly in good agreement with those given in earlier publications [5]. The oxide matrix consists of UO_2 in solid solution with isomorphic oxides of the lanthanides as well as with the stabilized cubic modification of ZrO_2. However, BaO and SrO are dissolved in this phase only to the extent that the chemical potential of $(Sr, Ba)ZrO_3$ precipitates permits (Fig.2).

There are two different phases containing metallic fission products. These can be distinguished by the great difference in their molybdenum concentrations (Table II). From the fluctuations observed in the compositions of the two distinct phases, it has been concluded that at least two solid-solution phases exist in the multicomponent Mo-Tc-Ru-Rh-Pd system under high burn-up conditions.

Palladium is transported through the pyrocarbon and is found in small quantities in the SiC coating/pyrocarbon interface zone. The results indicate that the vapour pressure of palladium is not sufficiently reduced by interaction with fission products to reduce migration. No second phase containing palladium was detected in the kernel.

The local distributions of solid fission products such as caesium, barium and strontium in the coating are determined primarily by their transport properties. Their enrichment in the pyrocarbon/silicon carbide interface zone was sporadically observed, and is attributed to interaction with the silicon carbide [3]. The rare-earth concentrations in the coating were below the limit of detection in the particles with UO_2 kernels (Table III).

2.2.2. Particles with carbide kernels (UC₂)

The composition of the carbide fuel matrix shows that zirconium is uniformly distributed throughout the matrix. Because zirconium is insoluble in the initial UC_2 phase, it is assumed that a monocarbide $(U, Zr)C$ phase exists under these conditions.

The distribution of metallic fission products molybdenum, technetium, ruthenium and rhodium is correlated with that of the fuel matrix within the kernel (Fig. 3). It was not possible to decide if this correlation is due to a finely dispersed mixture of known complex carbides [6], e.g. $U_2(Ru, Rh)C_2 + U(Mo, Tc)C_2$, or to a solid solution. BR2-P16 was the first in a series of irradiation experiments with carbide fuel particles that showed differentiation in the fission-product distribution, so that separate phases other than the matrix were visible in the X-ray images (Fig. 3).

The alkaline earth metals barium and strontium were distributed in a separate carbon-rich phase, possibly as the dicarbides, BaC_2 and SrC_2. A comparison with the caesium distribution suggests that a portion of the caesium exists as an intercalation compound coexisting with the dispersed barium and strontium carbides.

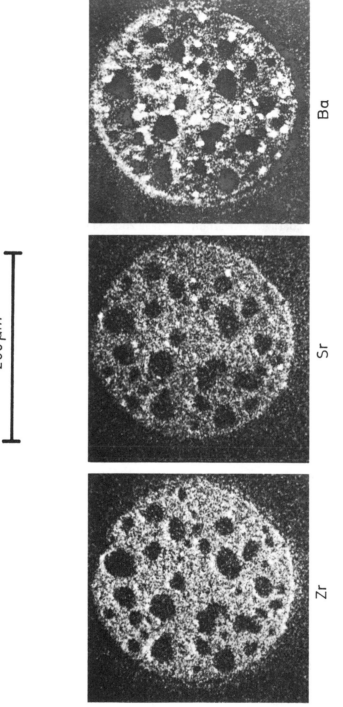

FIG. 2. Electron microprobe analysis of an irradiated UO₂ kernel. The X-ray scanning images show (Sr, Ba)ZrO₃ precipitation.

200 µm

Ba

Sr

Zr

TABLE II. MEAN ATOMIC RATIOS OF METAL
FISSION PRODUCTS IN TWO DIFFERENT
PRECIPITATION PHASES OF IRRADIATED
UO_2 KERNELS

Phase	Mo	:	Ru	:	Tc
I	10.2	:	1.8	:	1
II	3.1	:	1.7	:	1

TABLE III. MEAN CONCENTRATIONS OF RARE-EARTH FISSION
PRODUCTS IN THE BUFFER LAYERS OF FUEL PARTICLES WITH
DIFFERENT KERNELS AFTER HIGH BURN-UP (~70% FIMA)

Kernel type	Cerium (ppm)	Neodymium (ppm)
UO_2	<60	<25
UC_2	3100 ± 600	4700 ± 1200
90% UO_2 + 10% UC_2	280 ± 80	400 ± 30

The lanthanides were found not to be dissolved in the UC_2 matrix, in agreement with the results of an earlier investigation [7]. These fission products existed in another carbon-containing phase, presumably a dicarbide, distinct from the alkaline-earth carbides. These results reflect in part the results of simulation experiments with UC fuels made by Lorenzelli and Marcon [8].

Measurements made on the carbide particles show high concentrations of rare earths in the pyrocarbon coatings. These fission products, which are poorly retained in the kernel, lead to a distinct corrosion of the SiC layers, in spite of the relatively low irradiation temperatures.

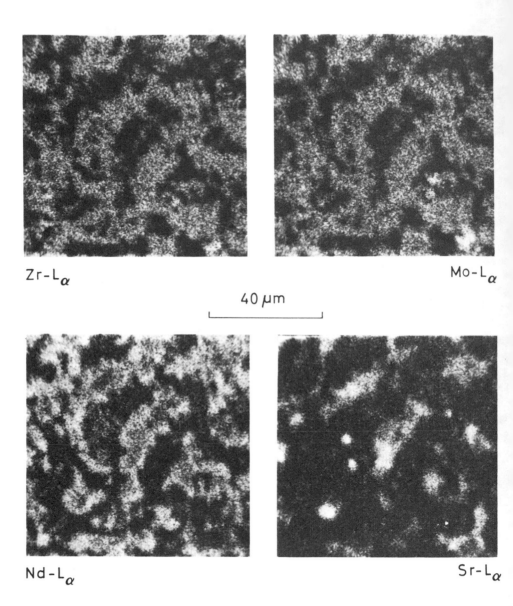

Zr-L$_\alpha$

Mo-L$_\alpha$

40 μm

Nd-L$_\alpha$

Sr-L$_\alpha$

FIG.3.   *Electron microprobe analysis of an irradiated* UC$_2$ *kernel showing X-ray scanning images of fission-product distribution.*

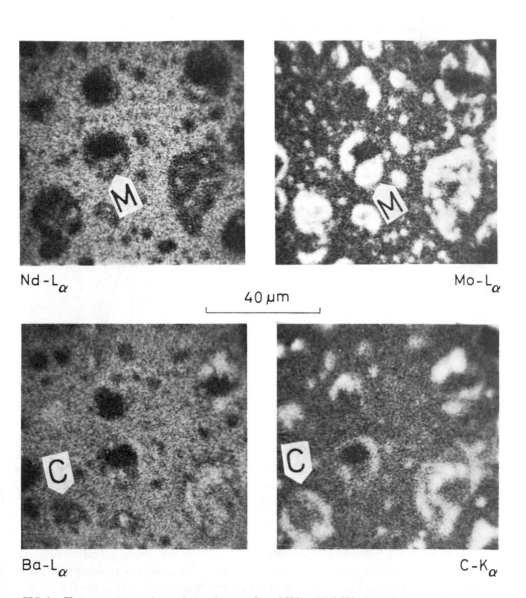

FIG.4. Electron microprobe analysis of an irradiated $UO_2 + 10\%$ UC_2 kernel showing X-ray images of fission-product distribution. (M: metal-rich phase; C: carbide-rich phase.)

2.2.3. Particles with carbide-containing oxide kernels (90% UO_2 + 10% UC_2)

Whereas the oxide matrix in the carbide-containing oxide kernel is similar to that of a pure UO_2 kernel in many respects, it differs in that no strontium or barium zirconate segregations were seen. However, barium was seen concentrated in a separate carbon-rich phase, probably as BaC_2 (Fig. 4).

The metallic fission products molybdenum, technetium, ruthenium, rhodium, palladium and zirconium appeared together with smaller amounts of rare earths and uranium in carbon-containing segregations. In the metallographic examinations at higher magnifications, these white inclusions were found to consist of multiphase segregations. The local resolving power of the electron microprobe is normally too low to determine the compositional differences between these phases, so that they appeared as monophase inclusions in the X-ray images of Fig.4.

Whereas the lanthanides were not released from the pure oxide kernels, they were detected in the pyrocarbon coatings of the 90% UO_2 + 10% UC_2 kernels, apparently as a result of the carbide added. The concentration, however, was one order of magnitude lower than that found with pure UC_2 kernels (Table III).

2.3. Discussion

The results show the effect of the initial composition of fissile kernels on the chemical behaviour of the various fission products and the consequences of such different behaviour for the particle failure mechanisms.

The release of the lanthanides and alkaline-earth metals from carbide kernels is appreciably greater than that from UO_2 kernels due to the lower solubility in the uranium fuel matrix and to the lower free enthalpies of formation of the carbides as compared with those of the oxides. This explains the appreciably greater concentrations of these fission products in the coating of carbide kernels. In the carbide-containing oxide kernels, the lanthanides occurred partly in a carbon-containing phase, which makes the detected release from such kernels understandable. The alkaline-earth metals do perhaps contribute to corrosion of the silicon carbide layer, but this is of minor importance when compared with the corrosion due to the rare earths.

In all of the types of kernels considered, there is only slight retention of palladium, and this is independent of composition. Because the fission product yield of palladium in highly enriched fissile kernels is relatively low, the SiC corrosion cannot be influenced by palladium release as adversely as with fuel particles of low enrichment.

The occurrence of alkaline-earth carbides such as BaC_2 is a good indicator of the effectiveness of adding carbide as a means of reducing the oxygen potential. Following the thermodynamical predictions of the distribution of fission products between the various carbide and oxide phases in a burned up U-C-O system [9], the presence of BaC_2 shows that the oxygen buffering reaction between BaC_2 and BaO is effective even in particles with 70% FIMA burn-up and a 10% carbide content. Hence, it is seen that the oxygen pressure has been so stabilized that free CO gas cannot be produced in larger quantities and no amœba effect occurs, which is in agreement with the results of the post-irradiation examination.

3. FISSION-PRODUCT BEHAVIOUR UNDER SIMULATED CORE HEAT-UP CONDITIONS

3.1. Materials

The behaviour of fission products at very high temperatures, in the range of 1800–2400°C, was investigated by using unirradiated coated particles in order to be able to distinguish between the following different effects:

● Interactions of heavy metals (uranium or thorium) with the coating;

● The influence of the CO pressure and CO permeation through the coating;

● Interactions of solid fission products with the coating.

Different types of unirradiated fuel particles were used to investigate the various parameters influencing the kinetics and mechanism of coated particle failure at very high temperatures. These particle types can be divided into two groups:

● Unirradiated coated particles with different kernel compositions and different coatings;

● Unirradiated fissile particles with oxide kernels and simulated burn-up.

3.1.1. Unirradiated coated particles

The kernel compositions used were UC_2, $UO_2+10\%$ UC_2, UO_2, ThO_2 and $(U, Th)O_2$; the coatings were TRISO (with SiC) coatings and BISO (without SiC) coatings. Details are given in Table IV, and in the last column the chemical compositions of the deposition gases are listed.

3.1.2. Fissile particles with simulated burn-up

The chemical reactions of solid fission products play an important role in determining the mechanisms by which coated particles fail at very high temperatures. It is nearly impossible to distinguish between the effects due to different groups of fission products by investigating *irradiated* particles under core heat-up conditions.

TABLE IV.　UNIRRADIATED COATED PARTICLES FOR CORE HEAT-UP SIMULATION EXPERIMENTS

Kernel type	Diameter (μm)	Coating type[a]		Deposition gas
UO_2	200	⎫	⎧ Buffer layer;	C_2H_2
90% UO_2+10% UC_2	200	⎬ TRISO	⎨ Inner LTI layer;	C_3H_6
UC_2	200	⎭	⎩ SiC layer;	CH_3SiCl_3
			Outer LTI layer	C_3H_6
ThO_2	500	⎫ BISO	⎰ Buffer;	C_2H_2
(Th, U)O_2	500	⎭	⎱ LTI or HTI layer	C_3H_6[b] or Cl

[a]　LTI: lower temperature isotropic process;　HTI: higher temperature isotropic process.
[b]　Sometimes as a C_3H_6/C_2H_2 mixture.

Therefore, solid fission products used for the simulation of burn-up were divided into three groups. These groups, which were introduced into fissile UO_2 kernels during the kernel fabrication process, are as follows:

- Metallic fission products (Mo, Ru, Rh, Pd);
- Rare earths (Y, La, Ce, Pr, Nd, Sm, Eu);
- Alkaline earths and zirconium (Sr, Ba, Zr).

Each group was chosen on the basis of its distinct chemical behaviour in oxide fuel materials, as described earlier in this paper (§ 2.2.1).

The concentration of fission-product elements in the kernels was calculated for a 90% ^{235}U enrichment at a burn-up of 75% FIMA. The calculations were made with the thermal neutron cross-sections, fission yields and fission product half-lives as published in the Karlsruher Nuklidkarte of 1974 [10]. An irradiation time of one year at a thermal neutron flux of 10^{14} cm^{-2}·s^{-1} was assumed for the radioactive fission products.

The basic kernel material was uranium dioxide. The kernels were fabricated by the Hydrolysis Process [11], which is based on a rapid solidification of droplets from concentrated uranyl nitrate solution containing urea and hexamethylene-tetramine, in silicon oil at 90°C. The metallic fission product elements molybdenum,

TABLE V. UO$_2$ KERNELS WITH A SIMULATED BURN-UP OF 75% FIMA

No.	Kernel dia. (μm)	density (g$\cdot$cm^{-3})	Fission products simulated	Element	Metal atom concentrations (at.%)
1	217	9.4	metals	Mo	18.2
				Ru	8.7
				Rh	2.2
				Pd	1.0
				U	69.9
2	218	8.9	rare earths	Y	5.7
				La	7.3
				Ce	17.3
				Pr	6.3
				Nd	19.2
				Sm	4.5
				Eu	2.2
				U	37.2
3	225	9.4	alkaline earths and zirconium	Sr	7.3
				Ba	4.9
				Zr	23.4
				U	64.4

ruthenium, rhodium and palladium were suspended as a fine powder in the starting solution. This powder was obtained from a homogeneous mixture of MoO$_3$ with soluble compounds of ruthenium, rhodium and palladium by reduction with hydrogen at 500°C. The rare earths, as well as strontium, barium and zirconium, were added to the starting solution as nitrates. During the heat treatment of the kernels in a reducing atmosphere (4% H$_2$ in argon) at 1350°C the first fission-product group formed metallic inclusions, the other two groups formed oxides. The compositions, diameters and densities are given in Table V.

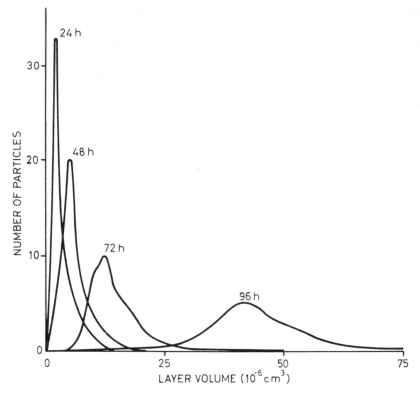

FIG.5. *Distribution of kernel/buffer reaction-layer volumes as observed microradiographically with 100 BISO-coated ThO₂ particles, of 500 μm kernel diameter after four successive annealings at 2300°C, each lasting for 24 h. The average growth rate of reaction-zone volume, as indicated by the most frequently found volume values, is roughly constant during the first two anneals (24 h and 48 h), but accelerates with time and nearly triples during the fourth annealing period (96 h).*

After sintering, the kernels were coated in a fluidized bed by deposition of four layers:[2]

- Buffer layer :   95-μm-thick porous pyrocarbon (from C_2H_2)
- Inner LTI :   30-μm-thick dense pyrocarbon (from C_3H_6)
- SiC layer :   30-μm-thick silicon carbide (from CH_3SiCl_3)
- Outer LTI :   35-μm-thick dense pyrocarbon (from C_3H_6)

---

[2] LTI: Lower temperature isotropic process.

3.2. Core heat-up simulation experiments

3.2.1. *Permeation loss of CO from coated oxide particles at very high temperatures*

When coated UO_2 or ThO_2 particles are heated to elevated temperatures, the following reaction between the kernel oxide and the surrounding pyrocarbon can occur:

$$MO_2 + 4C \rightleftharpoons MC_2 + 2CO \qquad\qquad (M = U \text{ or } Th) \qquad\qquad (I)$$

The carbon monoxide produced progressively accumulates in the gas-tight outer coating, and eventually builds up a pressure sufficient to retard Reaction (I) by mass-action effects. Thus, the excellent high-temperature stability of pyrocarbon-coated oxide particles is generally attributed to the containment of the CO within the particles at pressures ranging from 0.04 bar at 1600°C, to 2.1 bar at 2000°C, or to 31 bar at 2400°C [12].

Nevertheless, Reaction (I) can be observed to proceed in a forward direction at a low but finite rate in coated particles heated to temperatures above 2000°C. Eventually the coatings fail, with an average life-time which is characteristic of the coating design [13]. After the equilibrium CO pressure has been established, further production of CO by Reaction (I) requires the release of the CO. Two possible mechanisms for release of CO from coated particles are (a) coating failure, and (b) permeation of CO through the coating.

Coating failure was excluded as a failure mechanism in the earlier [14] and in the present experiments on the basis of microscopic examinations of sections of annealed particles. By exclusion, the CO release necessary for propagation of Reaction (I) must be attributed to permeation through the coatings. The initially gas-tight coating can become permeable when the metal in the kernel oxide MO_2 penetrates the buffer layer to cause a structural change that induces porosity in the outer high-density coating layers. The kernel oxide, however, is very immobile in pyrocarbon and cannot migrate as such to the outer coating. Naoumidis and Schenk [14] have shown that the failure rate of TRISO-coated particles is about 10 times greater at 2300°C in particles with UC_2 kernels than in those with UO_2 kernels, from which it follows that the diffusion of uranium through the pyro-carbon layers as UC_2 must be more than 10 times as fast as that of uranium as UO_2.

In order to examine the CO permeation of coated particles in greater detail, the time dependence of volume growth of the interaction zone between the buffer pyrocarbon and UO_2 and ThO_2 kernels was determined microradiographically. In each experiment, a sample series of 100 particles was embedded in a graphite matrix shaped into a disc 2 mm thick. Typical smoothed histograms obtained with commercially produced BISO-coated ThO_2 particles after annealing for successive 24 h periods of time at 2300°C are illustrated in Fig. 5. The distribution

1 mm

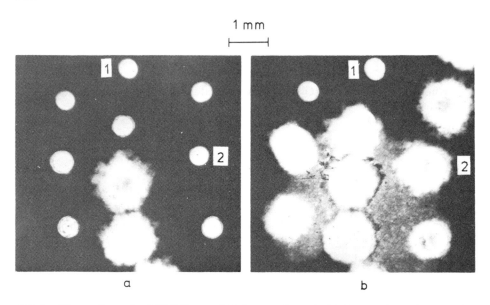

a   b

FIG.6. *Microradiograph of TRISO-coated ThO$_2$ particles embedded in graphite, after annealing.*
(a) 20 h at 2400°C: Three particles have failed and ThC$_2$ is seen to have migrated into the
surrounding graphite matrix; (b) 40 h at 2400°C: The particles numbered 1 and 2 in Fig.6a
exhibit similar states of progress of reaction and are not expected to fail during the second
anneal; however, as seen in Fig.6b, particle 2, a "nearest neighbour" to a failed particle,
did fail.

curves show that the average growth rate of the reaction-zone volume and the most
frequent reaction volume (the volume corresponding to the distribution peak)
was initially very low and nearly constant. This is to be expected for Reaction (I)
proceeding under near-equilibrium conditions at a rate controlled by a constant
CO pressure and, therefore, with a constant rate of permeation. With longer
annealings, the reaction-volume growth rate accelerates and the scatter in the
results increases. The accelerated rate in the more advanced stages of the reaction
evidently stems from an increased permeability brought about by the heavy metal
present at concentrations too low to be seen in microradiographs, but sufficient
to cause catalytic restructuring of the pyrocarbon and corrosion of the outer
coatings.

The effect of small quantities of released thorium on coating failure is vividly
seen in a "nearest neighbour" type of failure, an effect which is commonly
observed. In a sequence of repeated high-temperature anneals, particles located
adjacent to a failed particle in the graphite matrix are often observed to fail
suddenly whereas the next-nearest neighbours do not. A typical example of this
nearest-neighbour failure effect is illustrated in Fig.6.

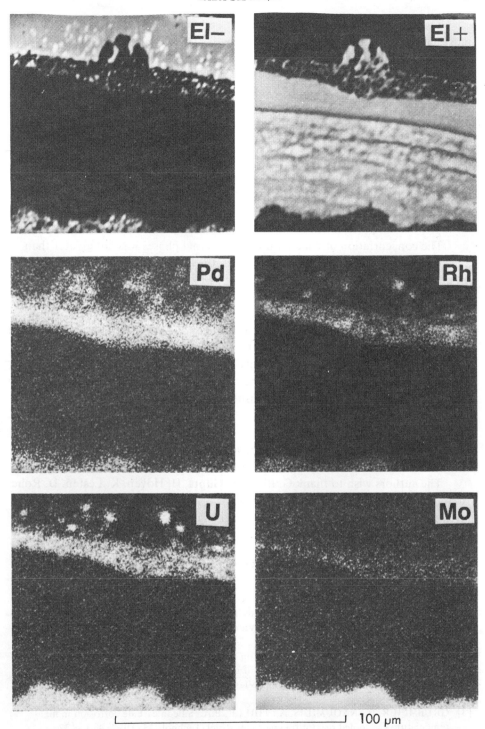

FIG.8.   Effect of uranium and solid fission-product metals on the corrosion of the SiC coating interlayer (2000°C, 6 h).

The second phase appeared to have the composition of $U(Ru, Rh)_3$ which must have resulted from chemical reaction with the UO_2 matrix during the annealing at 1800°C (15 min). After annealing at 2000°C (6 h), the second phase was found with another composition, with a metal ratio of $(Ru_{0.4}Rh_{0.1})(Mo_{0.4}U_{0.1})$.

The reaction can be formulated as:

$$UO_2 + C = U \text{ (metal solution)} + 2CO \qquad\qquad\qquad (II)$$

where the formation of the metal solution by the fission-product metals can help drive the reaction forward. However, uranium is also found in the fission-product inclusions formed in $UO_2 + UC_2$ fuel under normal irradiation conditions at lower temperatures (Table I). Reaction (II) is believed to occur in irradiated fuels mainly if the CO pressure is buffered to low values by the UC_2 addition.

The concentration of palladium in both kernel phases was not greater than 2 at.%. Thus palladium was transported from the kernel more rapidly than the other metals.

The transport of fission products in the buffer layer and their interaction with SiC were studied using X-ray images, the results of which are illustrated in Fig.8. The most striking features are that uranium is not only transported through the buffer layer but is also involved in the chemical reaction with the SiC layer. The distribution of the other metals molybdenum, ruthenium rhodium and palladium more or less demonstrated the same behaviour as did uranium. However, reaction of molybdenum with SiC seems to be a less important feature.

ACKNOWLEDGEMENTS

The authors wish to thank G. Blass, A. Gupta, H. Hoven, K. Lesten, D. Rohe, A. Schirbach and W. Tirtey for their careful work performing the experimental investigations.

REFERENCES

[1] LINDEMER, T.B., de NORDWALL, H.J., OLSTAD, R.A., "Analysis and measurement of mass transport in coated UO_2 and other HTGR oxide particles", Thermodynamics of Nuclear Materials 1974 (Proc. Symp. Vienna, 1974) Vol.1, IAEA, Vienna (1975) 163–171.

[2] NAOUMIDIS, A., THIELE, B., "Studies on the amoeba effect: carbon transport in coated fuel particles under the influence of a temperature gradient", Thermodynamics of Nuclear Materials 1974 (Proc. Symp. Vienna, 1974) Vol.1, IAEA, Vienna (1975) 173–186.

[3] GRÜBMEIER, H., NAOUMIDIS, A., THIELE, B.A., Silicon carbide corrosion in high-temperature gas-cooled reactor fuel particles, Nucl. Technol. 35 (1977) 413–427.

[4] DRESCHER, H.E., DUWE, R., PETERSON, K., SCHENK, W., UHLENBUSCH, L.,
     "Neuere Ergebnisse zur Freisetzung von Spaltprodukten eines 3000 MW$_{th}$-Kugelhaufen-
     reaktors beim hypothetischen Störfall ungehinderter Coreaufheizung", Proc. Reaktor-
     tagung (Hannover, April 1978) pp. 585—588.
[5] FÖRTHMANN, R., GRÜBMEIER, H., KLEYKAMP, H., NAOUMIDIS, A., "Chemical
     behaviour and improved retention of fission products in irradiated HTGR fuels",
     Thermodynamics of Nuclear Materials 1974 (Proc. Symp. Vienna, 1974) Vol.1, IAEA,
     Vienna (1975) 147—162.
[6] HOLLEK, H., Ternäre Carbide und Nitride der Actiniden und Lanthaniden mit anderen
     Übergangsmetallen, Ges. für Kernforschung mbH, Karlsruhe, Rep. KFK-1726 (1972).
[7] NOMURA, T., NAOUMIDIS, A., NICKEL, H., Phasenuntersuchungen an den U-Nb-C-
     und U-Ce-C-Systemen, Kernforschungsanlage Jülich GmbH, Rep. JÜL-1432 (1977).
[8] LORENZELLI, N., MARCON, J.P., "Evolution de l'activité du carbone en fonction du
     taux de combustion dans un combustible carbure", Behaviour and Chemical State of
     Irradiated Ceramic Fuels (Proc. Panel Vienna, 1972), IAEA, Vienna (1974) 99—113.
[9] HOMAN, F.J., LINDEMER, T.B., LONG, E.L., TIEGS, T.N., BEATTY, R.L., Stoichio-
     metric effects on performance of high-temperature gas-cooled reactor fuels from the
     U-C-O system, Nucl. Technol. 35 (1977) 428—441.
[10] SEELMANN-EGGEBERT, W., PFENNIG, G., MÜNZEL, H., Karlsruher Nuklidkarte,
     4. Auflage, Gesellschaft für Kernforschung mbH, Karlsruhe (1974) ISBN 3 87253 084 4.
[11] FÖRTHMANN, R., Die chemischen Grundlagen des Hydrolyseverfahrens zur Herstellung
     sphärischer Kernbrennstoffteilchen, Kernforschungsanlage Jülich GmbH, Rep. JÜL-950-
     RW (1973).
[12] PIAZZA, J.R., SINNOT, M.J., J. Chem. Eng. Data 7 (1962) 451—457.
[13] SCHENK, W., NAOUMIDIS, A., "Failure mechanism and fission product release in HTR
     fuel under conditions of unrestricted core heatup events", Nuclear Power Reactor Safety
     (Proc. ENS/ANS Topical Meeting, Brussels, Oct. 1978).
[14] NAOUMIDIS, A., SCHENK, W., "Das Verhalten von unbestrahlten und bestrahlten
     Brennstoffteilchen in HTR-Brennelementen bei hohen Temperaturen", in Kernforschungs-
     anlage Jülich GmbH Rep. JÜL-1533 (1978) 23—29.
[15] BRAMMAN, J.I., SHARPE, R.M., THOM, D., YATES, G., J. Nucl. Mater. 25 (1968)
     201—205,
[16] O'BOYLE, D.R., BROWN, F.L., DWIGHT, A.E., J. Nucl. Mater. 35 (1970) 257—266.
[17] HAINES, H.R., POTTER, P.E., RAND, M.H., "Some phase diagram studies of systems
     with fission-product elements for fast reactor fuels", these Proceedings, Vol.1, paper
     IAEA-SM-236/42.

DISCUSSION

U. BENEDICT: In your presentation you stated that you had detected
barium and strontium in the UO_2-base phase, and that you had not included
them in the UO_2-base solid solution but had listed them as a separate (Ba, Sr)
oxide phase. Does this mean that you had micro-inclusions containing barium
and strontium which you were unable to distinguish from the matrix phase in the
microprobe?

Have you found a true solid solution of barium and strontium in a UO_2 matrix
phase? If so, did you find large differences in solubility between the two elements?

This is interesting in view of an early UKAEA AERE report by McIver which states that one of the two elements is soluble and the other insoluble in UO_2.

A. NAOUMIDIS: Our experience with phase relationship studies on the solubility of SrO in UO_2 is similar to that of McIver. However, these solutions are unstable at T > 1400°C and decompose to form SrO gaseous species. Although we have not performed corresponding investigations on the $BaO-UO_2$ system, we conclude from BaO vapour pressure measurements[3] of UO_2 with BaO additives that BaO can be dissolved in the UO_2 phase. So we cannot rule out the existence of a solid solution of barium and strontium in the oxide matrix, even in the presence of the zirconate phase. No comparison can be inferred from this regarding the solubilities of strontium and barium in the UO_2 matrix.

H. GRÜBMEIER: I would add a brief comment in reply to the question concerning the resolving power of our electron microprobe work. We are indeed not able to distinguish between the solid solution of barium and strontium in the oxide matrix and separate oxide phases. This is possible only in the case of the larger inclusions.

---

[3] HILPERT, K., FOERTHMANN, R., NICKEL, H., Study of the vaporization of Ba from UO_2 nuclear fuel particles by high-temperature mass spectrometry, J. Nucl. Mater. **52** (1974) 89–94.

CHAIRMEN OF SESSIONS

General Chairman	P.A.G. O'HARE	United States of America
Session I	H.R. IHLE	Federal Republic of Germany
Session II	P.E. POTTER	United Kingdom
Session III	A. PATTORET	France
Session IV	D.D. SOOD	India
Session V	H. BLANK	Commission of the European Communities
Session VI	V.V. AKHACHINSKIJ	Union of Soviet Socialist Republics
Session VII	Z. MOSER	Poland
Session VIII	J. FUGER	Belgium
Session IX	M.G. ADAMSON	United States of America

SECRETARIAT OF THE SYMPOSIUM

Scientific Secretary:	J.D. NAVRATIL	Division of Research and Laboratories, IAEA
Administrative Secretary:	G. SEILER	Division of External Relations, IAEA
Editor:	E.R.A. BECK	Division of Publications, IAEA
Records Officer:	S.K. DATTA	Division of Languages, IAEA
Liaison Officers:		
Government of the Federal Republic of Germany	G. HERRMANN	Bundesministerium für Forschung und Technologie. Bonn
Local	O. RENN	Kernforschungsanlage Jülich GmbH

HOW TO ORDER IAEA PUBLICATIONS

 An exclusive sales agent for IAEA publications, to whom all orders and inquiries should be addressed, has been appointed in the following country:

UNITED STATES OF AMERICA   UNIPUB, 345 Park Avenue South, New York, NY 10010

 In the following countries IAEA publications may be purchased from the sales agents or booksellers listed or through your major local booksellers. Payment can be made in local currency or with UNESCO coupons.

ARGENTINA	Comisión Nacional de Energía Atomica, Avenida del Libertador 8250, RA-1429 Buenos Aires
AUSTRALIA	Hunter Publications, 58 A Gipps Street, Collingwood, Victoria 3066
BELGIUM	Service Courrier UNESCO, 202, Avenue du Roi, B-1060 Brussels
CZECHOSLOVAKIA	S.N.T.L., Spálená 51, CS-113 02 Prague 1
	Alfa, Publishers, Hurbanovo námestie 6, CS-893 31 Bratislava
FRANCE	Office International de Documentation et Librairie, 48, rue Gay-Lussac, F-75240 Paris Cedex 05
HUNGARY	Kultura, Hungarian Foreign Trading Company P.O. Box 149, H-1389 Budapest 62
INDIA	Oxford Book and Stationery Co., 17, Park Street, Calcutta-700 016
	Oxford Book and Stationery Co., Scindia House, New Delhi-110 001
ISRAEL	Heiliger and Co., Ltd., Scientific and Medical Books, 3, Nathan Strauss Street, Jerusalem 94227
ITALY	Libreria Scientifica, Dott. Lucio de Biasio "aeiou", Via Meravigli 16, I-20123 Milan
JAPAN	Maruzen Company, Ltd., P.O. Box 5050, 100-31 Tokyo International
NETHERLANDS	Martinus Nijhoff B.V., Booksellers, Lange Voorhout 9-11, P.O. Box 269, NL-2501 The Hague
PAKISTAN	Mirza Book Agency, 65, Shahrah Quaid-e-Azam, P.O. Box 729, Lahore 3
POLAND	Ars Polona-Ruch, Centrala Handlu Zagranicznego, Krakowskie Przedmiescie 7, PL-00-068 Warsaw
ROMANIA	Ilexim, P.O. Box 136-137, Bucarest
SOUTH AFRICA	Van Schaik's Bookstore (Pty) Ltd., Libri Building, Church Street, P.O. Box 724, Pretoria 0001
SPAIN	Diaz de Santos, Lagasca 95, Madrid-6
	Diaz de Santos, Balmes 417, Barcelona-6
SWEDEN	AB C.E. Fritzes Kungl. Hovbokhandel, Fredsgatan 2, P.O. Box 16356, S-103 27 Stockholm
UNITED KINGDOM	Her Majesty's Stationery Office, Agency Section PDIB, P.O. Box 569, London SE1 9NH
U.S.S.R.	Mezhdunarodnaya Kniga, Smolenskaya-Sennaya 32-34, Moscow G-200
YUGOSLAVIA	Jugoslovenska Knjiga, Terazije 27, P.O. Box 36, YU-11001 Belgrade

 Orders from countries where sales agents have not yet been appointed and requests for information should be addressed directly to:

 Division of Publications
International Atomic Energy Agency
Wagramerstrasse 5, P.O. Box 100, A-1400 Vienna, Austria

80- 02776